Forests, People and Power

The Earthscan Forestry Library

Jeffrey A. Sayer, Series Editor

Forests, People and Power

The Political Ecology of Reform in South Asia

Edited by
Oliver Springate-Baginski and Piers Blaikie

EARTHSCAN
London • Sterling, VA

First published by Earthscan in the UK and USA in 2007

ISBN: 978-1-84407-347-4

Typeset by FiSH Books, Enfield, Middlesex.
Printed and bound in the UK by TJ International, Padstow
Cover design by Susanne Harris

For a full list of publications please contact:

Earthscan
8–12 Camden High Street
London, NW1 0JH, UK
Tel: +44 (0)20 7387 8558
Fax: +44 (0)20 7387 8998
Email: earthinfo@earthscan.co.uk
Web: **www.earthscan.co.uk**

22883 Quicksilver Drive, Sterling, VA 20166–2012, USA

Earthscan publishes in association with the International Institute for Environment and
Development

A catalogue record for this book is available from the British Library

Library of Congress Cataloging-in-Publication Data

Forests, people and power : the political ecology of reform in South Asia / edited by
Oliver Springate-Baginski and Piers Blaikie.
     p. cm.
   Includes bibliographical references and index.
   ISBN-13: 978-1-84407-347-4 (hardback : alk. paper)
   ISBN-10: 1-84407-347-5 (hardback : alk. paper)
   1. Forest policy—India. 2. Forest policy—Nepal. 3. Forest management—India—
Citizen participation. 4. Forest management—Nepal—Citizen participation. I.
Springate-Baginski, Oliver. II. Blaikie, Piers M.
   SD645.F674 2007
   333.750954—dc22

                                                                    2006039460

The paper used for this book is FSC-certified and totally chlorine-free.
FSC (the Forest Stewardship Council) is an international network to
promote responsible management of the world's forests.

Contents

List of Contributors

Jagannath Adhikari works for Martin Chautari, a leading social development and advocacy NGO in Kathmandu, Nepal. He obtained his PhD from the Australian National University in Canberra. His main research interests are agrarian change, rural and urban development, labour migration and the remittance economy, food security, and natural resource management. He has also undertaken research and advocacy work on environmental justice in relation to natural resources management.
Email: jadhikari@mos.com.np

Madhusudan Bandi is a PhD Research Scholar jointly at the Centre for Economic and Social Studies (CESS) in Hyderabad, India, and the School of Development Studies, University of East Anglia, UK. He has an MPhil in Political Science and a degree in Law. His research interests include governance issues in community forestry, backward classes in local governance and secularism in Indian politics.
Email: madhusudan-bandi@rediffmail.com

Ajit Banerjee is an independent consultant in Kolkata, India. He received his Diploma in Forestry from the Indian Forest College, Dehradun, and served in the Indian Forest Service for 22 years. He completed his PhD in forestry at the University of Toronto, Canada. From 1984 to 1995, Ajit Banerjee was a senior forestry specialist at the World Bank. After his retirement from the bank he worked in China as a consultant promoting participatory forestry on behalf of KfW, the German Development Bank. He currently lives in Kolkata.
Email: akbanb@vsnl.net

Binod Bhatta is director with Research and Forestry and Natural Resources Management (F/NRM) in Kathmandu, Nepal, and a specialist with Alliance Nepal. He completed his PhD in silviculture and forest influences at the University of the Philippines in Los Banos. Binod Bhatta has worked at Winrock International Nepal (1999 to 2005), the Gesellschaft für Technische Zusammenarbeit (GTZ)-supported Nepal–German Churia Forest Development Project (1996 to 1999), as head of natural forest silviculture and management section at the Forest Research and Survey Centre, Ministry of Forest and Soil Conservation, Government of Nepal (1993 to 1996), and as training officer at the Training Division of the Ministry of Forest and Soil Conservation (1984 to 1993).
Email: binod@winrockint.wlink.com.np

Piers Macleod Blaikie is a professorial fellow at the Overseas Development Group, University of East Anglia, UK. He retired from the School of Development Studies, University of East Anglia, Norwich, UK in February 2003, but continues to teach and undertake research as a professorial fellow. His research interests include environment–society relations in the developing world; political ecology; and environmental policy in South and South-East Asia and in Southern and Central Africa.
Email: p.blaikie@uea.ac.uk

Om Prakash Dev is a founder member of the Resources Development and Research Centre (RDRC) in Kathmandu, Nepal. He has worked as a community forester since 1983, including with the UK Department for International Development (DFID)-funded Nepal–UK Community Forestry Project (1991 to 2001) and the Livelihoods and Forestry Programme (2001 to 2003). He completed his PhD at the Geography Department, Leeds University, UK, and is currently coordinating action research on various aspects of participatory forest management in Nepal.
Email: omprakash.dev@gmail.com

Akhileshwar L. Karna is a forest officer with the Ministry of Forests and Soil Conservation in Nepal. He has been working for the Ministry of Forests and Soil Conservation for more than 18 years in various capacities. He has an MSc in forestry from Reading University, UK, and is currently a PhD research student at the School of Development Studies, University of East Anglia, UK. His research and development interests include strengthening stakeholder partnerships in forest management.
Email: a.karna@uea.ac.uk

V. M. Ravi Kumar is a lecturer in the Department of History, Babashed Bhimrao Ambedkar University, Lucknow. His main research interests are environmental history and political ecology.
Email: mr_vejandla@yahoo.com

M. Gopinath Reddy is a reader and fellow with the Centre for Economic and Social Studies (CESS) in Hyderabad, India. A political science and public administration specialist, his current research interests include decentralized governance, livelihoods and poverty analysis, and institutional approaches to natural resource management.
Email: mgrjl@yahoo.com

V. Ratna Reddy is a professor and senior fellow at the Centre for Economic and Social Studies in Hyderabad, India. An economist by training, Ratna's current research interests include natural resources and environmental economics, new institutional economics and agricultural policy.
Email: vratnareddy@cess.ac.uk

M. Srinivasa Reddy has a PhD in Rural development and currently works as research officer at the Centre for Economic and Social Studies, Hyderabad. His research interests include rural development, agriculture and resource management, and recent work relates to watershed management, joint forest management and rice cultivation intensification systems.
Email: sreenivasdrreddy@yahoo.com

Sushil Saigal is a programme manager at Natural Resource Management and Coordinator, Resource Unit for Participatory Forestry (RUPFOR), Winrock International India, New Delhi, India. He holds an MBA in forestry management from the Indian Institute of Forest Management, Bhopal, India, and an MSc in forestry and land use from the University of Oxford, UK, and is currently a PhD research scholar at the University of Cambridge, UK. Prior to joining Winrock, Sushil Saigal worked with the Society for Promotion of Wastelands Development (SPWD), New Delhi, where he coordinated the National Network on Joint Forest Management and was also closely involved with the Forest, Trees and People Programme of the United Nations Food and Agriculture Organization (FAO).
Email: saigal.sushil@gmail.com

Kailas Sarap is professor of economics at Sambalpur University, Orissa, India. His research interests include agrarian relations, land and labour markets, natural resources, access to credit, and social securities. Kailas Sarap has been a visiting fellow at the University of Namur, Belgium, and a Commonwealth fellow at the University of Oxford, UK.
Email: sarap_k@rediffmail.com

Madhu Sarin is an independent adviser on social development and collaborative natural resource management in Chandigarh, India. Originally an architect, she switched to urban development planning at the Development Planning Unit, University College London, UK. She is currently exploring the links between the devolution of power and authority to gender-equal and democratic community institutions, promotion of environmentally sustainable livelihoods and the forest and land-use policy frameworks conducive to promoting them, and, in particular, linking local field practice to policy change. Madhu Sarin is a board member of a number of Indian non-governmental organizations (NGOs) and of the International Institute for Environment and Development (IIED) in London, UK.
Email: msarin@sancharnet.in

Oliver Springate-Baginski is a senior research fellow at the Overseas Development Group, University of East Anglia, UK. His research interests include community-based forest and other natural resource management, collective action and local government.
Email: oliver.springate@uea.ac.uk

List of Figures, Tables, Maps and Boxes

Figures

Tables

Acknowledgements

This book, an outcome of a research project funded through the UK Department for International Development (DFID) Forest Research Programme, has benefited from the involvement of a wide range of contributors across South Asia. Our greatest thanks go to the numerous local people who have given their time so generously so that we could gather information and develop an understanding of their local situation. We are particularly grateful to the local people who have helped us in Nepal over the last three years during the extreme political difficulties there. We wish them a peaceful, prosperous and democratic future.

We would also like to thank the many foresters at central, state, district and field levels who have helped us to understand forest management policies, practices and outcomes from their perspective.

In India we would particularly like to acknowledge the contributions of Sushil Saigal to the research project and the text of this book, and the tireless support, advice and guidance of Madhu Sarin, who has been closely involved throughout the project and has contributed extensively throughout this book. Thank you! We would also like to recognize the contribution of Sushil's colleagues at Winrock International India, particularly Mamta Borgoyary and Neeraj Peters.

In West Bengal, Dr Ajit Banerjee wishes to thank his research team, particularly Sukla Sen, Satyananda Das and Professor Tapan Misra, and also Asish Kr Das, Shakti Pada Haldar, Ali Bandyopadhyay, Sandipa Ghosh, Sukanta Das, Sunandan Saha, Joy Talukdar, Sudipta Maity P. S. Banerjee, S. C. Dutta Heerak Nandy and Arunoday Chakraborty. We also send our best wishes to Dr Debal Deb for success in his courageous legal stand in defence of community-based forest biodiversity protection.

In Orissa, Professor Kailas Sarap wishes to thank Sri Prasant Kumar Das, Murali Gartia, Sambhu Sahu, Somanath Sethy and Tapas Kumar Sarangi for collection and analysis of data, and the many forest officials at Bhubaneswar and in different areas of the study districts.

In Andhra Pradesh, Professor Ratna Reddy and Dr Gopinath Reddy wish to thank their research team and colleagues at the Centre for Economic and Social Studies (CESS), particularly Professor Mahendra Dev for supporting their involvement in the project.

Our Nepal research coordinator, Dr Om Prakash Dev, wishes to thank his field team, in particular, Ajeet Karn, Eak Raj Chatkuli, Damyanti Pandey and Hemant Yadav. We also wish to acknowledge the guidance and support throughout this project of Dr Damodar Parajuli of the Ministry of Forests and Soil Conservation, Government of Nepal, and to express our condolences to his family for his tragic recent death.

We would also like to acknowledge the help of the many field enumerators who have assisted the field researchers in data collection and analysis.

We thank our institutional colleagues who have helped with administrating the logistics for the research study. At the Overseas Development Group, particular thanks go to Jane Bartlett, Jo Jones and Karen Parsons for their efficiency, precision and patience.

Our editors! We wish to thank Jonathan Cate, who drew the maps and diagrams to exacting standards, and Ashish Kothari of Kalpavriksh for his incisive comments on an earlier draft of the book. Also last, but by no means least, Sally Sutton, who edited the final version of the book, without whose professionalism and stamina we would have found it difficult to complete the manuscript on time.

Finally, thanks to DFID's Forestry Research Programme, and particularly to John Palmer for agreeing to support this project, and to Katelijne Rothschild Van Look for her kind support throughout the project.

Note: this publication is an output from a research project funded by DFID for the benefit of developing countries. The views expressed are not necessarily those of DFID.

Oliver Springate-Baginski, Shimla, India
Piers Blaikie, Norwich, England
January 2007

Note on *Panchayats* in India

Panchayat (or *gram panchayat*) is a Hindi term meaning a committee of five village elders or leaders, charged with decision-making on village affairs and dispute resolution (*gram* meaning village, *panch* meaning five). The *panchayats* were historically selected bodies, hence retaining local hierarchies, typically dominated by upper class/caste males. *Panchayati Raj* (village self-government) became a nationalist cause in India during the struggle for independence, and subsequently the state's promotion of *panchayati raj institutions* (or PRIs) in keeping with the Gandhian ideal of *Gram Swaraj* (village republic) progressed gradually. However in 1992, through the 73rd Amendment to the Indian Constitution, the *panchayats* became constitutionally mandated to facilitate strengthened and more representative decentralized local government. An elections process was also mandated with reservations to ensure the inclusion of women and other marginalized groups. Two higher administrative levels were also introduced: *panchayat samitis* at *tehsil* ('block') level; and the *zilla parishad* at district level. Below the *gram panchayat* there is another level, the *gram sabha,* the customary village forum where all members of each hamlet or village are expected to deliberate and hold their representatives accountable. Incidentally the term *panchayat* is also used to refer to local village committees that are not part of *panchayati raj. Van panchayats* for instance are village forest committees created in the 1930s in hill areas of Uttar Pradesh (now Uttaranchal) to manage village forests there.

List of Acronyms and Abbreviations

ADB	Asian Development Bank
AKRSP	Aga Khan Rural Support Programme
ANSAB	Asia Network for Sustainable Agriculture and Bioresources
AP	Andhra Pradesh
APO	annual plan of operations
APP	Agriculture Perspective Plan
Ausaid	Australian Agency for International Development
BC	backward caste
BINGO	big international non-governmental organization
BISEP-ST	Biodiversity Sector Programme for *Siwalik* and *Tarai*
BIWMP	Bagmati Integrated Watershed Management Project
BZCF	buffer zone community forest
CARE	Cooperative for Assistance and Relief Everywhere
CBD	Convention on Biological Diversity
CBNRM	community-based natural resource management
CBO	community-based organization
CBS	Central Bureau of Statistics (Nepal)
CEC	Centrally Empowered Committee (*otherwise known as* National Level Committee on Forestry)
CESS	Centre for Economic and Social Studies (India)
CEW	community extension worker
CF	community forest/community forestry
CFDP	Community Forestry Development Project
CFM	collaborative forest management
CFMG	collaborative forest management group
CFUG	community forest user group
ChFDP	Churia Forest Development Project
CIAA	Commission for Investigation into Abuse of Authority (Nepal)
cm	centimetre
CollFM	collaborative forest management
CPM	Community Party of India
CPNUML	Communist Party of Nepal – United Marxist and Leninist
CSE	Centre for Science and Environment
DANIDA	Danish International Development Agency
dbh	diameter at breast height
DDC	district development committee
DFCC	district forests coordination committee
DFID	UK Department for International Development
DFO	district forest officer (Nepal)/divisional forest officer (India)
DFPSB	District Forest Product Supply Board
DFRS	Department of Forest Research and Survey (Nepal)
DoF	Department of Forests (Nepal)

EC	executive committee
EFEA	Environment and Forest Enterprise Activity
EIA	environmental impact analysis
EU	European Union
FAO	United Nations Food and Agriculture Organization
FCA	1980 Forest Conservation Act
FD	forest department
FDA	forest development agency
FDI	foreign direct investment
FECOFUN	Federation of Community Forest Users, Nepal
FINNIDA	Finnish International Development Agency
FMUDP	Forest Management and Utilization Development Project
FORESC	Forest Research and Survey Centre
FPC	forest protection committee
FPDB	Forests Product Development Board
FRI	Forest Research Institute (Dehra Dun)
FSCC	forest-sector coordination committee
FSI	Forest Survey of India
FUG	forest user group
FWC	Firewood Corporation (Nepal)
GB	general body
GCC	Girijan Cooperative Corporation
GDP	gross domestic product
GEF	Global Environment Fund
GO	government organization
GoI	Government of India
GoWB	Government of West Bengal
GPS	global positioning systems
GTZ	Gesellschaft für Technische Zusammenarbeit (German Agency for Technical Cooperation)
ha	hectare
HMGN	His Majesty's Government of Nepal (as of 2006, Nepal government)
HP	Himachal Pradesh
IAS	Indian Administrative Service
IBRAD	Indian Institute of Bio-Social Research and Development
IFA	1927 Indian Forest Act
IFI	international funding institution
IFS	Indian Forest Service
IGA	income-generating activity
IGNFA	Indira Gandhi National Forest Academy
IIED	International Institute for Environment and Development
IIFM	Indian Institute of Forest Management
ITDA	Integrated Tribal Development Agency (India)
ITK	indigenous technical knowledge
IUCN	World Conservation Union (*formerly* International Union for the Conservation of Nature)
JBIC	Japan Bank for International Cooperation
JFM	joint forest management
JFMC	JFM committee
JICA	Japan International Cooperation Agency
JPC	Joint Parliamentary Committee

JTRC	joint technical review committee
KBK	Koraput, Bolangir and Kalahandi districts
KFC	Kathmandu Forestry College
kg	kilogram
km	kilometre
km²	square kilometre
KMTNC	King Mahendra Trust for Nature Conservation
LAMPS	Large-Scale Adivasi Multi-Purpose Society
LFP	Livelihoods and Forestry Programme (DFID-Nepal funded)
LHF	leasehold forestry
LNGO	local non-governmental organization
LPG	liquid petroleum gas
LSGA	1998 Local Self-Governance Act
m	metre
m³	cubic metre
MAI	mean annual increment
MAP	medicinal and aromatic plant
MC	management committee
MDG	Millennium Development Goal
MIS	management information systems
MLA	member of the legislative assembly
mm	millimetre
MoAC	Ministry of Agriculture and Co-operatives
MoEF	Ministry of Environment and Forests (Government of India)
MoF	Ministry of Finance (India)
MoFSC	Ministry of Forests and Soil Conservation (Government of Nepal)
MoLD	Ministry of Local Development
MoPE	Ministry of Population and Environment
MoU	memorandum of understanding
MP	member of parliament
MPFS	Master Plan for Forestry Sector
MTO	mass tribal organization
n	total population sample size
NACRMLP	Nepal–Australia Community Resource Management and Livelihood Project
NAEB	National Afforestation and Eco-Development Board
NAP	National Afforestation Programme
NARMSAP	Natural Resource Management Sector Assistance Programme
NBSAP	Indian National Biodiversity Strategy and Action Plan
NCA	National Commission on Agriculture
NCS	National Conservation Strategy
NEPAP	Nepal Environment Policy and Action Plan
NFAP	National Forestry Action Plan
NFC	National Forest Commission
NFP	1988 National Forest Policy
NGO	non-governmental organization
NGSP	non-government service provider
NP	national park
NPC	National Planning Commission
NRC	Nepal Resettlement Company
NRM	natural resource management

NRSA	National Remote Sensing Agency
NSCFP	Nepal–Swiss Community Forest Project
NTFP	non-timber forest product
NWDB	National Wasteland Development Board
ODA	Overseas Development Administration (UK Government)
ODI	Overseas Development Institute
OECD	Organisation for Economic Co-operation and Development
OFD	Orissa Forest Department
OFDC	Orissa Forest Development Corporation
OFMP	Operational Forest Management Plan
OP	operational plan
PCC	project coordination committee
PCCF	principal chief conservator of forests
PESA	*Panchayat* Extension to Scheduled Areas
PF	*panchayat* forest
PFM	participatory forest management
PIL	public interest litigation
PMC	project management committee
PPF	*panchayat* protected forest
PRA	participatory rural appraisal
PRI	*panchayati raj* institution
PRSP	poverty reduction strategy paper
RCDC	Regional Centre for Development Cooperation
RD	revenue department
RDRC	Resources Development and Research Centre (Nepal)
RECOFTC	Regional Community Forestry Training Centre (Thailand)
RF	reserved forest
RFCC	regional forests coordination committee
RO	range officer
RP	range post
RRAFDC	Rural Region Agro-Forestry Development Centre
RUPFOR	Resource Unit for Participatory Forestry (India)
RSU	regional support unit
SAMARPAN	Strengthening the Role of Civil Society and Women in Democracy and Governance project
SAP	structural adjustment programme
SATA	Swiss Agency for Technical Aid
SDC	UK Sustainable Development Commission
SDC	Swiss Development Cooperation
SF	social forestry
SFP	Social Forestry Project/Programme
SGVY	integrated village afforestation and eco-development (*Samanvit Gram Vanikaran Samiriddhi Yojana*)
SHG	self-help group
SICFPG	Self-Initiated Community Forest Protection Group
Sida	Swedish International Development Agency
SIFPG	self-initiated forest protection group
SNV	Netherlands Development Organization
SPWD	Society for Promotion of Wastelands Development (India)
ST	scheduled tribe
TAL	*Tarai* Arc Landscape project

TCFDP	*Tarai* Community Forestry Development Project
TCN	Timber Corporation of Nepal
TDCC	Tribal Development Co-operative Corporation
THED	Theory of Himalayan Environmental Degradation
TII	Transparency International India
UK	United Kingdom
UN	United Nations
UNDP	United Nations Development Programme
UNEP	United Nations Environment Programme
US	United States
USAID	US Agency for International Development
VDC	village development committee
VFPC	village forest protection committee
VIKSAT	Vikram Sarabhai Center for Development Interaction
VP	*van panchayat*
VSS	*Vana Samarakshyan Samiti* (forest protection committee)
WBFD	West Bengal Forest Department
WBFDC	West Bengal Forest Development Corporation Limited
WBSG	West Bengal State Government
WIMCO	Western Indian Match Company
WLPA	1972 Wild Life Protection Act
WLR	Wildlife Reserve
WTLBP	Western *Tarai* Landscape Building Project
WWF	World Wide Fund for Nature (*formerly* World Wildlife Fund)

Introduction

Setting Up Key Policy Issues in Participatory Forest Management

Piers Blaikie and Oliver Springate-Baginski

A guide to this book

This book examines the issue of reform in forest management policy in India and Nepal, considering in detail three major Indian states (West Bengal, Orissa and Andhra Pradesh) and two regions of Nepal (the mid-hills and the plains, or *tarai*). A central issue of reform on which much current policy debate revolves is the role of local people in forest management. Participatory forest management (PFM) – a label used to describe a range of policy reform measures related to this issue – is examined in detail in this book. The term is employed here as a focus of debate, and we use it to refer to any policy that *claims* to be participatory in whatever terms and that applies whatever criteria the user chooses. Thus, the authors make no prior claim to what constitutes '*genuine*' participation, although the concluding chapters outline the purposes of participation in different contexts and the obstacles and facilitative forces that shape the policy outcomes of 'participation'. It will quickly become apparent that many claims to 'participation' are made for many different reasons. Assessments of these claims have been made by numerous different actors, including policy-makers; activists; politicians; international funding agencies; forest users of all kinds, from landless tribal people to village elites; and functionaries of the forest administrations at different levels.

There is a large and lively literature on PFM in Nepal and India, and although some very interesting data and innovative analyses have been presented, much of it covers similar ground in the sense that, since the mid 1990s, there has been a pattern of rather pessimistic conclusions that the promise of PFM reform has not been fully realized. Indeed, it is difficult to escape the well-trodden path of rehearsing an historical analysis of the colonial origins of forest management policy, especially in India, and its longevity and durability; of considering the limited changes that have occurred; and once again of rehearsing the continuing case for reform to bring justice and democratization for local forest users. This book follows this path some of the way, but also seeks to take different approaches. It makes the central argument that participation in forest management is primarily justified on the basis of social justice and common law because forests have, until relatively recently, provided major support for rural livelihoods before this was gradually undermined by patterns of state aggrandizement of the forest estate at the expense of local people and their customary rights of access and usufruct. This process of states undermining local rights still continues – indeed, in some places even under the guise of PFM. The later chapters in Part I (especially Chapters 2 and 3) address why policy broadly continues on this path and why reform (with

PFM at its centre) continues to be so difficult. Chapter 4 outlines in theoretical detail the issue of rural people's livelihoods and how PFM should, and actually does, affect the livelihoods of different people. Part II consists of detailed studies of the three Indian states and two Nepalese regions (the middle hills and the *tarai*), including an examination of the regional policy genealogy and the emergence of PFM in comparative context. Quantitative analysis of the forest's contribution to the livelihoods of different groups (men, women, wealthy, landless, tribal people and so on) in different ecological conditions is presented, with a focus on the differences that the introduction of PFM has made to these disparate groups. Finally, in Part III, policy conclusions are drawn, and the political, economic and administrative feasibility of reform is assessed.

This chapter is a summary of our approach to this complex subject and reports the main findings in outline. The basis of our approach rests on the assumption that all statements about forests and forest lands are intrinsically political. The policy process simply cannot be understood or reformed (either radically or piecemeal) without understanding this. The approach of political ecology is introduced and promoted to reach this end. A brief summary of the established arguments concerning the direction of the struggles over forests and forest lands – pessimistic and optimistic – is provided later in this chapter in order to set the scene for the rest of the book and to point out the well-trodden discursive paths referred to earlier. For the busy reader, the last part of the chapter outlines the main research questions and findings. For the *very* busy reader, key words and concepts are italicized throughout the text. Key references only are given in this chapter, and fuller referencing can be found in subsequent chapters.

Issues of 'people and forests' in India and Nepal

There has been a *long struggle* between the state and different sections of civil society and local people over the control, management and use of India and Nepal's land resources, particularly over what have been officially classified as forest lands and forest resources. Indeed, the state's right to control forests is asserted in the earliest South Asian texts on statecraft (see Kautilya, 1992). Over the last century, particularly due to colonial and economic expansion, these struggles have intensified. Certain issues constantly re-emerge: the appropriate role of the state in managing what has been defined as a national resource; environmental justice; and rights and entitlements, especially for populations who rely substantially on the forest for subsistence or small-scale commodity production. In some areas these issues are particularly relevant to indigenous peoples since forest management engages with issues of resource rights and ancestral domains, livelihood systems and cultural identity. Additionally, the issues of environmental conservation and protection (of watersheds, soil and water), biodiversity, vegetation and forest cover, and wildlife conservation have all become either subsumed by or at least overlap in complementary or contradictory ways a broadly defined forest policy. All of these issues emerge at different scales (the federal, state, district and local levels).

Insofar as the management of forests is partly a 'war of words' and partly policy argumentation between different protagonists, the formal institutions of state have exerted a powerful influence on the outcome of these struggles through the production and deployment of powerful and persuasive policy narratives. The present circumstances of these struggles are, in part, new, as well as being of long historical standing. They are, as always, *highly political*, even if certain parties wish to define them as technical (and under the control of those formal institutions that claim a monopoly of technical – and authoritative – knowledge). For example, the imposition by a forest department of an 80-year rotation of specified commercial species in a working plan is both a technical choice and a socio-political one since it asserts one group's priorities while displacing those of existing users' and rights-holders' preferences for a different (probably shorter rotation) management system and a different and broader species mix. The working plan may also restrict access to the forest in specific ways and therefore reduce access to forest products that support people's livelihoods. Indeed, local users may not necessarily be interested in any kind of rotation at all, but only in selective needs-based extraction from natural growth. During the 19th century, the promotion of teak 'improvement planting' for British shipbuilding and export was a political decision. In the post-independence period, the deforestation of extensive areas of Dandakaranya area took place (now occurring in forests in Orissa, Andhra Pradesh and Chhattisgarh) in order to rehabilitate large numbers of refugees from East Pakistan. More recently, during the 1970s, it has been claimed that politicians encouraged local people to cut forests for economic gain, in this way seeking to win their votes. To take another example, socio-political choices over forest use have often been made in the name of national development, rather than promoting sustainable public use of the resource. In Nepal, the ejection of *Sukhumbas*i (literally squatters, but commonly historically settled tribal people not granted legal rights to their land) from forests in the *tarai* has been encouraged by politicians who used forest land and the trees on it as patrimony for favoured clients (Ghimire, 1998). These political issues also intersect with broader concerns, such as the formation of new states in the Indian union, tribal politics, an inequitable political economy, non-inclusive political representation, and armed struggle in a number of forested tracts in India, as well as the Maoist insurrection in Nepal.

New issues have also appeared in forest policy reform debates over recent decades, championed by international agencies, intellectuals within both India and Nepal, and the increasingly politicized and vocal marginalized sections, including Adivasis, Dalits and scheduled castes in India and Nepal. Albeit with different emphasis, all have focused on the assurance of providing basic needs from the forest, the participation of local forest users (especially poorer groups) in forest management (hence, participatory forest management), gender equity, the democratization of environmental knowledge, and a sharper poverty focus on social and environmental justice. PFM has been the main focus of official policy reform although, in different guises. The idea is not new, as the historical account of policy in Chapter 1 shows. PFM policy often coexists uncomfortably with local people's informal and customary forest management practices (most explicitly, as we see in Chapter 8, in Orissa's self-initiated forest protection groups). In India, PFM is known as joint forest management (JFM); in Nepal the terms used are community forestry (CF) and the less widespread models of leasehold forestry (LHF) and collaborative forest management (CollFM).

Different policy actors, international funding agencies, activists and other commentators make a variety of claims of PFM. Calls have been made for environmental justice and environmental equity for forest fringe and forest dwellers, and for women; for more effective management of forests to achieve various conservation objectives, as well as improved income streams, especially for the poor; and for the improved provision of a range of forest products that underpin the livelihoods of local people. Furthermore, in much of the international literature, the process of participation itself is claimed to empower people, increase their sense of becoming citizens rather than remaining subjects (to use Mamdani's phrase in Mamdani, 1996), improve their political and organizational skills, and bring the advantages of more coordinated collective action that uncoordinated individual effort could never achieve. However, more recent critiques of participation have also been made, which are discussed and elaborated on in the context of PFM in Part III of this book. Participation for some implies accommodation of local people's wishes in forest management; but for others (usually forestry professionals), it implies a loss of centralized control, a dilution of management objectives at a national scale, and a disregard of scientific knowledge and research in making informed and sound technical decisions. The latter group is often observed to favour what might be called the 'I decide, you "participate"' interpretation of participation whereby local people are marshalled to achieve predetermined objectives in exchange for some promised share of benefits as an incentive. Thus, the issue of PFM is central to forest policy in India and Nepal, and informs a wide range of policy debates.

Issues of justice, equity between classes, ethnic groups and gender, sustainability, human rights and the purposes of forests are expressed in all manner of ways – as policy statements for public consumption, as laws and circulars, as manuals of standard operating procedures for foresters, as media items, and as verbal dialogue between local people and other actors at the village level. The individual words used and the argumentation are interwoven and used inter-textually, meaning that the same words and ideas may be employed by different actors to draw different and sometimes contradictory conclusions. As we discuss later, the term 'forest' itself has different meanings for different people, as do other terms, such as 'scientific forestry', 'forest protection', 'participation' and the involvement of local people in decision-making and activities. 'Participation' as used in PFM can be advocated in the service of competing – and sometimes contradictory – narratives. For example, participation can be said to be desirable *instrumentally* because it reduces management and policing costs on forest lands and, hence, reduces the responsibilities of the forest administration without undue risk of over-harvesting of an unprotected resource. At the same time, participation may be advocated *as an end in itself* on the grounds of social criteria such as empowerment, social justice and the development of social institutions. Alternatively, it can be *pragmatically* argued that the participation of forest users in forest management may be beneficial in some circumstances for sustaining forest quality, whereas – this argument goes – without the participation of forest users in planning, managing and protecting the forest, it becomes an open-access resource and forest quality will (continue to) decline. Thus, the different meanings of specific words used in policy argumentation is one of the central ideas of the book, returned to mainly in Chapter 3 (see Roe, 1994; Hajer, 1995; and Apthorpe and Gaspar, 1996, for a more abstract discussion, and in an applied Indian context, see Agrawal and Sivaramakrishnan, 2001)

The variety of meanings attributed to participation has been encouraged as a result of the wide variety of actors who use the word in forest policy debates. There have also been convergences of their concerns. To use Hajer's term, 'discursive coalitions' (Hajer, 1995, p13) can be identified around the term 'participation', in which different actors connect formally and informally with forest policy groups around specific storylines even though they may never have met and strategized together, and even though they may have, in other aspects, divergent agendas. Pressures and incentives have come both from the *international sphere* (multilateral organizations, bilateral donors and international groups of foresters), and from local social movements and in-country *activists and intellectuals*, often with good access to senior policy circles, and have been pressing for similar things developed in a variety of narratives for many years.

The debates over participation *per se* have now evolved into wider debates concerning forms of government, democratic decentralization, devolution and other governance issues. Therefore, participation has become a broader matter of contesting exclusive state ownership and control of forests, rather than simply their management with or without 'participation'. However, the forest narratives from within and outside India and Nepal cross an important political divide. Sovereign countries do not have to listen to, or at least take seriously, those narratives and policy styles promoted by international funding agencies (IFIs). They can pick and choose to incorporate those aspects that suit them for a wide variety of political and fiscal purposes, but may only acquiesce to these narratives of participation or pay lip service to them in order to attract donor funds. Here, the relative leverage of different international and bilateral institutions in India compared to Nepal is important in helping to explain how policy is formulated and implemented. The challenge in making sense of all this is that forest management is political and subject to many different and changing representations by different parties. 'Facts' are malleable and often fiercely contested, and there is a remarkably wide gap between rhetoric, the intention of the law, guidelines, policy documents and what really goes on in the field. As with some other state initiatives, (e.g. education, health provision, metropolitan plans, transport and irrigation developments), the outcomes are similarly difficult to agree on and have been highly contested. In all cases, it has been difficult to build a consensus on what monitoring criteria for participation in forest management might be. This is particularly so for forests where there has not been a consensus over the role of the forest sector in rural development and poverty alleviation, on what 'participation' actually means and is supposed to deliver, and on where the definition of 'participation' is enforced (i.e. whether the '*I* have decided that *you* participate' model or some other more democratic one is the norm).

Our approach

Our approach may not be as straightforward as many readers might like. A book about forest policy, in many people's view, may best be carried through by first identifying the problem clearly, and second, by presenting new and unequivocal evidence, seeing beyond and discounting political posturing, interrogating existing evidence and presenting facts, and then making a list of policy recommendations. In other words, the book should assume that 'truth can talk to power' (Wildavsky, 1979). On the other hand, as mentioned above, there has been a stream of publications following this model that have attempted to 'talk to power', but whose appeals to reason have had limited impact due to the entrenched position of the forest administration establishment (Gadgil and Guha's *Ecology and Equity*, 1995, being one of the most cogent). An alternative route, in view of the limited purchase of the above approach, may appeal more to academic than to professional audiences. This is the post-modernist route, involving deconstruction of different socially constructed legitimating narratives around the claims of the powerful to the forest and their divergence from different everyday practices of coercively enforced control of the forest (similar to Sivaramakrishnan's *Modern Forests*, 1999, for example). However, the authors of this book believe that there is a pragmatic middle way that combines a discursive approach (focus on words, narratives and argumentation and, essentially, political in nature) with empirical evidence and reasonably rigorous hypothesis testing. There are crucial assumptions in many of the competing narratives which *are* amenable to evidence-based testing – for example, changes in forest condition following the start of PFM in an area, different people's access to the forest before and after PFM, and whether species choice in the micro-plan for a village took account of local people's preferences.

All of the questions asked in this book are political, and the framing of hypotheses is clearly shaped by a particular political stance. However, at the same time, hypotheses can be tested in a clear, positivist manner. Therefore, analysis is not all about talk and meaning, but also, in a carefully circumscribed way, about proof and what is 'true' or 'false'. To take further examples, what are the impacts of different kinds of official restrictions on the future livelihoods of poor people and on forest quality? How do different land tenures, both *de jure* and *de facto*, impact on forest conditions and on the distribution of access to different groups of people, particularly the poor? These questions can be answered in fairly straightforward ways. The key issue here is the *impact of forest policy on poor people*. In the Indian case, many entire indigenous communities, not all of whom were originally deprived, have been made poor through the disenfranchisement and appropriation of their ancestral resources by forest administration (see, for instance, Singh, 1986). There are counterarguments that do not deny these historical processes but consider them necessary and unavoidable in the drive towards a modern society (which is discussed later). The book takes responsibility for choosing this issue as central in framing the research questions; but we have to acknowledge that it may not be the way in which other parties in forest management frame the 'problem'.

The title of the book includes the word 'power'. The preceding discussion has introduced the idea of *discursive power*. The rhetorical weight of policy argumentation, the presentation of evidence, persuasive language, claims of scientific authority and so on are key aspects of discursive power, but they are linked to other aspects of power. The first is the means by which knowledge is produced and disseminated. Therefore, large and relatively well-funded institutions such as the Indian forestry administration or the bigger IFIs can finance and undertake research and dictate their terms and objectives, while local forest users have their own knowledge and management techniques concerning their local forest, but do not have the means to represent them authoritatively as a feasible alternative to 'official' knowledge (see Banuri and Marglin's book with the self-explanatory title, *Who Will Save the Forests? Knowledge, Power and Environmental Destruction*, 1993, and, more recently, Sundar's 'The construction and destruction of indigenous knowledge in India' in Ellen et al, 2000). Our book has many examples of what we term the politics of forest knowledge, and it follows through its implications for forest sustainability and people's livelihoods.

Finally, other aspects of power concerning forests are exercised by different people. These involve other actors with particular interests in the forest, such as both international and indigenous conservation non-governmental organizations (NGOs) and individual conservationists who have, at various points, had a significant influence on forest policy and practice. Overall, interests in the forest across the range are usually exclusively material and specifically pecuniary. There are struggles over who defines the value of forest land and its constituent parts (for example, timber of different species, fodder and non-timber forest products, or NTFPs), who controls them and who can capture them. There are forest contractors, local landlords and their clients, and organized crime and corruption, together with politicians for whom forests may be part of their means of funding election campaigns, exerting control and getting people to do what they want, including using coercion, threats and acts of violence. There are other civil society actors such as the press, the insurgents in many parts of our study area, and formal institutions (e.g. the judiciary). The latter, in the form of the Supreme Court, has been especially influential in its public interest litigation judgments on forestry matters in India. Finally, there are also forest dwellers for whom the forest is home, culture, habitat and the material basis of their livelihoods. There is nothing specific to South Asia about these aspects of power – they are widespread internationally; but an effective analysis of forests and people must take account of them. Otherwise, analysis is confined to the debating chamber, where the written word, usually found in formal policy documents, becomes the focus of discussion and the messy reality of policy on the ground is overlooked.

This book sets out to answer some of the key questions about how and why policy is made and implemented, and about the impacts of PFM on forests and on different groups of people. It provides evidence and analysis and makes an effort to outline transparent argumentation in such a way that readers can make their own judgements. Nonetheless, no analysis of social and political issues such as forestry can successfully claim to be neutral. As we have already said, this book is political in the sense of prioritizing certain issues for investigation, being watchful of arguments and key words that carry heavy baggage and rhetorical statements aimed at non-participant audiences rather than at those who affect and are affected by what actually happens on the ground.

Throughout this book the issue of the power wielded by the forest administrations is central. It must not be assumed that this power only derives from the dominance it enjoys in policy argumentation. It also derives from a range of other factors. In India at least, the forest administration is one of the most well-established, durable and powerful clusters of civil institutions in South Asia. The *forest departments have enjoyed long-established control of huge areas of forest land* or 'land designated as forest' (over 22 per cent of the land area in India and 39 per cent in Nepal), and have constantly sought to extend and deepen that control ostensibly in order to fulfil their goal of conserving forests, halting their degradation and diminution in the name of the national interest, providing materials for national development, and, until recently, providing revenue generation. Chapters 1 and 2 discuss the historical antecedents and present the implications of this state of affairs. Territorial control is a major objective in its own right for the power it provides, and in India and Nepal the forest estate even includes large areas of 'forest' lands that neither have trees growing on them nor are suitable for growing timber in the future. Examples include high mountain areas above the tree line, such as Spiti district, Himachal Pradesh, or Ladakh in Kashmir, India, Mustang in Nepal and grasslands adjacent to villages throughout India.

We also examine the extent and nature of the forest services in India and Nepal on the ground, and here the picture is much more complex and, in some cases, surprising. Many of the works published during the late 1980s and early 1990s project the forest administration of India as a powerful entity able to enforce its stringent rules and regulations (Shiva, 1987; Gadgil and Guha, 1992). Recent literature on forestry shows that the forest administrations are not simply autonomous and powerful agencies, able to enforce their agenda unambiguously. Their policies are significantly constrained and altered by powerful political parties, leaders and other departments, not least the revenue department (Saberwal, 1999). Local elites also exercise a powerful influence on the nature of forest policy outcomes. Although a huge number of forest staff are notionally in place for the entire notified forest area, an increasing number of posts now remain vacant due to fiscal limitations on fresh recruitment. In many areas, there are severe constraints to even minimal surveillance, policing, extension and forest management. Thus, the territory formally under the management of the forest administrations may not demonstrate the power of the administration as it may seem when viewed through the lens of a map in a district office. The forest administrations of both India and Nepal (or, in some cases, individuals acting in their own informal capacity within them), and also often communities, have powerful *local* allies, such as forest contractors, sawmillers, manufacturers and traders of wood-based and other NTFPs, and the building, medical and aromatic plant industries. However, as Part II will show, these alliances have an equivocal role in the exercise of power by forest administrations. Usually, they have diverse interests in obtaining access to various forest land products that are not necessarily congruent with the objectives of official forest management policy. Furthermore, policies such as PFM provide new opportunities and constraints to improving access to these products, and their different strategies to improve their position may not necessarily assist in reaching official policy goals at all.

While the power of the forest administrations to implement policy on the ground as it is written in policy documents is often equivocal and dissipated, their discursive power (particularly in India) is dominant. They have control of the production of official knowledge about forests. They have a number of distinguished colleges to train forest officers and undertake technical research. They can map forests, claim new territories, decide on the criteria for designating forest land, and control the drawing-up of forest working plans. In India, these are *only* prepared by the forest administration, and in the case of PFM micro-plans, they are usually dominated by the divisional forest officer. All of this can be argued by forestry professionals to be entirely necessary, requiring prudent and carefully thought out plans in the name of modern, sustainable and scientific practice, and it remains a central part of the official policy narrative. Perhaps this is a central claim of *all* official policy narratives!

Examining these claims and the overwhelming bureaucratic power that both draws on and constructs them requires that the authors of this book interrogate these powerful policy narratives, including other narratives from forest dwellers and activists. For example, Nepal's forest administration is different from that of India because although historically based on the Indian model, the forest department has never been able to establish effective control of forest areas in remote hill and mountain regions even though it has had a strong presence in accessible areas of the *tarai* and areas surrounding the main towns in the hills. It has been more accommodating to ideas from civil society and international donors. Government budgetary pressures in Nepal, particularly in recent years, have meant that donor-funded projects have a less troubled passage from conception to implementation on the ground. For these reasons therefore, the power to enact legislation, write manuals and shape the practice of forest management on the ground is more diffuse and less concentrated in Nepal's forest administration than it is in India. Other narratives from both national and local institutions are heard clearly in Nepal, as in some instances in India, too.

Since the late 1960s, both countries have been characterized by the spread of insurgencies, particularly in the impoverished forested areas and more remote rural areas, involving a range of groups that have become known as *Naxalites* in eastern and central India, and the more unitary *Maobadi* in Nepal (which has spread across Nepal since the 1990s and is currently destabilizing the state structures). This has precipitated a renewed crisis in state governability. The insurgent's cause is viewed with varying levels of legitimacy by rural communities in whose name they act, particularly where armed conflict has descended into a dismal cycle of tit-for-tat brutality and opportunistic gangsterism or has brought repressive onslaughts by state security agencies. Nevertheless, the various groups' manifestoes generally involve demands for revision to iniquitous land and forest rights regimes, as well as redress of the alleged corruption of the forest administration. Their relations with existing NGOs, bilateral forestry projects and community-based organizations (CBOs) have brought about a variety of outcomes (elimination, accommodation, 'business as usual' and adaptation). These, too, are examined in this book. Here again, other political considerations, not primarily centred on forests and people, impinge strongly on forest policy and its outcomes.

Finally, in the discussion of our approach, there is the issue of accounting for policy impacts. A *diversity of policy and outcomes* in terms of forest management and livelihoods operates on a number of different scales and is familiar to statisticians, geographers and other social and natural scientists. There is also a diversity of policy in both space (different policies in different states, or across other jurisdictional borders) and time. Policies are not set in stone, but are dynamic. They resemble an amoeba, slippery and mobile, and without clear and definable boundaries where the policy effect can be unambiguously differentiated from other causal factors, background noise and contingencies. It is therefore usually difficult to identify and measure the policy effect (see Long and Van der Ploeg, 1989, for a generic discussion on policy, and Blaikie and Sadeque, 2000, for examples in the Himalayan region). This book engages continually with this problem. Spatial and temporal diversity also have troublesome implications for policy-makers everywhere. Universalizing and reductionist blueprint policies and laws are generally applied to stabilize policy-makers' expectations and make complex realities apparently understandable and governable (see Roe, 1994, and Scott, 1998, for analyses of the ways in which bureaucracies handle and govern complexities and uncertainties). However, they inevitably cause mismatches with local conditions that produce unintended outcomes over a variable political and ecological terrain. Indeed, this difficulty is one of the arguments for decentralization of forest management and PFM, where local conditions can be matched to locally appropriate management plans overseen by those who have a strong interest in their being effective. For example, Nepal's *tarai* have a completely different forest ecology, settlement history and socio-economic structure, as well as disparate politics, than the hills. To take another example, the political environments of West Bengal and Andhra Pradesh are profoundly different and the same (or similar) forest policy may mean completely different things and have different impacts in each of the state capitals and on the ground. There are varying settlement histories, local agrarian political economies and forest types, all of which combine to produce different outcomes and different impacts. Even differences between the personality of one district forest officer (DFO) and another, and a transfer of personnel between a division, district or different supporting bilateral donor-supported projects and NGOs at the local level, can lead to major differences in outcomes for the condition of the forest, as well as for livelihoods. The very existence of diversity has policy implications, too, in terms of blueprinting or flexibility in policy and manuals for daily practice (e.g. the writing of a micro-plan for a local forest) and the discretion wielded by local officers in writing and implementing such material. The literature promoting the decentralization of environmental management by local communities is extensive; a few key references are given here and others are discussed throughout the book, especially in Part III (Berkes, 1989; Bromley, 1991; Cleaver, 1999, 2002). The other side of the decentralization coin is the increasing expropriation of indigenous common property and locally managed forests over the past 150 years (Singh, 1986).

Political ecology: Understanding the politics of the environment

Political ecology is a useful and rapidly developing conceptual approach, and here we briefly introduce how it is used in this book. This section is not intended as a full literature review of the burgeoning field. It provides an outline of the ways in which political ecological analysis may contribute to progressive, just and technically sound forest policy and practice, and gives some key references.

This book follows *four main strands of political ecology. The first strand* concerns the contested ways in which biophysical ecology is interpreted and negotiated. We are specifically concerned here with the ecology of 'forests', the inverted commas here are used to imply that the category itself is socially constructed and contested, and not intrinsically self-evident. The two words in the term 'political ecology' suggest two rather different approaches to knowledge and, more specifically, to truth (that is, what can be proved and disproved and what we can know). On the one hand, ecology and, more generally, environmental science are conventionally assumed to be separate and epistemologically different from politics. According to this assumption, ecology is objective, rational and empirically justified through experimentation, wherever possible, and politics is subjective and socially constructed by different persons or groups with their own beliefs, cultures and strategies. To return to the rhetorical question 'what is a forest?', there are a number of contending definitions. To tribal women, the word has a specific connotation, involving habitat, identity and, in material terms, specific products. For strategic planners, on the other hand, the category of forest is an administrative category implying a desired land use and it need not include any trees at all, but is useful in making claims to extend the control of the forest service over new areas in order to fulfil its mission of achieving a national target of a minimum percentage of green cover (see Robbins, 2003, on the politics of a seemingly technical issue of land categorization in Rajasthan). Different actors put different values on nature (Kothari et al, 2003); but these tend to be overlain in official, formal and policy arenas by the rationalist claims of scientific forestry and the research that informs it. Agrawal (2005) has coined the term 'environmentality', which traces the connections between power, knowledge, institutions (particularly those of the state) and subjectivities, with field examples from forest policy in the Kumaon region of the Himalayas.

Thus, political ecology (re-)integrates environmental science and politics, and acknowledges that there is a politics of science (in this book, 'scientific forest management'), as well as of other forest knowledge, such as indigenous forest knowledge, popular knowledge in the media and so on (Stott and Sullivan, 2000, pp15–116; Forsyth, 2003, pp1–23). Two books with the same title discuss this issue (Agrawal and Sivaramakrishnan, 2001, with case studies from India; and Castree and Braun, 2001, focusing on more theoretical issues and a wider geographical spectrum). Saberwal and Rangarajan (2003) provide a number of case studies of the political ecology of ecological science and the conservation of fauna and flora, with an emphasis on India. The issues raised in these and other publications will be returned to later.

Our approach therefore develops a critical stance with regard to *all* competing accounts of the forest given by people who are in some way or another linked to it. It treats with scepticism the assumption that policy develops when new scientific evidence is presented to policy-makers who accept the new truth and adapt or adopt policy accordingly – in other words, when 'truth talks to power'. Thus, political ecology combines issues of power, the construction of knowledge, argumentation and the narratives in which these are embedded. It does not take for granted powerful and (for some) attractive narratives that are seemingly based on a single truth. Political ecology also has a wider focus on the ways in which natural resources (forest, grazing land, etc) are understood and represented in policy and civil society arenas. Nor does it leave unexamined populist assertions of the virtue and truth of other knowledge, such as local and indigenous knowledge. This, therefore, implies that the book focuses on the claims made by different actors and their resort to truth as verified by scientific testing, to natural justice, to equity, human rights or other means of persuasion. New narratives about forest, based on what Foucault called 'an insurrection of subjugated knowledge' (Foucault, 1980, in Forsyth, 2003, p157) should be acknowledged and given space, but also critiqued with the same rigour as dominant accounts. There are many explorations into what the production of subaltern knowledge might entail in terms of learning, syllabi of forest officers, new institutions, spaces for deliberation and so on (Campbell in Hobley, 1996). Back in the (more abstract) academy, Escobar (1998) calls for new and sometimes hybrid accounts of life and culture (and, in this context, forests and people). Peet and Watts (1996) describe their approach as 'Liberation Ecologies', in which they call for subaltern peoples to be allowed to speak for themselves, to be free to talk about their experiences – in this case, of forests and the politics of control and use – and to be heard by other more powerful actors who have a near monopoly on the production and dissemination of knowledge about the environment (and, in this case, forests). Chapters 2, 3 and 4 discuss the different actors engaged in forest management and their narratives (by now the list will be familiar, even to those new to forest management: scientific management, participation, cultural survival of forest dwellers, and livelihood needs of forest-adjacent and forest-fringe people and 'distant' forest users).

In the second strand, political ecology provides more structural explanations of the ways in which different groups gain access to the 'forest': who becomes marginalized; who gains and who loses; how (that is, strategies of interested parties, and who succeeds in carrying them out); and why (e.g. the exercise of differential economic power, coercion and violence). In this book, understanding access to natural resources, especially the forest, is central to understanding how forest policy and the current state of forests impact on society. Forest policy does not work itself out on a blank canvas, but is embedded in state, regional and local political ecologies. A particularly unequal agrarian political economy with marginalized and vulnerable forest users will shape the impact of a forest policy in a different manner from a more egalitarian and politically aware agrarian society subject to the same forest policy. Following some political ecological studies, we have considered more strictly political economy issues of class and social stratification, capital accumulation and the role of the state (see Blaikie, 1985; Blaikie and Brookfield, 1987; Watts, 1993, for a more structural approach). This has been termed the 'environmental politics' or the 'politicized environment' aspect of political ecology (Zimmerer and Bassett, 2003, p3).

In the third strand, political ecology addresses the dialectic relationship between ecology and society. A constantly evolving dynamic is at work. Forests shape people (their habitat and material practices, technology, identity and culture), and, at the same time, people shape forests – through ongoing livelihood use, as well as policy development and formal management practices, and their intended and unintended outcomes. For example, one might examine forest-dwelling refugees expelled by dam construction and follow through the environmental consequences of these refugees in terms of forest use and their adaptation to new patterns of earning a livelihood. Another example would be high-value monoculture teak forests planted to replace natural forest – a political decision taken over a century ago – which has created valuable single-species forests from which the state is anxious to exclude all livelihood uses. In this case, today's exclusions result from the political ecology of more than a century ago, but result in very contemporary and immediate environmental and social consequences.

Thus, political ecology, with its strong historical sense, explains present-day relations between the agrarian political economy and the forest (distribution, composition, quality, commercial value, diversity, etc) in terms of settlement history, class structure and local nexus between the forest service, local elites and politicians. These explanations put agrarian political economy and its interactions with forest policy into a context of spatial variation through time, and emphasize the variations of people's relations with the forest and forest land. It is useful to characterize a number of political ecological zones in each state since these relations between people and trees vary, sometimes markedly. For example, central West Bengal has little, often very degraded, forest, and limited interest is shown in current standing timber by commercial actors. However, opportunities for plantations with longer rotations will obviously attract attention from commercial and industrial actors in the future. In contrast, North Bengal has plenty of commercially very valuable timber, high levels of cross-border activity, some smuggling, old tea estates with particular timber demands, and many very poor out-of-work tea estate workers who rely on the forest for subsistence – a different policy challenge altogether. But in the south-west of the state (studied in detail in Chapter 7), there are large forested tracts and concentrations of tribal people, which presents a particular set of socio-economic relations between the agrarian political economy and forests. In conclusion, an historically rooted political ecology makes sense of variations in people–forest links and of the outcomes of forest policy by understanding how policy outcomes are mediated at lower geographical scales.

In Nepal, there are similar or, in some areas, even more marked variations in political ecology and the histories that produced them. The northern Himalayan region covers alpine and high altitude forests, shrubs and rangelands where population is sparse. While forests are important to livelihoods, they are linked to pastoral systems, medicinal herb collection and higher altitude agriculture. In the mid-hills, farmers are particularly engaged in forest protection and management since their livelihoods rely much more intensively on forests, which provide materials for subsistence farming and the sustenance of household livelihoods (primarily as a source of firewood, fodder for stall-fed livestock, soil nutrients for privately cultivated agricultural land, and construction timber). In the inner *tarai* (the Siwalik and Churia hills and inner valleys) and in the *tarai* proper, the situation is again markedly different. Forests have recently undergone rapid felling since standing timber is commercially very valuable, and there has been significant migration to the region from the hills. Local populations, many of whom are tribal, have been expelled by more powerful settlers and commercial fellers backed by the state itself through both legal and illegal means, often involving serious violence, killings and burning of forest dwellers' houses (Ghimire, 1998). The political ecology of the *tarai* and the rest of Nepal is so different from that of the middle hills that the book divides the case study of PFM in Nepal into two separate chapters (Chapters 5 and 6).

In the fourth strand, a political ecology approach leads to critical understanding of how environmental policy is made, the exercise of power, practices on the ground and the discourses that shape them at different levels. Such an understanding throws light on how the participation of local people in forest management, particularly the poor in an already inegalitarian agrarian society, might be pursued – not only by policy reform 'from above', but via other routes taken by a variety of different actors that bypass some of the roadblocks which stand in the way of justice and a reduction in rural poverty.

Pessimistic and optimistic stereotypes

Two contrasting assessments of PFM can be characterized. The first tends to radical pessimism and is often generated by a structuralist explanation in which human agency and political dynamics are given less prominence than deterministic economic and political economic forces. Pessimism also arises from activists and others with an informed historical sense, who interpret the history of struggle in which the state and its class allies have won most of the battles, and who have experienced the entrenched and well-defended positions of the Indian and, to a lesser extent, Nepalese forest administrations.

A pessimistic view sees PFM, as implemented in practice, as detrimental to the livelihoods of the poor. This is because it is characterized by the persistence of long-term state-dominated forest policy frameworks that have extended and deepened control by the forestry services through defining new tracts of land as forest and specifying expanded regulatory and exclusive management programmes. Historically, this has had the result of reducing or extinguishing local rights and decision-making control through traditional management systems, thereby criminalizing local subsistence use, which may have been an established customary practice for generations. Contrary to initial expectations, PFM has, in many cases, led to a tightening rather than a loosening of central control and has not resulted in a real devolution of power. PFM also takes place in an already inegalitarian agrarian political economy in both India and Nepal (Panday, 1999; Timsina, 2002). The economic and political power of elites in rural areas enables them to take advantage of the disturbances in established practice provided by the introduction of PFM, and to influence and benefit from 'the rules of the game' governing who gets what. In some cases, PFM actually *increases* the leverage they possess over the poorer section of the population – quite contrary to the pro-poor intentions of PFM.

We use two metaphors to describe the main thrusts of the pessimistic view of the impact of PFM's policy, legal and administrative frameworks on forests and people. The first is that PFM is a 'wolf in sheep's clothing'. The appearance of democratic reform of forest management (the sheep's clothing) belies the 'wolf' that lurks beneath and turns out to devour the rights, produce and incomes of poor rural people – quite contrary to the skin-deep promises of PFM. Some would say it is no more than a distraction from the real power play on forest land control, where the forest department and other agencies have been asserting monopoly control of many areas, including tribal ancestral lands. Furthermore, PFM may represent a serious risk: adopting it results in the destruction of pre-existing local institutions through hitherto unrecognized rules and regulations and, *de facto*, the exercise of new powers over the majority of local people. Lastly, PFM institutions set up by government have not been embedded within community socio-political structures, nor have they been given any legal basis. Rather, they are more akin to transient conscription or 'company unions' coopted to achieve local peoples' compliance (see Hobley, 1996, p245, who summarizes the 'cynic's view of participation').

·   The second pessimistic metaphor is based on the argument that PFM has been little more than tokenistic 'oil on the squeaking wheel' of the remnant colonial forest management system, temporarily buying time to diffuse calls for a more drastic overhaul of forest governance. It is a discursive strategy to withstand international fashion, financial pressure and national clamour for post-colonial democratization of 'forest' land control. Furthermore, PFM has been a recent ahistorical and, as yet, unproved distraction of short duration compared to other processes of forest policy. PFM may therefore have served in some areas as a Trojan horse concealing subversive elements within its apparently benign form. Any progressive policy trends in forest administration are attributed to external pressure mostly from within India, but also less pressingly from international funding organizations – so the pessimistic story goes. The only reason the Ministry of Environment and Forests (MoEF) in India is now talking about 'historical injustice' to tribal communities in the consolidation of state forests at all is due to phenomenal pressure from the grassroots through politicians' mass protests. But even at the time of writing, the MoEF is fighting to retain control over the process instead of letting a new actor, the Ministry of Tribal Affairs, gain a recognized role in forest management in tribal areas.

The more *optimistic view* is that compared to the pre-PFM situation, there has been substantial improvement. PFM is part of what can only be a gradual process of reform in the management and use of forests in which local people, including the poor and marginalized, have, indeed, seen an improvement in the contribution of forests to their livelihoods and benefited from improved representation of their forest needs in micro-plans and working plans. PFM demands widespread and fundamental change in the work practices of forest administrations, in training, in attitudes of frontline staff, and in the financing of the forest service, its relations with politicians and its institutions. To expect reform of all these aspects in a matter of a generation is unrealistic. Moreover, there have been widely agreed on encouraging trends, particularly in forest condition:

> *Nepal's community forestry [Nepal's form of PFM] has proved that communities are able to protect, manage and utilize forest resources sustainably. The community forestry approach is therefore a source of inspiration to all of us working for sustainable forest management and users' rights. Nevertheless, further innovation, reflection and modification in community forestry are needed according to local context to address the social issues, such as gender and equity.* (Pokharel, 2003, p6)

Here, difficult negotiations about the form of forest micro-plans have shifted from colonial-style timber extraction and 'fortress conservation' towards democratic, devolved and intensified management that still preserves production objectives, biodiversity conservation and watershed protection. Progressive developments, policy learning exercises and win–win outcomes can all be found. Progress is bound to be slow, the argument runs; but the momentum has now become unstoppable. Furthermore, forestry is being increasingly politicized and has become a matter of winning potential votes in state legislative assemblies in India. In Nepal, networking and alliances of forest user groups for different purposes have been formed. Social movements, CBOs, NGOs and new alliances (even with forestry staff of all levels who are favourably inclined towards more devolution of forest management) are beginning to shape PFM in a more accountable and democratic manner in the mid-hills of Nepal and also in West Bengal and Andhra Pradesh. There is considerable evidence that the adverse effects suggested by pessimistic interpretations of current trends are being officially noted and more openly discussed, and that a sustainable policy learning process has been established (Kanel, 2004).

The explanation of *how and why* these policy, legal and administrative frameworks have evolved and the possible direction of further developments requires a deeper analysis of the policy process itself and is discussed below. There are a variety of frameworks for forest management from within the forestry services in India and Nepal, from foreign donors and a variety of existing, customary or self-initiated forest management institutions that predate the introduction of PFM. All of these receive critical attention. Both the more critical and the optimistic views have implications for policy and intervention. The positive view would encourage gradualist interventions working towards reform within the general current structures. A more critical view would suggest that the current structures are dysfunctional, often unable to resist the vested interests of the more powerful, and require drastic overhaul, which is unlikely to come from within the forest service itself. Therefore, a strategy of working for change from outside the forest service is also indicated. The view of this book is that both strategies need to be followed.

Key research questions and summary findings

This section presents five basic research questions that the authors believe are central to the issue of forest policy reform, at the centre of which is participatory forest management. They are initially discussed here and expanded on in detail at the national level for India and Nepal in Part I of the book, and then in Part II within India for the states of Andhra Pradesh, Orissa and West Bengal, and in Nepal for the Nepalese hills and *tarai*. The implications of these findings for policy reform are discussed separately in Part III.

1 What have been the livelihood impacts of the different implementation strategies of PFM in varying areas of India and Nepal?

This is the first and most important empirical research goal of the book. This is because there is a fundamental difference of view between many in the forest services of India and Nepal and local forest users, activists and consultants over the extent to which forests actually underpin the livelihoods of people who live in or near them, and the importance of this issue. If the forest provides negligible contributions to people's livelihoods, then exclusions and strictures on the use of forest resources in the name of forest protection and production-oriented forest management will not have a serious social or economic impact on forest-adjacent people and are therefore justifiable in the name of national economic development, modernization and environmental conservation. On the other hand, where forest use forms a significant part of livelihoods, particularly of poorer groups, the enhancement of livelihood-oriented use of forest land and forest resources by forest-adjacent people should form an important goal of forest management. A third view accepts the current dependence of forest-adjacent users, but seeks a different solution. Forest-dependent livelihoods can and should be minimized through alternative livelihoods, such as 'eco-development'.

We are particularly concerned to understand how *de facto* forest resource management and access opportunities have changed under PFM. Findings predictably reflect wide variations in both initial conditions and as influences on the implementation process due to the large number and diversity of physical, social and institutional factors, as mentioned earlier. The approach suggested is to recognize that there is a wide variety of political ecological regions that shape the outcomes of not entirely uniform PFM policies. However, although the jury is still out in areas where forest regeneration or continuing degradation as a result of PFM has had insufficient time to affect livelihoods, there *are* discernable patterns of environmental and social change as a result of the implementation of PFM in some areas.

Impacts. The impacts of PFM on livelihoods have been very varied, both at the local and intra-household, as well as at the regional and state levels. It is not surprising that the wealthy use the forest less for subsistence purposes and petty commodity sales than the medium-rich marginal farmers and the landless, but that they are often in a position to gain more from the new opportunities which PFM offers them regarding both access to and distribution of forest products and in terms of defining forest management priorities to suit their economic needs. The contribution of total income derived from the forest varies between about 10 per cent and over 35 per cent for sample households by village. However, it is the poor and those with little or no private land who rely on the forest most. The forest also provides essential wild foods and income opportunities in the difficult dry season before cultivated crops mature. Tribal people with a long history of forest habitation and those who have been marginalized to the least productive lands have the greatest reliance on the forest.

Most groups have done better from PFM in terms of improved and legitimated access to the forest for their livelihoods. Off-take of the varied products of the forest (e.g. fuelwood, fodder, wild foods and NTFPs) and water security have improved overall, although at a much lower level than might have been the case if livelihood-oriented forest management systems had been introduced. Usually, access to fuelwood tends to decline for an initial period of several years due to closure for protection. This may extend to the longer term, when restrictions are placed on the quantity or quality of fuelwood that may be collected – for example, only dead and fallen twigs and branches, of which there are few in a degraded forest.

However, there are considerable exceptions to this optimistic finding, mostly concerning poorer and politically weak groups. Earlier studies, including one by the World Bank, have indicated that groups such as head loaders end up as major losers since most PFM groups ban collection for sale. Similarly, with regeneration of tree growth and the establishment of plantations, many NTFPs and fodder grasses of value to the poor may decline due to canopy shade. Many user groups ban grazing altogether, which places households dependent on wage labour at a serious disadvantage since they can hand-harvest fodder only at the cost of losing wages. Studies in Andhra Pradesh suggest that traditional grazier communities have totally lost access to their grazing lands and have often been excluded from *Vana Samarakshyan Samiti* (forest protection committee, or VSS) membership (thereby even losing wage labour opportunities and future entitlement to shares of income from PFM). There are some politically marginal groups whose access to the forest has further deteriorated (for an earlier review, see Sarin et al, 2003). In Nepal, the rural poor have been better able to position themselves in the new PFM dispensation than in India (although, again, there are some studies indicating the opposite; see Malla, 2000; Malla et al, 2003).

In both countries, these cases of improved access to the forest have far more important livelihood implications for the landless and the poor, who have little or no private land-based resources. However, the changes have been very modest compared with what might have been anticipated from earlier claims for PFM, with 'final harvest' benefits turning out to be minimal. There has also been a considerable loss of cultivatable land for groups in some areas due to the imposition of the 'forest' category on land under *de facto* cultivation, forest fallows and grazing land through PFM (especially in tribal upland areas in Andhra Pradesh). This has resulted in loss of access for livelihood uses, such as subsistence crop production critical for food security and grazing for the poor due to its perception as being inimical to 'tree' growth.

2 How do different policy, legal and administrative frameworks of forest management affect livelihoods, especially those of the poor?

While the research answers to question 1 review the impacts of policy on livelihoods in general terms, question 2 focuses on the impacts of different policy, legal and administrative frameworks on outcomes.

Impacts. The findings about the impacts of different policy, legal and administrative frameworks on livelihoods tell a complex story. The official implementation of PFM in degraded areas has usually proved more effective in improving forest conditions than the pre-PFM 'fence-and-fine' approach to forest management, especially in India (although in Nepal this previous approach had been comparatively much less in evidence). However, contrary to this general finding, many PFM village forest micro-plans have been written without genuine and wide consultation with local people and with a lack of transparency.

The selection of tree species, rotations and protection measures were considered to be contrary to the wishes of local people, who were usually sidelined, the tree species being determined by default in the selection made in forest department 'macro' working plans. In many areas of the three Indian states, local DFOs did not take the participatory and consultative processes of writing the micro-plan seriously at all – they did not see the point of it since the micro-plan had already been written at the district level using criteria and goals consistent with the working plan. Thus, while participation in micro-planning existed on paper, it was seldom realized to any meaningful extent. Furthermore, in some (particularly tribal) areas, PFM is implicated in a wider effort by the forest service to extend and deepen controls over lands hitherto protected by Schedules 5 and 6 of the Indian Constitution or subject to ongoing filing of claims for customary rights, with the result that many, particularly tribal, people have ended up with more restrictions and less access to their customary lands and forests than before (the 'wolf in sheep's clothing' metaphor applies in these cases).

In India, the introduction of the JFM programme through administrative orders rather than through changes to the law in most states has been a major constraint. The terms of partnership between government and local communities are not based in parliamentary legal process and law, but rather in discretionary bureaucratic orders. Local forest management groups have no independent legal existence and are not, as yet, linked to the decentralized local government system. In many states, government orders have been changed several times through changes in administrative orders, leading to confusion among both forest department field staff and local communities. For instance, in Orissa, JFM groups formed on the basis of the 1988 and 1990 government resolutions were declared null and void by the resolution of July 2003 (Pattanaik, 2004). Land and forest tenure issues have been an overriding and contentious problem caused by the *de facto* precedence of forest department reservation of forests over the constitutional protection of tribal resource rights and of recognized formal processes of rights settlement.

Regarding the special issue of forest policy and legal and administrative frameworks for tribal areas, there have been serious problems of inter-sectoral confusion and coordination in the implementation of PFM (e.g. the coordination of multi-stakeholder processes, integrating PFM with local government, including *panchayats* and other departments such as revenue, rural development, the Integrated Tribal Development Agency (ITDA), public works and the intersection of different PFM policies). There has also been frequent bypassing of the Ministry of Tribal Affairs by the Indian forest administration. This has resulted in the forest management objectives of the IFS being in direct conflict with the constitutional objectives of safeguarding tribal resource rights and cultures, and has skewed the policy process towards the exclusive objective of forest protection.

In Nepal, devolution of forest management has been more effective because of heavy donor support (both financial and advisory), with most donors promoting PFM, a larger cadre of reform-minded forest officers and, compared with India, a shorter history of centralized state control. There have been enabling acts and regulations that have promoted PFM (both community forestry and, to a lesser extent, leasehold forestry and collaborative forest management), with much more generous conditions concerning sharing forest produce with local forest users. However, the district forest officer still generally takes a dominant role in conducting forest inventories, writing local operational plans (OPs) and monitoring the activities of the community forest user group (CFUG). In the hills, the formation of user groups has been rapid and we can estimate that at least half, if not two-thirds, have remained active despite the decline in field support due to *Maobadi* activities in the majority of districts. Forest condition has improved in the hills in the majority of CFUGs, and income from the sale of forest products has been spent on infrastructural improvements and has provided the capital for CFUGs' own credit provision. However, with a few exceptions, the sums involved were small per group. There has been some exclusion of poor people from membership of the newly formed CFUGs, and executive committees and local management are firmly in the hands of men (not women), the literate and the wealthy. Forest management in the *tarai* has been beset by malpractice, encouraged by the high commercial value of timber there, the high rate of immigration and forest clearance for agriculture, and competition between forest conservation and clearance. Distant users of the forest and, again, the poor and members of marginalized groups have been heavily penalized and excluded from both access to the forest and arrangements for the sale of timber through CFUGs. Relations between the district forest officer and CFUGs are significantly more difficult and tense than in the hills, where the former is restricted to a protection role often supported by firearms. There is evidence of widespread illegal felling with the connivance of CFUGs and the forest service.

3 How far have the claims and aspirations for PFM by different actors been fulfilled and what have been the main opportunities and constraints to their achievement?

These are two linked key questions and answers require an evaluation of outcomes in the field according to forest administrative staff and forest users of different types, as well as a comparison with what is stated in policy documents. The successes and shortfalls of policy goals are explained in terms of factors that favour or constrain 'participation' in forest management.

Outcomes. Turning to India first, community management of forests has a long lineage (see Chapter 1), and although there have been a number of cases of state-supported PFM during the 1930s onwards in the western Himalayas and Madras Presidency, the roots of the recent PFM programme in India can be traced back to experiments with community participation in the early 1970s – most notably in two key experiments in the villages of Arabari in West Bengal and Sukhomajri in Haryana. Building on these positive experiences, the National Forest Policy (NFP) issued in 1988 retained the focus on forest conservation; but livelihood requirements of forest-fringe communities were also mentioned as one of the basic objectives of forest management. More significantly, the policy document stated that a 'massive people's movement' with the active involvement of women should be created to meet the country's forest management objectives. Subsequently, in 1990, the Ministry of Environment and Forests (MoEF) issued a circular that led to the formal launch of the PFM programme in the country.

The 1990 circular clearly stated that the PFM programme was to be limited to 'degraded' forests. From the 1990 JFM notification:

> The National Forest Policy, 1988, envisages people's involvement in the development and protection of forests. The requirements of fuelwood, fodder and small timber, such as house-building material, of the tribals and other villagers living in and near the forests are to be treated as first charge on forest produce. The policy document envisages it as one of the essentials of forest management that the forest communities should be motivated to identify themselves with the development and protection of forests from which they derive benefits.

In 2000, the MoEF issued another set of guidelines stating that PFM may be extended to 'good' forests, although most states have not yet included 'good' forests under the programme. The exclusion of 'good' forests from the PFM programme makes it clear that the state's main objective has been, and continues to be, the regeneration of degraded forests and the extension of forest cover, and community involvement is seen as an effective strategy for achieving this objective. This has become even more explicit with the launch of the National Afforestation Programme in 2000, which aims to bring one-third of the country's area under forest and tree cover, mainly through PFM.

Impacts. The claims and aspirations of the forest administration in India regarding the introduction of PFM (specifically, the JFM programme) are difficult to gauge since the attitudes of the majority to PFM are ambivalent or, more openly, opposed. There has been little in the way of efforts to mobilize 'a massive people's movement', or to involve people in the development and production of forests other than by offering financial incentives in the longer term and (in some cases of externally funded projects) employment in forest management for a few years. However, some of the aspirations of the Indian forest administration to improve forest quality and ground cover have been achieved insofar as JFM has facilitated this. Only degraded forests have been given over to JFM by state administrations, and although it has also been the intention to make over 'good' forests, state forest administrations have been reluctant to do so despite a range of experiences. In Maharashtra, for instance, community struggles relating to forests (e.g. in Mendha-Lekha, Gadchiroli) led the state government to accept that standing forests can be given for JFM. The aspirations of local forest users in the face of JFM are frequently framed by informal institutions that existed at the time of the introduction of PFM. In some states (e.g. in Orissa, studied in Part II), there is a long history of such institutions, and JFM often undermined and confused existing arrangements. In addition, the 'deal' which members could expect was often much less flexible and generous than that given in JFM.

It is important to be able to gauge the level of practical commitment as opposed to rhetorical strategies on the part of India's forest administration. The findings throughout this book speak of ambivalence, public versus private and professional agenda, and widespread resistance to JFM within the service from the majority. On the other hand, from the state's perspective, the programme has been successful in regenerating degraded forests in many parts of the country. The Forest Survey of India's *The State of Forest Report* (FSI, 1999) in 1999 showed that the overall forest and tree cover in the country had increased by 3896km^2 and dense cover by 10,098km^2, compared to the assessment made in 1997. One of the reasons cited for this improvement was implementation of the PFM programme (FSI, 1999). However, there is widespread scepticism of these statistics, as is discussed later in this book.

Local communities, on the other hand, have often seen PFM programmes as a means of greater access and control over forest resources near their villages; decriminalization of forest produce extraction and reduction in harassment by local forest department officials; wage employment opportunities; village development works; and enhancement of their cash income, at least in the short term. Several NGOs and activists view PFM as the first step towards the devolution of power and control over resources to the local level. Local communities' experiences seem quite varied depending on the local context in which the programme was implemented, especially in instances where pre-existing forest management institutions existed. In many places, relations between forest department staff and community members seem to have improved. There has been significant employment generation, as well as the creation of infrastructure (e.g. check dams) in many areas, especially in states where the programme was funded through an externally assisted project. However, the sustainability of this inducement to cooperate with JFM is suspect when funding stops, and many shifting cultivators expressed their intention of returning to such cultivation as soon as paid employment ceases. There have also been reports of increased intra- and inter-village conflicts due to JFM. A study reported a large number of inter-village conflicts in Andhra Pradesh due to JFM (Samatha and CRYNet, 2001). There have also been reports of forest produce being extracted by industries after forests were regenerated by the local community due to pre-existing leases (Sarin et al, 2003).

In Nepal, the Forest Policy of 1989 (revised in 2000) emphasized that the community forestry (CF) programme (the main form of PFM in Nepal) would take priority over all other forest management strategies. This policy clearly stated that the priority of the PFM programmes would be to support the needs of the poorer communities or the poorer people in the community. The 1995 Forest Regulations made special provision for people living below the poverty line through another form of PFM called leasehold forestry (LHF). However, the impact of the LHF programme in reaching and benefiting poor people was limited because the Forest Act clearly mentioned that community forestry would have priority over LHF, and therefore the specific poverty focus was overtaken by CF, which had a much less poverty-focused brief.

About 14,000 CFUGs incorporating about 1.5 million households have been formed to date, and about 1.1 million hectares of forest area have been handed over as community forest. Despite its large coverage, the CF programme has had difficulty in addressing the needs of poorer people, especially in the *tarai* (Winrock, 2002; Kanel, 2004). One of the main factors facilitating the success of the CF programme was the 1993 Forest Act, which provided a legal basis for the implementation of CF, simplified the handover process and recognized CFUG as a self-governed, autonomous institution to manage and use community forests according to the operational plan. However, some subsequent amendments to the Forest Act and statutes and circulars issued in relation to CF since the act have contributed to widespread controversy, highlighted especially by the Federation of Community Forest Users, Nepal (FECOFUN). These included the imposition of a tax on the sale of forest products to non-users and portraying collaborative forest management (CollFM) as an alternative form of PFM in parts of the *tarai* districts. In addition, since the enactment of the Local Self-Governance Act (LSGA) in 1998, there have been counterclaims about forest resource management and the right to collect taxes by local government authorities (village and district development committees – VDCs and DDCs). Therefore there is now considerable uncertainty over the future of CF, exacerbated by the present political turmoil in the country.

4 What have been the most important factors in facilitating or inhibiting the sort of PFM that enhances livelihoods, especially the livelihoods of the poor?

Factors facilitating PFM. The most important factors regarding successful PFM implementation at the local level were found to be:

- wide and inclusive representation, combined with informed participation in decision-making by all sections in the writing of the micro-plan and the working plan, and (for local management) an understanding of overall policies, schemes and programmes;
- a long-term deliberative relationship between forest department staff and wide representation of local people (often a matter of the personality and professional motivation of the district forest officer, combined with the availability of capable leadership and public spiritedness within the membership of local forest users);
- the influence of donor involvement both in providing policy incentives for change and in acknowledging the difficulties of sustaining policy initiatives when donor funding stops;
- a favourable local 'political ecology' of forest, people and politics (including a useful forest for local users; political awareness and adequate representation of local users, especially the poor and poor women; a consensus on the values and desirable uses of different forest products; and the means to protect the forest from outsiders and from those in the user group who infringe on a widely accepted and understood set of rules);
- state politics that, at the minimum, do not interfere in policy and implementation for political favour and advantage; and
- long experience of local management of forests by villages and their committees in cases of customary and self-initiated village protection that predate PFM (mostly in Andhra Pradesh, Orissa and in the middle hills of Nepal).

Factors inhibiting PFM. Those factors inhibiting PFM are the absence or opposite of the favourable factors listed above, and a number of more pervasive and general factors. As mentioned earlier, the Indian forest administration is, at all levels, ambivalent towards PFM, although a significant minority favours it in some form. Foresters are trained to be authoritative and scientific protectors of the forest, and not social engineers or local facilitators. Nor do district forest officers (DFOs) have sufficient resources, time and skills to get involved in time-consuming negotiations, follow-up and monitoring. In the case of India, PFM is also seen by some as a betrayal of the forest administration's historic mission of forest protection and extension of green cover through scientific management. These attitudinal aspects of most (but by no means all) forest staff in India are an important inhibiting factor in the implementation of the participatory aspects of PFM.

In Nepal, there has not been the same long historical experience of state responsibility in forest management, and local responsibility for forest management has not experienced the same intensity and reach of state control as in India. Forest user groups have been able to assume forest management to a greater extent than in India. However, there is still an ongoing struggle between two different forest policy styles. The first is the *older style* of centralized conservationist policy, linked in the *tarai* to the possibility of valuable forest revenues from timber being threatened by the formation of user groups, more transparency, and dilution of revenue claims for the state, some of its well-placed employees and their informal allies. The second style is much more *participatory*, although balanced against professional training, which prioritizes conservation (often with very conservative rates of off-take) and forest inventories (which favour species of commercial value at the expense of species and practices important for a wide range of subsistence needs). While the second style is certainly more conducive to a PFM that supports livelihoods, it is also overshadowed by more traditional professional priorities. There are also political initiatives to recentralize the management of forests and to increase taxation on CFUGs' forest produce.

Finally, a more pervasive problem militating against participation in PFM is the manner in which forest knowledge is created and acknowledged. Nepal's forest administration has the power to 'frame and name' the landscape, and to author manuals of forest management, including forest rotations, planting, coppicing and felling techniques. Its historic management objective of commercial timber production and, later on, of green cover and afforestation has resulted in an emphasis on long-term rotations of high-value species monocultures. If local people were able to assert their customary practices, integrating livelihood objectives within multifunctional and multiple-use local resource management, this would be more likely to generate greater opportunities for remunerative work, shorter rotation or selective felling for multiple products for subsistence and market. While statistical description, cartographic recording and manuals of accepted practice are all essential management tools, they imply authoritative knowledge and exclude alternative technical and practical knowledge held and used by forest users themselves. In addition, knowledge of rights, procedures and political paths to resist the erosion of liberties or to reassert rights to the forest; information about local meetings and negotiations; the ability to take part effectively in negotiations about forest plans; the monitoring of results and audits – all of these were asymmetrically distributed at the interface between Nepal's forest administration and a dispersed and often politically disorganized and differentiated rural population. This is not surprising bearing in mind the centralized, regulatory role that forest administrations have exercised for over 150 years along with all of the management tools to implement it.

5 What have been the ecological impacts of PFM?

Impacts. The general impact of PFM implementation has been an improvement in forest cover and condition. In four of the five study areas, forests have unambiguously improved

through increased protection and restricted product extraction. However, state PFM implementation – especially in India – has promoted state forest priorities, often leading to
reduced biodiversity and, therefore, the range of livelihood products. It has also led to the
promotion of exotic plantations of species irrelevant to livelihood use (e.g. eucalyptus and
silver oak). In the unique conditions of Nepal's *tarai*, the impact is more ambiguous (see
Chapter 6). Forest extraction has been reduced, although whether forests have improved, or
whether their decline has slowed, is a difficult question to answer, as we shall see.

References

Agrawal, A. (2005) 'Environmentality: Community, intimate government, and the making of environ-
    mental subjects in Kumaon, India', *Current Anthropology*, vol 46, no 2, pp161–190
Agrawal, A. and Gibson, C. (2001) *Communities and the Environment: Ethnicity, Gender and the State
    in Community-Based Conservation*, New Brunswick, Rutgers University Press
Agrawal, A. and Sivaramakrishnan, K. (eds) (2001) *Social Nature: Resources, Representations, and Rule
    in India*, New Delhi, Oxford University Press
Apthorpe, R. and Gasper, D. (1996) *Arguing Development Policy: Frames and Discourses*, London,
    Frank Cass
Banuri, T. B. and Marglin, A. (1993) *Who Will Save the Forests? Knowledge, Power and Environmental
    Destruction*, London, Zed Books
Berkes, F. (ed) (1989) *Common Property Resources: Ecology and Community-Based Sustainable
    Development*, London, Belhaven
Blaikie, P. M. (1985) *The Political Economy of Soil Erosion in Developing Countries*, London, Longman
Blaikie, P. M. and Brookfield, H. (eds) (1987) *Land Degradation and Society*, London, Methuen
Blaikie, P. M. and Sadeque, Z. (2000) *Policy in High Places: Environment and Development in the
    Himalayan Region*, Kathmandu, Nepal, International Centre for Integrated Mountain
    Development (ICIMOD)
Brandis, D. (1897; reprinted 1994) *Forestry in India*, Dehradun, India, Natraj Publishers (reprint by
    WWF India)
Bromley, D. W. (1991) *Environment and Economy: Property Rights and Public Policy*, Oxford, UK, and
    Cambridge, US, Blackwell
Castree, N. and Braun, B. (eds) (2001) *Social Nature: Theory, Practice and Politics*, Malden,
    Massachusetts, Blackwell Publishers
Cleaver, F. (1999) 'Paradoxes of participation: Questioning participatory approaches to development',
    Journal of International Development, vol 11, pp597–612
Cleaver, F. (2002) 'Institutions, agency and the limitations of participatory approaches to development',
    in Cooke, B. and Kothari, U. (eds) *Participation: The New Tyranny?*, London, Zed Books, pp36–55
Debroy, B. and Kaushik, P. D. (eds) (2005) *Energising Rural Development through Panchayats*, New
    Delhi, Academic Foundation
DFID (Department for International Development) (1999) *Shaping Forest Management: How
    Coalitions Manage Forests*, London, DFID
Divan, S. and Rosencranz, A. (2001) *Environmental Law and Policy in India: Cases Materials and
    Statutes*, New Delhi, Oxford University Press
Escobar, A. (1998) 'Whose knowledge, whose nature? Biodiversity, conservation and the political ecol-
    ogy of social movements', *Journal of Political Ecology*, vol 5, pp53–82
Forsyth, T. (2003) *Critical Political Ecology: The Politics of Environmental Science*, London, Routledge
FSI (Forest Survey of India) (1999) *The State of Forest Report*, Dehradun, Ministry of Environment and
    Forests, Government of India
Gadgil, M. (2001) *Ecological Journeys: The Science and Politics of Conservation in India*, New Delhi,
    Permanent Black
Gadgil, M. and Guha, R. (1992) *This Fissured Land: An Ecological History of India*, New Delhi, Oxford
    University Press
Gadgil, M. and Guha, R. (1995) *Ecology and Equity: Use and Abuse of Nature in
    Contemporary India*, London, Routledge
Ghimire, K. (1998) *Forest or Farm? The Politics of Poverty and Land Hunger in Nepal*, New Delhi,
    Manohar Publication

Grove, R. H., Damodaran, V. and Sangwan, S. (eds) (1998) *Nature and the Orient: Essays on the Ecological History of South and South East Asia*, New Delhi, Oxford University Press

Hajer, M. A. (1995) *The Politics of Environmental Discourse: Ecological Modernization and the Policy Process*, Oxford, Clarendon Press

Handmer, J., Norton, T. and Dovers, S. (eds) (2001) *Ecology, Uncertainty and Policy: Managing Ecosystems for Sustainability*, Harlow, Prentice-Hall

Hannigan, J. (1995) *Environmental Sociology: A Social Constructionist Perspective*, London, Routledge

Hobley, M. (1996) *Participatory Forestry: The Process of Change in India and Nepal*, London, Overseas Development Institute, Rural Development Forestry Study Guide No 3

Jeffery, R. and Sundar, N. (eds) (1999) *A New Moral Economy for India's Forests? Discourses of Community and Participation*, New Delhi, Sage

Kanel, K. (2004) 'Twenty-five years of community forestry: Contribution to Millennium Development Goals', in *Proceedings of the Fourth National Workshop on Community Forestry*, 4–6 August, Kathmandu, Nepal

Kautilya (1992) *The Arthashastra* (ed L. N. Rangarajan), Delhi, Penguin

Kothari, S., Ahmad, I. and Helmut Reifel, H. (eds) (2003) *The Value of Nature: Ecological Politics in India*, New Delhi, Rainbow Publications

Long, N. and Van der Ploeg, J. D. (1989) 'Demythologizing planned intervention: An actor perspective', *Wageningen Studies in Sociology*, vol 29, no 3/4, pp226–249

Malla, Y. B. (2000) 'Impact of community forestry policy on rural livelihoods and food security in Nepal', *Unasylva: International Journal of Forestry and Forest Industries*, vol 51, no 202, pp37–45, www.fao.org/documents/show_cdr.asp?url_file=/DOCREP/X7273E/x7273e07.htm

Malla Y. B., Hari, N. and Branney, P. (2003) 'Why aren't poor people benefiting more from community forestry?', *ODI Rural Development Forestry Newsletter/Journal of Forests and Livelihoods*, London, ODI

Mamdani, M. (1996) *Citizen and Subject: Contemporary Africa and the Legacy of Late Colonialism*, Princeton, Princeton University Press

O'Riordan, T. and Stoll-Kleeman, S. (2002) 'Deliberative democracy and participatory biodiversity', in O'Riordan, T. and Stoll-Kleeman, S. (eds) *Biodiversity, Sustainability and Human Communities*, Cambridge, Cambridge University Press

Panday Devendra Raj (1999) *Nepal's Failed Development: Reflections of the Mission and Maladies*, Kathmandu, Nepal, Nepal South Asia Centre

Pattanaik, M. (2004) 'Orissa' in Bahuguna, V. K., Capistrano, D., Mitra, K. and Saigal, S. (eds) *Root to Canopy: Regenerating Forests through Community–State Partnerships*, New Delhi, Commonwealth Forestry Association (India Chapter) and Winrock International, India

Peet, R. and Watts, M. (eds) (1996) *Liberation Ecologies: Environment, Development, Social Movements*, London, Routledge

Pokharel, B. K. (2003) *Contribution of Community Forestry to People's Livelihoods and Forest Sustainability: Experience from Nepal*, Montevideo, World Rainforest Movement, www.wrm.org.uy/countries/Asia/Nepal.html

Robbins, P. (2003) 'Beyond ground truth: GIS and the environmental knowledge of herders, professional foresters, and other traditional communities', *Human Ecology*, vol 31, no 1, pp233–253

Roe, E. (1994) *Narrative Policy Analysis: Theory and Practice*, Durham, NC, Duke University Press

Roy, S. B. (ed) (1995) *Enabling Environment for Joint Forest Management*. New Delhi, Inter-India Publications

Saberwal, V. K. (1999) *Pastoral Politics: Shepherds, Bureaucrats and Conservation in the Western Himalaya*, New Delhi, Oxford University Press

Saberwal, V. and Rangarajan, M. (eds) (2003) *Battles over Nature: Science and the Politics of Conservation*, Delhi, Permanent Black

Saigal, S., Arora, H. and Rizvi, S. (2002) *The New Foresters: The Role of Private Enterprises in the Indian Forestry Sector*, London, International Institute for Environment and Development

Samatha and CRYNet (2001) *Joint Forest Management: A Critique Based on People's Perspectives*, Unpublished report

Sarin, M. with Singh, N. M., Sundar, N. and Bhogal, R. K. (2003) *Devolution as a Threat to Democratic Decision-Making in Forestry? Findings from Three States in India*, London, ODI Working Paper 197, February (also in Edmunds, D. and Wollenberg, E. (eds) (2003) *Local Forest Management: The Impacts of Devolution Policies*, London, Earthscan Publications)

Scott, J. C. (1998) *Seeing Like a State: How Certain Schemes to Improve the Human Condition Have*

Failed, New Haven, Yale University Press

Shiva, V. (1987) *Forestry Crisis and Forestry Myths: A Critical Review of Tropical Forests: A Call for Action*, Penang, World Rainforest Movement Publications

Singh, C. (1986) *Common Property and Common Poverty: India's Forests, Forest Dwellers and the Law*, New Delhi, Oxford University Press

Sivaramakrishnan, K. (1999) *Modern Forests: Statemaking and Environmental Change in Colonial Eastern India*, Delhi, Oxford University Press

Stocking, M., Helleman, H. and White, R. (2005) *Renewable Natural Resources Management for Mountain Communities*, Kathmandu, Nepal, International Centre for Integrated Mountain Development, and London, UK Department for International Development

Stott, P. and Sullivan, S. (eds) (2000) *Political Ecology: Science, Myth and Power*, London, Arnold Press

Sundar, N. (2000) 'The construction and destruction of indigenous knowledge in India', in Ellen, R., Parker, P. and Bicker, B. (eds) *Indigenous Environmental Knowledge and its Transformations: Critical Anthropological Perspectives*, London and New York, Routledge

Timsina, N. (2002) 'Empowerment or marginalization: A debate in community forestry in Nepal', *Journal of Forest and Livelihood*, vol 2, no 1, pp29–33

Victor, M., Lang, C. and Bornemeier, J. (eds) (1998) *Community Forestry at a Crossroads: Reflections and Future Directions in the Development of Community Forestry*, Proceedings of an International Seminar held in Bangkok, Thailand, 17–19 July 1997, RECOFTC Report No16, Bangkok

Watts, M. (1993) *Silent Violence: Food, Famine and Peasantry in Northern Nigeria*, Berkeley, University of California Press

Wildavsky, A. (1979) *Speaking Truth to Power: The Art and Craft of Policy Analysis*, Boston, MA, Little Brown and Co

Winrock International (2002) *Emerging Issues in Community Forestry in Nepal*, New Delhi, Winrock International India

Zimmerer, K. and Bassett, T. (eds) (2003) *Political Ecology: An Integrative Approach to Geography and Environment-Development Studies*, New York, Guilford

Part I

Key Issues and Approaches

In this first part of the book we consider the context of participatory reform in forest management in South Asia, as a preparation for the detailed analysis of the actual regional situations which are presented in Part II. In Chapter 1 we consider the historical background to the emergence of participatory forest management (PFM) in both Nepal and India. We turn our attention to considering the 'policy process' and the difficult path of policy reform in Chapter 2. We then, in Chapter 3, look in greater detail at the actors involved in forest management and their structural positions and engagement in the policy process. Lastly, in Chapter 4, we present an analytical approach to considering how local people's livelihoods are affected by forest management reform.

1

Annexation, Struggle and Response:
Forest, People and Power in India and Nepal

Oliver Springate-Baginski and Piers Blaikie
with Ajit Banerjee, Binod Bhatta, Om Prakash Dev, V. Ratna Reddy,
M. Gopinath Reddy, Sushil Saigal, Kailas Sarap and Madhu Sarin

The issues addressed in this book are not new. Conflict over the control and use of forests and forest lands has been in the political arena for as long as local people and the state have shared an interest in the same resource. We, therefore, need to examine the historical pattern of annexations of forest and land by the state, the struggles of the people in defence of livelihood-related forest access, and policy responses that have led to the emergence of participatory forest management (PFM). These must be understood in terms of long-term historical processes, including, in particular, the emergence of powerful and centralized state forestry agencies.

Over the past 20 years, PFM has, however, led to new paths in forest management – mainstreaming local peoples' involvement in forest management, especially those who had been increasingly challenged and marginalized by previous forest policies. In India, these date from before the 1878 Forest Act until the present time, and in Nepal, from the Nationalization of Forests in 1956/1957 until the 1970s.

The new direction of PFM emerged during a period of increasing academic attention to the parallel 'subaltern' history of local forest management – which has focused on both contestation and conflict against the imposition of colonial forest management (e.g. Guha, 1983, 1989; Gadgil and Guha, 1992, 1995; Chaudhury and Bandopadhyay, 2004). This academic attention was accompanied by grassroots protests and rebellions against commercial forest management by the state – for example, the Chipko and Jharkhand movements and widespread protests in Bastar, which refocused attention once again on issues of the rights of local forest users (GOUP, 1921; Guha, 1989; Saxena, 1995; Munda and Mullick, 2003).

In Nepal, recognition of the customary rights of local hill people to protect the forest and decide on resource use, coupled with innovative lessons from forest handover in Sidhupalchok district in the mid 1970s, laid the foundations of community forestry policy from 1976 to 1993. In 1956/1957, all forests in the country were nationalized with the intention of establishing state control over forest resources. Prior to the return of a constitutional monarchy and the overthrow of the ruling Rana family, forests were under feudal management control (mainly focused on large timber revenues from the valuable *tarai* forests). However, in the process of nationwide nationalization the Department of Forests (DoF) also acquired control over the forests in the hills. The DoF, at the time, was neither prepared nor equipped to shoulder responsibility for managing all of the country's forests. As a result, it channelled its efforts into the *tarai* region (with the most valuable timber) and neglected most of the hill forests altogether. The impact of forest nationalization in the hills was mixed. In some places, the forests were degraded or cleared since the DoF was unable to protect and manage them because nationalization had reduced what had been common property managed by local people to an open-access resource, where trust, rules of use and access to users and exclusion of outsiders were completely undermined. In other places, where people did not experience any difference due to nationalization (since DoF presence was non-existent anyway), the forest was used as before. However, even with strong protection and guarding, the *tarai* forests became increasingly fragmented and degraded. Malaria eradication, the construction of the East–West Highway, and official and illegal clearing of forest land for resettlement of hill migrants in the 1960s and early 1990s were important factors. Illegal trade in valuable *tarai* timber within Nepal and with adjoining India through the open border also contributed significantly to widespread felling. Such heavy degradation as a result of incompetent and corrupt forest management lent weight to those who believed that it was almost impossible to manage forests without some kind of people's participation and local protection of the forest by those who had a stake in it. Whether people's participation can provide an effective institutional alternative to corrupt and inefficient state-run alternatives remains to be seen, and is discussed in Chapter 5.

The emergence of participatory forest management in India

This section briefly introduces the long history of conflict over forest management in *India,* from which the recent PFM policies have emerged. The main elements have been well rehearsed in recent literature (Ravindranath and Sudha, 2004; Guha, 1983, 2001; Gadgil and Guha, 1995; Hobley, 1996; Grove et al, 1998; Jeffrey and Sundar, 1999; Ravindanath et al, 2000; Sundar et al, 2001; Edmunds et al, 2003). What concerns us here are the historical origins of the recent efforts to re-orientate forest management towards a more participatory style. At all times, attention is focused on the impacts of forest policy; changes to the rights and obligations of different parties; who gained access to what; who was represented at various levels; and what actually happened. There have been many excellent and invaluable research documents on the emergence of PFM that show that, in different guises, PFM has been an important issue for many years. These accounts usually have a strong agenda for promoting PFM and identify the main actors who have done so. For example, Jeffrey and Sundar (1999) provide an account of the emergence of PFM. It is interesting to note that the discussion of 'the impact of research and documentation' (Jeffrey and Sundar, 1999, p34f), as well as of that of donors, lays out their representations and findings; but there is no mention of the response from politicians, non-governmental organizations (NGOs) and the forest administrations at national and state level. This book acknowledges the obstacles in the path of PFM and goes further than many academic and policy-relevant accounts to ask *why* they are there, what are the causes of their seeming durability, and what strategies might be followed to improve sustainable livelihoods and forest quality. Table 1.1 provides a timeline of forest management in India since colonial times.

Records of forest management practices are available from as far back as the Mauryan Empire (circa 300 BC), indicating that timber has been a valued and traded commodity for millennia. The Mauryans created reserved forest areas for elephants, maintained by state employees. Emperor Asoka's edicts mention massive tree plantation activities by the state (Sagreiya, 1994). By the time of the Moghul era, a timber market had penetrated much of the Deccan and northern belt, and accelerated clearance of plains forests for agricultural land to increase state revenue had begun (Singh, 1996).

Table 1.1 *Evolution of participatory forest management in India*

Year	Event	Significance to/impact on participation
1864	Indian Forest Service constituted	Beginning of planned state forest management.
1865	Indian Forest Act	First attempt at legislation.
1878	Indian Forest Act (revised)	Negative: process of forest reservation started; alienation of many rural communities, protests and rebellions.
1882	Madras Presidency Forest Act	More elaborate and sensitive mechanisms for settling rights included.
1895	Forest Policy	Agriculture given priority over forestry.
1927	Indian Forest Act	This legislation still governs Indian forest administration; it included provision for village forests – but this was not implemented.
1930	Separation of Indian Forest Department into state-level forest departments	
1920s–1930s	Kumaon Forest Grievance Committee (1921) Punjab Garbett Commission (1937)	The two investigations in response to anti-forest reservation, anti-forest department agitations and burnings led to the inception of the *van panchayats*, withdrawal of the forest department from large reserve forest areas and restoration of customary rights in Uttar Pradesh and forest co-operative societies in Kangra.
1952	Forest Policy (India's first independent policy)	Emphasis on industrial and commercial needs; local needs labelled secondary to 'national' interest; *ad hoc* adoption of objective of 33% forest cover.
1950s	Nationalization of forests of princely states, *zamindars* (landlords) and private owners	Large-scale felling by owners before handing over to the government.
1950s–1970s	Continuing forest degradation and conflicts between the forest department and the rural communities due to prioritizing commercial exploitation at the cost of local livelihoods	Evidence of ineffectiveness of the forest policy and growing unrest; Chipko and Jharkhand movements, Bastar protests and unrest in Andhra Pradesh.
Early 1970s	Experiments with community participation on forest lands	These led to the later emergence of the JFM programme.
1972	Wildlife Protection Act	Creation of national parks and wildlife sanctuaries.
1976	National Commission on Agriculture (NCA) report Forests moved to concurrent list Ministry of Environment and Forests (MoEF) set up	Focus on replacing natural forests with commercial plantations; advent of social forestry on non-forest lands to reduce livelihood dependence on forests.
1980	Forest Conservation Act (with amendments in 1988)	Central permission for diverting forest land to other uses becomes mandatory. The leasing of forest land to any private party banned – which requires the timber and pulp industry to meet its needs from private lands; existing leases not to be renewed on expiry. A fundamental turning point from the economic exploitation of forests and forest land to conservation.

Table 1.1 *continued*

Year	Event	Significance to/impact on participation
1980 onwards	Green felling bans in many states	A 15-year ban on green felling in Uttar Pradesh hills by order of India's then Prime Minister Indira Gandhi, after the Chipko Andolan. Further green felling bans emerge as a result of many other states' self-imposed moratoria, mainly through chief minister's environmental initiatives (e.g. Himachal Pradesh in 1984; Gujarat in 1986; Orissa in 1992; and many others). Later, the National Level Committee on Forestry, popularly known as the Centrally Empowered Committee (CEC), imposes a range of restrictions, including a ban on green fellings above 1000m or in fragile areas or without an approved working plan (1996).
1985	National Wasteland Development Board (NWDB) set up	Large-scale afforestation programme starts.
1988	New Forest Policy	Focus on conservation and subsistence needs, as well as protection of rights.
1990 – 1 June	Government of India joint forest management (JFM) notification	Formal acceptance of the JFM approach.
1990 – Sep. 18	Government of India (GoI) guidelines for regularizing encroachments, settling disputed claims over forest lands and converting forest villages into revenue villages	Based on 1987–1989 report of the scheduled castes and scheduled tribes; commissioner points out widespread unrest in tribal areas due to faulty forest settlements under which rights are not recognized. Non-recognition of land leases given by revenue department due to faulty land records; dismal conditions in 'forest villages' under forest department control. All states asked to constitute committees of forest department, revenue department and tribal officials to resolve problems.
1992	73rd Amendment to the Constitution – *panchayati raj*	Made decentralized governance through elected three-tier *panchayati raj* (local government). Institutions with a list of 25 functions/activities to be assigned to them.
1996	*Panchayat* Extension to Scheduled Areas (PESA)	Under the 73rd Constitutional Amendment, PESA enacted specifically for Schedule V areas, permitting greater space for continuation of traditional systems. Revisions to central guidelines on JFM through an empowered *gram sabha* (village assembly).
1995 onwards	Supreme Court orders in T. N. Godavarman *versus* Union of India and CEL WWF-India *versus* Union of India and Others cases	The marathon Godavarman case (which began in Tamil Nadu but escalated to the Supreme Court) continues to be the biggest judicial intervention in forest administration in India. It led to the creation of the CEC under the Environment Protection Act. CEC directives have intervened in a range of issues, including state forest department felling and tribal land rights. The CEL case on settlement of rights in protected areas addresses the government's failure to settle rights before declaring protected areas. Directives to settle rights immediately have had a major impact on forest policy relating to protected areas.

Table 1.1 *continued*

Year	Event	Significance to/impact on participation
		Both the Godavarman and the CEL–WWF cases have led to fundamental changes that have had a wide impact on forest management. For example:
		• New authorities, committees and agencies such as the CEC and the Compensatory Afforestation Management and Planning Agency have been established.
		• No forest, national park or sanctuary can be de-reserved without the approval of the Supreme Court.
		• No non-forest activity is permitted in any national park or sanctuary even if approval had been obtained under the 1980 Forest (Conservation) Act.
		• An interim order in 2000 prohibited the removal of any dead or decaying trees, grasses, driftwood, etc from any area comprising a national park or sanctuary. It was also directed that if any order to the contrary had been passed by any state government or other authorities, that order should be stayed.
2000 and 2002	MoEF revision to central JFM guidelines	Updating the 1990 guidelines, the revised 2000 guidelines call on states to increase the participation of women, extend JFM to 'good' forest areas, contribute to regeneration and forest resources, recognize self-initiated groups, and promote conflict resolution. The 2002 revised guidelines further recommend clear memoranda of understanding (MOUs) to be signed, local forest protection groups to strengthen their link with *panchayats*, and capacity-building in local non-timber forest product (NTFP) marketing.
2002	National Forestry Action Plan (NFAP)	The NFAP, prepared by the MoEF, Delhi, and supported by the United Nations Development Programme (UNDP) is claimed to be 'a comprehensive, strategic long-term plan for the next 20 years to address the issues underlying major problems of [the] forestry sector in India in line with the National Forest Policy, 1988. The objective of the NFAP-India is to bring one third area of the country under forest/tree cover ... to enhance the contribution of forestry and tree resources to ecological stability and people-centred development through qualitative and quantitative improvements in the forest resources.' It was prepared over the 1990s (http://envfor.nic.in/).
2006	National Forest Commission (NFC) publishes report	Established in December 2003, the seven-member NFC was set up with former Chief Justice B. N. Kirpal as chairperson to review and assess the impacts of existing forest policy and legal frameworks according to ecological, scientific, economic, social and cultural viewpoints, as well as the current status of forest administration and institutions at national and state levels to meet the emerging needs of civil society. The resulting report is lengthy but somewhat orthodox, and includes a dissenting minority report from C. P. Bhatt (2006).

In pre-colonial feudal periods, forests, pastures and grazing lands close to rural habitation were under common use and management, and were subject to a variety of customary regulatory practices. Further from villages, local rulers set aside specific areas for their own recreational use (e.g. hunting reserves) and also applied varying levels of controls and taxes on the use or trade of forest products. Baden Powell argued that:

> *There never was a time when the government could not issue an edict reserving certain valuable trees, such as teak, sandal, black wood and other valuable trees, as royal trees, nor any time when the chieftain of the province would have hesitated to enclose a large area of the wasteland as a hunting preserve.* (Baden-Powell, 1892).

The early colonial period was characterized by extensive exploitation and plunder of forests by private contractors, primarily to feed demand for maritime construction timber, although the land revenue objectives of the British Raj were also served by encouraging the conversion of forest land to agricultural use (Gadgil and Guha, 1995, p120). The forests were also felled to meet timber and fuelwood needs for cantonments and urban centres. The advent of railways in India in 1853 led to further large-scale felling to fulfil the need for railway sleepers and, initially, for fuel for steam engines.

This unregulated clear felling of extensive areas led to alarm that strategically important timber supplies were threatened. The India Navy Board stressed the need for timber conservation policies as early as 1830 to save the forests from devastation (Hobley, 1996). This concern gave rise to the formation of the Indian Forest Department for the 'orderly exploitation' of India's forests, and the associated legal and rights structures that continue to this day, most of them diluting, modifying and sometimes totally curtailing the rights of local livelihood-oriented forest users.

The Indian Imperial Forest Service was set up in 1864, headed by Dr Dietrich Brandis, a German forester, as the first Inspector General of Forests from 1864 to 1883 (Guha, 1983). Its functioning required a legal basis from which to assert its authority, which was provided by the hastily drawn up first Forest Act of 1865. This, however, was never fully implemented. A draft revised Forest Act was circulated in 1869 to strengthen the state's control over forests, and the ensuing debate foreshadowed current PFM policies and practices. The final act established the legal and administrative architecture of the forest bureaucracy, which has largely persisted to the present day. The fundamental issue concerned the customary livelihood-oriented forest use of local people to adjacent forest, and the extent to which their rights should be recognized, commuted or extinguished. The colonial state argued that forest use had been based on the agreement of the raja and therefore was a privilege rather than a right, and since the colonial government was the successor to the rajas, it now had the prerogative to extinguish these privileges where it saw fit (Ribbentrop, 1900, p97). Voices of dissent emerged from officers in the Madras presidency:

> *The provisions of this Bill infringe the rights of poor people who live by daily labour (cutting wood, catching fish and eggs of birds) and whose feelings cannot be known to those whose opinions will be required on this Bill and who cannot assert their claims, like [the] influential class, who can assert their claims in all ways open to them and spread agitation in the newspapers.* (Venkatachellum Puntulu, cited in Guha 2001, p215)

It is interesting to see how the local revenue officials in the Madras presidency reflected on the relation between communities and forests. Venkatachellum Puntulu, the deputy collector of Bellary district, further argued that:

> ... *it is known fact that all the jungles in this part of country are the common property of the people and that the poor persons who live near them enjoy their produce from immemorial time.* (Board of Revenue Proceedings, Madras, 1871, cited in Guha, 2001).

Dietrich Brandis was also strongly opposed to the 'annexationist' approach and consistently urged finding a middle way between systematic forest management in extensive valuable tracts and accommodating the needs of local people. This could be achieved, he suggested, by creating a local administrative structure for the facilitation of village forest management, eventually leading to the village assuming management responsibility:

> *Not only will ...[communal] forests yield a permanent supply of wood and fodder to the people without any material expense to the State, but if well managed, they will contribute much towards the healthy development of municipal institutions and local self government.* (Brandis, 1884)

Despite these appeals, the 'annexationist' position advocated by Baden-Powell prevailed, and the resultant Indian Forest Act of 1878 led to the expansion of commercial exploitation of the forest and the inevitable removal of important livelihood materials from forest-adjacent peoples (Poffenberger and McGean, 1996, p59). It introduced a system of categorizing forests into three classes. State or 'reserved' forests were set aside where forests were of commercial value. Customary rights here were 'settled', meaning that they were generally converted to 'privileges' to be exercised elsewhere or totally extinguished. The second class of forest was 'protected', wherein rights and privileges were recorded, but not settled, although all valuable tree species were even here reserved by the government, and 'damaging' practices such as grazing could be restricted (Rangarajan, 1996). The Forest Act also provided for 'village forests', but since their formation first required their reservation by the forest department, local people became suspicious and this provision was hardly implemented. The *van panchayats* (VPs), the so-called forest villages in montaine Uttar Pradesh, were created not under the Indian Forest Act, but under the Scheduled Districts Act, which was later repealed. The first revision of the original VP rules was made under the Indian Forest Act in 1976; but some in Uttaranchal even today question their legality since the rules applied to forests were created under a different act. The only post-independence use of Section 28 of the 1927 Indian Forest Act (which provided for the creation of village forests) was by Orissa, and only for social forestry wood-lots, mostly on non-forest land and with highly exclusionary rules that did not empower the community at all. A retrospective revision of the rules for these woodlots has, under JFM, introduced 50:50 'benefit-sharing' between village and FD, although the land affected by these rules and their revisions does not even belong to the Orissa Forest Department.

Thus, the colonial administration declared proprietary governmental rights over forests. To what extent these were limited by existing rights of private persons or communities was to be subsequently determined (Ribbentrop, 1900). However, the onus of reporting and proving infringement of rights was on the village people, who were ignorant of Western concepts of property (Vira, 1995) and most of whom, moreover, were illiterate. It has been suggested that a majority of users did not register their claims and thus failed to secure their rights legally (Singh, 1986). Consequently, in many places the local communities' rights over forests were either extinguished or transformed into privileges (and, later, 'concessions') that could be withdrawn at the will of the government.

In any case, the Indian Forest Department was able to extend its power and control, as well as its territories. The initial 1865 Forest Act pronounced that 'wherever it is expedient that rules having the force of law should be made from time to time for the better management and preservation of forests wherein, rights are vested in her Majesty for the purpose of the Government of India' (Stebbing, 1926, p8). This was the starting point of state intrusion into the complex customary rights and resource-use patterns then existing in India. It was followed by intense debate over the appropriate balance of rights between local people and the forest department, leading to the 1878 Forest Act, which provided the legal framework for the states' reservation and demarcation of 'valuable' forest tracts and extinguishment of local rights. It also allowed flexibility for further reservations as time went by. The 1927 Forest Act is still in force today, leading to current legal powers concerning the establishment and demarcation of boundaries, trespass, and cutting and control of movement of forest products. The subsequent Forest Conservation Act stipulates that even when forest land needs to be diverted to other uses, an equivalent amount of other land should be made available for compensatory afforestation and notification as forest in order to prevent the decline of the notified area. Although the aim, at the inception of the forest department, was a response to unregulated cutting by forest contractors, the extension of 'scientific forest management' legitimized the takeover of the uncultivated commons from previous local users. The reservation and composition of the 'forest estate' (defined as land under standing forest or any land 'recorded' as forest in any government record irrespective of whether it had, or could have had, any forest) subsequently took place. It required the revision of complex local land tenures in favour of the state under the guise of 'legal process'. In many cases, local customary rights were never formally settled at all and reservation was put on a provisional basis, which then became, *de facto*, permanent.

Although the main objective of reserving forest land has been to exercise exclusive access to timber areas, many areas of the uncultivated commons – often without any standing forests but labelled 'wastes' due to their inability to yield any revenue – were also taken over by the FD (particularly 'wastelands' – i.e. grazing lands) and were subsequently put under forest plantation, then often planted with unbrowsable exotics. The composition of the forest estate reflected a cultural preference for settled agriculture and a negation of pastoralism. For instance, the severe impact of the Punjab famines of the 1870s have been largely attributed to the disruption of the pastoralist–agriculturalist interaction due to the forest department annexation of grazing lands (Chakravarty-Kaul, 1989).

In this way, the state laid its claim on most of the commons. While some areas were selectively reserved due to the valuable forests that they harboured, in other cases huge areas were declared state forests without any vegetation surveys (for example, the 1893 notification in Uttarakhand), so the areas reserved did not necessarily have any trees on them. The lands sought for conversion to state forests often had pre-existing communities who used them for various livelihood needs and had diverse communal property resource management regimes, often with the acceptance of local rulers. In many places, people resisted the forest reservation policy and even burned some reserved forests in protest (Guha, 1983, 1989). Some of these protests compelled the colonial government to reduce the area to be reserved significantly (as in Uttarakhand and Bastar) and provide for, or at least promise, village *nistari*. The term '*nistar*' literally means satisfaction of subsistence needs and was restricted to household consumption. Officially, these were called '*bona fide* needs', although this term has not been defined anywhere, thereby leaving enormous discretionary powers of interpretation in the hands of forest officials.

There are a few cases where the extinguishment of customary rights did not take place. In Jharkhand, for example, special tenancy acts were enacted and the government even recognized the traditional *Mundari Khuntkatti* system of the Munda tribe, under which the original families settled in the village enjoy ancestral rights over the village area, including the right to clear existing forest land to settle male heirs. Many tribal areas were declared agency areas with special administrative arrangements, which gave some variable degree of protection from the reduction or extinguishment of customary rights (Sarin, 2003).

Thus, the Indian Forest Department was rapidly developed in terms of staffing levels to fulfil the role set out for it. Forests were to be demarcated; the rights of local people were either commuted to privileges or were extinguished, where deemed necessary, during reservation; working plans were drafted; and, in a few cases, rights were established. The late 19th and early 20th centuries saw a period of extensive reservation of forests, and by 1900, there were 81,400 square miles (211,00km²) of reserved forests and 8300 square miles (21,500km²) of protected forests. By 1927, a new Forest Act was introduced to modify and revise the 1878 act. This act retained the provision for village forests, although this remained unimplemented.

The reservations and gradually increasing enforcement of restrictions on local people's customary use of forests had a massive and complex impact on rural societies. There is an extensive literature on the effects in every region of India. Centuries-old practices were suddenly obstructed or criminalized (Singh, 1986; Gadgil and Guha, 1995; Munda and Mullick, 2003). For instance, in mountain areas transhumant pastoralists, accustomed to seasonal long-distance grazing, travelling as far as Afghanistan and the alpine pastures of the Himalaya, had their use of grazing areas, often already reduced in extent by conversion of land use to agriculture or plantation, subjected to increasing restrictions, gradually rendering their livelihoods unworkable (Chakravorty-Kaul 1996). Even the customary seasonal timing of grazing in forests came under restriction. To take another example, in West Bengal, after the private forests were acquired by the West Bengal Estates Acquisition Act in 1960, the Indian Forest Department started digging deep 'cattle-proof trenches' around the forest blocks, thus depriving local people of customary grazing. The trenches were frequently breached; nevertheless, long-established customary practices were obstructed and criminalized.

Another forest policy was announced in 1894. This policy was greatly influenced by the report on the improvement of Indian agriculture written by Dr Voelcker, which suggested that forest policy should serve agricultural interests more directly and recognize the importance of forests to livelihoods. Accordingly, the policy gave emphasis to the links between agriculture, forests and, in modern parlance, livelihoods. It noted:

> *It should be remembered that, subject to certain conditions to be referred to presently, the claims of cultivation are stronger than claims of forest preservation... Accordingly, wherever an effective demand for culturable land exists and can only be supplied from forest area, the land should ordinarily be relinquished without hesitation.* (GoI, 1894)

These policies of forest reservation led to the Indian Forest Department becoming 'unquestionably the most unpopular arm of the British Raj' (Guha, 2001, p217). Both widespread resistance (e.g. breaches and arson) and outright rebellion occurred across the country (including, but not only, Chhotanagpur in 1893; Gudem-Rampa in 1879–1880 and 1922–1923; Bastar in 1910; Midnapur in 1920; Uttarakhand in 1915–1920; and Adilabad in 1940). By the 1920s and 1930s, agitations against forest settlements and restrictions to local people's rights, as well as the high-handed behaviour of forestry personnel, had become intense, converging with the independence movement. For instance, in 1910, the *adivasis* (tribal or indigenous peoples, although the term is contested) of Bastar again revolted against the forest policy of reservation since it threatened their traditional rights. This movement lasted for a decade before being crushed by the colonial government (Behar, 2002; for a discussion of these uprisings, see Panikkar, 1979; Arnold, 1982; Pati, 1983; Murali, 1984; and Tucker, 1984).

Acknowledgement of local people's grievances did emerge in some areas as a strategic accommodation to quell the unrest. The formation of the Madras Forest Grievance Committee and forest *panchayats*, and the *van panchayats* in Kumaon and British Garwal on the recommendation of the Kumaon Forest Grievance Committee, together with the withdrawal of the Indian Forest Department from a large part of the reserve forest and restoration of all people's rights, were some of the actions taken to reduce political agitation. Other examples of the colonial state's acknowledgement of the grievances arising from the withdrawal of forest users' rights of the forest included the establishment of the Punjab Forest Grievance Committee, which in 1937 recommended the creation of the Kangra Forest Co-operative Societies; a significant reduction in forest area reserved in Bastar; the creation of *nistari* forests; recognition of *Mundari Khuntkatti* forests in Jharkhand; the Chhotanagpur and Santhal Parganas tenancy acts; and Wilkinson rules in Jharkhand. A noticeable characteristic of these policies was that the management of forests near human habitation was given to local forest users via the creation of local bodies. For instance, in the Madras presidency, independent *panchayats* were created to manage forests. Under this scheme, local people were given usufruct rights on grazing and non-timber forest products (NTFPs), and the forest department was restricted to an advisory role only. These institutions were dismantled after the 1950s under the claim that a more effective management of natural resources for national development was needed.

This brief account shows that forest governance as an administrative entity has often been refashioned by urgent political demands. However, the main issue was the restoration of local resource rights and control over their management for the livelihood functions of local people, albeit as 'oil on a squeaky wheel' to diffuse conflict in specific areas, rather than an overhaul of the entrenched legal structures. Nonetheless, these compromises that the state had to make to avoid political upheaval and, in some cases, the wilful destruction of forests by incendiarism did involve the same issues as those at stake in PFM today.

After independence, the 1952 Forest Policy restated the main provisions of the 1927 Forest Act and reasserted that its fundamental element (orderly timber production and extraction for national needs) held good. The objective became 'national development' and industrialization, and local needs were explicitly stated to be secondary to the 'national' interest. Between 1951 and 1988, the 'net' state forest estate was further expanded by 26 million hectares (from 41 million hectares to 67 million hectares), largely through 'vesting' the non-private lands of ex-princely states (merged with the Union of India after independence) and of *zamindars* as state forests (Saxena, 1995, 1999). In poorly surveyed, predominantly tribal areas such as in Orissa, this was largely done through blanket notifications with the help of an amendment to Section 20 of the Indian Forest Act, declaring them 'deemed' forests. In many of these areas, the procedure for enquiring into and settling local people's rights was not followed. The 1952 policy stressed 'national needs' such as 'public benefit', as did the 1894 policy, but was much less sympathetic to forest-fringe communities' requirements. It noted:

> *The accident of a village being situated close to the forest doesn't prejudice the right of the country as a whole to receive the benefits of a national asset.* (GoI, 1952)

This is an example of the principle of 'eminent domain' that justified the rights of the state to a resource for national benefit over and above any claimed rights of local people who may have ancestral claims on that resource (see also Chapter 3 for further discussion). The 1952 policy listed the 'paramount needs' of the country that were to provide the basis of forest management. The policy-makers, however, included everything from environmental services to industrial raw materials, and from rural subsistence requirements to revenue for the government under the list of 'paramount needs'. There was no prioritization of these needs and it is obvious that all of these could not be met simultaneously. There were, however, no guidelines as to how choices were to be made between these competing claims (Vira, 1995).

Large tracts of forests were nationalized through the issuing of notifications, often unaccompanied by any surveys or settlements, which remain the major cause of conflict even today in states such as Orissa, Madhya Pradesh, Chhattisgarh, Andhra Pradesh and elsewhere. The nationalization of *Zamindari* forest led to large-scale deforestation as *zamindars* felled most of the standing stock before handing over the land.

Industrial and revenue considerations dominated Indian forestry during the years after India's independence. The forests were viewed as raw material and revenue sources for this economic development programme (Vira, 1995). Such mining of resources for revenue often decimated the livelihood resource base of local communities. In contrast, the investment made in the forestry sector was extremely small compared to other sectors of the economy. The share of forestry in public-sector outlay averaged only 0.53 per cent from 1951 to 1980 (GoI, 1986).

In 1970, the National Commission on Agriculture (NCA) was appointed, which also looked into the issues related to forests. The NCA gave its final report in 1976 and set the course for Indian forestry for the next two decades. It noted that while forests occupied 23 per cent of India's landmass, their contribution to the gross national product (GNP) was less than 1 per cent. The NCA, however, ignored the non-monetized forest-based economy of rural and tribal communities, as well as the economic value of the protective functions of forests (GoI, 1976). It came to the conclusion that mixed vegetation has no commercial value and therefore these forests should be felled and replanted with fast-growing commercially important plantation species (NCA, 1976). It suggested the creation of corporations to manage forests on business principles and to attract finance from institutional and other sources. Consequently, quasi-autonomous forest corporations were set up in most states. Thus, the period of recovery of local rights to forest users had been truly eclipsed.

The NCA viewed local communities' dependence on the forests as a major cause of forest destruction and a major obstacle for production forestry. Thus, in order to free the forest lands for the latter purpose, it suggested that local communities' needs should be met by a social forestry programme on *non-forest* lands, such as village commons, government wastelands and farmlands (GoI, 1976). The NCA's other concerns were the cultivation of wastelands and increasing agricultural productivity. Both of these objectives could also be met through social forestry. The idea was that the plantations would be raised on degraded and marginal lands, thereby improving their productivity. It was hoped that the increased supply of fuelwood from these plantations would meet local needs and even generate surplus for the market. This, in turn, would reduce the use of cow dung as fuel so that it could then be used as manure in agricultural fields. According to estimates, over 458 million metric tonnes of wet dung were being used annually as fuel. If this was used in agricultural fields, it could potentially fertilize 91 million hectares and increase food output by 45 million metric tonnes (Srivastava and Pant, 1979).

It is clear that social forestry started as an attempt to keep rural communities out of existing forest lands, rather than as a participatory rural development programme, and acquired the connotation of a people's programme only much later (Pathak, 1994). The following quotes from the NCA report clearly show the government's earlier thinking:

> *Free supply of forest produce to the rural population and their rights and privileges have brought destruction to the forest, and so it is necessary to reverse the process. The rural people have not contributed much towards the maintenance or regeneration of the forests. Having overexploited the resources, they cannot in all fairness expect that somebody else will take the trouble of providing them with forest produce free of charge... One of the principal objectives of social forestry is to make it possible to meet these needs in full from readily accessible areas and thereby lighten the burden on production forestry. Such needs should be met by*

> *farm forestry, extension forestry and by rehabilitating scrub forests and degraded forests.*
>
> *Production of industrial wood would have to be the raison d'être for the existence of forests. It should be project oriented and commercially feasible from the point of view of cost and return.* (GoI, 1976)

The fundamental shortcomings of the social forestry programme took some time to gain recognition. The programme was generally criticized by national NGOs and environmentalists on four grounds: that the programme remained confined to large farmers only; that it did not produce fuelwood and fodder as promised; that it resulted in a reduction in areas devoted to food crops; and that it promoted monocultures of eucalyptus, which were environmentally undesirable (Saxena and Ballabh, 1995a).

In 1982, the Government of India circulated a draft Forest Act that led to intense mobilization against it by activists and intellectuals who challenged the 'centralizing thrust and punitive orientation' of the 1878 Forest Act on which it was based (Guha, 2001, p231). However, Guha suggests that after the withdrawal of the act, over the 1980s the 'politics of blame' evolved into the 'politics of negotiation' as foresters and their critics began a process of dialogue. This led to the approval of the 1988 National Forest Policy, which changed the emphasis from industrial production to ecological protection and the satisfaction of local people's needs. However, the 1988 document remained a policy and not an act, and as such was no more than a statement of intent unaccompanied by legislative changes. The policy was adopted by parliament, however, and there was a reasonable amount of discussion. Still, it can be argued that not bringing the policy forward for negotiation and drafting allowed ambiguity and latitude for discretion and manipulation on the part of the forest administration. However, there have been repeated attempts to bring back a commercial focus. Proposals have been made three times in the last two decades to lease forest lands to industry for growing raw materials, including the latest move in 2006, which was cleverly disguised as an industry–community collaboration.

Recent forest movements, particularly the seminal Chipko and other subsequent movements, have played a major role in mobilizing local people to challenge the commercial orientation of state forest management. Impacts have been felt both directly on the ground and also through the movements' wider effects on policy. The internationally famed Chipko Andolan movement of the 1970s in Uttaranchal has been amply discussed (Guha, 1989). It led to a major victory in 1980 with a 15-year ban on green felling above 1000m above sea level in the Himalayan forests of that state by order of India's then prime minister, Indira Gandhi. The leaders (C. P. Bhatt and Sunderlal Bahuguna) continue as active voices for reform. Chipko also inspired the villagers of the Uttara Kannada district of Karnataka Province in southern India to launch a similar movement to save their forests, and in September 1983, the people of Salkani also 'hugged the trees' in Kalase Forest. The local term for 'hugging' in Kannada is *appiko*, and so the *Appiko Andolan* movement emerged. The *Andolan* (or movement) mobilized local people across the Western Ghats to agitate against the state forest management systems that have excluded local people and their priorities from the forests, leading to clear felling and exotic monocultures. The agitation led to state recognition that local people should be involved in biodiversity protection and therefore forest management. The people's movement had a major effect in generating pressure for a natural resources policy more sensitive to people's needs and the natural environment.

Hence, a third stage, the 'politics of collaboration', gradually emerged (Hobley, 1996, p59). In three states, Haryana, West Bengal and Gujarat, different approaches to involving local communities in forest and local resource management were being experimented with. In West Bengal, innovative local foresters had begun forming village forest protection committees and encouraging them to play a role in forest regeneration over the 1980s. In

Haryana, a different approach was being tried whereby communities were supported to develop an integrated natural resource management strategy to improve their livelihoods through sustainable use managed by their own institutions. This involved not only the forest sector, but also water and grass in a broader focus on livelihoods (Banerji, 1996; Sarin, 1996). The impressive progress of the West Bengal experiments led to the national-level JFM circular of 1990. With a persuasive set of bureaucrats at the central level exhorting other states, most state forest departments adopted JFM, issuing their respective JFM notifications over the next 10 to 12 years. Currently, virtually all states in India have adopted JFM, especially after significant funds for JFM became available under a national scheme.

At this point, the conception of JFM by the Indian forest administrations must be carefully rehearsed, since it laid the foundations of current debates, and many of the initial assumptions about contemporary PFM made by the forest administrations have not been sufficiently examined. The core principle was that a local 'village forest protection committee' should be constituted by the forest department and comprised of local people living around a 'degraded' forest. They are permitted, as privileges (JFM does not grant rights), NTFPs for their own use and are allowed to exclude non-members from forest use. In return, they are given a mix of additional benefits that may include wage labour opportunities and a part of the net revenue from timber harvesting. Micro-plans are mainly drafted by forest department staff for the JFM areas and are only then opened for discussion with the village (this is further investigated in Part II). Thus, the local people's 'participation' is simply at the level of agreeing to and then implementing forest department plans, rather than negotiating their own. So-called entry point activities are decided with village views taken into account; but these generally have little to do with the forest and are seen as incentives to the village to get involved in areas of welfare and village infrastructure, such as wells, community halls and so on. Even these entry points are often seen as an intrusion into the domain of local government institutions. For further detailed discussion on JFM orders for each state, see Sarin (1998, 1999), Khare et al (2000) and Shah (2003).

PFM in India has been described as a 'care and share' approach. If the local committee satisfactorily protects the area for five to ten years, it is promised a share of the timber or other major produce if, but only if, forest department working plans permit such harvesting. No rights are granted and the villagers have no legal authority to enforce protection. In most states, a 'participatory' micro-plan is supposed to be prepared for the JFM forest. While, in many cases, no such plans are ever prepared, where they are, they are written by forest department staff. The forest department has the right to unilaterally terminate a JFM agreement if the village committee is considered to have failed to honour its responsibilities. If the forest department fails to honour *its* responsibilities or the promised share of benefit, there is little the villagers can do about it. Many forest departments have also unilaterally revised their JFM orders, in some cases *reducing* the major product shares originally promised. The structure, composition and names of the village institutions have also been unilaterally changed by the forest department many years after the initiation of JFM by the villagers of Madhya Pradesh, Rajasthan, Gujarat and Andhra Pradesh.

During the early 1990s, bilateral and multilateral donors began to support PFM projects, and this certainly acted as an incentive for forest departments to adopt at least the rhetoric of participatory practice. Donors had become involved in supporting 'social forestry' projects in the 1970s, which promoted non-forest supply of livelihood tree products and ignored the fundamental issue of the criminalization of communities' livelihood-oriented forest use. By the early 1990s, some donors felt that they had 'learned from their mistakes' and promoted the new, more participatory, JFM approach, whereas others, such as the Swedish International Development Agency (Sida), withdrew altogether. The World Bank was the first to provide large-scale support to the emerging model of JFM in West Bengal, followed by similar support to Andhra Pradesh, Kerala, Madhya Pradesh, Maharashtra and

Uttar Pradesh. The pioneering *Sukhomajri* approach from Haryana was not adopted, apparently because large donors preferred to support the simpler sectoral benefit-sharing model of West Bengal, rather than the more holistic multi-sectoral approach evolved in *Sukhomajri*. The main donors have been the World Bank, the UK Department for International Development (DFID), the Japan Bank for International Cooperation (JBIC) and Sida. JBIC has emerged as the largest donor to the forestry sector, with a cumulative commitment of US$1.06 billion. Although the projects funded by JBIC have a large afforestation component, it claimed in 2005 that its projects have resulted in the creation of about 8000 JFM committees (JFMCs), with another 2500 in the pipeline. The total area covered by these JFMCs is over 1.5 million hectares (National Afforestation and Eco-Development Board, 2005). Another major donor, the World Bank, has supported eight state-level and one national project over the 1990s, with a total outlay of US$528 million (World Bank, 2005). In the tenth Five-Year Plan (2002 to 2007), the central government started a scheme to support the JFM programme through forest development agencies (FDAs), which are ostensibly federations of JFMCs. However, they are constituted and controlled by forest departments, with the conservator of forests for the forest circle in question as the chairperson and the DFO as the member secretary of each FDA, and the local forest guards as the member secretaries and joint account holders of each JFMC affiliate of the FDA. Until 31 March 2005, 620 FDA projects had been sanctioned by the MoEF. The financial data available from 397 FDA projects indicate that these have a combined outlay of 9.63 billion rupees (GoI, 2005).

Other institutions have also played a major role, even if not a financial one. The Ford Foundation performed an innovative facilitative and networking role, especially in the emergence and early days of JFM in India, led by staff members with strong activist agendas (such as Jeff Campbell and Mark Poffenberger), backed up by study and implementation support given by a large number of in-country NGOs. The Aga Khan Rural Support Programme (AKRSP, India), a Gujarat-based NGO (with its chief executive at that time Anil C. Shah, a retired senior Indian Administrative Service (IAS) officer with excellent reach within state and national governments), adopted a form of JFM. The chief executive had supported plantations on degraded forest lands in some AKRSP project villages even without legal permission prior to JFM and was one of the drafters of the 1990 JFM order, as well as of the Gujarat order in 1991. Many other NGOs were involved in a pioneering way at this time, including Gujarat-based VIKSAT (Vikram Sarabhai Center for Development Interaction) and West Bengal-based IBRAD (Indian Institute of Bio-Social Research and Development). In most cases, the success of these initiatives depended on the collaboration of those forestry officials who were supportive of PFM.

At the state level, some states have been more dynamic than others in adopting and implementing JFM, with some such as Andhra Pradesh even claiming to have moved from JFM to community forest management, although the claim is probably more a matter of re-labelling than of substance, as Chapter 9 discusses. Why the policy change to JFM came about in the states at the time depended on a number of convergent factors, not least the informal coalitions of progressive bureaucrats, activists and civil society leaders with international agency support. Furthermore, there was a combination of civil society protests as well as growing environmental concerns in government over forest degradation, and increasing recognition (even in some forest departments) of the non-viability of a policing approach to forest protection.

The *de facto* forest control by a large number of organized local communities in many states such as Orissa (as discussed in Chapter 8) and Maharashtra has also had a major influence on forest department policy, forcing the states and centre to recognize the validity of their involvement in forest management and obliging them to find mechanisms through which their role could be recognized.

The promulgation of the 1988 Forest Conservation Act has had a number of significant impacts on forest governance. Its main provision is that forest land cannot be converted for other uses without central MoEF clearance, and its introduction signified a major shift of policy towards environmental protection against the risks of uncontrolled industrial development and agricultural clearance of forests.

There has, to date, been a lack of explicit linkages between the official JFM form of PFM in India and the 1993 and 1996 decentralization laws and constitutional amendments, contributing to inconsistency between the supposed jurisdiction of local government bodies over village resources and their effective control by the forest departments.

Currently, the focus of many donors is moving from specific forestry projects to broader forest-sector reforms, and the language is increasingly shifting towards livelihoods and poverty alleviation. For example, DFID has been supporting forest-sector reform projects in Himachal Pradesh and Orissa. In Andhra Pradesh, the JFM programme has now evolved into what is termed community forest management (CFM), wherein much greater powers are supposed to have been devolved to the local communities than in the JFM programme. In government circles, there is also a renewed focus on afforestation (through JFMCs, as well as *panchayat*s) to achieve the national target of 33 per cent tree and forest cover by 2012. Recent debates have centred on the Scheduled Tribes (Recognition of Forest Rights) Bill and community-based management of protected areas, especially the recommendations made in the draft National Biodiversity Strategy and Action Programme (TPCG and Kalpavriksh, 2005).

The emergence of participatory forest management in Nepal

The emergence of PFM in Nepal has been significantly different from the Indian case, but also bears some similarities. One of the initial stimuli in Nepal came from international concerns over the theory of Himalayan environmental degradation that deforestation in the hills of Nepal was leading to downstream flooding (also claimed in the Indian and Chinese cases; see Eckholm, 1976, pp74–100, who championed the theory; Ives and Messerli, 1989, and Ives et al, 2002, who severely criticized it in later years; and Blaikie and Muldavin, 2004, who review the enduring impact of the theory in spite of its severe criticism in academic circles). Donors and the international community put increasing pressure on the Nepalese government to take drastic measures. At the same time, the Nepali Foresters Association came to recognize its inability to protect forests under the command-and-control model. This section briefly outlines the historical context of policy change.

Prior to the uniting of Nepal by the Shah dynasty in 1743, the region was made up of many small principalities whose feudal rulers built their power through developing patronage relations with local elites and functionaries. These local elites were responsible for collecting tax revenues and were rewarded through *jagir* (temporary) and *birta* (permanent) grants of agricultural and forest land. After 1743, the Shah kings continued to use land grants. During the subsequent Rana period (1846 to 1951), the land grant policy gradually became confined to close family members and key officials (Malla, 2001). Many farmers became tenants on the *birta*-holders' land, often obliged to participate in exploitative contract farming arrangements. During this time the Rana government emphasized the extension of agriculture to further expand its tax base. The tenure systems varied across areas, the main type being *birta,* under which the local landlords had the responsibility to manage the forest and granted rights, in turn, to local households to use the forest. Feudal lords arranged for the regulation of access to forests and forest products through the local land revenue administrators *talukdar* (western hills)/*jimmawal* (eastern hills) and their *chitaidar* (forest watchman), paid through the *manapathi* system in grain or kind for protecting the forest. Although timber extraction was regulated, local people generally had free access to non-commercial forest products. Cultivation of millet in temporary forest plots (shifting cultivation, or *khoriya*) was widespread, undertaken partly as a means to evade agricultural taxation.

With the fall of Rana rule in 1951, the new government nationalized all forests of Nepal in 1957 with the intention of taking over the ownership and control of the forest resources from a few wealthy and influential families. It put forest management under the Department of Forests and issued a new Forest Act in 1961. The rapidly expanded forest department assumed an enlarged role for forest protection and management, as well as forest product marketing and the management of forest industries. Yet, with little capacity and with very limited infrastructure, the situation on the ground could change only very gradually. Feudal elites were best positioned to take advantage of the new opportunities and often became political leaders, and alliances commonly developed between them and the new forestry staff who were struggling to assert their authority. Farmers' rights in the forests improved very little. Nationalization also completely undermined customary local forestry management institutions, turning what was an effective system of community-based natural resource management institutions into an open-access resource, allowing government, illegal forest contractors, outsider villagers and entrepreneurs to fell trees. Nationalization of forests, coupled with a national cadastral survey and land registration in the names of individual owners, triggered the conversion of forest to agricultural land in many places in order to assure users that it was they, finally, who were awarded private ownership of the land.

Later, with the realization that forests were continuing to be felled, people's participation in forest management appeared in academic papers and project documents. Following pilot project in Thokarpa and Banskharka villages in Sindhupalchok (a hill district of Nepal), new participatory policy guidelines were proposed in the 1975 National Forest Plan, followed by *panchayat* forest (PF) and *panchayat* protected forest (PPF) regulations in 1978. This was the first formal legal initiation for PFM in Nepal. Over time, the management authority of the *panchayat* was found to have several limitations for livelihood-oriented forest management: partiality, favouritism and nepotism were shown by the people holding the power in *panchayats*, the lowest-level political body. These behaviours seriously marginalized the poor and many others. Often the benefits were taken by the political allies of the *panchayat* leaders. Another limitation was the boundaries of the *panchayat* and forests. In several instances, the actual traditional users of the forest were in one *panchayat*, whereas their forest was in another, and with PF and PPF regulations they could not access these forests. Later, with the formulation of the Master Plan for the Forestry Sector (1989), these shortcomings of PF and PPF regulations were recognized, and PFM took a new direction under the present concept of community forest user groups (CFUGs). These reforms were undoubtedly helped by the resumption of the multi-party political system in 1990; although the forest user group (FUG) concept was already realized and would have been there with the master plan and new forest policy of 1989 (even without the multi-party political system), the political change in 1990 helped to formulate FUGs as more democratic and independent institutions.

The community forestry policies were strengthened through the 1993 Forest Act and the 1995 Forest Regulations; but these primarily emphasized the reinforcement of forest protection capacity rather than developing wider livelihood opportunities for local people. The high-value *tarai* forests were hardly considered in the programme, and the monopoly on forest product marketing there remained with the Department of Forests – hardly surprising in view of the threat which community forestry might pose to the revenues from valuable timber that flowed to the state and individuals employed by the state and in civil society.

During the late 1990s, the debate began to shift away from the instrumental promotion of 'participation' as a means to the end of forest conservation towards livelihood and poverty issues, local governance and gender equality, mostly at the promptings of donors. Reform of

Table 1.2 *Genealogy of forest policy in Nepal*

Year	Event	Significance to/impact on participation
1927	Establishment of *Kath Mahal*, a proto-department of forests	*Kath Mahal* created (a state institution to control forests and generate revenue).
1939	Establishment of eastern wing and western wing	Focus on revenue collection from *tarai* forests.
1942	Establishment of Department of Forests (DoF)	Establishment of DoF and expansion of forest administration all over *tarai* and some parts of the hills.
1957	Forest nationalization	Nationalization of forests controlled by feudal powers.
1960	Establishment of party-less *panchayat* system	Party-less *panchayat* system of government headed by the king.
1961	Forest Conservation Act	Provision made for national Five-Year Plan period to harvest *tarai* forests on sustained yield basis under strict forest governance by the DoF. Beginning of DoF policing role; but DoF unable to manage forests on sustained yield basis and deforestation continues.
1967	Forest Protection Act	Period of DoF's strengthened policing role through strict legal provisions in which the district forest officer (DFO) is invested with immense authority and judicial powers to act as judge to look after forest offences.
1973	National Park and Wildlife Conservation Act	Declaration of forest areas as national park and wildlife reserve in *tarai* and high mountains in the Himalayan region. In the *tarai*, beginning of era of restriction on people's customary forest-use rights.
1974	Various acts	Several other acts between 1959 and 1974 (such as the 1959 Land Grant Abolition Act; the 1963 Birta Abolition Act; the 1964 Land Reform Act; and the 1974 Pasture Land Nationalization Act) further increase the power of the DoF.
1975	Concern about Himalayan environmental degradation	Issues of environmental degradation raised and widely advocated for urgent action to avert the predicted catastrophe; this draws significant international attention and support.
1976	National Forestry Plan	The plan recognizes that forests are integral to rural communities and acknowledges the role of forests in economic development. It also recognizes that despite all the *de jure* powers of the DoF, its *de facto* inability to enforce them renders it ineffective in most hill areas. This is also the first formal policy document to recognize the need for people's participation in forest management. Establishment of Ministry of Forest (later the Ministry of Forests and Soil Conservation, or MoFSC).
1977	Amendment of 1961 Forest Act	The 1961 Forest Act is amended based on the 1976 National Forest Plan and seeks local participation. This broadens the concept of basic need policies to include conservation measures for the public good, in addition to service provisions for individuals.

Table 1.2 *continued*

Year	Event	Significance to/impact on participation
1978	1978 *Panchayat* forest (PF) and *panchayat* protected forest (PPF) regulation	Community Forestry Rules 1978 (PPF/PF rules) incorporate people's participation in forest management; but since forests are handed over to the local government and then to the *panchayats*, the participation of locals is not ensured. Artificial regeneration plantations emphasize increased forest resource. Technical forestry (large-scale plantation) carried out with the support of the World Bank and bilateral projects.
1982	Decentralization Act	Provides for decentralization and devolution of power to local government. User group concept is introduced for development projects and efforts; but forest handover to the *panchayat* continues. The MoFSC continues to expand its administration and opens regional directorates in 1983.
1984	Private and leasehold forest rules	The need and importance of private forest and leasehold forestry in the sector are recognized; but very limited industrial leasehold forests are given and private forest registration is also limited as people do not understand the value of such registration with the DoF for small patches of private forest under their control.
1989	Master Plan for Forestry Sector (MPFS)	The Master Plan for Forestry Sector prepared for the balanced development of the forestry sector as a major economic sector by managing forests under different regimes in order to fulfil the needs of the economic sector, for poverty alleviation and for environmentally sound use of forests. Its priority programme is Community and Private Forestry.
1990	Restoration of multi-party democracy	Restoration of democratic government, which enhances the formation of civil society groups, NGOs and the involvement of the private sector in development.
1992	District Development Committee (DDC) Act	Local government (village development committee, or VDC), and District Development Committee (DDC) Act devolves power and authority for development using natural resources and own governance systems.
1993	Forest Act	Ensures the rights and access of people in forests through participatory forest management (e.g. community forestry and leasehold forestry).
1995	Forest Regulations	Supports and enhances the implementation of community forestry, which recognizes the customary right of people to use forests through community forest user groups (CFUGs) as independent institutions.
1995	Community Forestry Operational Guidelines	Issued by the DoF to implement community forestry uniformly across the hill region. These guidelines are based on earlier experiences of community forestry. Uniform procedures proposed for CFUGs hitherto adopt different procedures developed by projects under the broad framework of national Community Forestry Guidelines. Due to heavy donor support and incentives, CFUGs are handed over hastily to local communities without proper preparation.

Table 1.2 *continued*

Year	Event	Significance to/impact on participation
1995	Agriculture Perspective Plan (APP)	The APP identifies four forestry priorities, including community forestry in the hills and mountains, and commercial forest management in the *tarai*.
1997	Political conflict	Beginning of Maoist activities in the country.
1999	Forest Act (first amendment)	Intended to control financial and other irregularities in the CFUGs through the involvement of district forest offices.
1998	Operational Forest Management Plan (OFMP) of *tarai* and inner *tarai* districts	Operational Forest Management Plan for 18 *tarai* and inner *tarai* districts prepared and approved, but not fully implemented due to several factors, including weak commitment by the government to the plan, insufficient availability of resources for its implementation, and protest from the Federation of Community Forest Users, Nepal (FECOFUN), which favours community forestry over other forms of forest management.
2000	Revised Forest Sector Policy	The policy recommends community forestry for hills and collaborative forest management (CollFM) (another form of PFM co-management, with the state taking a greater controlling role than in community forestry) in *tarai* and inner *tarai* districts. It also suggests that small isolated forest patches in the *tarai* could be handed over as CFUGs. Leasehold forestry promoted for poverty alleviation.
2003	Collaborative forest management directives approved by the Ministry of Forests and Soil Conservation (MoFSC)	CollFM guidelines focusing on *tarai* and inner *tarai* districts are put forward to pave the way for the pilot implementation of CollFM. The CollFM policy is contested by FECOFUN and civil societies.
2004	Fiscal ordinance for fiscal year 2004–2005	Tax lowered on CFUG revenues from timber sale. In 2000, a directive (legalized through ordinance in 2003) is issued by the MoFSC imposing 40% tax on sale of timber by CFUGs outside their group.
2005–2006	Civil turmoil	The political situation of the country worsens in 2005 with the King assuming emergency powers and curtailing civil rights, which affects some functions and operations of CFUGs. After major public agitations in spring 2006, constitutional democracy is restored.

Source: adapted from Dev (2003) and Parajuli (2003)

forest management in the *tarai* also assumed an increasing urgency in policy debates. With handover to communities slow or stalled, the continued deterioration of high-value forest has been driven by both corrupt and illegal practices in timber management and formal exclusion of 'distant' forest users' access to livelihood-related forest products. Overall, inequity in access opportunities remain a continuing issue in the new 'second generation' agenda of community forestry according to Malla (2001, p288):

> *In general, local elites and government officials continue to work together to limit the access of poor forest users to the forest resources they need, often with the tacit support from donors and non-governmental organizations (NGOs). The result has been an uneven greening of Nepal's land, with continued hardship for most forest users.*

Despite the impetus for reform over the 1990s, Malla (2001) maintains that forest department officials have still been able to work in alliance with local elites, mostly illegally, to 'limit peasant access to forest, preserve for themselves the authority over the resource ... maintaining opportunities for patronage for local elites' (Malla, 2001, p299). Political instability over the 1990s due to widespread dissatisfaction with the pace of reform and the perceived corruption of Kathmandu political cliques developed into widespread Maoist insurgency across the country. This has clearly caused extensive disruption (see Chapters 5 and 6 for further discussion).

The 1993 Forest Act represented a major transformation of the role of the DoF from policing towards facilitation. At the same time, the 1998 Local Self-Governance Act provided the basis for the overall devolution of power, whereas previously different sectors often had overlapping and conflicting policies and legislations. However, the process has yet to bring decision-making closer to the people, mainly because institutional reforms to support it have not been put in place. District development committees (DDCs) do not really govern districts at all, but are only generally informed of what is happening at the sectoral 'development' offices, which really are the central driving force of developmental activities. Local people thus face frustration in successfully representing their interests in district-level programme planning and implementation since it is the sectoral offices that make the major decisions and seldom have a forum for local people. There remain important contested issues concerning the rights to collect taxes from timber products and coordination issues between the DoF and the DDCs and village development committees (VDCs). It is likely, if not inevitable, that CFUGs will become incorporated within the VDC apparatus for development planning and implementation (NPC, 2003). This conflict between the forest department and local devolved government institutions also finds many similarities with the situation in India, discussed above.

Community forestry policy may have dropped down the government agenda due to the recent political crisis, yet it remains highly controversial and politically sensitive. The Federation of Community Forest Users, Nepal (FECOFUN) has been increasingly active in recent years. It has particularly challenged the method of unilateral development of proposed revisions to the Forest Act by the MoFSC. Amendments have, nevertheless, been introduced in some areas – for instance, imposing taxes on forest user groups' timber marketing. In 2004, the tax imposed by the government on CFUG timber sales was reduced from 40 per cent to 15 per cent, primarily due to the lobbying of FECOFUN. This dramatic change of policy reflects the change in the balance of power around policy-making, and the maturity of the political process in Nepal with the emergence of a federation of FUGs.

Policy in the Nepalese *tarai*

The contrasting political ecology in the hills and *tarai* has had different implications for implementing PFM (this is discussed in more detail in Chapter 6). While forest policy has, in many respects, been officially the same (acts and circulars are – with a few exceptions, found in Table 1.2 – identical for both *tarai* and the rest of the country), the pace of implementation, the degree and nature of international and bilateral aid and the detail of user groups has been quite different. While hill forests, usually in patches close to settlements, have historically been managed primarily by local people for their subsistence use under various land tenure systems with little market or state penetration, *tarai* forests have been used by feudal families for patronage and revenues for many years, and are still a source of patronage for local politicians who allow squatting in return for votes. Smuggling fuelwood and timber to India and Kathmandu is commonplace, while long-established associations between forest and local users are absent (Hobley, 1996, p164). Although policies are formally very similar for both regions, there are clear divergences between the *tarai* and hills regarding their implementation. Largely because *tarai* forests are far more commercially valuable, community forestry has not been rigorously applied in the *tarai* because the government (and private individuals working within it) fears that it would lose a great deal of revenue, which has led to a virtual halt to handing over forests as community forestry. The problem of identifying long-established and rightful users of the forest and rights of access to the forest remains unresolved. Table 1.3 shows the extent of community forest handover in the hills and *tarai*, and the slow rate of handover in the latter region.

Thus, the *tarai* forests are still rapidly becoming degraded while there is confusion over who should be the rightful stakeholders (Ojha, 2000). There are serious doubts over the DoF's ability to protect *tarai* forests, given the fact that 570,000 hectares of forestland were lost between 1964 and 1985 (HMGN, 1988a). There are still 500,000 hectares of productive forests which need to be protected through proper policy, implementation strategies and honest commitment (DFRS, 1999; Khadka, 2000).

Table 1.3 *Status of handover to, and characteristics of, community forest user groups in the hills and* tarai *(including inner* tarai*)*

Community forest (CF) and forest user group (FUG) characteristics	Hills	Tarai	Total
Total CF land area (hectares)	1,107,614	77,210	1,184,824
Number of FUGs	12,723	1478	14,201
Total households in FUGs	1,323,941	309,467	1,633,408
Average households per FUG	104	209	115
Average area per FUG	87.00	52.24	83.43

Source: CPFD (2005)

The rules about revenue-sharing between CFUGs and the DoF have been amended; but some of the previous rules about CFUGs' rights to invest fund money for forests and other development activities have been kept intact. Previously, there was no compulsion to invest any money for forest development, so that its use was based on FUGs' own decision-making. The new provision is that CFUGs have to spend at least 25 per cent of their group funds on forest development. Most of the CFUGs and FECOFUN consider this heavy handed on the part of the government. Moreover, according to the 1999 Local Self-Governance Act, the DDC also has the right to levy tax on natural resources such as forest products, water, stone and sand. This shows that there are still contradictory interests between FUGs, the DoF and the DDCs regarding revenue-sharing,

According to CFM directives for *tarai* forests only, 75 per cent of the revenues collected from the sale of the forest products (timber and fuelwood) go to the government and 25 per cent to local bodies. The distribution of 25 per cent to local bodies has been decided by the District Forests Coordination Committee (DFCC) (MoFSC, 2003). Similarly, for their own purposes, CFM groups and sub-groups manage the distribution of timber, fuelwood and other forest products. However, for business purposes, there is a need to follow auction rules and the selling rates must not be less than the government's royalty rate. The situation in the *tarai,* where there were very few projects (and these focused on establishing plantations and a few other minor forestry development activities) is different. One of the reasons for the lack of participatory approaches in the *tarai* was that use rights are contested and do not have a long-established history as in the hills.

While hill forests are intermixed with settlements and are mainly used for subsistence economy, *tarai* forests could provide both better livelihood opportunities for local populations and important revenues for the government (Pokharel, 2003). However, the Operational Forest Management Plan (OFMP) did not develop management approaches that provide for both sets of management objectives. Most recently, Collaborative Forest management (CollFM) has been launched, but is still in the pilot phase in three districts (Bara, Parsa and Rauthat), and it is too early to review implementation.

Inconsistencies in forest policy in Nepal

The community forestry programme has provided implementation experience on participatory development in the hills, where serious attempts have been made at grassroots-level empowerment, at least since the 1980s (Gilmour and Fisher, 1991; Baral, 1999; Baral and Subedi, 1999). Several anomalies between the Forest Act and the Forest Regulations (which apply both to hill and *tarai* forests) have, however, added to the challenge. Contradictions and overlaps also exist between forest-related acts and other acts, such as the Local Self-Governance Act, the Land Revenue Act, the Public Roads Act, the Nepal Mines Act and the Mines and Minerals Act, the Water Resources Act, the Soil and Water Conservation Act, and the Environment Protection Act (Chapagain et al, 1999).

Furthermore, the DoF and the MoFSC frequently change forest rules by means of circulars that are neither publicly distributed, nor debated in parliament. The practice of sending undebated circulars to implement policy decisions has strong negative impacts. Circulars issued by the government should also reach all stakeholders at the discussion stage, such as FUGs, DDCs and VDCs. Following the publication of the National Conservation Strategy, plans for biodiversity conservation are currently being prepared, but with very limited consultation. Recently, an ordnance was passed that specifies that the management of national parks can be leased to NGOs which meet certain stringent criteria. Unfortunately, the King Mahendra Trust for Nature Conservation (KMTNC) is the only NGO deemed to meet these criteria. The forest management criteria of the forest department and of the KMTNC conflict, and this presents ongoing problems.

Donor projects in Nepal

Community forestry in Nepal is influenced more strongly than in India by foreign donors in both conception of policy and implementation for reasons examined in more detail in Chapter 2. There has been a wide variety of donors active in PFM in Nepal, and the division of territorial responsibilities by different donor projects has, at times, covered the whole country. In Table 1.4, various donors involved in community forestry and their coverage are listed. Most districts of Nepal have been supported by donors in implementing PFM programmes. In the following section, the implementation approaches of donor PFM projects are discussed.

These donor-funded projects require the cooperation and also collaboration with service providers, who may also be government bodies (such as the Ministry of Forests and Soil Conservation (MoFSC) including DoF and district forest offices), NGOs and private firms.

The ways in which donor-funded forestry projects work often reflect predispositions of the respective donor countries with respect to their overall aid administration in Nepal. The projects exhibit an apparent variation in the modalities of their implementation, which, in turn, defines how deeply they engage with the MoFSC and the DoF, local government and

Table 1.4 *Major donors in participatory forest management and district coverage*

Donor	Starting year	Current projects/ programme	Project-supported districts and forest user groups (FUGs)	
			Hill districts	***Tarai* districts**
Australian Agency for International Development (Ausaid)	1966	Nepal–Australia Community Resource Management and Livelihood Project (NACRMLP) (2003– 2006)	2	–
World Bank	1979–1997	–	Phased out	Phased out
Danish International Development Agency (DANIDA)	1997–2005	Natural Resource Management Sector Assistance Programme (NARMSAP)	38	0
UK Department for International Development (DFID)	1993	Livelihoods and Forestry Programme (LFP) (2001–2011)	11	4 (including Dang)
Swiss Development Cooperation (SDC)	–	Nepal–Swiss Community Forestry Project (NSCFP)	3	0
German Agency for Technical Cooperation (GTZ)	1992	Churia Forest Development Project (ChFDP) (phased out in 2005)	1	2
Netherlands Development Organization (SNV)	2002	Biodiversity Sector Programme for *Siwalik* and *Tarai* (BISEP-ST)		8 (including Makwanpur)
US Agency for International Development (USAID)	1978	Environment and Forest Enterprise Activity (EFEA) (1996–2002) (phased out)	5	3
CARE Nepal/USAID		Strengthening the Role of Civil Society and Women in Democracy and Governance project (SAMARPAN)	6	6
World Wide Fund for Nature (WWF)	2002	*Tarai* Arc Landscape project (TAL)	0	3

civil society groups, as well as (eventually) the beneficiaries. At one end of the spectrum lie the Biodiversity Sector Programme for *Siwalik* and *Tarai* (BISEP-ST) and the Natural Resource Management Sector Assistance Programme (NARMSAP, phased out in 2005), which operate within the administrative structure of the MoFSC. Apart from their own administrative or staff-related activities, they support almost all programme activities through relevant 'counterpart' structures within the MoFSC/DoF, and thus have a much more circumscribed space in which to engage civil society groups for policy advocacy. Around the centre of the continuum lie such projects as the Livelihoods and Forestry Programme (LFP), the Nepal–Swiss Community Forestry Project (NSCFP), the Nepal–Australia Community Resource Management and Livelihood Project (NACRMLP) and the Churia Forest Development Project (ChFDP), which work with the MoFSC/DoF, as well as with NGOs and civil society groups. At the other end of the spectrum lies aid from USAID, which is implemented independently of the MoFSC through NGOs or private firms in liaison with the MoFSC.

Since 1990, donor policy and programmes have been dominant in policy-making in the forestry sector in Nepal. Although some mechanisms for coordination between different forestry-sector donors exist at ministry level, it is common to find a range of donors with different policies, strategies, approaches and procedures working more or less independently. Some of the main variations in their approaches have been:

- the intensity of support and funding level;
- the focus on community forestry alone or on a more integrated approach with wider community development and livelihoods;
- the level of institutional involvement of different stakeholders, such as local governments, NGOs and CBOs, and the private sector;
- the focus on developing new models and practices; and
- the support of the MoFSC on an extensive sectoral basis to implement plans and programmes more effectively, rather than to support area-specific projects in the field.

In general, donor-aided project personnel (both Nepalese nationals and visiting foreign consultants and project personnel) have a difficult interface to negotiate with government counterparts with regard to the purposes to which funds should be put. The multidimensional nature of poverty and livelihoods necessitates a far wider approach to development support, while DoF personnel resist any attempt by forestry projects to escape narrowly defined 'forestry' concerns and concern themselves with wider areas of activity (for example, gender mainstreaming, poverty and livelihood issues). In particular, the 'basic needs approach' that emerged as a distinct development approach towards the end of the 1970s (Blaikie et al, 1976) continues to haunt the MoFSC and to reoccur within policies and their interpretations at both national and district levels. For example, community forestry and forest inventory guidelines continue to emphasize the necessity for basic needs fulfilment from the forest; but the links to practical policy change to facilitate this have not been made explicit.

Recent and current policy and implementation issues in India and Nepal

The history of forest policy sets the scene and shapes the current changing policies and practices of PFM in important ways. The following generalizations about significant changes in policy focus and style, and their relevance to the emergence of PFM, can be made.

First, a *crisis in command-and-control forest management* has manifested itself in increased conflict, deteriorating forest conditions and forest protection movements (from the late 1960s onwards in India). Further attempts to stem these adverse symptoms of policy were made by increasing centralized control of forests and environment/wildlife protection systems. In turn, the recent emergence of tribal rights campaigns, the increasing politicization of forest issues, with attention now being paid in India by members of parliament (MPs) and members of the legislative assembly (MLAs) to forest rights issues, as well as the activities of mass tribal organizations (MTOs), federations of self-initiated community forest management groups and associations of forest users in both countries, have all increased pressures on the Indian forest administration. Simultaneously, there have been strong NGO campaigns against policies to accelerate the commercialization of forests. There is a complex three-way relationship among environmental conservation NGOs and individuals, the forest department and communities (or NGOs backing community interests). Different alliances are formed depending on the context – for example, in India, environmental conservation NGOs and communities come together against mining or the leasing of forest land to industry, while NGOs and the forest department are loosely in alliance against local communities on the issue of forest-based community rights (the Scheduled Tribes and Forest Dwellers (Recognition of Forest Rights) Bill being a topical instance). It is within the context of these pressures that PFM must be analysed. However, as the examples illustrate, these pressures have varied significantly through time and in different areas. In India, the main focus of struggle has centred on the recovery of rights lost in the past to the forest administration. Does PFM represent an advance to recover these rights, or is it a 'wolf in sheep's clothing' that will continue to devour remaining rights of access and management?

In Nepal, without a history of a colonial (and, it may be argued, neo-colonial) forestry service as in India, the crisis of 'fortress forestry' is less marked. The state has never been able to exercise the same control over forests as in India. However, in the hills, nationalization of forests has produced a crisis of legitimacy of different forms of local self-management. Thereafter, struggles between technically minded forest staff with forest conservation as their priority and a differentiated local rural populace with a variety of livelihood priorities emerged. Struggles existed between the Nepalese state and other actors (such as patronage-wielding politicians, smugglers, armed forest mafia and local forest users) in an attempt to hang onto the revenues from the *tarai* forests. The threat of their loss to more accountable, dispersed and politically aware user groups contributed to confusion, foot dragging by the state and its officials, and continuing loss of access to forests by many politically marginal groups and distant users. Thus, the question arises: is there a crisis of 'fortress conservation' in Nepal as there is in India? The answer is a heavily modified affirmative – there *are* key issues that set the forest administration against local people and the CFUGs in Nepal, but in different ways and with different members of the public than those in India.

Second, there has been a wide and dynamic *variety in the exploration of alternatives to command-and-control forestry over a long period of time, one of which is self-initiated conservation by communities in both India and the Nepalese hills.* As the accounts of policy history in this chapter have shown, there has been a long history of restoring (or partially restoring) the rights of forest users in India (e.g. the Sukhomajri and Arabari experiments, acknowledgement of grievances, social forestry, and various policy statements advocating more participation by local forest users in management). There have also been counter-strategies for seizing back control of forests and following the management objectives of commercial forestry, conservation and increasing green cover. In Nepal, a series of forest acts and guidelines have promoted PFM; but here, also, there have been counter-policy shifts to curtail PFM.

Third, *selective acceptance and accommodation of PFM by some key policy-makers has occurred both in India and in Nepal,* leading to widespread implementation of PFM, at least as a label. A critical examination of what 'acceptance' implies, and whether this is a rhetorical accommodation of convenience or a change in behaviour that reshapes processes and social relations on the ground, is one of the main research objectives of this book. In other words, to what extent do pressures from social movements, widespread protests, activists and donor-initiated policy, supported by a minority of Indian and Nepalese nationals within both forestry administrations, *actually affect* the outcomes of policy on the ground? Rhetorical acceptance of the general principles of PFM without precise supporting measures regarding land tenure and the legal restoration of the rights of forest users can be an infinitely extendable strategy. Mere lip service and formal authorization while dissipating PFM through foot dragging, hiding behind regulations, and adapting bureaucratic policies to follow 'business as usual' procedures is common, both in India and Nepal.

Fourth, *policy history in both countries has laid the terms of reference for current PFM strategy.* 'Participation' has very different meanings to different actors (as Chapters 3 and 11 show in more detail). To most in the Indian forestry administration, participation implies that the state already possesses the overall rights to controlling forests, and 'participation' is merely an offer to local people to protect forests (almost always 'degraded' rather than healthy forests) in return for a share of produce and fixed-term cash for labour inputs. Attention is therefore drawn to the degree to which local people have had a say in the management objectives of a forest's management plan and the implications of this management for people's livelihoods. The more radical alternative challenges whether the forest administration alone should have acquired the rights of forest management and revenue collection in the first place. The Nepalese case is less acutely contradictory since the state has been unable to control forests to the same extent as in India. However, in the *tarai*, there is an alliance of state employees, other entrepreneurs and local landed elites which has provided a powerful opposition to PFM and which can also shape the way in which policy is made and implemented. In the hills of Nepal, the meaning of 'participation' is perhaps less dictated by powerful bureaucratic fiat and less controlled by a near monopoly of the production of knowledge (and data) about forests.

References

Acharya, K. P. (2001) *A Review of Foreign Aid in Nepal*, Kathmandu, Citizen's Poverty Watch Forum

Adhikari, B. (2002) 'Forest encroachment: Efforts to overcome problems', in *Hamro Ban*, Kathmandu, DoF

Adhikari, J. (2003) 'Aid in developing Nepal', *Himal South Asia*, July, pp63–64

Adhikari, J. and Ghimire, S. (2002) *A Bibliography of Environmental Justice in Nepal*, Kathmandu, Martin Chautari

Amatya, S. M. and Shrestha, K. R. (2003) *Nepal Forestry Handbook*, Kathmandu, Nepal Foresters' Association

Anderson, R. S. and Huber, W. (1988) *The Hour of the Fox: Tropical Forests, The World Bank and Indigenous People in Central India*, New Delhi, Vistaar/Sage

APROSC (Agriculture Projects Services Centre) (1995) *Nepal Agriculture Perspective Plan*, Kathmandu, APROSC

Arnold, D. (1982) 'Rebellious hillman: The Gudem-Rampa raisings, 1893–1924', in Guha, R. (ed) *Subaltern Studies,* vol I, New Delhi, Oxford University Press

Baden-Powell, B. H. (1892) *A Manual of Jurisprudence for Forest Officers*, Calcutta, Government Press

Bajracharya, K. M. (2000) 'Intensive management of the *tarai* and inner *tarai* forests in Nepal', in *Management of Forests in Tarai and Inner Tarai of Nepal,* National Workshop Organized by the Nepal Foresters' Association, 11–12 February 2000, Kathmandu

Bajracharya, K. M. and Amatya, S. M. (1993) 'Policy, legislation, institutional and implementation problems', in Gautam, K. H., Joshi, A. L. and Shrestha, S. (eds) *Forest Resource Management in Nepal: Challenges and Need for Immediate Action*, The National Seminar, 31 March–1 April, Kathmandu

Banerjee, A. K. (1996) *Joint Forest Management: The Haryana Experience*, Ahmedabad, Environment and Development Book Series, Centre for Environment Education

Banerjee, A. K. (2004) 'Tracing social initiatives towards JFM', in Bahuguna, V. K., Mitra, K., Capistrano, D. and Saigal, S. (eds) *Root to Canopy*, New Delhi, Winrock International India and Commonwealth Forestry Association

Baral, J. C. (1999) *Government Intervention and Local Process in Community Forestry in the Hills of Nepal*, PhD thesis, University of Western Sydney, Hawkesbury, Australia

Baral, J. C. and Subedi B. R. (2000) 'Some community forestry issues in the *terai*, Nepal: Where do we go from here?' *Forests, Trees and People*, vol 42, pp20–25

Bartlett, A. G. and Malla, Y. B. (1992) 'Local forest management and forest policy in Nepal', *Journal of World Forest Resource Management*, vol 6, pp99–116

Behar, A. (2002) *Peoples' Social Movements: An Alternative Perspective on Forest Management in India*, Working Paper 177, London, Overseas Development Institute

Bhatia, A. (1999) *Participatory Forest Management: Implications for Policy and Human Resources Development in the Hindu Kush-Himalayas, vol V, Nepal*, Kathmandu, International Centre for Integrated Mountain Development

Bhatt, C. P. (2006) *Report of National Forest Commission*, Delhi, MoEF

BISEP-ST (Biodiversity Sector Programme for *Siwalik* and *Tarai*) (2002) *Biodiversity Sector Programme for Siwalik and Tarai*, Programme Document, Kathmandu, His Majesty's Government of Nepal

Blaikie, P. M., Cameron, J. and Seddon, J. (1980) *Struggle for Basic Needs: A Case Study in Nepal*, OECD Monograph Series, Paris, OECD Development Centre

Blaikie, P. M. and Muldavin, J. S. S. (2004) 'Upstream, downstream, China, India: The politics of environment in the Himalayan region', *Annals of the Association of American Geographers*, vol 94, pp520–548

Brandis, D. (1884) 'The progress of forestry in India', in *The Indian Forester*, vol 10, no 11, pp508–510, November 1884, cited in Guha, R. (1997) 'Dietrich Brandis and Indian forestry', in Poffenberger, M. and McGean, B. (eds) (1997) *Village Voices, Forest Choices*, Delhi, Oxford University Press, p95

Branney, P. and Yadav, K. P. (1998) *Changes in Community Forest Condition and Management 1994–1998*. Kathmandu, Nepal–UK Community Forest Project

CARE-Nepal (2002) *Impact Study Report on New Policy in Community Forestry with Reference to Banke, Bardia, Kailali and Dang Districts*, Kathmandu, CARE-Nepal

Carney, D. (1998) *Sustainable Rural Livelihoods: What Contribution Can We Make?*, London, Department for International Development

Central Bureau of Statistics (2001) *Statistical Pocket Book*, Kathmandu, His Majesty's Government of Nepal

Central Bureau of Statistics (2002) *Population Census 2001*, Kathmandu, His Majesty's Government of Nepal

Chakravarty-Kaul, M. (1996) *Common Lands and Customary Law: Institutional Change in North India over the Past Two Hundred Years*, Delhi, Oxford University Press

Chapagain, D. P. (1999) *Current Policy and Context of the Forestry Sector with Reference to the Community Forestry Programme in Nepal*, Working Paper, Kathmandu, SEEPORT and PROPUBLIC

Chapagain, D. P., Kanel, K. R. and Regmi, D. C. (1999) *Current Policy and Legal Context of the Forestry Sector with Reference to the Community Forestry Programme in Nepal: A Working Overview*, Kathmandu, Nepal-UK Community Forestry Project

Chaudhury, B. B. and Bandopadhyay, A. (eds) (2004) *Tribes, Forest and Social Formation in Indian History*, New Delhi, Manohar Publishers

ChFDP (Churia Forest Development Project) (2003) *Report on a Progress Review Mission*, ChFDP and GTZ, Nepal

Convention on Biological Diversity (1992) Convention on Biological Diversity, www.biodiv.org/doc/legal/cbd-en.pdf

Corbridge, S. E. (1991) 'Ousting Singbonga: The struggle for India's Jharkhand', in Dixon, C. and Heffernan, M. J. (eds) *Colonialism and Development in the Contemporary World*, London, Mansell, pp152–182

CPFD (2005) Community Forestry Database of the Community and Private Forest Division (Department of Forests), Government of Nepal, Kathmandu, November

CSE (Centre for Science and Environment) (1982) *The State of India's Environment, 1982: A Citizen's Report*, New Delhi, CSE

Deb, D. (2006) 'Sacred ecosystems of West Bengal', in Ghosh A. K. (ed) *Status of Environment in West Bengal: A Citizens' Report*, Kolkata, Society for Environment and Development (ENDEV)

Dev, O. P. (2003) *Collective Action for Forest Management And Livelihood Improvement: An Analysis of Community Forest User Groups in Nepal*, Leeds, UK, University of Leeds, School of Geography

DFRS (Department of Forest Research and Survey) and MoFSC (Ministry of Forests and Soil Conservation, Government of Nepal) (1999) *Forest Resources of Nepal (1987–1998)*, Kathmandu, Forest Resource Information System Project Publication No 74

DoF (Department of Forests) (2002) *Hamro Ban*, Kathmandu, Department of Forests, Government of Nepal

Dogra, B. (1985) 'The World Bank *vs* the People of Bastar', *The Ecologist*, vol 15, no 1/2, pp44–49

Eckholm, E. P (1976) *Losing Ground: Environmental Stress and World Food Prospects*, Oxford, Worldwatch Institute, and New York, Pergamon Press

Edmunds, D. and Wollenberg, E. (2001) 'Historical perspectives on forest policy change in Asia: An introduction', in *Environmental History*, www.lib. duke.edu/forest/Publications/EH/ehapr2001.html, pp190–212

Edmunds, D. and Wollenberg, E. (eds) (2003) *Local Forest Management: The Impacts of Devolution Policies*, London, Earthscan

Edward, D. M. (1996) *Non-Timber Forest Products from Nepal: Aspects of the Trade in Medicinal and Aromatic Plans*, FORESC Monograph 1/96, Kathmandu, Ministry of Forests and Soil Conservation

Eswaran, V. B. (2004) 'Genesis of JFM in India', in Bahuguna, V. K., Mitra, K., Capistrano, D. and Saigal, S. (eds) *Root to Canopy*, New Delhi, Winrock International India and Commonwealth Forestry Association

Falconer, J. (1996) 'Developing research frames for non-timber forest products', in Perez, M. R. and Arnold, J. E. M. (eds) *Current Issues in Non-Timber Forest Products Research*, Bogor, Indonesia, Centre for International Forestry Research

FAO (United Nations Food and Agriculture Organization) (1978) *Forestry for Local Community Development*, Rome, FAO Forestry Paper No 7, FAO

Farrington, J. and Baumann, P. (2000) *Panchayati Raj and Natural Resource Management: How to Decentralize Management over Natural Resources*, Research Proposal submitted to the Ford Foundation, New York

Fisher, R .J. (1989) *Indigenous Systems of Common Property Forest Management*, Hawaii, East-West Centre

Ford Foundation (1998) *Forestry for Sustainable Rural Development*, New York, Ford Foundation

Friedmann, J. (1992) *Empowerment: The Politics of Alternative Development*, Oxford, Blackwell

FSI (Forest Survey of India) (1988) *The State of the Forest Report, 1987*, Dehradun, Ministry of Environment and Forests, Government of India

FSI (1999) *The State of Forest Report*, FSI, Dehradun, Ministry of Environment and Forests, Government of India

Gadgil, M. and Guha, R. (1992) *This Fissured Land: An Ecological History of India*, New Delhi, Oxford University Press, and Berkeley, LA, University of California Press

Gadgil, M., and Guha, R. (1995) *Ecology and Equity: Use and Abuse of Nature in Contemporary India*, London, Routledge

Gadgil, M., Prasad, S. N. and Ali, R. (1983) 'Forest management and forest policy in India: A critical review', *Social Action*, vol 33

Gautan, M. (2002) 'Women's participation in forest development', in *Hamro Ban*, Kathmandu, DoF

Ghimire, K. (1992) *Forest or Farm? The Politics of Poverty and Land Hunger in Nepal*, New Delhi, Oxford University Press

Gilmour, D. A. and Fisher, R. J. (1991) *Villagers, Forests and Foresters: The Philosophy, Process and Practice of Community Forestry in Nepal*, Kathmandu, Sahayogi Printing Press

Gilmour, D. A. and Fisher, R. J. (1997) 'Evolution in community forestry: Contesting forest resources', in *Community Forestry at a Cross Road*, 1977 Seminar Proceeding, Bangkok, RECOFTC

GoI (Government of India) (1894) Circular No 22-F, 19 October, Calcutta, Department of Revenue and Agriculture, GoI

GoI (1952) *National Forest Policy Resolution*, 12 May, New Delhi, Ministry of Food and Agriculture, GoI

GoI (1961) *Third Five-Year Plan*, New Delhi, Planning Commission, GoI

GoI (1976) *Report of the National Commission on Agriculture: Forestry, vol IX*, New Delhi, Ministry of Agriculture and Irrigation

GoI (1986) *Report of the Inter-Ministerial Group on Wood Substitution*, New Delhi, Ministry of Environment and Forests, GoI

GoI (1999) *National Forestry Action Programme – India*, New Delhi, Ministry of Environment and Forests, GoI

GoI (2002) *Joint Forest Management: A Decade of Partnership*, New Delhi, Ministry of Environment and Forests, GoI

GOUP (Government of Uttar Pradesh) (1921) *Report of the Forest Grievances Committee for Kumaon*, Lucknow, Forest Development Agency, GOUP

Griffin, D. M. (1988) *Innocents Abroad in the Forests of Nepal: An Account of Australian Aid to Nepalese Forestry*, Canberra, Anutech

Grove, R., Damodaran, V. and Sangwan, S. (eds) (1998) *Nature and the Orient: The Environmental History of South and Southeast Asia*, New Delhi, Oxford University Press

Guha, R. (1983) 'Forestry in British and post-British India: A historical analysis', *Economic and Political Weekly*, vol 18, pp 1882–1897, 1940–1947

Guha, R. (1989) *The Unquiet Woods: Ecological Change and Peasant Resistance in the Himalaya*, Delhi, Oxford University Press

Guha, R. (2001) 'The prehistory of community forestry in India', *Environmental History*, vol 6, pp213–238

HMGN (His Majesty's Government of Nepal) (1988a) *Master Plan for the Forestry Sector, Nepal, Country Background, Kathmandu*, HMGN/ADB/FINNIDA

HMGN (1988b) *Master Plan for the Forestry Sector, Nepal, Main Report*, Kathmandu, MoFSC

HMGN (1995) *The Forest Act 1993 and Forest Regulations 1995*, Kathmandu, HMGN

HMGN (1998) *The Ninth Plan (1997–2002)*, Kathmandu, National Planning Commission, HMGN

HMGN (2000) *Revised Forest Policy*, Kathmandu, MoFSC

Hobley, M. (1996) *Participatory Forestry: The Process of Change in India and Nepal*, Rural Development Forestry Study Guide No 3, London, Overseas Development Institute

Hobley, M. and Malla, Y. B. (1996) 'From the forests to forestry – the three ages of forestry in Nepal: Privatization, nationalization, and populism', in Hobley, M. (ed) *Participatory Forestry: The Process of Change in India and Nepal*, London, Overseas Development Institute, pp65–82

Human, J. and Pattanaik, M. (2000) *Community Forest Management: A Case Book from India*, Oxford, Oxfam

IISc (Indian Institute of Science) (2004) *Joint Forest Management: Performance and Impact Studies: State Reports – Gujarat, Karnataka, Rajasthan, West Bengal*, Bangalore, Centre for Ecological Sciences, Indian Institute of Science

Ives, J. D. (1998) 'The Himalaya environmental change and challenge in the Himalaya – misguided attempts at development: Population growth and poverty', *Geography Institute Papers*, vol 40, no 1, Tokyo, Nippon University, pp34–51

Ives, J. D. and Messerli, B. (1989) *The Himalaya Dilemma: Reconciling Development and Conservation*, London, John Wiley and Sons

Ives, J. D., Messerli, B. and Jansky, L. (2002) 'Mountain research in South-Central Asia: An overview of 25 of UNU's mountain projects', *Global Environmental Research* vol 6, no 1, pp59–71

Japan Bank for International Cooperation (2005) 'JBIC's assistance to forestry sector in India', Presentation by Vineet Sarin, JBIC representative at the National Consultative Workshop on JFM, 14–15 July, New Delhi

Jefferey, R. (1988) *The Politics of Health in India*, Berkeley and London, University of California Press

Jeffrey, R. and Sundar, N. (1999) *A New Moral Economy for India's Forests?*, New Delhi, Sage Publications

Jerram, M. R. K. (1982) *A Text-Book on Forest Management*, Dehradun, International Book Distributors

Jewitt, S. (1998) 'Autonomous and joint forest management in India's Jharkhand: Lessons for the future?', in Jeffery, R. (ed) *The Social Construction of Indian Forests*, Edinburgh, Centre for South Asian Studies, and New Delhi, Manohar

Jodha, N. S. (2001) *Life on the Edge: Sustaining Agriculture and Community Resources in Fragile Environments*, New Delhi, Oxford University Press

Joshi, A. L. (1993) 'Effects on administration of changed forest policies in Nepal in Warner', in Warner, K. and Wood, H. (eds) *Policy and Legislation in Community Forestry*, Proceedings of a Workshop, RECOFTC, Bangkok

JTRC (Joint Technical Review Committee) (2000) *Community Forestry Issues in Tarai: Proceedings of Workshop on Community Based Forest Resource Management*, Nepal, Joint Technical Review Committee (JTRC)

Kanel, K. (2000) 'Management of the *tarai*, inner *tarai* and Churia forest resources: A reflection and perspective', Paper presented at *Tarai* Management Workshop, Nepal Foresters Association

Karna, A. L. (1998) *Critical Examination of Current Approaches to Participatory Community Forestry Planning in Nepal*, Dissertation Prepared in Partial Fulfilment of the Requirements for the MSc in Forestry Extension, University of Reading, UK

Khadka, M. (2000) 'Past and present forest management experiences along with prospects of sustainable forest management in the *tarai* and inner *tarai* forests of Nepal', Paper presented at Tarai Forest Management Workshop, Nepal Foresters Association

Khare, A., Sarin, M., Saxena, N. C., Palit, S., Bathla, S., Vania, F. and Satyanarayanan, M. (2000) *Joint Forest Management: Policy, Practice and Prospects*, London, IIED

Kondas, S. (1985) 'Social forestry in India', *Indian Forester* (special issue), November, pp887–898

Lal, J. B. (1992) *India's Forests: Myth and Reality*, Dehradun, Natraj Publishers

LFP (Livelihoods and Forestry Programme) (2001) *Livelihoods and Forestry Programme Document,* Kathmandu, DFID

LFP (2002) *Livelihoods and Forestry Programme Prospectus 2002*, Kathmandu, LFP and DFID

LFP (2003) *Data Complication on Community Forestry and Potential Community Forestry in Kapilbastu District*, Kathmandu, Innovative Development Associates (IDEA)

Ludden, D. (2002) *India and South Asia: A Short History*, Oxford, Oneworld Publications

Mahat, T. B. S., Griffin, D. M. and Shepherd, K. R. (1986a) 'Human impact on some forests of the middle hills of Nepal. Part 1: Forestry in the context of traditional resource of the state', *Mountain Research and Development,* vol 6, pp223–232

Mahat, T. B. S., Griffin, D. M. and Shepherd, K. R. (1986b) 'Human impact on some forests of the Middle Hills of Nepal. Part 2: Some major human impacts before 1950 on the forests of Sindhu Palchok and Kabhre Palanchok', *Mountain Research and Development,* vol 6, pp325–334

Mahat, T. B. S., Griffin, D. M. and Shepherd, K. R. (1986c) 'Human impact on some forests of the middle hills of Nepal. Part 3: Forests in the subsistence economy of Sidhu Palchok and Kabhre Palanchok', *Mountain Research and Development,* vol 7, pp53–70

Mahat, T. B. S., Griffin, D. M. and Shepherd, K. R. (1987) 'Human impact on some forests of the middle hills of Nepal. Part 4: Detailed study in south-east Sindhu Palchok and north-east Kabhre Palanchok', *Mountain Research and Development,* vol 7, pp111–134

Malla, Y. B. (2001) 'Changing policies and the persistence of patron–client relations in Nepal: Stakeholders' responses to changes in forest policies', *Environmental History*, vol 6, no 2, pp287–307

MLJ (Ministry of Law and Justice) (1999) *Local Self-Governance Act*, Kathmandu, Ministry of Law and Justice, Law Books Management Board

MoFSC (Ministry of Forests and Soil Conservation) (2000) *Memo: The Ministerial Concept Paper for Collaborative Forest Management (CoFM)*, Kathmandu, Unofficial Translation of MoFSC

MoFSC (2003) *Collaborative Forest Management Directive*, Ministry of Forests and Soil Conservation, Kathmandu

MoPE (Ministry of Population and Environment) (2000a) *Nepal Population Report, 2002*, Kathmandu, HMGN

MoPE (2000b) *The State of Population*, Nepal, Ministry of Population and Environment

Mosse, D. (2003) *The Rule of Water: Statecraft, Ecology and Collective Action in South India*, New Delhi, Oxford University Press

Mosse, D., Farrington, J. and Rew, A. (eds) (1988) *Development as Process: Concepts and Methods for Working with Complexity*, New Delhi, India Research Press

Munda, R. D. and Mullick, S. (eds) (2003) *The Jharkhand Movement: Indigenous Peoples' Struggle for Autonomy in India*, IWGIA document no 108, Copenhagen, International Work Group for Indigenous Affairs

Murali, A. (1984) 'Alluri Sitaramaraju and the Manyam Rebellion of 1922–1924', *Social Scientist*, vol 12, no 4, pp3–33

Nadkarni, M. V. (1989) *The Political Economy of Forest Use and Management*, New Delhi, Sage Publications

National Afforestation and Eco-Development Board (2005) 'Salient features of FDA projects', Presentation by Dr Sanjay Kumar, NAEB representative at the National Consultative Workshop on JFM, 14–15 July, New Delhi

Natural Resources Management Sector Assistance Programme (1997) *Programme Document*, Kathmandu, Embassy of Denmark

NCA (National Commission on Agriculture) (1976) *Report of the National Commission of Agriculture*, New Delhi, Ministry of Agriculture and Irrigation

Nepal–Australia Community Resource Management Project (2003) *Nepal–Australia Forest Users Equity and Livelihoods Project*, Kathmandu, Nepal–Australia Community Resource Management Project

Nepal–UK Community Forestry Project (1994) *Community Forestry in Nepal: A NUKCFP Briefing Document,* Kathmandu, NUKCFP, DFID

Neupane, H. (2003) 'Contested impact of community forestry on equity: Some evidence from Nepal', *Journal of Forest and Livelihood,* vol 2, no 2, Forest Action, pp55–61

NPC (National Planning Commission) (2003) *Tenth Five-Year Plan (2002–2007)*, Kathmandu, NPC

Ojha, H. (2000) *Terai Forestry and Possible Strategies for Management,* Kathmandu, Asia Network for Sustainable Agriculture and Bioresources, pp33–36

Overseas Development Group (2003) *Social Structure, Livelihoods and the Management of Community Pool Resource in Nepal*, ODG Report, University of East Anglia, Norwich, ODG

Pal, S. (2000) 'Community based forest management in Orissa – a new way forward', *Forest Trees and People Newsletter* no 42, Uppsala, Forest Trees and People Network

Panday, D. R. (2002) *Corruption, Governance and International Cooperation: Essays and Impressions on Nepal and South Asia*, Kathmandu, Transparency International

Panday, K. K. (1985) *Some Tenurial Aspects of Environmental Problems in Nepal in Land, Trees and Tenure*, Proceedings of an International Workshop on Tenure Issues in Agroforestry Land Tenure Centre (LTC) and International Council for Research in Agroforestry (ICRAF), Kenya

Panikkar, K. N. (1979) 'Peasant revolts in Malabar in the nineteenth and twentieth Centuries', in Desai, A. R. (ed) (1979) *Peasant Struggles in India*, Bombay, Oxford University Press

Pant, M. M. (1979) 'Social forestry in India', *Unasylva*, vol 31, no 125, pp19–24

Parajuli, D. P. (2003) *Evolution of Forest Policy in Nepal*, Kathmandu, Unpublished monograph

Pathak, A. (1994) *Contested Domains: The State, Peasants and Forests in Contemporary India*, New Delhi, Sage Publications

Pathak, A. (1995) 'Law, private forestry and markets', in Saxena, N. C. and Ballabh, V. (eds) *Farm Forestry in South Asia*, New Delhi, Sage Publications

Pati, B. (1983) 'Peasants, tribal and national movements in Orissa (1921–1936)', *Social Scientist*, vol 7, no 32, pp25–49

Poffenberger, M. (1990) 'Forest management partnerships: Regenerating India's forests', Executive Summary of the Workshop on Sustainable Forestry, in Bhatia, K. and McGean, B. (eds) (1990) *Forest Management Partnerships: Regenerating India's Forests: Executive Summary of the Workshop on Sustainable Forestry*, New Delhi, Ford Foundation

Poffenberger, M. (ed) (2000) *Communities and Forest Management in South Asia*, Gland, Switzerland, IUCN

Poffenberger, M. and McGean, B. (1998) *Village Voices, Forest Choices: Joint Forest Management in India*, New Delhi, Oxford University Press

Pokharel, B. K. (2003) *Contribution of Community Forestry to People's Livelihoods and Forest Sustainability: Experience from Nepal*, World Rainforest Movement, www.wrm.org.uy/countries/Asia/Nepal.html

Pokharel, B. K. and Amatya, D. (2001) 'Community forestry management issues in the *tarai*', in *Community Forestry in Nepal: Proceedings of the Workshop on Community Based Forest Resource Management*, 20–22 November 2000, Godawari, Lalitpur, Kathmandu, Joint Technical Review Committee, pp167–188

Raina, R. (2002) *Study on Networks in Community Forestry in India*, Unpublished PGDFM Report, Indian Institute of Forest Management, Bhopal

Rangarajan, M. (1996) *Fencing the Forest: Conservation and Ecological Change in India's Central Provinces, 1860–1914*, Delhi, Oxford University Press

Rangarajan, M. (2003) 'The politics of ecology: The debate on wildlife and people in India, 1970–1995', in Saberwal, V. and Rangarajan, M. (eds) *Battles over Nature: Science and the Politics of Conservation*, New Delhi, Permanent Black

Rao, G. B., Goswami, A. and Agarwal, C. (1992) *Trends in Social Forestry in India*, Sweden, Report prepared for the Swedish International Development Agency (Sida)

Rao, V. S. (1961a) 'The old forest policy: One hundred years of Indian forestry', *Indian Forester*, vol II, Appendix IV

Rao, V. S. (1961b) *One Hundred Years of Indian Forestry 1861–1961*, Dehradun, Souvenir Forest Research Institute

Ravindranath, N. H., Murali, K. S., and Malhotra, K. C. (eds) (2000) *Joint Forest Management and Community Forestry in India: An Ecological and Institutional Assessment*, New Delhi and Oxford, IBH Publication

Ravindranath N. H. and Sudha, P. (2004) *Joint Forest Management in India: Spread, Performance and Impact*, Hyderabad, Universities Press

Regmi, M. C. (1978) *Land Tenure and Taxation in Nepal*, Kathmandu, Ratna Pustak Bhandar

Ribbentrop, B. (1900) (reprinted 1989) *Forestry in British India*, New Delhi, Indus Publishing Company

Riley, J. M. (2002) *Stakeholders in Rural Development: Critical Collaboration in State-NGO Partnerships*, New Delhi, Sage Publications

Rjal, B. and Petheram, R. J. (2001) *Extension for Community Forestry Development in The Midhill Zone of Nepal*, International Union of Forestry Research Organizations: Proceedings of the Extension Working Party (S6.06-03) Symposium 2001, Institute of Land and Food Resources, University of Melbourne, Australia

Sagreiya, K. P. (1994) *Forests and Forestry*, New Delhi, National Book Trust

Sangawan, S. (1999) 'Making of a popular debate: The *Indian Forester* and the emerging agenda of state forestry in India, 1875–1904', *The Indian Economic and Social History Review*, vol XXXVI(2), pp187–237

Sarin, M. (1997) 'Grassroots initiatives versus official responses: The dilemmas facing community forest management in India', in *Community Forestry at a Crossroads: Reflections and Future Directions in the Development of Community Forestry*, Proceedings of an International Seminar held in Bangkok, Thailand, 17–19 July 1997, RECOFTC report no 16, Bangkok

Sarin, M., Ray, L., Raju, M. S., Chatterjee, M., Banerjee, N. and Hiremath, S. (1998) *Who Is Gaining? Who Is Losing?: Gender and Equity Concerns in Joint Forest Management*, Delhi, Society for Promotion of Wastelands Development

Sarkar, S. (1980) 'Primitive rebellion and modern nationalism: A note on forest Satyagraha in the non-cooperation and civil disobedience movements', in Panikkar, K. N. (ed) *National and Left Movements in India*, New Delhi, Vikas Publication

Saxena, N. C. (1990) *Farm Forestry in Northwest India*, New Delhi, India, Ford Foundation

Saxena, N. C. (1994) 'Forests, people and profits: New equations for sustainability', Paper presented to the Workshop on Policy and Implementation Issues in Forestry, New Delhi, Dehradun, Centre for Sustainable Development and Natraj

Saxena, N. C. (1995) 'Forest policy and the rural poor in Orissa', *Wastelands News*, vol II(2), Delhi, SPWD, pp9–13

Saxena, N. C. (1999) 'Participatory issues in joint forest management in India', Foundation Day Lecture, Delhi, SPWD

Saxena, N. C. and Ballabh, V. (eds) (1995a) *Farm Forestry in South Asia*, New Delhi, Sage Publications

Saxena, N. C. and Ballabh. V. (1995b) 'Farm forestry and the context of farming systems in South Asia', in Saxena, N. C. and Ballabh, V. (eds) (1995) *Farm Forestry in South Asia*, New Delhi, Sage Publications, p250

Sayer, J. A. and Maginnis, S. (eds) (2005) *Forests in Landscapes Ecosystem Approaches to Sustainability*, London, Earthscan

Seddon, D. and Hussein, K. (2002) *The Consequences of Conflict: Livelihoods and Development in Nepal*, London, ODI Livelihoods and Chronic Conflict Working Paper Series no 185

Seeland, K. (1997) 'What is indigenous knowledge and why does it matter today?', in Seeland, K. and Schmithusen, F. (eds) *Local Knowledge of Forests and Forest Uses among Tribal Communities in India*, Zurich, Department Wald-und Holzforschung

Seeley, J. (2003) *Livelihood Labelling: Some Conceptual Issues*, Norwich, UK, University of East Anglia

Shah, A. (2000) *Emergence of Joint Forest Management in India: A Journalistic Documentation in the Hands of the People*, Ahmedabad, Gujarat Institute of Development Research

Shah, A. (2003) 'Fading shine of golden decade – the establishment strikes back', Paper presented at the National Seminar on New Development Paradigms, Ahmedabad, Gujarat Institute of Development Research

Sharma, A. R. (2002) 'Community forestry development programme' in *Hamro Ban*, Department of Forests, Kathmandu, His Majesty's Government of Nepal

Sharma, A. and Ramanathan B. (2000) *Joint Forest Management in Jhabua: A Preliminary Documentation*, New Delhi, WWF-India

Sharma, S. (1998) *Decentralization and Local Participation for Development: Policies and Realities in Nepal*, Kathmandu, Annapurna Offset Printing Press

Shivaramakrishnan, K. (1999) *Modern Forests: Statemaking and Environmental Change in Colonial Eastern India*, Delhi, Oxford University Press

Shrerstha, G. (2002) 'Demand and supply of timber and fuelwood in Nepal', in *Hamro Ban*, Department of Forests, Kathmandu, His Majesty's Government of Nepal

Shrestha, H. S. (2002) 'Monitoring and evaluation situation of Forest Development Programme', in *Hamro Ban*, Department of Forests, Kathmandu, His Majesty's Government of Nepal

Sigdel, H. (2002) *Profiles of Programmes/Projects under Tarai Arc Landscape (TAL)*, Kathmandu, World Wildlife Fund Nepal Programme

Singh, Chetan (1996) *Natural Premises: Ecology and Peasant Life in the Western Himalaya*, New Delhi, Oxford University Press

Singh, Chhatrapati (1986) *Common Property and Common Poverty: India's Forests, Forest Dwellers and the Law*, New Delhi, Oxford University Press

Singh R. V. (2002) *Forests and Wastelands: Participation and Management – The Ford Foundation 1952–2002, Celebrating 50 Years of Partnership*, New Delhi, Ford Foundation

SPWD (Society of Promotion of Wastelands Development) (1992) *Joint Forest Management: Concepts and Opportunities*, Proceedings of the National Workshop at Surjkund. Society of Promotion, New Delhi, SPWD

SPWD (1993) *Proceedings of the National Level Meeting of the SPWD National JFM Network 1993*, New Delhi, SPWD

SPWD (1998) *Joint Forest Management Update*, New Delhi, SPWD

Srivastava, B. P. and Pant, M. M. (1979) 'Social forestry on a benefit-cost analysis framework', *Indian Forester*, vol 105, no 1, pp2–35 (quoted in Pant, 1979)

Statz, J. (2003) *Community Forest Management Demonstration Programme, Integrated Planning Processes for Natural Resource Management and the Distant User Approach*, Draft Report of Churia Forest Development Project, Lahan

Stebbing, E. P. (1926) *The Forests of India*, vol 11, London, John Lane, Bodley Head Limited

Sundar, N., Jeffery, R. and Thin, N. (2001) *Branching Out: Joint Forest Management in India*, New Delhi, Oxford University Press

Sundar, P. (1995) *Patrons and Philistines*, New Delhi, Oxford University Press

Takimoto, A. (2000) *Impact of Community Forestry in Banke and Bardia Districts of Forestry and Partnership Project*, Kathmandu, CARE-Nepal

Tamang, D., Gill, G. J. and Thapa, G. B. (eds) (1993) 'Indigenous management of natural resources in Nepal', Policy Analysis in Agriculture and Related Resource Management Project, Kathmandu, Ministry of Agriculture/Winrock International

Thin, N., Neeraj, P. and Prafulla, G. (1998) *Muddles about the Middle: NGOs as Intermediaries in JFM*, Edinburgh, Centre for South Asian Studies

Timsina, N. P. (2002a) *Political Economy of Forests Resource Use and Management*, PhD thesis, University of Reading, UK

Timsina, N. P. (2002b) 'Empowerment or marginalization: A debate in community forestry in Nepal', *Journal of Forest and Livelihoods*, Forest Action, vol 2, no 1, pp27–33

Timsina, N. P. (2003) 'Viewing FECOFUN from the perspective of popular participation and representation', *Journal of Forests and Livelihoods*, Forest Action, vol 2, pp67–71

Tiwari, S. (2002) 'Access, exclusion and equity issues in community management of forests: An analysis of status of community forests in mid-hills of Nepal, in Winrock International' (2002) *Policy Analysis of Nepal's Community Forestry Programme: A Compendium of Research Papers*, Kathmandu, Winrock International

TPCG (Technical and Policy Core Group) and Kalpavriksh (2005) *Securing India's Future: Final Technical Report of the National Biodiversity Strategy and Action Plan*, Pune and Delhi, Kalpavriksh

Tucker, R. P. (1984) 'The historical roots of social forestry in the Kumaon Himalayas', *Journal of Developing Areas*, vol 13, no 3, pp341–356

UNDP (United Nations Development Programme) (2001) *Nepal Human Development Report*, UNDP, www.undp.org

Valkeman, G. (1997) *Community Forestry Working Plan 1997–2000*, Natural Resource Management Series, December, vol 1, Kathmandu, Makalu-Barun Conservation Project

Vira, B. (1995) *Institutional Change in India's Forest Sector, 1976–1994 – Reflections on State Policy*, Research Paper No 5, November, Oxford, Oxford Centre for the Environment, Ethics and Society

Voelcker, J. A. (1893) *Report on the Improvement of Indian Agriculture*, London, Eyre and Spottiswoode

Westoby, J. (1962) 'Forest industries in the attack on underdevelopment', *Unasylva*, vol 16, no 4, pp168–201

Winrock (2002) *Emerging Issues in Community Forestry in Nepal*, Nepal, Winrock International

World Bank (1978) *Forestry Sector Policy Paper*, Washington, DC, World Bank (quoted in Hobley, M., 1996)

World Bank (1999) *Poverty in Nepal at the Turn of Twenty-First Century*, vols I and II, Washington, DC, World Bank

World Bank (2005) 'World Bank forestry programs in India', Presentation by Ms Reena Gupta, World Bank representative at the National Consultative Workshop on JFM, 14–15 July, New Delhi

WWF (World Wide Fund for Nature) (2004) *Tarai Arc Landscape Programme*, WWF, www.world-wildlife.org/tigers/pubs/Tarai_Arc2004.pdf

Yadav, N. P. (2003) *Community Forest User Group Impacts on Community Forest Management and Community Development*, UK, University of Leeds, School of Geography

2

Understanding the Policy Process

Piers Blaikie and Oliver Springate-Baginski

Linkages between policy and livelihood impacts

Policy may be briefly defined as a set of stated intentions and resultant practices in the name of the public good. The policy process is the means by which policy is conceived, negotiated, expressed and, perhaps, brought into law, and the procedures of implementation and practice. This and the following chapter examine the ways in which forest policies and, particularly, participatory forest management (PFM) have come about. This task is necessary because policy reform does not emerge as a linear response to 'truth talking to power', as the Introduction has discussed – in other words, as a result of facts from research or other sources that reveal new truths and support alternative rational arguments for a policy change. Changes to policy are much more complex than this simplistic rationalist model. Thus, the traditional question 'How can research be transferred to the policy sphere?' is currently replaced by the question 'Why are some of the ideas that circulate in the research/policy networks picked up and acted on, while others are ignored and disappear?'. In the case of forest policy in India and Nepal, there has been a long and distinguished literature on forest policy reform; and yet, the forest administrations responsible for reform have, by and large, been slow to pick up appeals for justice and a more pragmatic policy that balances the needs of rural forest users with those of commercial forestry and 'green cover' imperatives. This and the next chapter seek the reasons for this.

As in most policy-making, multiple actors, often with divergent versions of the 'truth' and competing objectives, are involved in negotiating formal policy. In many cases, outcomes of policy and what happens on the ground may bear little resemblance to the intentions of those who shape and draft policy documents of various types. Some authors have taken a highly sceptical view of the conventional rationalist position on the policy process. Apthorpe (1997), for instance, says that 'the plainer and clearer a policy is painted, the more it is driven by evasion and disguise'. While our approach to policy acknowledges evasion and disguise, it is also tempered by an acceptance of the instrumental intentions of the authors of policy, in which there is often good faith, professionalism (as interpreted by the forester or administrator) and a rational application of policy, as they see it, towards achieving stated outcomes. There are other voices that may not be heard and are unable to either join the negotiating table or whose knowledge is deemed by more powerful actors to be worthless or illegitimate. This focus may sound like a contradiction in terms – to listen to voices that do *not* contribute to formal policy-making. However, this focus identifies the quality and degree of representation and 'participation' in the policy process and interrogates the claims made for PFM that it is in some way 'participatory'; thus, voices that are not heard (or are disregarded) in the policy process must be a central concern. There is also a need to take into account the decisions and actions being made by others on the ground outside the formal policy process. These actions help to shape a *de facto* policy on the ground, but may only affect the formal policy process as a reaction or response to a previous cycle of policy-making. Therefore, we consider it necessary to explain the policy process itself – evasions, good faith, ambiguities and strategies of main actors included (see also Apthorpe and Gaspar, 1996; Scott, 1998; Keeley and Scoones, 1999; Sutton, 1999; Shankland, 2000).

This chapter analyses the policy process in terms of the *structures, institutions and actors involved and their relationships*. Policy is not made only by political leaders in conjunction with the senior bureaucrats of the ministry concerned, but is profoundly affected at all stages by a whole cast of other actors, including other ministries, international funding institutions (IFIs), 'street-level bureaucracies' (i.e. the field staff at the lower levels of the policy process), the judiciary, and also by civil society (e.g. social movements, non-governmental organiza-tions (NGOs), prominent scientists and intellectuals, local politicians, state politics and entrepreneurs). Therefore, enquiry into forest policy should not confine itself to the head offices and ministry buildings of New Delhi and Kathmandu. This chapter discusses the policy process, but also the structures and formal procedures insofar as they are significant in shaping these policy processes. An excursion into the administrative detail of the forest administrations of India and Nepal may seem a diversion from the path to understanding policy process; but we consider it necessary from time to time as they essentially shape the process. To give a brief example, the job descriptions of district/divisional forest officers (DFOs) are important in explaining what the officers do from day to day, and also the way in which they form attitudes to new PFM initiatives, which may require a different set of practices, professional values and attitudes.

At all stages of the policy process, and particularly in implementation, the politics of knowledge production are an important element. They concern the production of 'authori-tative' knowledge about forests (e.g. the classification of land as 'forest', statistics on forest cover and quality, technical manuals on forest management, and the role of forests both in the economy and in the environment), as well as what is deemed unacceptable and worth-less knowledge, such as the use and management of a wide range of forest products for subsistence use that are not commercially attractive. The dissemination of forest knowledge to specific audiences and the ways in which this type of politics is played out have a profound impact on forest management and livelihoods. These elements are linked to the historical origins of forest policy and to the inception of PFM policy as discussed in Chapter 1. In this sense, policy history is important to current policy analysis.

How knowledge is produced and communicated to others takes the discussion from the structural and institutional aspects of forest policy in this chapter to forest discourses. In Chapter 3, we examine in more detail the substance of the *major discourses* used in forest policy, and how these are deployed and practically implemented in the policy process. Discourse here is defined as an ensemble of ideas, concepts and categories. Discourses frame certain problems and ignore or 'brush out' others. There are key strings of propositions made by different actors in the policy process that are amenable to analysis. The concerns of this chapter (actors, structures and institutions) are very closely linked to the discourses produced by these actors and institutions. However, the organization of the discussion into two chapters (policy process and actors in Chapter 2 and discourses in Chapter 3) should not indicate a separate theoretical domain or a separation between actors in their structural and institutional positions and what they say.

In Chapter 4, the focus turns to the potential *impacts of the implementation of PFM policy on livelihoods*. As we trace the policy process to the impact of policy on livelihoods, implementation becomes an integral part of the policy process. There are dangers in sepa-rating implementation from policy formulation, as decision-makers can then attribute any perceived policy failures to 'poor implementation', 'lack of political will', 'absence of adequate scientific research capabilities' and so on, rather than to the policy itself (Clay and Schaffer, 1986). A conceptual separation of policy-making and implementation might allow policy-makers, therefore, to abrogate responsibility for their policy and pass the blame down the administrative line to intermediate administrators or field operators (e.g. forest rangers or beat guards). As an illustration of statements of intent in policy statements without a clear indication of the necessary mechanisms for implementation, the fourth principle of the 1988 Indian Forest Policy states that the policy will involve 'creating a massive people's movement with the involvement of women, for achieving these objectives [listed] and to minimize pressure on existing forest[s]'. The policy statement must be judged not only by the nobility (or otherwise) of its sentiments, but also by the means by which it will be implemented. The policy puts forward no mechanisms at all to implement this creation of a mass people's movement, and therefore a degree of scepticism about this aspect of the policy seems to be in order. This book looks at the absence of institutional mechanisms for translating policy goals into practice.

In order to link policy process and discourse to actual impacts on livelihoods on the ground, we use a framework that charts the major determinants of livelihoods and the distribution of different livelihoods to different groups of the population within agrarian political economy, in general. Having set up the framework, we treat the livelihood impacts of PFM as a particular and additional set of factors to those already operating in the agrarian political economy and to those shaping livelihoods, but also as a factor that potentially alters the ongoing processes of agrarian political economy. Part II of the book elaborates on these links in the different national, regional and local conditions, and analyses the empirical data on livelihood impacts of PFM and on the policy process at the field level. Thus, explanations of the livelihood impacts of PFM must incorporate, where directly relevant, the whole sequence of policy history and specific policies through to implementation, and how these impacts work themselves out on the ground within existing social and environmental conditions – the whole sequence of analysis from how policy is formed at the central level (discussed in this chapter) through to the state level (discussed in Part II).

Participatory forest management in the policy process

Figure 2.1 provides a framework for the policy process and maps what the authors believe to be the main relationships between structures, institutions and actors in forest policy process. The policy process takes place at different levels, with feedback and reflexive relations between them. There are the international, national (in India there is also the state level), district and local levels. The assumption here is that policy is best judged as a whole process from initial negotiations in the context of an ongoing policy history right through to outcomes on the ground. Policy is not *made* in the capital city and *implemented* at the district and local levels. Therefore, understanding policy requires that analysis is focused at different levels.

Figure 2.1 maps an analysis that is intended as generic and can be made in general terms to analyse the policy process in forestry and the role of PFM in any country. However, here it focuses first on India and then Nepal. The figure is intended to be used as a conceptual map of a complex process occurring at these different geographical scales and levels of administration (e.g. national, state, regional and local). The development of formal policy is a process in which many actors take part, and policy is shaped in many different ways, with the result that explanation becomes complex. As a result, points of entry for reform cannot be identified simply by policy recommendations at the higher levels alone. Therefore, opportunities and constraints to reform can best be understood by an account of all the most important policy 'drivers' (key processes and influences) operating at a variety of levels. Figure 2.1 and accompanying text are a means of dealing with the complexity involved.

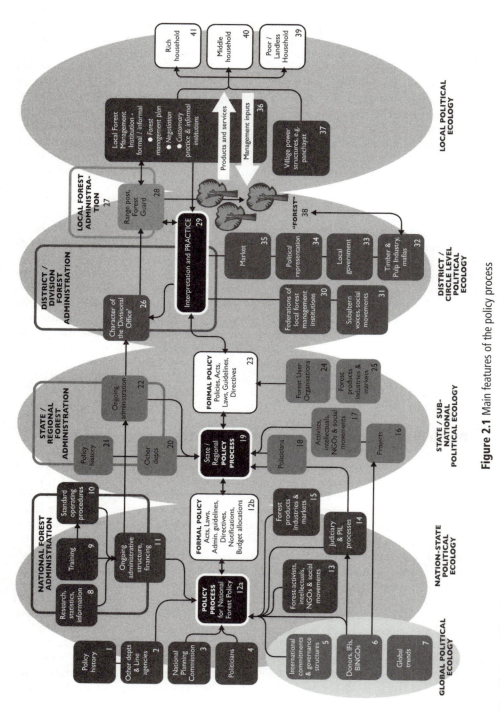

Figure 2.1 Main features of the policy process

Source: Piers Blaikie and Oliver Springate-Baginski (original material for this book)

There are a few caveats about the interpretation of Figure 2.1. The first is that it aims to simplify without doing violence to complexity, uncertainty and ambiguity. Readers will no doubt be able to identify linkages that are not charted. We argue that inclusion or exclusion is decided according to the importance we attach to the central arguments and 'storyline' of the book. Regarding uncertainty and ambiguity, with which actors in the policy process live as a part of daily life, these are verbally described in the commentary. Arrows linking different elements (cells) in the figure are easy enough to draw, but sometimes require detailed explanation as to the type of linkage. Also, Figure 2.1 may give an impression of linearity in the policy process, whereas many policy processes (and forestry is no exception) are reflexive or iterative, with feedback to a number of elements in the system. At the national level, for example, there are a wide number of linkages, reflexive flows of information, and personal interactions between different institutions and the individuals within them, and these are discussed in detail. Different elements in the process are numbered in the figure to facilitate cross-reference between text and diagram. The reader may like to treat Figure 2.1 as a small-scale road map to facilitate a view of the larger terrain and a simplified representation of the location of a particular actor or institution.

Figure 2.1 identifies four linked levels of the policy process, starting on the left at the highest level – the global political ecology, in which global commitments and dominant issues in development ('global trends', Cell 7 in the diagram – e.g. neo-liberalism, governance and participation). The institutions involved are bilateral and multilateral donors, labelled IFIs (Cell 5). For example, the World Bank is funding a significant PFM programme in Andhra Pradesh, as well as in other states (Cell 16, at the state level). Other big international non-governmental organizations (BINGOs), such as the World Wide Fund for Nature (WWF) and the World Conservation Union (IUCN) (Cell 6) are significant actors in the forest policy process. There are also international environmental agreements (Cell 5) that may affect forestry policy in complex, albeit so far minor, ways. India is a signatory to the Convention on Biodiversity (CBD), but was shamed at a recent CBD conference of parties in Brazil for not honouring its commitments (including not having an official National Biodiversity Strategy and Action Plan, or mechanisms for ensuring benefit-sharing, protection of indigenous knowledge or involvement of indigenous/tribal communities in the management of protected areas by NGOs). Probably even more important is the Clean Development Mechanism, under which all kinds of projects are being agreed to, with India having the largest number of such projects of any country to date.

Policy-making at the national level in India

The policy analysis starts at the *national level* since this has become the locus at which the overall strategy and structures for the state's management of the forest are decided. In societies where forest users have a more powerful role in policy-making, the starting point may well be better situated at multiple locations at the grassroots level. Here we examine the nature of the policy process from the top down, not as a normative statement of how to analyse the process, but in acceptance of the fact that in India, especially, forest policy is a top-down process. Despite some states in north-east India enjoying greater constitutional autonomy in forest policy formulation, national policy has invariably reshaped policy even there, in part due to ambiguities in the legal boundaries separating the two. At the state/district level, in tribal-dominated areas, there is a parallel jurisdiction of tribal departments or integrated tribal development agencies (ITDAs), which is discussed in more detail below.

Thus, for both India as a whole and the three states chosen for detailed study and for Nepal, a range of actors engage and negotiate in the production of formal policy. Acts, laws,

guidelines and notifications (see Cell 12a in Figure 2.1), as well as administrative arrangements, are required to implement these, involving budget allocations for the administration
to operate right down to the household level. Most of the discussion in this chapter focuses
on the national level, since Part II of this book deals with forests and forest policy at the state
level, and also discusses how policy ultimately affects user groups and the households within
them.

At the core of the policy process at the national level is the legislature (see Cell 12a in
Figure 2.1) and its passage of permanent, legally binding legislation (acts, byelaws, etc; Cell
12b in Figure 2.1) and recurrent allocation of resources to ensure its execution, implementation or enforcement by the administration and executive. The production of a formal
policy document represents a temporarily stabilized moment of relative power configuration
in an ongoing dynamic and, often, very fluid process until it is debated, negotiated and acted
on again, then interpreted and fought over. Therefore, policy-making is seldom initiated on
a *tabula rasa*, but has an established policy history (see Cell 1 in Figure 2.1) and a whole environment of standard operating procedures, bureaucratic norms and sets of expectations of
the different actors involved. The policy process therefore involves modifying what has
already been established according to the prevailing priorities and expedients of political
actors and alliances of the past. The *policy history* is a powerful pre-existing *discursive
resource* within which current reforms are sought by particular actors for particular reasons
(see Chapter 3). As we have seen in Chapter 1, forest management has a great deal of
momentum in South Asia, where very large, well-established and multifaceted administrative
structures have assumed the control and management of forests for many years.

In Figure 2.1, *forest product industries and markets* (see Cell 15) are important actors, at
least in India. The state forest administration has historically dominated this area, initially
through contractors and through leasing forest areas to forest-based industries, and later
through parastatal forest corporations. Here, there are three main groupings. The first and
most obvious is the timber and wood product industries (including plywood); second, the
pulp industry; and, third, the non-timber forest product (NTFP) traders and industries.
Examples include the Western Indian Match Company (WIMCO), the Bhadrachalam Paper
Company for pulpwood and the Titaghur Paper Mills in Orissa. As might have been
expected, during recent negotiations these companies and others were not concerned with
the participation of local forest users or producers, local profit-sharing, co-operatives or
other forms of collaborative ventures with local organizations. These industries were major
forces in policy influence until the 1980 Forest Conservation Act banned the leasing of forest
land to any private party asking industry to meet its needs from private lands. Existing leases
were not to be renewed on expiry. This, combined with green felling bans in many states, has
reduced the importance of forests for such industries. Industry has been trying, since the
1980s, to gain access to 'degraded' forest lands for captive plantations, which has met with
vehement resistance from NGOs. During recent years, the Ministry of Environment and
Forests (MoEF) has again begun talking about the need to attract private investment to
afforest degraded forest lands, and under the changed context of economic liberalization has
worked out a 'multistakeholder partnership on forests' scheme in collaboration with the
Confederation of Indian Industry. If implemented, this would likely convert PFM in India
into a mechanism for producing raw material for industry on terms effectively decided by
forest departments and industry. Other major stakeholders are mining companies and dam
builders (which are not interested in forest products at all, but rather in what lies *beneath* the
land or other potential uses, including submergence).

It is useful to take a wider perspective of the political environment that determines forest
policy. *Democratic constitutional processes*, which the executive administration (as 'public
servant') is expected to translate into practice, set the national policy for forest management
in general political terms. Although the forest administrations' power is, theoretically, limited

by their legal mandate, monitoring of the use and abuse of considerable discretionary powers enjoyed by administrators/foresters under the law is poor. The dissonance between policy and law also needs to be recognized. Policy articulates the government's intent, but cannot be translated into executive action unless matching legislation is enacted. In India's case, although the 1988 Forest Policy (note 'policy' – not an act, which has had full scrutiny in parliament) radically changed forest management objectives contained in the 1952 Forest Policy; but no changes were made to the 1927 Indian Forest Act (IFA). In the major Godavarman public interest litigation (PIL) case, popularly known simply as 'the forest case', which has been continuing in the Supreme Court since 1995, the court has passed a wide range of orders based on interpretations of conservation laws: the IFA; the 1980 Forest Conservation Act (FCA); and the 1972 Wild Life Protection Act (WLPA), which violate the spirit of the 1988 policy with respect to the rights of tribal and other forest dwelling communities by prioritizing the objectives of forest conservation (Sarin, 2005a, 2005b).

In India, the executive ministry (administration) consists of two echelons – namely, the technically qualified chief forester, with his field executives (often called the directorate, or the executive arm, headed by the principal chief conservator of forests, or PCCF), and the Indian Forest Department (the administrative arm headed by the forest minister; the PCCF is also the executive head of the Indian Forest Department, below the minister) (see Cell 11 in Figure 2.1). Policies are effected through the collaboration of the chiefs of both echelons in formulating and implementing the policy. Normally, however, forest policies originate from the national or, sometimes, the state forest directorate. A policy may also emanate from the field offices; but the chief decides whether to formulate it and sends it to the administrative office for agreement. The secretary, if he agrees, then obtains draft proposals vetted by the minister before the policy is issued as a government order for implementation under his signature. Recently, policies have been put up for comment on ministry websites, and depending on how proactive the concerned ministry is, it may also organize consultations with different stakeholder groups to gain their views. This happened recently with the new draft environment policy prepared by the MoEF. Most environmental NGOs protested against the fact that the draft was only put up on the website and only in English, thereby making it inaccessible to local communities and members of *gram sabhas* and *panchayats*, who are most directly dependent on environmental resources. The MoEF also organizes meetings with state forest departments, with industry and with a small select group of NGOs. Many NGOs are able to send their critical comments directly to the ministry. The problem remains, however, that although the revised draft environment policy has been finalized, it has not been made public because it must first be presented to the Cabinet for approval. Any document to be presented to the Cabinet continues to be labelled secret, and any wider sharing of the draft for consultation with other stakeholders could be termed a breach of parliamentary privileges, leading to the rejection of an entire bill. After a new policy document has been cleared by the Cabinet (consisting of all ministers of the central government), it is then tabled in parliament, at which point it also becomes a public document. Many of these documents are passed in parliament with little debate; but active groups concerned about their content can get members of parliament (MPs) to raise questions about particular provisions in order to get them changed. Therefore, although a policy is not legally enforceable, it does go through legislative screening and approval, and is not just a matter of one minister approving what his ministry's secretary has invented.

The real problem in achieving a democratic policy process remains how to practically organize a genuinely consultative process for developing a new policy in which the large numbers of people likely to be directly affected at the grassroots are able to participate in meaningful ways (the drafting of the Indian National Biodiversity Strategy and Action Plan (NBSAP) was commissioned by the MoEF from a leading environmental NGO, Kalpavriksh, who developed it through a consultative mechanism, although the final submission in 2003 has so far been rebuffed by the MoEF). This is altogether a different matter from, for example, the revision of the *van panchayat* rules or state joint forest management (JFM) orders. In these cases, institutions such as *van panchayats* and specific JFM groups will be directly affected by any revision, and invariably have much clearer views about the changes (although these are sometimes unilaterally revised by state forest departments anyway). However, even in these situations, very little attention is paid to specific groups affected by the revision. This problem is all the more intractable when a new policy affects a much more amorphous and differentiated section of the population, as discussed in Part III. Nevertheless, the style of the whole procedure is commonly very top down, with undue dominance from the executive and the administration rather than from local forest users, who should play the dominant part in democratic processes.

A. K. Mukherjee, a retired inspector general of forests, gave a summary of the actual process through which the 1988 Forest Policy was formulated in his keynote address at a national workshop on JFM at the Indian Institute of Forest Management (IIFM) in 2006 (Mukherjee, 2006). His account was instructive since many of the usual channels of communication between different actors were bypassed. He pointed out that some of the most significant changes in forest and environmental policies were made by Prime Minister Indira Gandhi. On her return from the Stockholm conference in 1972, she called a meeting at Dehra Dun and asked why forests were so rapidly becoming degraded. When the foresters told her that they had no control over the diversion of forest lands to other uses, she ratified the 42nd Constitutional Amendment, which put forests on the concurrent list (i.e. the list of issues to which the Indian national government must give assent to state's decisions), and so since then both the central government and the states must approve changes in land use and forest clearance. The 1980 Forest Conservation Act was later enacted to prevent state governments from clearing forests for other uses without central government approval. During this time, politicians, foresters and NGOs were represented on the Central Board of Forestry, which was chaired by the prime minister. The draft of the 1988 Forest Policy was deliberated on in five meetings of the board over a period of ten years and finally approved in 1987. The board thereafter became non-functional and was effectively dissolved in 1990. One of the biggest hurdles faced in getting approval for JFM, particularly its extension to healthy and intact forests in the circular of February 2000, came from the Ministry of Finance, which was concerned with the loss of revenue to the government from sharing forest benefits with communities.

Secretaries of the Indian Forest Directorate (the executive arm of the forest administration) report to politically elected ministers (in the administrative arm). However, ministers' interest in and knowledge of forestry are highly variable; some have questionable personal integrity and playing to state-level political constituencies is not uncommon. For example, the last two forest and environment ministers of India have been from the Dravida Munnetra Kazhagam party and have used their ministry to give rapid clearance for environmentally damaging development projects that they felt would increase the party's standing in the electorate.

Other departments and line agencies (see Cell 2 in Figure 2.1), especially the revenue departments in India (but also in Nepal, to a lesser degree), have an historic rivalry with the forest administration. During the mid to late 19th century, many tax incentives were offered to promote the conversion of the forest frontier to agriculture, and the forest departments had to justify the reservation and protection of forest land against the revenue departments, which charged that tax revenues were being lost as the land would be more productive if brought under sedentary agriculture. The origin of the term 'wasteland' lies in the categorization of lands not yielding any revenue (most of which were uncultivated common grazing lands) as 'waste' by the colonial administration. In recent forest policy, the line agencies dealing with revenue and mines have wielded considerable influence over both land uses that are alternative to forest and the revenue implications of PFM, as opposed to a more centralized forest management. While the forest administration (both department and directorate) has been the main state agency for controlling and managing forests, increasingly it is the writ of the Cabinet Committee on Economic Affairs (see Cell 2 in Figure 2.1) which seem able to assert its will in ways that are likely to be contrary to those of the forest administration. The main concerns are the speeding up of environmental and forest clearances for mining and other projects attracting foreign direct investment (FDI). While policy is framed by the ministry, its execution is by the department. In practice, administrators in ministries rely on technical foresters to frame policy, which reduces the significance of their different roles. It is extremely difficult for administrators to act counter to the wishes of professional foresters, except where there is an even stronger political issue, as in the case of the present demand to speed up forest clearances for mining. A topical case in point is that of the Saranda forests, which are under threat of felling in order to open them up to mine their rich iron ore deposits. The forest administration of Jharkhand is urgently seeking to have the Saranda forests declared 'virgin forests': C. P. Khanduja (DFO, Saranda) was quoted as saying 'Since mining can take place in virgin forests only after exploring possibilities in non-virgin areas, this is one way to protect biodiversity and the variety of species in the area from destruction' (*Central Chronicle*, 2006). Even the foresters concerned are extremely worried about the mining threat facing some of the best remaining forests.

The *National Planning Commission* has also played an important part in shaping forest policy (see Cell 3 in Figure 2.1). It writes India's Five-Year Plans and monitors their performance. Preparation of the 11th Five-Year Plan is already at an advanced stage. The plan contains fairly detailed reviews of past policies, and the achievements and problems of different sectors, and lays down the direction for the coming five years. The National Planning Commission wields very considerable power and stipulates broad policy approaches, funding and inter-sectoral issues; but the major role remains in budgetary allocations. The forest administration has often complained that allocations for forestry have been miniscule compared to the revenue that they were earlier asked to generate. The National Planning Commission has also approved a number of centrally sponsored schemes, such as the ongoing National Afforestation Programme (NAP), under which forest development agencies (FDAs) are being formed to receive direct funding for JFM by all states. National Planning Commission allocations have often depended on the dynamism of the individual member looking after forestry and environmental affairs. During the last government, due to a very active member (himself a forester), several large projects were framed and were approved by the commission (e.g. the NAP, the Bamboo Mission, the Medicinal Plants Board, etc).

The *judiciary* (see Cell 14 in Figure 2.1) has played an increasingly important role in forest policy in India, primarily due to the growing importance of public interest litigation. This, in particular, has led to the Godavarman case, or 'forest case', which has had far-reaching implications for PFM in tribal areas and in the north-eastern states of India. Forestry is, at present, on the concurrent list, which authorizes both the central government and the states to legislate (giving both opportunities for flexibility, as well as contradictions and confusions). While there are national laws (e.g. the IFA and the FCA), there are also 'policy frameworks' (e.g. the NFP) and other guidelines and notifications without any formal legal power. Protected forests can only be notified under the IFA, but the settlement of rights required under the IFA has simply not been carried out in many cases. The IFA empowers the government to classify any government wasteland as 'forest'. Large areas, including communal *jhum* (shifting cultivation) lands in the north-eastern states have been classified in this way and are recorded as 'unclassed state forests'. This is a strange category as 'un-classed forests' are not notified under the IFA at all. It is unclear how and through what formal and accountable process such a classification or recording is achieved. This seems to be a clear case of forest officials being able to exercise unaccountable power in classifying rotational cultivation lands as forests. Although not legally notified as forests, under a 1996 interim order of the Supreme Court they have been brought under the purview of the FCA. Until the judicial interventions under the PIL forest case, the legal status of these 'forests' remained notional and ambiguous since they do not belong to the government and, in fact, are of a special type of land use under which the land alternates between short periods of cultivation followed by longer periods of fallow during which natural regeneration of secondary forest growth takes place. Classifying them as 'forests', as officially understood as a single-use category, is itself erroneous, besides the legal anomalies in the process of such classification. The interim order by the court in December 1996 ruled, however, that the FCA would apply to all lands 'recorded' as forests in any government records and conforming to the dictionary definition of forests irrespective of ownership. This judgement has increased forest department and MoEF control of such lands considerably.

Donors, IFIs and BINGOs (see Cell 6 in Figure 2.1) have played an important role in shaping forest policy in both India and (especially) in Nepal. They are able to offer significant incentives (loans and grants) for policy reform of various sorts, including PFM, and implementation. They may also exercise disincentives when states are in fiscal difficulties, such as unwillingness to renegotiate or 'roll over' loans without other conditions being fulfilled. Donor funds are actually spent in a diversity of ways. They may fund state government budgets (as in the case of DFID), or provide technical and financial support to forest departments to implement forest projects; or they may directly fund projects that are independently implemented by donor-appointed project leaders through directly hired staff, NGOs and private organizations. However, it is rare for IFIs to fund current expenditure of forest administrations without conditions attached since this does not carry with it the leverage to induce reform. Donor-funded project support to states in India requires the approval of the Department of Economic Affairs in the Ministry of Finance, which, some have commented, sometimes influences project design more towards fund disbursement than 'reform' objectives.

There are a number of major criticisms of foreign donors and lenders that have a bearing on the promotion of PFM in both Nepal and India. First, they undermine the authority and capabilities of the nation state (the sovereignty issue), and confuse, undermine and divert senior policy-makers and their policies. Second, both multilateral and bilateral donors and lenders have national, regional and global agendas (stated and implicit), both in terms of policy and geopolitics, which may be politically unacceptable to national elites and senior policy-makers. Third, they tend to de-skill administrators and decision-makers through local leadership of a sector's management by means of projects and by poaching the more able forest staff from government service (to which they often do not return after the project is finished, but rather take up appointments in IFI-funded projects, private consultancy and higher education opportunities abroad). In the case of forestry programmes, this is more common in Nepal where independent project 'fiefdoms' are managed and directed in most sectors, largely by expatriate staff and local consultants, including former DoF personnel. In India, the forest department itself manages donor-supported projects with expatriate staff in supporting roles, and criticism of the confusion and diversion of policy cannot be raised so easily. Fourth, the arrival of highly paid foreign technocrats, ignorant of national and local cultures and politics, might elicit *swadeshi* (literally 'own country' or nationalistic) feelings in the most even-tempered of national policy-makers.

Donor projects are frequently accused of having the effect of 'queering the pitch' between the state forest administration and civil society groups. Donor funding support and technical support from foreign 'experts' can lead to forest departments insulating themselves from local pressures for reform or from improving interaction with local civil society groups. Furthermore, cash allocations for PFM-related service provision (e.g. local-level facilitation) can lead to selective patronage by forest departments of the more compliant civil society organizations and the exclusion of independent-minded ones, as has been observed, for instance, in Andhra Pradesh under World Bank-funded projects. On the other hand, it may be argued that donor projects can disrupt 'business as usual' and may be a mixed blessing – for civil society activist groups – by providing fora for them to challenge forest department practice and to apply pressure in terms of the gaps between the new PFM rhetoric and the field reality.

There are other broadly environmental interests that overlap the conservation aspects of forest policy. Typically, these are BINGOs, such as the IUCN, the WWF and others. Among the factors that have enabled IFIs to engage in policy discussions on PFM has been the acknowledgement that earlier models of social forestry programmes have failed to stem the degradation of forests. Second, in certain states in India an economic crisis has been emerging in which the policing of forests by the forest administration has proved too costly, especially in the face of local poaching following the undermining of pre-existing forest protection institutions and the Green Felling Ban in India. The destruction of common property management regimes followed, and the forests were reduced to *de facto* open access, which, in turn, increased the cost of central policing and simultaneously reduced revenue from the forest to pay for it. This threat to the forest administration (although very unevenly distributed throughout the country) was, it is claimed, one of the main background factors that made the alternative of PFM more attractive. It was the recognition of the non-viability of effective policing, combined with an awareness of the livelihood impacts of exclusionary forestry on forest dwellers, that provided extra leverage for IFIs to push their policies more effectively. This factor serves to accentuate the funding problem in relation to maintaining the forest administration, as discussed above.

It is difficult to take a consistent view of the real influence of IFIs in introducing PFM. For example, in India donors were pressing the Government of India for a more participatory approach at the same time that the 1988 Forest Policy was being circulated and widely discussed. The 1988 policy draft was more or less ready during the early 1980s, with Indira Gandhi's approval, but was abandoned after she lost the election and was subsequently assassinated. Although there was a major campaign against the new forest bill, then drafted by the MoEF, and the preparation of an alternative NGO draft bill, this had little influence on adoption of the new policy, which remained little changed from the original. However, as the process unfolded, there was next to no IFI presence. As Chapter 3 and Part II discuss in detail, donors were sometimes able to negotiate entry points into forestry policy, usually through establishing projects that privileged participatory forestry, although in the Indian case this was not universal. Whether donors were able to bring about a more participatory approach either in the forest administration, as a whole, or even, in practice, within their own project areas, is a much debated point. Anecdotal evidence (and this is the only 'evidence' available) suggests that bilateral donors and even big multilateral lenders need to complete projects once started and to spend funds. Therefore, when national policy-makers, politicians and bureaucrats resist pressure for reforms, the implicit threat of their not accepting loans or grants, project stalling or even closure generally prevents the donor from escalating pressure over reforms.

The most powerful institutions, such as the World Bank, are able to push their agenda more forcibly, but not necessarily in the direction of PFM. For example, the World Bank almost withdrew from a loan once the issue of unsettled rights and massive evictions in the Narmada Dam construction controversy surfaced, and the MoEF made it clear that removing encroachments was non-negotiable. There were large protests against the Madhya Pradesh forestry project by mass tribal organizations (MTOs) regarding the premise on which World Bank forestry projects were based (i.e. that local people are the main cause of degradation), which compelled a serious review within the Bank. The World Bank has now begun over, with 'pilot' projects in Jharkhand and projects under negotiation in Assam,

Madhya Pradesh and Chhattisgarh. These projects are being preceded by legal studies to understand existing legal rights and to ensure that JFM agreements do not undermine them. At a more discreet level, however, the World Bank is now exerting a strong influence on the drafting of a new environment policy and is re-engineering environmental impact analysis (EIA) procedures. Both have been attacked by environmental activist groups for being non-consultative and for making environmental protection subservient to economic development. If implemented, the new policy and procedures will have a far greater negative impact on forests and PFM in terms of the growing number of mining memoranda of understanding (MoUs) being signed. There are also efforts to promote the entry of private corporate interests in forests through structural adjustment programmes (SAPs), which again will imply a reversal to keeping the private sector out of national forests as set out in the 1988 Forest Policy.

It is difficult to produce evidence on the often hidden and ambiguous issue of the degree of influence that institutions or individuals hold in such a complex and drawn-out process as policy-making. This is partly due to the fact that decisions are frequently made behind closed doors and partly to the fact that decisions may be ambiguous, with a half-life of only a few days. But it is sometimes possible to discern the ways in which the policy turns out on the basis of wide precedent and strong imputed structural cause and effect. In this case, we argue that IFIs, even such large and powerful institutions as the World Bank, have lending targets and cannot afford to be too censorious over the lack of participation, compensation, the rights of rural people and other ideological agendas that are ostensibly important to them. Therefore, leverage in the name of reform is transient and seldom effective.

Forest activists, intellectuals and NGOs in India (see Cell 13 in Figure 2.1) have played a crucial role in representing forest users against injustices perpetrated by the state. There is a long history of conflict and struggle against the imposition of colonial forestry in India, and during the 1970s and 1980s revisionist historiographies sought to document subaltern voices (Ranajit Guha, 1989; Ramachandra Guha, 1989; Sarkar 1989; Arnold and Hardiman, 1996). Activists drafted an alternative Forest Act, the third edition of which was published in 1995. There have been many other examples of civil society action to prevent the handing over of degraded forest land and other 'wastelands' to industry on the grounds that these lands should remain common lands critical for the survival of the poorest (Hiremath, 1997). However, it is difficult to assess the impact of activists and intellectuals on policy reform. The usual strategy of any powerful bureaucracy confronted with a group of activists without mass political backing is one of polite reception, some argumentation, vague commitments and then business as usual – unless activists manage to form more powerful coalitions with politicians, social movements or a cadre of like-minded individuals within the target administration itself.

Moving now to the *forest administration* itself (see Cell 11 in Figure 2.1), it is necessary to give some detail of the structural characteristics of the administrations since they play a major role in shaping policy, procedures and the day-to-day activities of forest personnel at all levels:

- The Indian forest administration is a *permanently established organization* with a long-term and historically established territorial responsibility, and a long length of service. It is also adept at using established arguments and procedures to counter or deflect voices for change (see Chapter 1). In common with other 'primary extractive industries', such as coal and other mineral mining, the organization and its resource management practices emphasize the long-term nature of the enterprise and, therefore, the need for continuity. The formal role and objectives of the forest department emerged from its inception, as stated in a number of key documents. These were sustained yield of timber (para 24 of the 1952 National Forest Policy); the protection of forest (para 10 of the 1952

NFP) and, later, revenue (para 33 of the 1952 NFP); and wider environmental protection (soil conservation) and green cover (a goal of 33 per cent forest cover was adopted in the 1952 Forest Policy, para 19) (GoI, 1952). Objectives have changed, often forced on the forest administration by fierce resistance from users. More recently, the basic needs of forest users, participation and poverty reduction have entered the lexicon, together with eco-tourism, markets for environmental services, carbon sequestration, biodiversity and wildlife conservation. Some of these objectives may contradict others and therefore give rise to issues of priority. A further objective, as with most bureaucracies and large organizations, is the self-generating means for own reproduction and expansion. The Indian forest administration has been no exception, and has continually and selectively redefined its role in the face of changing circumstances. Indeed, during recent years, it has even sought to assume a rural development role, paradoxically citing its extensive field capacity as reason enough to annex more territory and powers.

- The Indian forest administration *has policing and quasi-judicial powers*, with powers to judge, fine and imprison offenders. Forest officers enjoy extensive discretionary powers, and there is frequent clamour from different states for field staff to be provided with better weapons to fight against organized timber smugglers and wildlife poachers. Some states have made such amendments. For example, forest officials in Assam have the power to use weapons, and wildlife conservationists have praised the killing of a number of alleged wildlife poachers by forestry staff. The authority enjoyed by forestry staff enables them to terrorize impoverished forest dwellers through beatings, burning their homes and crops, abusing women, filing offence reports against them and locking them up. Law and order issues and paramilitary policing have all become part of forest department culture (forest staff are trained in gun use and were until recently also trained in bayonet use and horse riding). Protection of the forest estate has involved the exercise of draconian laws. Policing has used the same legal provisions and practices against both illicit timber trading by organized illegal gangs and local people whose collection of fuelwood and other forest products for personal use has been criminalized. Whereas powerful timber smugglers and forest land grabbers usually escape penal action, hundreds of thousands of cases are filed against poor villagers for the pettiest violations of forest law. Part of this imbalance can be explained by the harsh conditions of the day-to-day life of frontline officers and the dangers of attempting to apprehend well-armed and well-patronized timber smugglers.
- The forest administration consists of a *very large staff*, compared with other government departments (over 90,000 in over 30 states), and is relatively well represented at all levels. Job descriptions are shaped by what the forest department sets out to do (see the following point). Additionally, many day wage labourers do not appear on the books. While the sheer size of an administration does not directly imply power, it certainly ensures that the many different interests of employees are expressed loudly and that any counter-moves to limit the remit and territorial jurisdiction of the administration will be met with a powerful response. The staffing structure of the forest administration involves:
  - the central Ministry of Environment and Forests (MoEF), under the minister of environment and forests, who is a member of the central cabinet of ministers;
  - the state forest departments under the forest/environment minister and the principal chief conservator of forests (PCCF);
  - training and research – for example, the Indira Gandhi National Forest Academy (IGNFA); the Forest Research Institute (FRI), with its several regional branches; Dehra Dun College; regional training centres for rangers; and forester training schools and, in some states, forest guard schools;
  - conservators at circle level and divisional forest officers (IFS trained);

  - field staff: range posts (rangers, beat officers, foresters, forest guards and forest labourers); and
  - units such as the PCCF wildlife and managing director of the Forest Development Corporation.
- The main *activities of staff* reflect what different employees are supposed to do and are specified in detailed job descriptions. Here there is a reflexive relation between official job descriptions and associated procedures, the objectives of the forest administration and the actual behaviour of staff. In other words, staff will do what they are trained to do and do not respond to innovations that require a different set of skills and standard operating procedures (as is usually the case with PFM) without scepticism and extreme caution. Within the staff a distinction is maintained, dating from colonial times, between forest administration staff in general and the IFS professional officer cadre. The IFS officers form part of the so-called 'All-India (civil) Services' along with the 'Indian Administrative Service (IAS) and 'Indian Police Service'. IFS is a specific title earned through 12 month training at Dehra Dun Forestry Academy. The main activities of the forest administration staff are as follows:
  - Surveying and reserving the forest estate, demarcating forests and facilitating actions that supposedly lead to the clear demarcation of government estate property. The 'forest estate' may be assumed to include not only high-value forests, but also any 'wasteland' (that is, non-private land, including village common land that the government has notified to be state forest) which may become liable to demarcation by the forest department.
  - Planning for forest management, involving the drawing-up of working plans according to management objectives. These plans primarily protect the state's exclusive use and ensure a sustained yield of timber – hence, the prioritization of technical silvicultural forest management. This necessarily involves long-term rotations and the exclusion of other forest users to ensure the protection of the timber species. Silvicultural management objectives imply the destruction of NTFPs, such as yielding climbers, bushes and trees considered inimical to timber yield by cutting and clearing. In turn, these priorities lead to loss of livelihoods, biodiversity and wildlife habitat.
  - Actual field management of forests, involving planting, maintenance, enumeration, inventory, thinning and harvesting, exclusive protection, and harvesting and marketing forest products – often by auction. Until the 1980s, forests were leased to industry at highly subsidized rates, and full commercial rates have only recently been made mandatory.

None of the above management practices and decisions involves local people in any capacity other than as labourers. Therefore, successful PFM requires a massive reorientation of job descriptions and responsibilities. As in any reform programme, there is usually a mismatch between new objectives and approaches with already-existing job descriptions and the very structures of the service, which are designed to reach totally different management objectives from those of PFM.

- The forest administration produces its own *technical and cultural knowledge system* (see Cells 8 and 9 in Figure 2.1) through which it controls the production of and access to forest knowledge (see the Dehra Dun Forest School and state forest academies in Cell 9 of Figure 2.1), leading to the elaboration of powerful policy narrative (see Chapter 3). A distinctive internal culture and knowledge system has emerged in the IFS over a period of over 150 years, reflecting and reproducing the main forestry management objectives and activities described above. The knowledge system claims to be 'modern', scientific and authoritative. 'Scientific forestry' has been the foundation of the professional repertoire of the forest administration; but under closer examination, 'scientific forestry' reflects subjective and political judgements (Banuri and Apfel-Marglin, 1993; Rittenbergen, 2001; Forsyth, 2003). Issues of local people's management needs and biodiversity and wildlife protection have not historically been included in textbooks such as Schlich's *Manual of Forestry* (1896) and Jerram's *A Textbook of Forest Management* (1892). The social choices over forest management objectives have already been decided at higher levels and have become part of the routine of foresters' professional repertoires. The centres for the development of these cultural and technical practices have been the Indira Gandhi National Forest Academy (IGNFA) and the Forest Research Institute (FRI) at Dehra Dun. The Indian Institute of Forest Management (IIFM) is a relatively new institution, established with a broader mandate to produce forest management graduates who must find jobs for themselves, in contrast to the trainees at IGNFA, who are selected for the IFS through national competitive examinations. The mandate of IGNFA is restricted to producing officers to manage the government forest estate for the IFS. IIFM graduates must have a broader perspective in order to find diverse jobs for the requirements of industry, donor agencies, NGOs, etc. There is a wide mix of faculty with economics, social science and technical backgrounds, in contrast to the more techno-centric training given in Dehra Dun.

* *Sources of revenue.* The forest departments have enjoyed considerable weight in policy-making and control over their policy agenda in the past due to the revenues they have generated, which are directed into the consolidated fund of the state, the budget allocated to forest departments being decided by government. Consequently, it has been a constant complaint of the MoEF and the forest departments that despite their being in charge of about 23 per cent of the country's territory, they have seldom been allocated more than 1 per cent of the national budget. Only the forest development corporations, or similar more autonomous agencies set up from the 1970s onwards, can retain their incomes as well as raise loans to finance their activities. The forest departments had until recently been generating surplus revenue for government. However, recurrent plus development costs have exceeded revenue expenditure over the past ten years, which puts administrations under extreme pressure, with the result that in many states there has been little maintenance or recruitment of new staff. The drop in forest maintenance has been due to the limited budgetary allocations, with even these allocations largely being diverted to PFM on 'degraded' forests, resulting in the management of healthy natural forests being neglected. Similarly, the lack of recruitment of new staff has no links with revenue generated by the department, but has been due to both central and state governments wanting to reduce their recurrent salary costs. Although IFIs are coming up with finance for capital expenditure, most states require funds for revenue expenditure in order to run the day-to-day activities of the service.

Two concluding points are important. The first is that the Indian forest administration is not monolithic, and policy outcomes, although shaped by the structural forces described here, can, from time to time, be given a substantially different character and style by individuals at all levels. Individual people make a difference, and no one is a helpless prisoner of the structural direction of flow against reform. The pressures to move towards a more participatory approach has variously affected different staff positions – some have responded with high levels of enthusiasm for the radical reorganization of field relationships, whereas others have sought continuity despite the changing circumstances.

The second point is that the forest administration is policy-maker, implementer, educator, producer and disseminator of knowledge, entrepreneur and policeman – all in one. It is debateable how far the forest administration actually 'makes policy' independently with a high degree of institutional discretion. It certainly is very influential, as argued above; but it is still subject to other influences and incursions from time to time (as evidenced, for example, in the role of Prime Minister Indira Gandhi). The 1988 Forest Policy was drafted during the early 1980s by the National Forestry Board, with representation from different ministries. At present, there is a clear intent to compel the MoEF to sidestep its environmental mandate in the interest of promoting a new liberal economy. Many sincere forest officers are finding themselves helpless to prevent some of the best forests from being opened up for mining or other development projects. There is also a challenge from the tribal forest rights lobby (see the Introduction and Chapter 1). In spite of these exceptions, we argue that the forest administration in India has become a formidable and stable institution with a high degree of autonomy. Therefore, most challenges to forest policy, if deemed to be contrary to the professional and pecuniary interests of key policy actors, can be ignored, in practice, or co-opted and altered by various strategies (see Chapters 1 and 3) in order to promote the long-established objectives of the service.

Countervailing pressures on the Indian forest administration

The previous discussion should not imply that the IFS is impervious to change, particularly to PFM. A number of countervailing forces continue to act as *pressures for change* in forest policy. Some have been recurrent over a long period, as Chapter 1 has shown.

First, the *credibility and professional reputation of the forest administration* have been put under pressure. The administration is held to be inflexible, inefficient and non-competitive by national government agencies (the planning commission, the treasury, and, particularly, the judiciary), other Government of India (GoI) institutions, international donors, intellectuals and civil society groups. Eminent authors have, for decades, repeatedly criticized the forest administration's fundamental lack of accountability (Saxena, 1994; Hobley, 1996; Sundar et al, 2001). Forest administrations of India have been accused of failing in their fundamental responsibility to protect forests: in many areas, rapid deforestation has become linked with wider environmental degradation. The administration is in the contradictory position of being charged with increasing forest cover to the target of 33 per cent while barely succeeding in protecting existing forests. In recent years, according to FSI statistics (FSI, 2003, p3), forest decline is claimed to have been halted. On closer analysis, however, the FSI has included commercial crop plantations on non-forest lands as 'forest cover', while natural forests still continue to decline in many states at an even higher rate than previously. Bose (2005) provides an evaluation of the current situation:

> *Since the last report of 2001 to 2003, the country's forest cover increased by about 2800 square kilometres (sq km), the report says. But this 0.41 per cent increase conceals an alarming figure: the net drop in the dense forest cover was 26,245 sq km. This means considerable expanses of dense forest areas (over 40 per cent tree cover) degraded to the 'open forest' category (10–40 per cent tree cover). The total addition in open forest area, from different categories, was 29,000 sq km. The 'change matrix' of types of forests reveals a bleak scenario: actually about 55,600 sq km dense forests have degraded into open forests, 420 sq km into scrub land and 27,821 sq km into non-forest areas. But other open scrub and non-forest areas have added to dense forests, making the net change low. Save a few states like Orissa and those of the north-east, all others recorded a loss in dense forests. The bulk of this, over 18,000 sq km, was in Karnataka, Kerala, Madhya Pradesh, Maharastra and Uttar Pradesh. The change matrix also raises another question: how can 46,177 sq km of open forests improve and become 'dense' in just two years...? FSI merely says its mandate is to present data, not explain it.*
>
> *There is a growing lack of confidence in FSI's ability to interpret data, coupled with advances made in the field of remote sensing and its increased use in India.*

The report doesn't identify land ownership while describing changes in forest cover, making it difficult to fix responsibility. FSI claims this is because these facts are not clear. But S. P. S. Kushwaha of Indian Institute of Remote Sensing, Dehradun, rubbishes the claim: 'In fact, over 70 per cent of this [information] is available and can easily be shown.' Also, despite the fact that technology permits distinguishing between natural forests and plantations, FSI doesn't do so. Kerala is shown with 40 per cent forest cover; that most of these are plantations is not registered.

Bose (2005) cites leading mapping experts who criticize the FSI report:

Eminent GIS and remote sensing expert Jagdish Krishnaswamy agrees: 'At present there is none of the accountability or transparency that comes with peer reviews. The process should be decentralized; FSI should work with other groups or ... hand the task over to experts in different regions.' The sense is that the crucial task shouldn't be handled by a single body, especially one under the government's pressure to assure the nation that its forests are safe.

The credibility of the Indian forest administration has also been under pressure from emerging evidence of its failure to successfully protect biodiversity, especially fauna in protected areas (biodiversity and flagship species conservation projects). The examples of the recent Tiger Task Force report and the Green Felling Ban (no green felling beyond 3000 feet (915m), no felling without a proper working plan approved by the MoEF) were both vigorously opposed by the forest administration; nevertheless, it was overruled. The Tiger Task Force was set up by the prime minister in 2005 after all the tigers in the Sariska Tiger Reserve were found to have disappeared. An interim order of the Supreme Court in December 1996 banned any felling without approved working plans across the country. The green felling ban in the hills, particularly in Uttaranchal, had been imposed during the 1980s by the government. There are also recurrent cases of forest departments being accused of corruption and being in league with timber smugglers. The famous *malik makbuja* case in Bastar is a key example where senior IAS and IFS officers were implicated in massive illegal felling from tribal lands. As a result, the Supreme Court banned all timber felling in Madhya Pradesh for several years (Sundar et al, 2001, p72). Another example was the shooting dead of four villagers in Dewas by police in 2001. Fact-finding reports revealed that villagers were catching illicit timber smugglers; but instead they were accused by the forest administration of stealing 'valuable' timber (Sarin et al, 2003).

Second, currently, there are *political challenges to the forest departments*. The most important is from tribal land rights, which challenge the legality of the forest reservation of extensive tracts of land in tribal areas and which takes the form of the Tribal Forest Rights Bill that threatens to contest the MoEF's exclusive control over lands designated as 'forest', particularly in majority tribal areas. The categorization of these lands has now been expanded into virtually all forest areas since the definition of forest dwellers has been likewise expanded under Schedule 5 of the Indian Constitution, which provides for special administration of such areas to ensure protection of tribal resource rights, livelihood systems and cultures. In an affidavit filed in the Supreme Court in July 2004 (MoEF, 2004), the MoEF admitted the following:

That, for most areas in India, especially the tribal areas, record of rights did not exist due to which rights of the tribals could not be settled during the process of consolidation of forests in the country. Therefore, the rural people, especially tribals who have been living in the forests since time immemorial, were deprived of their traditional rights and livelihood and, consequently, these tribals have become encroachers in the eyes of law.

   That these guidelines, dated 5 February 2004, are based on the recognition that the historical injustice done to the tribal forest dwellers through non-recognition of their traditional rights must be finally rectified. It should be understood clearly that the lands occupied by the tribals in forest areas do not have any forest vegetation. Further, that because of the absence of legal recognition of their traditional rights, the adjoining forests have become 'open access' resource as such for the dispossessed tribals, leading to forest degradation in a classic manifestation of the tragedy of commons.

In response to widespread protest against brutal evictions of tribal and other forest dwellers from forest lands as alleged 'encroachers', and to the demands of the broad-based Campaign for Survival and Dignity, in January 2005 the prime minister asked the Ministry of Tribal Affairs to urgently draft a bill entitled the 2005 Scheduled Tribes and Forest Dwellers (Recognition of Forest Rights) Bill, for early tabling in parliament. The draft was initially attacked by wildlife conservationists, the Indian Forest Officers' Association and even by the MoEF, despite the latter having submitted the affidavit in the Indian Supreme Court that the rights of tribals had not been recognized during the consolidation of state forests. The MoEF contended that 'forests' fall within its mandate and that the Ministry of Tribal Affairs had no jurisdiction over the matter, despite the fact that large areas declared forests are actually majority tribal areas under Schedule V of the Indian Constitution. The Prime Minister's Office organized several meetings to address MoEF and conservationists' concerns, but held firm on its decision to get the bill tabled by the Ministry of Tribal Affairs. It was eventually tabled in parliament on 13 December 2005 and, after being examined by a joint parliamentary committee of both houses of parliament, and subject to several significant revisions, it was finally passed on 18 December 2006. This signifies a momentous change of policy which now remains to be implemented.

   Third, another political challenge of a less direct nature is provided by the current *focus on poverty reduction* as a central goal of the United Nations Millennium Development Goals (MDGs) (see Cell 5 in Figure 2.1). The poverty reduction strategy papers (PRSPs) present the forest departments with difficult challenges. Once the connection is made between rural poverty and the reliance of many of the poorest on forest resources for survival and the exclusionary policies by the forest departments, poverty reduction becomes a potent weapon for reform, as well as for a more participatory approach. Both donors and the planning commission have made the connection between poverty and forest policy. Most donor-funded forestry projects now talk about livelihoods and poverty alleviation, and the recent 2006 World Bank report (*India: Unlocking Opportunities for Forest Dependent People*) has gone further than most donors in recommending tenure reform and institutional strengthening for local PFM bodies. Funding for the National Afforestation Programme of the National Afforestation and Eco-Development Board (NAEB) was approved by the planning commission during the tenth Five-Year Plan largely as a means of generating employment for the poor. One of the most important forest policy issues of poverty reduction is the NTFP management regime, which, we claim, could reduce poverty and increase livelihood security in many forest areas.

   Fourth, the *decentralization of government*, although not initially a problem for forest administrations, now threatens to shift accountability, management and revenue collection of forest lands from forest administrations to *panchayats*. This is particularly acute in the case of the *Panchayat* Extension to Scheduled Areas (PESA), which presents a very potent challenge to the pre-eminence and lack of accountability of local forest administrations. In West Bengal, for example, forest rangers already report their activities to the *panchayat* officials (co-opted in sub-committees). Furthermore, a court case currently pending in Punjab will decide whether revenue from JFM village forests should be shared with the *panchayat*.

Finally, in India it has been well recognized in recent years that *the electorate can be treated less and less as a passive 'vote bank'* to be controlled and relied on by dominant parties. Increasing educational opportunities and media exposure have ensured that the electorate is much more informed and independent in its voting patterns. Tribal groups, in particular, have become more politicized, and some tribal parties have emerged – one particular concern being tribal resource rights. Demands were made of the MoEF to settle land rights prior to recent national elections and political pressure was brought to bear by political representatives in order to win votes through policy concessions.

Policy-making at the national level in Nepal

The history of the forest service in Nepal is closely related to that of India, since systematic state forest management was initially an offshoot of the IFS, as described in Chapter 1. The forest administration (see Cell 11 in Figure 2.1), consisting of the Ministry of Forests and Soil Conservation (MoFSC) and the Department of Forests (DoF), is both the main state agency for controlling and managing forest and the main set of actors in the policy process. Directives and guidelines are also given to DFOs by the MoFSC and the DoF; but these do not have to be scrutinized by parliament, as Chapter 1 has described. Sometimes forest policy has been, and continues to be, made unilaterally by cabinet and the minister, which was not the intention of the Forest Act. For example, forest policy established in 1989 and revised in 2000 stated that green felling should be banned in the *tarai* national government-managed forests, directly contradicting the 1993 Forest Act, which stipulated that green trees can only be harvested if a working plan for a forest block has been approved by the ministry. This example reflects the inconsistency in forest policy-making and confusion in major policy decisions.

As Chapter 1 has mentioned, Nepal's local self-governance policy (under the Ministry of Local Development) conflicts with the 1993 Forest Act in a similar manner as the panchayati raj constitutional amendment in India. The Local Self-Governance Act (LSGA) contains 22 points that contradict other sectoral acts, including the 1993 Forest Act, and these contradictions have played out in the field depending on the relative strengths of line agencies and the coordination capabilities of DDCs and VDCs. The contradictions in forestry between VDCs and CFUGs are currently a central dilemma in Nepal's forest policy. The DoF and forest user groups, too, claim that they were the first to adapt to decentralized PFM and therefore should not have to give up the management and revenues of forests to devolved local governments.

For the fiscal year of 2005–2006 the government's aim was to include the forestry sector under local government (at DDC level) together with agriculture, horticulture and livestock sectors, starting in selected districts. However, under the devolution plans district-level civil service staff would remain under the supervision of their parent department and ministries, and so will remain accountable to them rather than to the DDCs for which they will be working.

Due to the insurgency and subsequent political transformation and constitutional reforms of 2006–2007 the process of the decentralization and amalgamation of district offices has not materialized. The new political objective has become making Nepal a federal state so that more effective devolution can be implemented, one in which forestry would become a sector in local government.

Two *para-statal organizations* (see Cell 11 in Figure 2.1), the Forest Product Development Board and the Timber Corporation of Nepal, were set up to fell timber from *tarai* forests and to fulfil the demand of growing urban centres and the Kathmandu Valley. However, there was widespread mismanagement and corruption, and illegal timber mafia from both Nepal and India thrived, often in association with forest staff. The 'policing' role continued as the responsibility of the DoF, but was not successful in stopping forest degradation, encroachment and smuggling of valuable timber through the leaky border with India, where it fetched high prices. In the hills the forest has been handed over to local communities, and now the DoF is providing legal and technical support to these CFUGs (see Chapter 1).

In Nepal, *the role of donors, IFIs and BINGOs in promoting PFM* (see Cell 6 in Figure 2.1) is significantly different from that in India. By the late 1980s, Nepalese forestry was visibly suffering from the ill-judged nationalization of forests following the 1957 Nationalization Act, whereby existing local institutions were expropriated of their customary rights and forests were reduced to open-access resources, with no resources allocated for the centralized policing which then became necessary. At a stroke, it turned foresters into policemen and licensing officers against the interests of almost all villagers (Hobley, 1985). The 1967 Forest Preservation (Special Arrangements) Act, for example, empowered DFOs to shoot wrongdoers below the kneecap if they in any way imperilled the life or health of forest officials (Talbot and Khadka, 1994). Forest policy therefore was in serious difficulties and could not be defended on the basis of evidence. This provided an opening for IFIs to promote early forms of PFM which emerged during the 1970s. After initial success foreign donors were able to increase their support for PFM promotion in the hills – for example, the Australian Agency for International Development (Ausaid); the UK Overseas Development Administration (ODA) (later to become the Department for International Development (DFID)); the German Agency for Technical Cooperation (GTZ); the Japan International Cooperation Agency (JICA); the Swiss Agency for Development and Cooperation (SDC); the United Nations Development Programme (UNDP); and the World Bank. The primary divide is between the hills and the *tarai*. While there has been considerable interest, progress and learning in the former (with important qualifications, as discussed in Chapter 1 and again in Chapters 5 to 6, the latter has proved that the political ecology of the *tarai* has been much more inimical to PFM. The main reason is that there is very valuable standing timber there, and participatory forestry threatens to increase transparency, open up contractual arrangements for more democratic scrutiny and share proceeds from timber sales with a wider public. Hitherto, tendering procedures for timber felling and sales have been made so complex and expensive that only the wealthy and well connected could bid for contracts (Iversen et al, 2006). Finally, there was (and remains) a high level of illegal felling and sale of timber across the 'semi-permeable' frontier with India (Blaikie and Seddon, 1978). For these reasons, donors have found that PFM has been a much more difficult problem to negotiate in the *tarai.* During the 1980s, the Tarai Community Forestry Project was funded by the World Bank. This project has been able to plant trees along road and canal sides, but has failed to form forest user groups due to lack of a coherent and well-implemented PFM policy for the *tarai*. The DoF made an Operational Forest Management Plan for most *tarai* districts; but this was not implemented due to lack of funding by either the Nepalese government or international funding institutions. Protests from forest user group networks and civil society demanded that community forestry be extended to the *tarai*. By 2000, none of the donor agencies except the GTZ were interested in financing projects in the *tarai* due to lack of a clear policy, powerful and hidden interests in illegal felling, reluctance of the government to facilitate any initiatives there that might actually become operational, and political manoeuvring for votes from illegal settlers (*sukhumbasi*). In short, the *tarai* forest issue was simply too hot to handle. DFID and CARE (Cooperative for Assistance and Relief Everywhere), with the US Agency for International Development (USAID), have now started the formation and support of CFUGs in the *tarai* and the government has put forward a new 'collaborative forest management' (CollFM) policy, although by 2006 this had already run into stalemate due to civil society opposition.

In Nepal, DFID and the SDC have funded programmes such as the Livelihoods and Forestry Programme (LFP) and work with local government, local non-governmental organizations (LNGOs), the Federation of Community Forest Users, Nepal (FECOFUN) and the DoF to strengthen local democratic structures, planning and decision-making. CARE, the Danish International Development Agency (DANIDA) and the Netherlands Development Organization (SNV) all have programmes with LNGOs and FECOFUN to improve governance and transparency, and overcome some of the problems associated with the conflict between the DoF and the DDCs and VDCs. At present, almost all donor agencies have shifted their priorities to poverty alleviation, welfare programmes and government reform. In the arena of IFI–DoF relations, the era of classic silvicultural 'fortress' forest policy in the name of soil and water conservation, economic development and modernization is coming to a close. This generalization cannot be made about Indian forestry policy.

The forest policy of Nepal may be summarized by the following seven characteristics:

1   *Inconsistency*. There are parallel policies in other sectors that contradict forest-sector policy, the most important being the potential conflict between the forest administration and the DDCs and VDCs. Additionally, land use change, especially the conversion of *tarai* forests to settlements, infrastructure, towns and industries has not been planned for within forest policy. Forest-related policy-making is also weak and non-transparent. Despite attempts to manage and regulate *tarai* forests through various Operational Forest Management Plans (OFMPs) and through the collection of taxes from the sale of major tree species (*Sal* and *Khair*) from *tarai* CFUGs, success has been very limited due to the almost non-existent cooperation and coordination between local forest users, donors and other stakeholders.

2   The DoF is a permanent government institution, with long-term mandates and responsibilities. The roles and objectives of the department were set out at its inception and are stated in a number of key policy documents (such as the 1976, 1989 and 2000 Forest Policies, as well as in forest acts, rules/regulations, circulars and directives. The 1960 Forest Act focused on forest protection and use for national economic development, and included strict provisions to exclude local people. Thus DoF personnel were mandated to carry out a policing role. Later, the 1976 Forest Policy, the 1989 Master Plan for Forestry Sector (MPFS), the 1993 Forest Act and associated 1995 Forest Rules promoted participation, demanding a reorientation in DoF personnel's role.

Under the MFSC, the DoF has 7070 personnel, who include 4927 (70 per cent) technical and 2143 (30 per cent) administrative staff members. There are 252 (4 per cent) gazetted officers and 6818 (96 per cent) non-gazetted staff in the DoF (DoF, 2002). The overall aim of the DoF is to:

- support the MFSC in preparing forest acts, legislation, rules, policy and strategies for the forestry sector;
- implement and coordinate forest development projects and programmes in the country;
- mobilize people's participation in forest management activities by providing information related to forest management and plantation;
- prepare plans for scientific forest management;
- collect revenue from forest products;
- bring uniformity in PFM programme implementation by providing guidelines;
- coordinate PFM projects and programmes under DoF policy strategies; and
- transfer the technology of private forest development through information dissemi-

nation.

The main job descriptions and activities of DoF staff are:

- surveying and mapping forest areas;
- planning forest management, including drawing up operational plans (OPs) for CFUGs and government forests according to guidelines and policy;
- advising CFUGs on technical forest management and silvicultural treatments;
- implementing and enforcing government policy and directives;
- ensuring that a forest inventory is carried out and that forest product harvesting is based on the annual allowable cut prescribed by the DFO; and
- marketing forest products, harvesting forests and selling products from government-managed forests (often at auction), including supporting sales from community forestry.

These job descriptions do not support the promotion of PFM at all, except in the drawing-up of plans for CFUGs, although documents referring to the role of the DoF include, in general terms, 'mobilization of people's participation'. However, this statement is not matched with any job description that may bring it about, as we have seen in the case of India; neither do the job descriptions mention PFM or training to support these activities. Nevertheless, officers in most donor-assisted districts are provided with various training courses related to technical aspects of forest management, CFUG institutional capacity development and reorientation. In order to address second-generation problems of CFUGs, such as governance, livelihoods, NTFP management and the demand-driven needs of CFUGs through participatory approaches, a massive rewriting of job descriptions, responsibilities and orientations is required, as in India.

3   The DoF has *extensive territorial control, but disappearing forests*. The DoF has jurisdiction and control over approximately 39.6 per cent of forest and shrub land (as well as mountain and open land), but has been unable to fully control and manage forests in either the hills or the *tarai*.

4   *Paramilitary policing in the tarai* is a salient characteristic of Nepalese forest policy. The DoF was formed in 1942 primarily in order to protect valuable forests in the *tarai* and to supply the timber and fuelwood needs of towns and cities. The DoF sought to assert its control and exclusive management over forest areas, requiring legal powers in order to achieve this. A paramilitary culture, fines and the forceful maintenance of law and order through strict laws have characterized its tenure. Forest officers were trained in gun and bayonet use and horse riding at the Dehra Dun Forest College in India, and ranger-level paramilitary training was initiated in Hetauda. Before 1976, Nepalese foresters were trained at Dehra Dun and were therefore acculturated to the Indian Forest Service. Armed forest guards were deployed in the *tarai*, but not in the hills, and their training centre was located at Tikauli in Chitwan district in order to train armed guards who were mainly retired military staff. Today, the DoF employs armed forest guards through an open competition; these guards are then trained by retired army staff already working under the DoF. Policing enforced the same legal provisions and practices against illicit timber trading by organized mafias as it did against local people, whose collection of fuelwood for personal use has been criminalized in the *tarai*. However, during the 1970s, Dr Mahat, a DFO who was also trained in Dehra Dun, recognized the inefficacy of the paramilitary approach, as well as its corrupt practice and the immense pressure for forest clearance from settlers, and became one of the original Nepalese instigators of PFM. His experience of the failures of paramilitary policy in the *tarai* was formative in establishing PFM.

5   *A dual role within the DoF*. After the policy of PFM began to be practised in IFI-funded

projects, and later by the DoF itself, the DoF had to provide technical advice, extension, institutional development and legal services to CFUGs, mostly in the hills, but also to fulfil its more traditional role of policing in government forests (largely in the *tarai*). This dual role leads to confusion and fragmentation in the overall responsibility of the MoFSC, which is to prepare and propose different acts, legislation and rules for forest and soil conservation sectors. The MoFSC formulates forestry-sector policy and strategies for the departments. However, these planning and legislative activities are still shaped by the traditional silvicultural 'fence-and-fine' and policing roles, as well as the newer PFM approach.

6   The *technical and cultural knowledge systems* (see Cells 8 and 9 in Figure 2.1) required for the implementation of PFM have not fully been developed and adapted. Despite the growing experience of PFM and the knowledge and skill of thousands of CFUGs, as well as of IFIs involved in PFM, the forest administration at the circle and district levels often operates in isolation from these new initiatives. Thus, strict technical models of management (silvicultural management objectives and very conservative limits of off-take imposed on CFUGs) are being adopted with only minor modifications and without adequate extension advice, negotiation and interactions with CFUGs. For example, the first forest inventory guidelines issued by the DoF to field staff were heavily biased towards the more technical aspects of forest sampling, forest inventory and calculation of annual yield. Similarly, academic institutions have tried to improve forest college curricula in relation to PFM; but technical forest management priorities remain unchanged. Donor-assisted projects and DoF personnel develop their knowledge and skills through training and field observation. However, the reach of this training is significantly reduced for more senior staff and those in the DoF in Kathmandu and regional centres, who receive very little instruction. In addition, there is no adequate mechanism to provide continuity to the knowledge and skill gained through PFM projects, especially in relation to CFUGs. One of the main reasons for the failure to incorporate PFM skills has been the lack of government funding: PFM funding relies overwhelmingly on IFIs, and IFI-funded projects have tended to operate largely in isolation with their innovations limited to their project territory, ending with the closure of the project. PFM knowledge creation and dissemination therefore tends to cease with the ending of IFI funding in an area.

Finally, local forest users' intellectual property rights receive no credit or assistance from the DoF despite the fact that forest protection and management is carried out by CFUGs, which use, when permitted, their own indigenous and customary knowledge and skills. For example, the local names and uses of medicinal and aromatic plants (MAPs) are collected from local experts within CFUGs, but are documented by the DoF under the officer who writes the report. Many local people have knowledge and experience of trees, NTFPs, MAPs and forests, as well as unique forestry skills; but these have not been used in developing forestry technology suitable for PFM.

7   *Overall transparency* of policy-making and implementation in formal practice does exist in the implementation and functioning of DoF and district forest offices. Despite the fact that the DoF is part of a unitary state, its district forest offices operate as do any other line agencies (such as agriculture, veterinary services and irrigation) under the overall premises of the Local Self-Government Act and the decentralization policy of the Nepalese government. All offices have to present their annual plan and budget. However, there has been a lack of transparency and accountability at a higher level within the sector over more strategic decisions in policy-making in the longer term.

A comprehensive and independently managed, and generally accessible, monitoring system is an important guarantee of transparency. The monitoring formats used by district forest officers are of three types:

1 a planning format to report progress against targets;
2 a DoF format to report needed information by different divisions at the centre; and
3 a project format for information required by that particular project.

Although a database of all the community forests in the country is maintained at the DoF, it does not have a common PFM monitoring system. Various PFM monitoring systems have been developed within projects with their own, often disparate, objectives; but these have not been harmonized in a common monitoring system. PFM projects do publish independent reviews and study reports; but these are not widely available and are sometimes even treated like the private and confidential property of the project. As a whole, more useful information is stored at the project and individual levels than in a common pool at the district and central level. In addition, the Institute of Forestry lacks up-to-date PFM-related information.

Countervailing pressures on the Nepal forest administration

Nepal, unlike India, does not possess an historically enduring forest administration that has managed to repel radical reform or to adapt to more threatening pressures. However, contradictory forces in Nepal have resulted in a number of changes in forest management, particularly in the introduction of PFM. These include, first, the long-established customary rights of local people, especially in the middle hills, which have not been abrogated and reduced by government action to the same extent as in many states of India. There is, therefore, an understanding about, and expectation of, local people that they will manage 'their' forest. In spite of the nationalization of forests in 1957, local technical and organizational skills still remain. Growing pressure has been applied by forest user group networks (e.g. FECOFUN) to extend and strengthen PFM in all areas and to safeguard the rights of forest user groups. Second, there have been widely discussed failures in managing forests in the *tarai*. Third, and more recently, self-governance and decentralization policy within the unitary structure of government have challenged the monopoly control of the DoF. Fourth, most donor assistance, either in the form of projects or institutional development, have poverty alleviation as a central goal, using CFUGs as an entry point. The tenth Five-Year Plan prepared a road map for poverty alleviation in its poverty reduction strategy paper, and all development sectors, including forestry, are to contribute to this goal. The forestry sector and PFM achievements are also being linked with the United Nations MDGs. How far words on paper will contribute to poverty alleviation through a more egalitarian and effective PFM is open to debate, and this is discussed in Chapter 3 and Part III. Fifth, insurgency applies pressure to change the mode of operation of all forest management, including PFM, and adversely affects its revenue collection, as well as the funds that community groups derive from the sale of forest products.

The state/sub-national level

The second level for policy-making is *the state* in the case of India. Nepal has a unitary state; but there are stark political ecological variations, particularly between hills and the *tarai*, which are discussed in the next section. Part II of this book discusses state-level (India) and hills/*tarai* (Nepal) forest policy (labelled 'state/sub-national political ecology in Figure 2.1) and impacts at lower levels: the 'district/circle political ecology', and the local and household levels. Most of the empirical findings, therefore, are discussed in Part II and only the generic policy drivers are rehearsed here.

The international and national political ecologies of the policy process, the main actors and their political relations have been described. Dynamic outcomes resulting from the interplay of these actors, formal policies, acts, laws, guidelines, directives and budget allocations are transmitted to the state level (Cells 16 and 25 in Figure 2.1). The second set of 'transformative steps' from intentions on paper to final outcomes in the field (Cells 36, 38, 39–41 in Figure 2.1) are initiated. At the state level (at both the state capital and its bureaucracies, and in civil society at state level), formal policy is reworked, interpreted and contested. There are struggles, media coverage, and attempts to co-opt and control within and between state and civil society institutions, and these are shaped by state politics, ecologies and history.

Clearly, it is important whether a PFM site is in, for example, West Bengal or across the border in Bihar. Twenty years of Left Front politics with widespread local mobilization and local party organization have changed the local situation markedly, while a more neo-feudal and ineffective centralism has ruled in Bihar. There have also been the forest policy experiments in West Bengal described in Chapter 1, with political mobilization, examples of administrative competence and a decentralization of power to *panchayat* level. These make a profound difference in West Bengal; but a less progressive environment has developed in Bihar. This means that any analysis and recommendations have to be politically embedded, and the more quantitative statistics we may collect at the village and household level will have to be interpreted within a broader but distinct state-wide context. For example, the issue of 'elite domination of forest user groups' can only be discussed in the context of a regional perspective. Our own studies indicate that irrespective of the oppression of forest dwellers by local elites, it is clear that when an open-access area is closed for protection, women and the poor usually lose out on critical access for livelihood needs – this outcome mirrors similar outcomes in an already inegalitarian political economy.

State administrations have a varying but always considerable legal and on-the-ground autonomy in the way that they receive, interpret and act on guidelines and notifications, and even laws. To take an Indian example, a public interest litigation, the Samata Judgment, was passed by the Supreme Court in 1997 involving the transfer of land to non-tribals in Schedule V areas of Andhra Pradesh, thereby depriving mining companies of access to tribal forested areas, with considerable implications for tribal rights in general and for PFM. Although the Supreme Court judgment asked other states with Schedule V areas to set up state committees to explore the enactment of similar protective laws as in Andhra Pradesh, none have done so. Recently, even the Andhra Pradesh government is exploiting a loophole in the judgment, which did not treat public agencies as non-tribals, therefore enabling it to open up tribal areas to mining. The large number of MoUs recently signed by Orissa's government with mining companies is among the biggest threats to PFM in the state. In addition, Orissa's government has also decided to lift the ban on green felling, which will destroy the *de facto* control that local groups have enjoyed over their forests. Another example of state-wise variations in forest policy concerns the notification of JFM, which allows each state to set its own objectives for JFM policy within a broad remit. A final example of differences between states' prioritization of PFM is the state of West Bengal, which had begun a form of JFM *before* the notification (the successful experiment in participatory forestry in the village of Arabari; see Bannerjee, 2004).

Another example of the differences between states' application of the JFM approach is the share of revenue between the state and the forest protection committee (FPC) (25 per cent net in West Bengal; 50 per cent in Gujerat; 50 per cent in Orissa; and 100 per cent in Andhra Pradesh, after marketing and other expenses have been deducted). A more significant issue is the very poor extent of actual sharing and disbursement of revenue to user groups and committees that has taken place in most states. West Bengal is the only state to have shared on a significant scale.

Interpretation and practice of forest policy at the district/division and local levels

The next level of transformation of policy to have an outcome on the ground is the *district/forest circle level*. We include both here, although their boundaries seldom coincide (in India the 'district' administrative category is not the same as 'forest circle', a forestry administrative category). This can lead to confusion because political ecologies (land use, settlement history and its interactions with land use and flora and fauna, agrarian political economy, and present forest characteristics) vary enormously across districts and forest circles. In some cases, the working plan can reflect these differences, although as Part II will show, it is likely that only the silvicultural variation will find its way into the plan and other considerations of political ecology will not be considered. For example, Dang district in Nepal includes both *tarai* and hill political ecologies. Similarly, in Visakapatnam district in Andhra Pradesh, there are both hill regions with shifting cultivation and tribal populations together with lowland paddy cultivation and low proportions of forest. These are ecological issues that shape the kind of forest, its extent, the distribution and kinds of livelihoods of people living there, and its politics both in relation to the state and at a more local level. Each also has a history of how the forests and other resources have been transformed by state forest management. Some circles have valuable timber, and PFM becomes politically entangled in struggles for control with forest contractors and forest officers, characterized by a variety of legal and illegal relations. Other forests have little commercial value (e.g. some Nepalese hill forests compared with *tarai* forests, the latter having a much tougher time in establishing PFM).

These variations on political ecology lead to inevitable prioritization by the district/divisional forest office, whereby more accessible areas and villages, and those with more valuable forest resources are chosen for interaction, with many of the more remote interior villages neglected. The setting-up of PFM usually requires lengthy negotiation, with well-publicized meetings including all of the more distant local forest users – not just a quick cup of tea and brief visits to the *pradhan panch* (the elected village council leader) and elites (as our research teams found to be closer to the truth in many instances). In Nepal, a period of one week is recommended, and CFUGs, once formed, complain that they have a need for 'post-formation support', particularly since they have not received proper formation support in the first place. Part II gives a number of examples of the paper formation of user groups in order to fulfil target numbers without planning, negotiation or even the minimum initial contact between staff and users in order to start the complex and lengthy process of PFM.

Policy and the local political ecology

Here we refer to the local level (meaning the settlement), both village and hamlet, where day-to-day interactions occur. In some ways, the political ecology of local policy is a micro-version of the political ecological variations at the district and division levels.

The most numerous group of all comprises the *local forest users* (Cells 39 to 41 in Figure 2.1). The word 'local' is an imprecise term, but here almost always refers to those people who are forest adjacent or who live in and relate to the forest. There are exceptions in the Nepalese *tarai,* where forest users whose homes are up to 15km from the present forest edge now travel increasing distances for grazing and fuelwood due to forest clearance, and there are also numerous examples of long-distance graziers in India.

The right side of Figure 2.1 provides a simplified diagram of the linkages between forest and people (here divided into three basic groups of 'poor and landless', 'middle' and 'rich' households; see Cells 39 to 41). The three-way categorization is, of course, an oversimplification, and detailed discussion in Part II reveals a range of households and their livelihoods. The satisfaction of their material needs varies between these groups in relation to the same forest. Generally, it is the poor and those who have no legal title to agricultural land at all who rely on the forest most; therefore, current poverty-focused programmes may be affected adversely by the exclusion of the poor and landless from the forest. Furthermore, as Chapter 1 has shown, a longer historical perspective reveals that exclusion of people from forests has impoverished large sections of the rural population, including the majority of agriculturalists, artisans who draw on the forest for their raw materials and pastoralists. Forest materials include a very wide range, such as wood fuel; leaf and grass fodder for stall-fed livestock; construction timber; NTFPs; wild foods; medicines; and so on. Details are given in the regional chapters in Part II.

Finally, we introduce the issue of the role played by rural people in the policy process. This is usually played out as collective action through federations or co-operatives, political activism through political parties, and the pursuit of wider agendas through armed insurrection. Other formal institutions, such as the *gram panchayat* in India and the VDCs and DDCs in Nepal, are also important in shaping the way in which PFM works on the ground. Rural people may have little influence on the process in the capital city, except indirectly when activists and intellectuals speak for them to opinion leaders and policy-makers (although there are exceptions – for example, where community self-initiated processes have forced state governments to give healthy standing forests for JFM in Maharashtra). They have been, and continue to be, *subjects* who are the recipients (and often victims) of policy, rather than *citizens* who have a voice in policy-making. In the absence of being able to make representations that are openly discussed and negotiated, alternative strategies to alter or derail official forest policy have been established through direct resistance (e.g. arson or violence against officials; see Tucker, 1984). Social movements, including the Chipko Andolan in Uttaranchal (Rangan, 2001), the Appiko movement in Karnataka and the Silent Valley Movement in Kerala (Gadgil and Guha, 1995), have been important, if often isolated, instances that have sometimes had longer-term impacts on the daily practice of forest policy. Although they have perhaps not had as significant an impact as much as the literature on them might suggest, the Chipko Andolan, for instance, was instrumental in bringing about a ban on green felling above 1000m, a significant policy change. Such movements, perhaps, also have more far-reaching inspirational and psychological impacts on both the general public and on policy-makers and policy influencers, which could eventually result in policy shifts.

The fifth level is the *household level* (Cells 39 to 41 in Figure 2.1). This also includes households acting in a collective fashion with local institutions and therefore could also be termed the 'village/household' level (see Cell 36 in Figure 2.1). The flows between forest products, household labour and consumption, agriculture and animal husbandry are depicted via a few suggestive arrows, and the detail is illustrated in Part II. Research at this level focuses on four linked issues in Part II:

1   an impact study (in summary form, the question: 'What impact has PFM had on your household and members within it?');
2   a distributional study that focuses on equity issues ('Who gets what and why?' and 'How has access and control of forests and forest land altered with PFM?');
3   the question: 'What collective action and empowerment has occurred with PFM, and how has it changed from pre-PFM periods?'; and
4   the question: 'What are the reflexive links between PFM and wider issues in local

politics?'.

Conclusions

1   Many diverse actors shape forest policy and the response to PFM; it is a multilevel and dynamic process with many feedbacks. The links between actors are reflexive and denote constant contention and renegotiation. Obviously, the forest service itself is a key institution in both countries, and the outcomes of PFM in the policy process, from conceptualization on, are shaped by a huge range of procedures, job descriptions, training schedules, bureaucratic repertoires and changing political relations between the main actors. Internal reform, therefore, must address this range, with each aspect linked to many others. Reform in training and the production of forest knowledge is linked to formal training and syllabi, and also to the way in which local knowledge can play a part in technical choice as a part of a process. Job descriptions and procedures have a strong historical momentum and simply will not be modified without other complementary changes ('contradictions and pressures', as this chapter terms them). Reform to policy and how it is made (the policy process) may be aided by identifying alliances in order to generate reform options and to be better able to exert political pressure. It is also essential to be aware that there are ambiguous strategies on the part of the forest services, such as rhetorical acceptance of PFM reform, but practical 'burial' of it in silence, and these should also be identified.
2   Strategies for PFM reform must take a broad and historically informed view in order to make sense of current policy processes. Many actors shape policy, and therefore a policy reform should be targeted not only at various sections of the forest service of India and Nepal, as well as other government departments, but at political parties, social movements, NGOs, the judiciary, BINGOs and IFIs. Coordination of purpose and the formation of national and international alliances, wherever possible, are an essential and key challenge.
3   A constitutional approach to forestry reform in which executive arms of the state should be responsive to the legislature and be representative of the wishes of citizens who express their choices through the ballot box may be effective. This democratic ideal is never reached, but may be worth presenting, providing an element in the case for reform.
4   PFM is part of a long-term transition in forest reform and cannot be expected to transform ingrained relations quickly, either in forest administrations or in agrarian societies. To expect reforms such as PFM alone to be instrumental in broader agrarian change is like expecting 'the tail to wag the dog'. The changes that PFM strives for require a long gestation period.

References

Agarwal, A., Narain, S. and Sen, S. (eds) (1999) *State of India's Environment: The Citizens' Fifth Report*, New Delhi, Centre for Science and Environment

Apthorpe, R. (1997) 'Writing development policy and policy analysis plain or clear: On language, genre and power', in Shore, C. and Wright, S. (eds) *Anthropology of Policy: Critical Perspectives on Governance and Power*, London and New York, Routledge Press, pp43–58

Apthorpe, R. and Gasper, D. (1996) *Arguing Development Policy: Frames and Discourses*, London, Frank Cass

Arnold, D. and Hardiman, D. (eds) (1996) *Subaltern Studies VIII: Essays in Honour of Ranajit Guha*, New Delhi, Oxford University Press

Banerjee, A. K. (2004) 'Tracing social initiatives towards JFM', in Bahuguna, V. K., Mitra, K.,

Capistrano, D. and Saigal, S. (eds) *Root to Canopy*, New Delhi, Winrock International India and Commonwealth Forestry Association

Banuri, T. and Apffel-Marglin, F. (1993) *Who Will Save the Forests? Knowledge, Power and Environmental Destruction*, London, Zed Books

Blaikie, P. M. and Muldavin, J. S. S. (2004) 'Upstream, downstream, China, India: The politics of environment in the Himalayan region', *Annals of the Association of American Geographers*, vol 94, pp520–548

Blaikie, P. M. and Sadeque, Z. (2000) *Policy in High Places: Environment and Development in the Himalayan Region*, Kathmandu, Nepal, International Centre for Integrated Mountain Development (ICIMOD

Blaikie, P. M. and Seddon, J. D. (1978) 'A map of the Nepalese political economy', *Area*, vol 10, no 1, pp30–31

Bose, S. (2005) 'In a sorry state', *Down to Earth,* 31 August, Delhi, Centre for Science and Environment

Central Chronicle (2006) *Central Chronicle*, Bhopal, 18 January, www.centralchronicle.com/20060118/1801305.htm

Chhettry, B., Francis, P., Gurung, M., Iversen, V., Kafle, G., Pain, A. and Seeley, J. (2005) 'A framework for the analysis of community forestry performance in the *tarai*', *Journal of Forest and Livelihood*, vol 4, no 2, pp1–16

Clay, E. J. and Schaffer, B. B. (eds) (1986) *Room for Manoeuvre: An Explanation of Public Policy in Agriculture and Rural Development*, London, Heinemann

Court, J. and Young, J. (2004) *Research and Policy in Development Programme (RAPID) Briefing*, Paper No 1, October, London, Overseas Development Institute

David, A. and Hardiman, D. (eds) (1994) *Subaltern Studies VIII: Essays in Honour of Ranajit Guha*, New Delhi, Oxford University Press

DFRS (Department of Forest Research and Survey) (1999) *Forest Resources of Nepal (1987–1998)*, Publication No 74, Kathmandu, DFRS

DFRS (2005) *Forest Cover Change Analysis of Tarai Districts (1990/91–2000/2001)*, Kathmandu, DFRS

DoF (Department of Forests) (2002) *Hamro Ban*, Kathmandu, Department of Forests, Ministry of Forests and Soil Conservation

Forsyth, T. (2003) *Critical Political Ecology: The Politics of Environmental Science*, London, Routledge

FSI (Forest Survey of India) (2003) *State of Forest Report*, Forest Survey of India, Dehra Dun

Gadgil, M. and Guha, R. (1995) *Ecology and Equity: Use and Abuse of Nature in Contemporary India*, London, Routledge

GoI (Government of India) (1952) *The National Forest Policy of India*, New Delhi, GoI

Government of West Bengal (2002) *Biodiversity Strategy and Action Plan for West Bengal*, Kolkota, Government of West Bengal

Guha, R. (1983) 'Forestry in British and post-British India: A historical analysis in two parts', *Economic and Political Weekly,* vol 18, no 45/vol 18, no 46, pp1882–1897, 1940–1947

Guha, R. (1989) *The Unquiet Woods: Ecological Change and Peasant Resistance in the Himalaya*, Delhi, Oxford University Press

Guha, Ranajit (ed) (1989) *Subaltern Studies VI*, New Delhi, Oxford University Press

Hiremath, S. R. (ed) (1997) *Forest Lands and Forest Produce – As if People Mattered*, Dharwad, National Committee for Protection of Natural Resources (NCPNR)

Hobley, M. (1985) 'Common property does not cause deforestation', *Journal of Forestry*, vol 83, pp663–664

Hobley, M. (1996) *Participatory Forest Management: The Process of Change in India and Nepal*, London, ODI

Iversen, V., Chhetry, B., Francis, P., Gurung, M., Kafle, G., Pain, A. and Seeley, J. (2006) 'High value forests, hidden economies and elite capture: Evidence from forest user groups in Nepal's Terai', *Ecological Economics,* vol 58, pp93–107

Ives, J. D. and Messerli, B. (1989) *The Himalaya Dilemma: Reconciling Development and Conservation*, London, John Wiley and Sons

Jerram, M. R. K. (1982) *A Textbook of Forest Management*, Dehra Dun, International Book Distributor

Kavita, P. (2003) *Civilizing Natures, Race, Resources and Modernity in Colonial South India*, New Delhi, Orient Longman

Keeley, J. and Scoones, I. (1999) *Understanding Environmental Policy Processes: A Review*, IDS Working Paper 89, University of Sussex, Brighton, Institute of Development Studies

Long, N. and Long, A. (eds) (1992) *Battlefields of Knowledge: The Interlocking of Theory and Practice in Social Research and Development*, London, Routledge

MoEF (2004) 21/7/2004 Affidavit No. 703 of 2000 in Writ Petition (Civil) No. 202 of 1995 (T.N. Godavarman Thirumalpad Versus Union of India and Others)

Mukherjee, A. K. (2004) 'Tracing policy and legislative changes towards JFM', in Bahuguna, V. K., Mitra, K., Capistrano, D. and Saigal, S. (2004) *From Root to Canopy: Regenerating Forests through Community State Partnership*, New Delhi, Winrock International India, pp35–44

Mukherjee, A. K. (2006) 'Evolution of good governance through forest policy reforms in India', in Workshop pre-prints from three-day workshop, 20–22 April, 2006, Bhopal, ICCF and IIFM, pp17–24

Narain, S. and Agarwal, A. (2003) *State of Forest Report 2003*, Delhi, Centre for Science and Environment

Peet, R. and Watts, M. (eds) (1996) *Liberation Ecologies: Environment, Development and Social Movements*, London, Routledge

Perkins, J. (2004) *Confessions of an Economic Hitman*, San Francisco, Berrett-Koehler

Philip, K. (2003) *Civilizing Natures, Race, Resources and Modernity in Colonial South India*, New Delhi, Orient Longman

Rahema, M. and Bawtree, V. (eds) (1997) *The Post-Development Reader*, London and New Jersey, Zed Books

Rangan, H. (2001) *Of Myths and Movements: Rewriting Chipko into Himalayan History*, New Delhi, Oxford University Press

Rangarajan, M. (1996) *Fencing the Forests; Conservation and Ecological Change in India's Central Provinces, 1860–1914*, New Delhi, Oxford University Press

Rangarajan, M. (2003) 'The politics of ecology: The debate on wildlife and people in India, 1970–95', in Saberwal, V. and Rangarajan, M. (eds) *Battles over Nature: Science and the Politics of Conservation*, New Delhi, Permanent Black

Rittenbergen, S. (2001) 'The history and impact of forest management' in Evans, J. (ed) (2001) *The Forests Handbook: Volume 2*, Abingdon, Blackwell Science

Saberwal, V. (1999) *Pastoral Politics: Shepherds Bureaucrats and Conservation in the Western Himalaya*, New Delhi, Oxford University Press

Sarin, M. (1996) 'From conflict to collaboration: Institutional issues in community management', in Poffenberger, M. and McGean, B. (eds) *Village Voices, Forest Choices: Joint Forest Management in India*, New Delhi, Oxford University Press

Sarin, M. (1999) *Policy Goals and JFM Practice: An Analysis of Institutional Arrangements and Outcomes*, Policy and Joint Forest Management Series 3, Paper written for the WWF–IIED project Policies that Work for Forests and People, New Delhi, WWF–IIED

Sarin, M. (2003) 'Bad in law: Analysis of forest conservation issues', *Down to Earth,* 15 July 15, pp36–40

Sarin, M. (2005a) *Laws, Lore and Logjams: Critical Issues in Indian Forest Conservation*, Gatekeeper series 116, London, IIED

Sarin, M. (2005b) 'Scheduled Tribes Bill 2005: A comment', *Economic and Political Weekly*, 21 May, pp2131–2134

Sarin, M. with Singh, N. M., Sundar, N. and Bhogal, R. K. (2003) *Devolution as a Threat to Democratic Decision-Making in Forestry? Findings from Three States in India*, Working Paper 197, London, ODI

Sarkar, S. (1989) 'The Kalki-Avatar of Bikrampur: A village scandal in early twentieth century Bengal', *Subaltern Studies,* vol VI, New Delhi, Oxford University Press

Saxena, N. C. (1994) *Policies, Realities and the Ability to Change: The Indian Forest Service – A Case Study*, London, ODI

Schlich, W. (1896) *A Manual of Forestry*, London, Bradbury, Agnew and Co

Scott, J. C. (1998) *Seeing like a State: How Certain Schemes to Improve the Human Condition Have Failed*, New Haven, Yale University Press

Shankland, A. (2000) *Analysing Policy for Sustainable Livelihoods*, IDS Research Report 49, University of Sussex, Brighton, Institute of Development Studies

Shiva, V. (1991) *Ecology and the Politics of Survival*, New Delhi, Sage Publications India Pvt Ltd

Shiva, V. and Bandyopadhyay, J. (1987) 'Chipko: Rekindling India's forest culture', *The Ecologist*, vol 17, no 1, pp26–34

Shiva, V. and Bandyopadhyay, J. (1988) 'The Chipko Movement', in Ives, J. and Pitt, D. (eds) *Deforestation: Social Dynamics in Watersheds and Mountain Ecosystems,* London, Routledge, pp224–241

Singh, S., Shastry, A., Mehta, R. and Uppal, V. (eds) (2000) *Setting Biodiversity Conservation Priorities for India*, Delhi, WWF

Srinidhi, A. S. and Lele, S. (2001) *Forest Tenure Regimes in the Karnataka Western Ghats: A Compendium*, Working Paper No 90, Bangalore, Institute for Social and Economic Change

Sundar, N., Jeffery, R. and Thin, N. (eds) (2001) *Branching Out: Joint Forest Management in India*, New Delhi, Oxford University Press

Sutton, R. (1999) *The Policy Process: An Overview*, ODI Working Paper No 118, London, Overseas Development Institute

Talbot, K. and Khadka, S. (1994) *Handing it Over: An Analysis of the Legal and Policy Framework of Community Forestry in Nepal*, Washington, DC, World Resources Institute

Tucker, R. P. (1984) 'The historical roots of social forestry in the Kumaon Himalayas', *Journal of Developing Areas*, vol 13, no 3, pp341–356

Vira, B. (2005) 'Deconstructing the Harda experience: The limits of bureaucratic participation', *Economic and Political Weekly*, vol 40, no 48, pp5068–5075

World Bank (2006) *India: Unlocking Opportunities for Forest Dependent People*, Delhi, Oxford University Press

3

Actors and their Narratives in Participatory Forest Management

Piers Blaikie and Oliver Springate-Baginski with Ajit Banerjee,
Binod Bhatta, Sushil Saigal and Madhu Sarin

Our approach

An important part of our approach to forest policy is the examination of 'policy narratives' or stories told by different protagonists. These are not 'just talk' or inventions for others' amusement, but persuasive constructions with a beginning (assumptions, problem framing, choice of issues, etc), a development (argumentation, supporting evidence, justifications, troublesome side issues and other relevant circumstances) and a conclusion (what should be done and policy recommendations). They use some facts, are ignorant of or deselect others, and interpret information in a particular manner in order to tell a persuasive and consistent story. They frame issues and problems in certain ways to focus on some issues and to exclude others. This may be done either consciously, as a strategy, or unconsciously, where the author has a particular set of facts and values that are not critically reflected on. Narratives are used in policy-making as much as in everyday life. They are a way of making sense of an uncertain, complex and contested world. In a more strategic sense, narratives may also be a means of persuading others. In no way is the labelling of an account as a 'narrative' meant to be derogatory or to imply falsehood or fantasy. On the other hand, however, we cannot assume that we know the actors' intention from our interpretation of what they say. As Chapter 2 has illustrated, forest policy is complex, with many competing political representations and political ecologies at different scales, and narratives fulfil important objectives for the actors involved. Narratives serve to stabilize their expectations and provide secure moorings in a shifting and sometimes threatening world; but they also perform representative and political purposes in the exercise of power by persuasion. Narrative analysis is therefore well suited for the treatment of policy (see Roe, 1994; Hajer, 1995; Apthorpe and Gaspar, 1996; Forsyth, 2003).

Policy narratives can be examined for logical consistency, and this book tests some of the main claims made for and against participatory forest management (PFM) in India (joint forest management, or JFM) and in Nepal (community forestry, or CF) on this basis. A major part of this book (mostly in Part II) considers empirical data to check these claims on the ground. There have been various statements and arguments from the forestry services, international funding institutions (IFIs) and other groups about the purposes of PFM and how to fulfil them. They are all amenable to logical and empirical examination. For example, in what ways is JFM 'jointly' managed in practice? How 'community oriented' is community forestry in Nepal? Do the benefits of participation logically follow from the premises outlined in policy documents? What really happens in daily practice, and does it logically follow the intentions in documents? Here we provide empirical evidence to answer these questions and to take further the analysis of the processes of establishing and practising PFM that have shaped the degree of participation and control by local people at the local level. The outcomes, in terms of changed livelihoods, are also discussed.

The book examines the main lines of argumentation used in narratives produced by the major actors. Many actors, apart from official sources, shape forest policy, which tends to be written and produced by, and presented to, audiences who are close to formal policy-making. Non-official narratives are also important, especially as the book focuses on the participation of a wide number of others outside formal policy-making. These include narratives from women's groups, tribal organizations, federations of forest users, activists and individual forest users, which are sometimes not written down or widely publicized. All narratives are often (but not always!) coherent, persuasive and 'common-sense' accounts; but they are usually competitive with others for the ear of particular audiences. Different stakeholders have different and often competing narratives. Examples include 'forest rights' and another similar narratives concerning 'natural justice' for the restoration of tribal people's rights over their ancestral domains (made by social movements, most in-country non-governmental organizations (NGOs), left-leaning politicians and tribal leaders). Another is the 'national interest' and the role of forests to provide raw materials for industrial growth, and 'scientific management of the forest', which claims authoritative knowledge over other sources of knowledge. A third is the imperative of longer-term soil and water conservation, and the improvement of agricultural productivity through afforestation and 'green cover'. There are others, too, that are relevant, such as biodiversity conservation, maintaining ecological balance and creating inviolate areas for wildlife, which affect forest management and forest-adjacent populations.

In other cases, however, narratives from different actors are not contested by all others, but find resonance or agreement, and those who tell them form a 'discursive coalition':

> *A discursive coalition is the ensemble of a set of storylines, the actors that utter these storylines, and practices that conform to these storylines, all organized around a discourse.* (Hajer, 1995)

The alliances themselves are sometimes temporary and strategic, often between the different actors who produce these narratives. In this chapter we discuss both the centralized 'classic' narrative (itself a discursive alliance) and, in broad opposition, the 'popular' discursive alliance, both of which are formed by quite diverse ideas and logics, but which can enlist mutual discursive support in policy argumentation and political action.

Narratives are usually amenable to critical review on the basis of empirical research, logical consistency and ethical principles. There are also framings of more specific lower-order forest issues, which are not full narratives but which also have strong implicit, rather than explicit, political weight because they contain assumptions that remain largely uninterrogated.

Finally, the uses (and abuses) of policy narratives for policy reform need to be discussed. A focus on narrative in the policy process is necessary but insufficient without addressing the reflexive relationship between the power of the author of the narrative and the narrative itself. This relationship may have structural aspects (e.g. funds, the means of broadcasting the narrative, networks of allies, the means of coercion, as well as the persuasiveness (and therefore the power of the narrative itself). Thus, it is not only the power of the narrative itself (ability to persuade target audiences), but also the power of the author deriving from access to other sources of power. For example, the World Bank may have a narrative of 'good practice' in PFM and may be able to publicize and fund it, while a small NGO (with 20 years of hands-on policy practice on the ground) may make similar policy recommendations and employ the same narrative. Which will have more influence on policy process? Larger and politically dominant partners have predominance in projecting their narratives. Hence, Chapters 2 (actors) and 3 (narratives) are closely related. A focus on policy narratives alone will also tend to privilege documents and printed matter, and the survival of the formal and well-financed over the spoken narrative by those who have poor access to a printing press or radio.

A whole chapter on narratives may be dismissed by some as 'all talk', implying that 'talk' is one thing, but practice and behaviour (which may or may not conform to the narratives) is quite another. Certainly, PFM seems to be a case in point – the rhetoric of participation and justice is included in policy documents, but may be diluted, lost or perverted, in practice. However, taking policy statements seriously, examining them carefully and critically, and identifying logical inconsistencies, selective use of slanted metaphors and the covert introduction of value judgements in the guise of rational argument are, we maintain, essential tasks in planning policy reform.

Policy narratives

This section introduces the main actors' narratives. Narratives are dynamic and are constantly being adjusted in response to external conditions and the changing circumstances of the actors. Therefore, a discussion of a narrative requires something of an historical perspective and an understanding of where the narrative may be heading in the future, although this will always be speculative.

First, we discuss the *state forest administrative narrative*, using a generic model – despite the fact that in India and Nepal, they have slightly different periods. The colonial narrative in India from the mid 1860s up until the mid 1980s is discussed first, and for Nepal we suggest that the model is applicable after the fall of the Ranas in 1950 up to the mid 1970s. In the next section we focus more carefully on the conflicts that this narrative has received from various actors in civil society (e.g. forest users, social movements and activists) from the 1980s onwards. The discussion then turns to the major areas of conflict over the substance of the narratives between the state and civil society in the policy process. Finally, the various engagements between state and civil society, leading to adaptations on both sides, are detailed.

Sometimes narratives contain words that carry a 'heavy freight' – which convey pivotal meaning that may be accepted by the audience of the narrative without question. These are powerful words because if accepted by an audience without critical examination they promote acceptance of the whole policy argument. In this sense, the argument is implicit in a word, rather than explicit, but, nonetheless, can be an important part of policy argumentation. In all cases, these words must be put into the institutional and historical context in which they are used and must also be attributed to authors. The words 'forest', 'participation', 'joint' and 'community' are some examples.

The 'state forestry' narrative

There has been a stable 'state forestry' narrative with a long historical precedent in both India and Nepal (see Chapter 1). In India, the narrative itself has engaged with both external events and reactions of people to forest policy (social movements, resistance, and 'weapons of the weak' (Scott, 1985), such as poaching, sabotage and arson). The state narrative also has a history of internal dialogue during the 1880s (see, for example, Baden-Powell, 1882, and Brandis, 1883, as discussed in Chapter 1) and, more recently, crises in the costs of policing, insufficient revenues to maintain the forest service and engagement with the narratives of some donors. In the case of Nepal, other changing external circumstances have constantly reshaped existing policy narratives. The emergence of Nepal from a semi-feudal past and the presentation of a modernization imperative were part of the thinking behind the nationalization of forests in 1957; but the realization that the state could not police the nationalized forest estate led to a readier acceptance of a more 'hands-off' and, lately, a more participatory approach.

Figure 3.1 identifies the main lines of argumentation in the classic and state-dominated policy narrative for Indian and Nepalese forestry. This is not meant to imply that the narrative is highly rigid and standardized, but to illustrate that it has been consistent and durable over time.

The figure is intended to be generic and, therefore, applicable to both India and Nepal. Although there are differences in emphasis for different parts of the narrative, there are also underlying similarities.

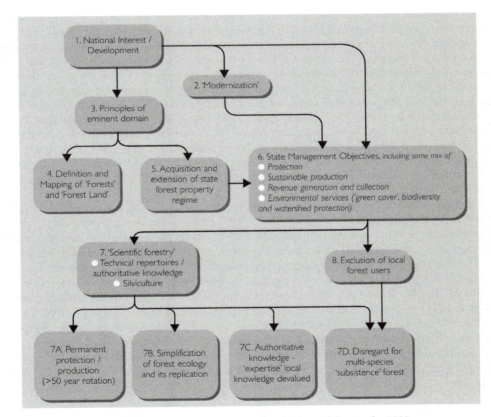

Figure 3.1 'State forestry' narratives: The classic model up to the 1980s

Source: Piers Blaikie and Oliver Springate-Baginski (original material for this book)

An additional narrative may be about the need for a centralized, trained bureaucracy – the notion that only such an agency could manage forests effectively has been strong through the past several decades and remains compelling even now, finding favour among a section of conservation/environmental groups and, of course, with the administration.

National interest/development

(Cell 1 in Figure 3.1)

After independence, national interest and 'national development' became key policy arguments. This has carried certain baggage in terms of nation-building, involving assumptions around prioritizing and trading off between rural and industrial development, as is apparent in the 'modernization' narratives. It has been argued that after independence civil society was already dominated by a relatively overdeveloped state and that the present autonomy of the Indian state is reflected more in its regulatory (and, hence, patronage-dispensing) role than in its development role (Bardhan, 1984). Certainly, the regulatory aspect of the classic forest department narrative conforms to this general characterization as authoritarian rather than developmental.

The implication for forests is that the national interest is primary and prevails over local concerns. Forests provide raw material inputs for industry, rather than for local subsistence users. Later on, 'green cover' objectives (see Cell 6), and, in certain areas, soil and water conservation and watershed management, were also assumed to be a proper role of the forest department (the first National Planning Commission of 1937 and the first Five-Year Plan). The Indian National Congress appointed various sub-committees to suggest recommendations to put India on the path towards 'modernization'. A sub-committee on soil and water conservation advocated that forest cover should be strictly protected in order to arrest soil erosion. It also advocated maximization of agricultural productivity by converting wastelands and forests into cultivated lands. Its main recommendation was maximization of forest productivity for use in industry. Thus, under the banner of national interests, local forest-dependent people's livelihoods were suppressed for the sake of national development. With the 1988 Forest Policy, ecological balance and environment protection, as well as local needs, became major priorities. The 1980 Forest Conservation Act (FCA) focused on maintaining forest cover, which implied the maintenance and, wherever possible, expansion of control over forest land.

Turning now to Nepal, a similar narrative emerged, but with some significant differences. After the fall of the Ranas, King Mahendra's government strongly pursued a policy of modernizing the state, and the nationalization of forests in the name of the national interest owed much to the desire to wrest their control from feudal elites for a modernizing state. The role of *tarai* forests, in particular, was to contribute to national development. The integration of hill people with those of the plains by out-migration from the hills, implying the clearance of *tarai* forests for agriculture, and the generation of revenue from the forests (Regmi, 1978), with legal and even illegal settlement in the *tarai*, were all supported. Donors such as the US Agency for International Development (USAID) supported malaria eradication and resettlement programmes. The establishment of a major sawmill at Hetauda run by the Timber Corporation of Nepal indicated the developmental imperative and the role of forests in promoting it. PFM could hardly be more contradictory to this part of the classic narrative.

Modernization

(Cell 2 in Figure 3.1)

In India, particularly after independence and as reflected in most Five-Year Plans, especially the first, the programme of industrial development was prioritized in planned development programmes. The notion of modernization has also been used by the forest administrations to justify the outlawing of non-modern ('primitive') practices practised by forest-dwelling communities, including a variety of forest fallows cultivation (*podu*). For example, here is a passage from the Forest Directorate of (then) Calcutta:

> *There is much need for introduction of improved logging techniques and mechanical extraction of timber from coupe to roadside. Large-scale use of gravity and portable power ropeways, Skyline cranes and mountain tractors are likely to solve this problem and planning on this line is already in progress. Locating sawmills and wood-based industries in the close vicinity of forests and use of portable power saws inside the forest may also go a long way to rapid development of forestry in this area. Future development of these forests has to be linked up with development of wood-based industries, including paper mills, so that the existing forests may be fully utilized, irrespective of the demand for conventional uses of timber.* (Ray, 1964, pp95–98)

The first Five Year Plan also emphasized agricultural expansion and irrigation, and the second, industrial development, with the exploitation of forests to assist in this goal (see the quote from Ray, 1964). Increased felling was planned to generate revenue and strategic timber supply for industry and to satisfy urban demand for construction and furniture. There was a push to reduce rotations and increase felled areas to supply factories during the late 1950s and early 1960s (Guha, 1983). In Nepal, the term 'modernization' was first used during the 1930s by Collier, a British adviser to Ranas, who suggested a rotational selection felling system (selecting trees with minimum diameter at breast height (dbh) of 30 inches, or 76cm, for felling) specifically for the *tarai*'s sal (*Shorea robusta*) forest, which was to be supplied for railway sleepers to British India and at the fall of the Ranas had become a central part of the vision of a new, post-feudal modern state.

Principle of eminent domain

(Cell 3 in Figure 3.1)

For millennia, there has been conflict between the state and people over rights to land. The principle of 'eminent domain' has been one way in which states have justified taking over land in the interest of the state. It maintains that in the name of the greater good, the rights of the state to a resource transcend local rights claimed by proximity and customary use (Commander, 1986). In colonial India, the state asserted its imperative through privileging the rights of the state above all others through a variety of tenurial arrangements. Baden-Powell argued that:

> *In old days, native rulers used often to set aside considerable areas of forest land as Shikarighar or hunting grounds, and these would be usually covered with thick and perhaps valuable forests. Such lands have now become the property of the British government following the principle of succession.* (Baden-Powell, 1882, p7)

Since independence, the IFS has upheld the premise that the forest is a 'national' resource and not a 'public good', and therefore national interests (as specified by the government and interpreted by the forest department) are paramount, and any customary rights are either abrogated or downgraded to a privilege. The 'pre-eminent domain' argument runs that the state's need for control of forest resources takes precedence over the needs of local people. Pre-eminent domain has been the legal basis for the forest department to define and reserve 'forests'. This implies that local people do not have rights, only privileges.

In Nepal, the main domains in the feudal regime during the Rana period used forests as private property and as a source of revenue for landlords, rather than for the state. Nationalization of private forests in 1957 made forests state property in order to contribute to the political, social and economic development of the state.

Definition and mapping of 'forests' and 'forest land', and acquisition and extension of state forest property regime

(Cells 4 and 5 in Figure 3.1)

Forest land is the land on which forest currently stands or putatively once stood. The Indian forest administration initially specifically reserved high-value timber stands, but gradually sought to reserve almost all non-private and non-village lands ('wastelands'), justifying its increasing estate in terms of the strategic colonial and, then, national interest, and in contemporary times increasingly in terms of forest cover (equated with tree cover) and wildlife habitat protection. There are three main issues regarding the definition of 'forests' and 'forest land'. First, the forest administration must categorize land that it wishes to annex as 'forest'. Although much of this land may have standing forest on it, this is not a necessary precondition – and, in fact, it may have been used for agriculture, pastoralism, forest fallows cultivation and other land uses, including customary community uses. Thus, 'forest' does not necessarily mean tree-dominated ecosystems. Even the existence of 'trees' does not define 'forests' – it is an administrative category rather than a description of current land use or vegetation.

The second point here is that 'forest cover' includes commercial tree plantations. This categorization can therefore fulfil other discursive objectives of the forest administration, such as contributing towards green cover objectives, increasing revenue and making a more significant contribution to 'national development' (i.e. pulp and wood-based industry, etc). This issue in India is related to the 'anti-eucalyptus monoculture narrative' (Saxena, 1994), a debate reflected internationally in contestations over monoculture tree plantations.

The third issue is the classification of 'degraded forest' and is important to the choice and allocation of potential forests for JFM in India. The system of classification seems very confusing and inconsistent. For example, the forest cover classes are officially given for the state of West Bengal as follows (bearing in mind that there are inter-state differences in classification):

- very dense forest: canopy density above 70 per cent (FSI, 2005);
- dense forest: 40 to 70 per cent canopy density (FSI, 2005);
- open forest: 10 to 40 per cent canopy density (FSI, 2005);
- mangrove: soil-tolerant ecosystem mainly in tropical and subtropical inter-tidal regions (GoWB, 2001);
- scrub: all lands with poor tree growth, mainly small or stunted trees having canopy density of less than 10 per cent (GoWB, 2001);
- non-forest: any area not included in the above classes (GoWB, 2001); and
- degraded forest: no official definition, but by common usage refers to open forests and scrub.

In practice, JFM is meant to be implemented only on 'degraded forest', yet, what 'degraded forest' is, is not specified. In some cases, it is taken to mean less than 40 per cent forest cover and in others, land areas classified as forest that have less than 10 per cent tree cover, which the Forest Survey of India (FSI) treats as 'non-forest'. Tree cover of 10 to 40 per cent is classified as 'open'; above 40 per cent cover is defined as 'dense' forest. In most states, JFM is only implemented in 'degraded forests', and this includes land that may not necessarily have been historically forest at all. Thus, there seems to be a high degree of political discretion in the classification of what constitutes a forest and its different categories.

In Nepal, there is also confusion (although not at the level of complexity that exists in India) over ownership of forest resources. Communal land is under village development committees (VDCs); but idle land near forests has not been claimed as private agricultural land for individuals' use since the Cadastral survey was undertaken during the 1950s to the 1980s. National forests are those forests that are not privately owned. Government forests have been defined by the 1993 Forest Act as areas 'fully or partially covered with forest'. The term national forest is defined to include barren lands or unregistered (*ailani*) land, foot trails, ponds, lakes or streams, and land across their banks within or in the vicinity of the forest. National forest includes community forest, leasehold forest, government-managed forest and protected area forest. The legal definition of forest is the main definition used by the Department of Forests (DoF) to distinguish government-owned from private forests. On the other hand, forest as a land use refers to areas that have tree cover present, even if degraded. Forest is categorized on the basis of the dominant species present in the forests, such as sal (*Shorea robusta*), khair (*Acacia catechu*), sisso (*Dalbergia sissoo*), pine (*Pinus roxburghii* and *P. wallichiana*), oak (*Quercus* spp), rhododendron (*Rhododendron arboretum*) and so on. Other categories are based on altitude, such as tropical sal forest, sub-tropical sal forest, temperate forest and alpine forest. On the basis of their management, the national forests have been categorized into government-managed forest, community forest, leasehold forest, religious forest and protected forest. Forests have been officially categorized according to two degradation statuses (full tree cover and shrub lands). As discussed earlier, land within or in the vicinity of forests, if not private property, is classified as 'forest' as defined by the 1993 Forests Act.

Returning now to both India and Nepal, claims by the forest service based on the contested categories discussed are recorded on surveys and maps. The boundaries, therefore, are specified as final, and land uses within them are exclusively classified according to unique categories (until such time as they may be modified at a later date). However, in both India and Nepal there remain large discrepancies within official statistics. In India, there is a difference of 9.13 million hectares in the area recorded as 'forest' in the Ministry of Agriculture's land records and in the Ministry of Environment and Forests (MoEF) records (FSI, 2005). In many areas, boundaries are not demarcated clearly and there are serious jurisdictional disputes between forest departments and revenue departments. In many areas, there are no good maps at all. In Orissa, rough blocks are marked on revenue maps as 'forest', and the forest department enforces the 'map' as if it conferred authority. When challenged in court, the forest department has been unable, in many cases, to prove its claims.

The main divergence between official forest development categories and those understood by local forest users is the exclusive land categories employed by line agencies versus inclusive and multiple categories through space (multiple uses and categories related to local livelihoods) and time (certain exclusions of uses at different seasons) employed by local forest users. Forest maps and plans are essential management tools, and this discussion does not deny their essential role. However, the focus of criticism is the claims made by official maps and the ways in which these claims are upheld in contradiction to local understandings, categories and management rules since these are used in a non-participatory way. This difference in 'mental cartography' also highlights the difficulties that face PFM. The practicalities of a participatory map with locally defined categories may, ideally, facilitate genuine PFM, and the construction of such maps has long been practised as part of participatory rural mapping (Campbell, 1995).

In Nepal, the effectiveness of forest mapping is also very poor, particularly in the hills. Cadastral survey maps attached to the operational plans (Ops) of community forest user groups (CFUGs) can refer to surveys made two to three decades ago. The technology used by the district forest office is still expensive and out of date (the equipment consists of chains, tapes and compasses, rather than the much more accurate and rapid global positioning systems, or GPS). The older survey systems have limited accuracy and precision, as indicated by the different results when the same parcel of forest land is resurveyed after some time lapse.

State management objectives

(Cell 6 in Figure 3.1)

Forest management strategies were adapted to serve the modernization paths which India and, to a lesser extent, Nepal followed. There was a need to supply appropriate timber and a few non-timber forest products (NTFPs) for infrastructural development, local commerce and export. The appropriate timber for shipbuilding, railways, urban construction, roadways, bridges and export from the 1860s until after India's independence were teak, sal, pine, *Acacia arabica* (now called *Acacia nilotica*) deodar, sandalwood and India rubber (*Ficus elastica*). Also, NTFPs such as pine resin (for turpentine) and *Caoutchouc* (exudations of the India rubber plant) were required by industry. This resulted in increased reservation of forest areas that contained these species and also their planting, where possible (plantations of teak, in particular, in south and central India and *Ficus elastica* in the north-eastern areas). There was also a pressing imperative to increase revenue for the exchequer. This was made possible by increasing the areas of reserved forest and by exploiting them with 'scientific' methods long developed by Europeans (mostly trained in Germany) but not indigenous to Indian conditions, and no other forestry system could be introduced except through the application of local research and the rewriting of forest manuals for Indian and Nepalese conditions. Furthermore, in Europe, especially Germany, France and the UK, forest reservation was the preliminary step to managing forests by the state, and this method was also followed in India.

After independence, the same state objectives for forests were followed even more rigorously. Forest and waste areas that were omitted from reservation before independence for village use were planted with industrially required species and protected by the state, thus excluding common use. It would be difficult for foresters of the time to imagine any alternative management system since they were all trained in classical forestry. They simply did not know any other system. They learned, for example, that the major elements of indigenous forestry (e.g. shifting cultivation) damage the forest; that transhumant pastoralists reduce regeneration; that home gardens and sacred groves are of minimal importance and thus do not need to be considered part of forest policy; and that local forest rights lead to the annihilation of forests – none of these activities of indigenous forest use serving the national purposes which forests were meant to serve. The foresters understood the purposes of state forestry as silviculture in the form of monoculture plantations for convenient and cheap exploitation, and to increase revenue. Hence, the forest officials continued with classical forestry as the role that forests were supposed to play in national development.

Thus, important elements of the narrative derive from these national priorities, which set the parameters for forest policy. The key ideas were that forests must provide for industrial demands; reservation of increased areas and plantations must be accelerated; and non-commercial and indigenous land uses are an impediment to be excluded from forests and forest land.

Other management objectives such as soil and water conservation, wildlife, 'green cover' and biodiversity were also stated and given priority at different times. Soil and water conservation as a major policy issue emerged in India during the 1930s, for example Hamilton writing for the Punjab Erosion Committee (Punjab Erosion Committee, 1931; see also Farooqi, 1997, for an historical overview). Immediately after the takeover of private forests during the 1950s and 1960s, soil and water conservation measures were built into afforestation and re-afforestation techniques. In Nepal, there has been a long period of concern about the impact of deforestation on soil erosion, about falls in the productivity of agricultural land and about downstream flooding and landslides, although the Theory of Himalayan Environmental Degradation (THED) has received some serious criticism, and with it the role of so-called deforestation (see Chapter 1 for discussion and references).

In 1952, the Indian Planning Commission initiated the idea that 'green cover' was necessary, and an arbitrary green cover target of 30 per cent was decided on and has become a powerful contribution to the Indian forest administration's claim to taking an expansionary approach and annexing and protecting more land that can be officially designated as forest. The unspoken assumption is that alternative management regimes, including PFM of various types, cannot be trusted to increase green cover.

Biodiversity narratives were a recent introduction of the late 1980s and 1990s, although 'wildlife' issues have a longer history going back until at least the 1970s. Wildlife protection was supported by many Indian NGOs and individual conservationists, including the then prime minister, Indira Gandhi. Biodiversity projects, in the more recent meaning of the term, mainly derive from initiatives from IFIs – for example, the World Conservation Union (IUCN), the World Wide Fund for Nature (WWF) and the Convention on Biological Diversity (CBD). These narratives usually imply highly exclusionary policies, in the case of Nepal, with responsibility for protection entrusted to the military. Proponents of biodiversity protection have frequently made alliances with the most conservative elements in the forestry administrations of both India and Nepal, promoting protected areas and excluding local people. Therefore, 'fortress' biodiversity conservation becomes an important ally in the production of the classic forestry narrative. However, this 'discursive alliance' is undoubtedly fragmenting, as the recent National Biodiversity Strategy and Action Plan debacle reflects. A leading environmental NGO, Kalpavriksh, was commissioned by the Ministry of Environment and Forests (MoEF) to develop a National Strategy and Action Plan; but after an exhaustive consultative process the pro-people recommendations were rejected by the MoEF. It appears that this could well be because the draft plan challenged the dominant state narratives in many ways.

Scientific forestry

(Cell 7 in Figure 3.1)
The claim that professional forest practice is based on 'scientific' principles primarily relates to empirical studies of the growth trends of specific species, leading to the development of 'growth tables' that allow a forest manager to predict the timber yield of the site in question under different management systems. Additionally, empirical study of variations in soil types, climatic variation, behaviour of different species and testing of various silvicultural systems all form the information base for the claims that the forest administration dispenses 'scientific' forestry. Whatever the quality of the evidence informing management planning, actual field practice in forestry remains as much an art as a science, although 'artistic forestry' as a descriptive label carries less discursive force and therefore has not caught on. Rackham (1990) has observed that 'Forestry is an art which, because of the long timescale, failures tend to be forgotten, not learnt from, and thus later repeated.'

'Scientific forestry' is authoritative in the sense that it claims 'truth' through being based on accepted methods of hypothesis testing. It overrides personal interpretation and political bias and puts statements beyond doubt and further argument. Alternatively, it is clear that any 'scientific forestry' has prior definitions of policy interest and always makes a series of decisions over the framing of scientific problems that select the testing of some hypotheses and deselect others. For example, the achievement of particular management objectives is a prior political decision about input into an agenda for scientific research according to assumptions of the overall purpose of the forest estate. Therefore, in order to provide inputs to industry, construction, paper-making and railway sleepers, scientific forest management tends to frame research problems around plantations, timber yield and quality maximization under alternative treatments. In turn, this usually results in long rotations, often in excess of 50 years (see Cell 7A, Figure 3.1) and the exclusion of local people (see Cell 8).

Scientific forest management leaves out of the frame of scientific research, for example, those research areas that are not clearly mainstream concerns and that have less commercial value (e.g. research into diverse product mixes from mixed species subsistence forest, or NTFPs with use values rather than commercial value). The authority derived from scientific forestry is based on the priority of silviculture, which then becomes the research focus and is 'in the frame' – while NTFPs and multi-species forests, which in part support livelihoods, remain out of the scientific frame (see Cell 7D, Figure 3.1). For example, before the commercial value of bamboo became evident there were plans to eradicate it as it was considered a weed!

The current scientific research undertaken in the forestry schools of both Dehra Dun and Pokhara is still affected by the historical inertia of the application of temperate principles of forestry practice to the tropics. The current scientific research undertaken at Dehra Dun and other regional institutes is heavily biased towards increasing the productivity of wood, resource inventory, afforestation, planting, the introduction of exotic species, botanical specifics and medicinal plants. There were some minor exceptions, such as the agro-forestry adaptation of *podu* (shifting cultivation) in 1919 to the *taungya* method (discussed below); but as yet there is no 'science' of participatory forest management, particularly concerning NTFPs and multi-species subsistence forests, in either India or Nepal. Indeed, there is considerable evidence that foresters in both countries do not have the necessary scientific knowledge, produced by appropriate research into what forest users would like to know, to contribute constructively to their local agenda.

In Nepal, the prediction of an impending energy crisis (related to the predictions made based on the consumption of firewood without considering the potentials for other alternatives and changes in technologies) led to research and trials on the adaptation of tree plantation in the 1960s in the Kathmandu Valley, with limited success. The prediction of a fuelwood deficit was made for the *tarai* region and big urban centres. This situation initiated immediate action in the form of the massive plantation of fast-growing species suitable for firewood. Scientific research made the case for rapid-growing species, and to support this initiative the Sagarnath Forest Development Project was initiated with a loan from the Asian Development Bank (ADB). The project selected eucalyptus as the main plantation species, justified by its fast growth and suitability to the site. At present, the claims made by Sagarnath plantations are only partially fulfilled and eucalyptus wood is being used for electric poles, fuelwood and particle board. However, the yield has not been at the level predicted (growth rate 24 cubic metres per hectare per year).

'Sustainable yield' is a key notion in the narrative, justifying long-term control of the resource. This necessitates predictability in planning (see Cell 7B, Figure 3.1) and therefore the exclusion of other interventions in the forest (that is, from local people). From this vantage point PFM is viewed as a disturbance in the long-term control of the resource (as we discuss later, local people have a wide variety of species preferences, product mixes and generally shorter, or even no, rotations). Scientific forestry as currently constituted leaves the multiple and diverse uses of forests and how they can be managed out of the frame and focuses on scientifically tested and replicable species choice and practice, largely limited to carefully selected species of commercial timber trees. In practice, timber production does not conform to the general model as it depends on numerous local factors (what foresters call the 'plot'), many of which (such as disease or market conditions) may be indeterminate or beyond the control of the planner. Therefore, the planner is faced with either accepting the limitations of the models on which forest management planning is based, or simplifying the 'real world' to conform to the model (see Cell 7B, Figure 3.1). PFM, on the other hand, is based on a completely different set of scientific assumptions and practice.

Authoritative knowledge and disregard for indigenous forestry

(Cells 7C and 7D in Figure 3.1)

The main elements of indigenous forestry are shifting cultivation (a form of agro-forestry); forest protection with occasional cutting to satisfy particular needs (e.g. house construction, ploughshares, bullock cart construction); dry fuelwood collection; grazing/fodder collection; hunting; protection of water sources and wetlands; transhumance; sacred groves; and knowledge and use of a wide range of NTFPs (including medicinal plants).

Shifting cultivation is an indigenous system that is particularly held in disdain by forestry administrations throughout South and South-East Asia. So long as the rural population was relatively small compared to present numbers, the forest area large and customary tenure secure, shifting cultivation worked well, as the fallow rotation was 15 to 20 years or even longer. However, it has to be understood that even the fallow period of 15 to 20 years was not conducive to forest biodiversity in the areas of forest where it was practised, although as a cultivation system it is far less deleterious than conversion to settled agriculture. Yet, it did not greatly affect the forests overall as the area under shifting cultivation was usually small compared to the total forest of the region. Shifting cultivation was sustainable for local people's agricultural sustenance and agricultural (as opposed to forest) biodiversity. With the doubling of the population every 30 years and the government's taking large portions of shifting cultivation area from the people for reservation and government use (see Chapters 1 and 2), the fallow rotation was drastically reduced, adversely affecting soil and agricultural productivity and the sustenance of the local people. Households needed to enlarge their annual cultivation area, which was not possible because of forest reservation and more families competing for the reduced area. This eroded and degraded many shifting cultivation forest areas and diminished their biodiversity (Ramakrishnan and Misra, 1981).

Around 1915, Dr Brandis introduced a variation of shifting cultivation that included some of the characteristics of classical forestry and shifting cultivation in one system. In Burma, this was called the *taungya* system. Forest villages were established and the villagers were allowed to raise agricultural crops for two years in clear-felled coupes between the lines of forest seedlings. As soon as the crops grew to shade the space between the tree seedlings, the villagers had to discontinue cultivation and move to a similar space in a new plantation. Each villager thus cultivated about 1 acre of space in a new plantation and 1 acre in a two-year-old plantation. In addition, they were granted some cultivation area for sedentary agriculture. This system continued in North Bengal until the 1960s when *taungya* was finally discontinued. The system was contrary to the inappropriate forestry systems developed in Europe and adopted in India, which has a range of tropical forest uses of a totally different nature. *Taungya* was a better proposition, but was not extensively used except in Bengal and eastern Uttar Pradesh. Instead of following up on *taungya* and refining it, the dominance of European forestry principles in Indian forestry finally extinguished it.

Another example of disregarded indigenous practice was the sacred grove. This is a protection management system without any specific provision for using any of the products constituting the grove and could be a very good system of management for biodiversity protection. Unfortunately, the idea was never discussed officially, nor developed or supported by forest departments. On the contrary, the forest departments have sought to include sacred groves as reserve forest and to manage them like any other forest category (see Box 7.2 in Chapter 7 on sacred groves in West Bengal).

A final example of the ignorance and disdain of indigenous knowledge that might have been recognized and facilitated by the forest administration is medicinal plants in Indian forest management systems (including in West Bengal). In fact, in East Bengal (now Bangladesh), Bankura, Midnapore and the Darjeeling Hills, knowledge of medicinal plants and their use in the Ayurvedic system of medicine was (and still is) prevalent. This idea could easily have been developed by the foresters for the management of some NTFPs. The idea was never recognized, although some development of NTFP management did take place (e.g. pine resins in the Uttar Pradesh Hills and the management of *Accessia catechu* for tannin extraction).

Exclusion of local forest users

(Cell 8 in Figure 3.1)

The tendency to exclude local forest users became a 'default position', and one that is logically consistent with the previous seven sub-narratives. It is now one of the key policy imperatives standing in the way of PFM. Many discursive paths lead to this practical result. Over time, with the additional implementation of silvicultural management plans for forests (see Cell 7, Figure 3.1), the forest administration sought to ensure the regeneration of desired species by excluding local people from cutting or grazing in managed areas of standing multi-species timber, as well as in plantations.

All of the foregoing forms a formidable narrative, with internal consistency and historical momentum. However, there are crucial alternatives to almost every aspect of this state narrative, and many of the assumptions and logical links on which it is based must be questioned. While exclusion by the forest administration may be necessary for some types of forest, it has become the general tendency for all types and for increasing areas appropriated by the forest administration.

The 'popular/civil society' narrative

This narrative is a linked set of hypotheses and propositions expressed by a diverse range of local forest users, activists, intellectuals, social movements, NGOs, community-based organizations (CBOs) and federations of forest users, often as members of alliances and discursive coalitions, as the opening section of this chapter has explained. As the narrative unfolds it will be clear that the actors themselves are extremely diverse, and the claims that each make are based on a wide range of factors such as morality, technical efficiency, equity, human rights and other socio-economic and political claims.

For example, the conservationist NGO narrative uses some, but not all, of the sub-narratives of the popular narrative. It is distinct from the state narrative and can be seen as a unique variant of the 'popular' narrative. While it aligns itself more with the state narrative in terms of wildlife conservation/protected area legislation and practice, it parts company with it over such issues as the role of civil society. In Figure 3.2, for instance, Cell 1 would be 'ethical imperatives' for the conservation NGO narrative, extending ideas of 'the public good' from a purely human focus to include those of a wider range of fauna and flora.

We call this narrative 'popular' ('of the people'); but it will be apparent that many of its aspects are produced by a very wide range of actors, some of whom are not 'of the people' at all, but whose own narratives resonate with popular sentiments. As with most popular narratives regarding the environment, they are produced in reaction to what is seen as outside incursions, restrictions and dispossession. They are common throughout the history of the last 300 years in Europe and North America, as much as in South Asia. They are primarily 'populist' in broad political terms (that is, anti-state and anti-big business), and claim to be the voice of the people celebrating local values, traditions and customs. They can also be quite conservative in the sense of endorsing existing inequalities and being reactive against what is seen as a powerful outside enemy, rather than proactive for disadvantaged sections of 'the people'.

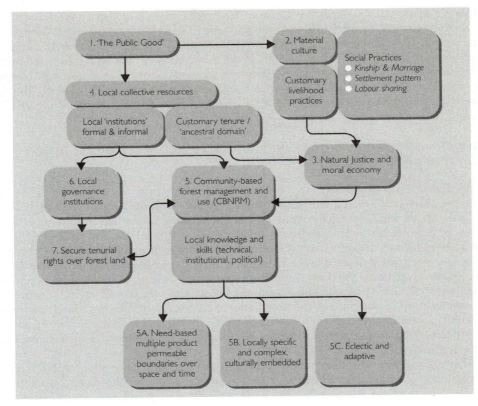

Figure 3.2 The 'popular/civil society' narrative

Source: Piers Blaikie and Oliver Springate-Baginski (original material for this book)

'The public good'

(Cell 1 in Figure 3.2)
The 'public good' refers to the common wealth of the local community and its constituent households, with reference to access to (and management control of) forest resources. The overriding argument for not severing links between people and forest is that the people should maintain their control and rights to use the forest in order to satisfy their welfare. Popular narratives about the 'public good' were articulated in response to other actors seeking to take over the forest. Three strands of popular narrative have been articulated as 'annexationist', 'pragmatist' and 'populist' (Guha, 1989).

'Material culture', customary livelihood and social practices

(Cell 2 in Figure 3.2)
The popular narrative of prompting local management of forests is not only based on technical production decisions about trees, but is also an inseparable part of local culture. For example, *podu, jhum, khoria phadani* (the main shifting cultivation systems in the region), agro-pastoral transhumance systems, forest protection and indigenous technical knowledge

(ITK) are all intimately linked to patterns of settlement, kinship and marriage, labour-sharing and inheritance. Also embedded in the customary use of forests is the moral economy of local people, by which is meant reciprocal relations, gift exchange, security nets and assurances between households. These notions of the moral economy are particularly relevant in the context of PFM, where the state assumes management control of the forest but abrogates any responsibility for ensuring alternative livelihoods for forest users, especially the very poor, now that the resource itself has been detached from the culture that nurtured and depended on it.

'Natural justice' and moral economy; 'local collective resources', local institutions and customary tenure

(Cells 3 and 4 in Figure 3.2)
Until the 1970s (and revived during the 1990s and 2000s with the Tribal Forest Rights Bill), the main actors who produced narratives based on natural justice were forest users themselves, in representations to forest authorities and courts, social movements and agitations and activism, as well as politicians promoting independence movements. The central core of these narratives was resistance to the abrogation of rights of use and management of forests and, in less severe cases, the downgrading of these into 'privileges'. The social and economic privations of excluded populations have been historically, and remain today, very serious. For example, up to one-third of the population of Kumaon in the Indian Himalayas had to emigrate as a result of British exclusionary policies (Tucker, 1984). The popular response to the state's seeking to take control of forests was to defend the legitimacy of rights in the name of natural justice.

The popular narrative rests on the assumption that the local is virtuous, supports collective action (e.g. communal labour such as the *parma* system in Nepal) and underpins a benign version of the moral economy, whereby individual greed and competitiveness is tempered by the moral imperative of the collective good (Kitching, 1982). An assertion of the right to the control and management of the forest, which represents an ancestral domain, is in direct conflict with the principle of eminent domain discussed above. It is a moral claim, and is made when the distinction between how the world is and how it ought to be becomes painfully apparent. Here, it is the exclusion of local people from the forest by the state (and, in the past, by local feudal rulers). A distinction can be drawn between two notions of morality. The first is a generalized morality that includes 'justice', which is generally the level at which political representation of forest issues is pitched at the state level, in party politics, and in negotiations between government and senior members of civil society. The second is a 'thick' local morality that expresses specific cultural meanings, rules and practice (Walzer, 1994). These local moralities underpin many of the local management practices, exclusions and inclusions, who gets what from the forest, and how adaptations are made to new incursions on the forest from outside. Moral authority becomes part of the new forest politics under PFM.

Community-based forest management and use, and local knowledge and skills

(Cell 5 in Figure 3.2)
From the 1980s onwards, the popular narratives were elaborated on by international academics in conjunction with a set of persuasive theories centring on community-based natural resource management theory (CBNRM). This group of closely linked theories was applied to a wide set of management issues, including forests, pastures, wildlife, and channel and tank irrigation. A range of theoretical benefits of CBNRM was elaborated on in academic writing and taken up by donors – for example, the Ford Foundation in India and the UK

Department for International Development (DFID) in India and Nepal. These benefits claimed to include:

- a pro-poor safety net;
- efficient management decisions that cope well with local ecological specificities and complexities (see Cell 5B in Figure 3.2);
- new institutional economics and public choice theory, which hypothesized that secure tenure (see Cell 7 in Figure 3.2) and mutual assurances of trust would be able to internalize externalities and secure self-regulated extraction;
- palliation of open-access problems (sometimes caused by the destruction of CBNRM and customary tenure by the state; see above and Cell 7 in Figure 3.2);
- transparency of management systems (it is easy to monitor what is going on since it is local and to obtain face-to-face recourse);
- capacity-building for good governance, democratic functioning of forest user groups (FUGs), conflict management, and FUGs as a model of real democratic institutions (see Cell 6 in Figure 3.2); and
- PFM as an entry point for rural development, community development and rural sustainable livelihoods (see Jodha, in Blaikie and Brookfield, 1987; Wade, 1988; Berkes, 1989; Ostrom, 1990; Bromley, 1992; Baland and Platteau, 1996; Brown, 1999; Adams and Hulme, 2001; Agrawal, 2001; Agrawal and Gibson, 2001).

CBNRM was promoted on the basis of two seemingly supportive sets of propositions; but on closer examination these have very different and fundamentally opposing ideological roots. The first was a neo-liberal critique of state engagement in environmental management (see the *World Development Report 1992: Development and the Environment*, World Bank, 1992), where properly functioning markets, clearly defined property rights and accurate economic information on the consequences of environmental decisions were promoted in order to overcome the worst excesses of expensive, often corrupt, non-accountable and inflexible bureaucracies. However, this neo-liberal strand of thinking had only very limited shared objectives with the second, 'popular', narrative. The latter bases its claim on (to take a brief listing as an illustration) equity, natural justice, a right to the means to secure a livelihood, and issues of community solidarity that private or state appropriation have fragmented.

There has been an upsurge of CBNRM initiatives around the world over recent decades, accompanied by a large literature that reflects the wide appeal of this approach. CBNRM embraces not just forest management, but also watersheds, soil conservation, biodiversity protection, artisanal fisheries, rangelands and wildlife. CBNRM in forestry in India and Nepal (and worldwide) has been established for a very long time, and the history of colonial forestry (after 1947, the Indian Forest Service and, to a lesser extent, the Nepalese administrations concerned with forests) has pursued policies that have encroached on and undermined these institutions. Historically, CBNRM has also been widespread in both Europe and the US, and in spite of the onslaught of commercialization and privatization, present-day examples can still be found, although they operate on a diminished scale and are often in decline. However, they are still widespread outside Europe and North America and, in a sense, have been rediscovered by policy-makers, NGOs and aid donors in Africa and Asia. The CBNRM thinking holds that local 'communities' should be mobilized to manage their local resources under a regulated management regime that has already been developed by local people. It constitutes the third major property regime after private property and state property. Many advantages, it is claimed, accrue to CBNRM institutions regarding the sustainability of the natural resource and the well-being of the people who manage it. However, it is also widely acknowledged that CBNRM institutions are increasingly meeting challenges from both state and civil society. These are political and economic in nature and include commercialization, growing inequality (internal and external to communities), population pressure causing overuse of common property, and a breakdown in solidarity and local authority for a variety of reasons. The state, too, has failed to recognize that resources are *already* being managed by local people under CBNRM, and that there are rules and regulations which have evolved and been proven by trial and error (Kothari et al, 1998).

The CBNRM model appeals to a wide variety of actors, and it is claimed that benefits to people and the environment alike appear in many forest narratives. Governments, for a different set of reasons, may be attracted to CBNRM as a low-cost means of achieving management objectives and international treaty obligations without fundamental structural change. CBNRM narratives have also been a highly influential rallying point for grassroots activists and advocacy groups. For instance, Narain and Agarwal (1989, pviii) say:

> *All rural settlements must have an active institution which has legal control over its immediate environment and access to funds. The role of the government must be that of an enabler of village-level planning and action rather than that of a doer.*

Despite this idealized conception, in practice, CBNRM in forestry, as in other sectors, has often failed to fulfil its promise. A number of explanations are offered for this. First, responsibilities are devolved by the state without power or rights (where the state has implicitly nationalized the CBNRM in the first place). Second, the 'community' as a homogenous group very seldom exists because of socio-economic differentiation, contradictory understandings of the resource and its uses, the breakdown of customary authority, in-migration and the illegal use of the resource by 'strangers' or free-riders; or it is beset by elites who corner common benefits. Third, there may be insufficient livelihood incentives because the resource in question is encroached on, privatized or significantly degraded in the first place.

However, in spite of these criticisms the popular narrative is able to form a discursive alliance with a whole range of non-local, national and global networks. These are brought to bear on the issue in PFM in India and Nepal by national intellectuals, authors and activists through reference to a huge international literature and many influential consultants and staff members of big international non-governmental organizations (BINGOs) and IFIs.

Another closely related aspect of the popular narrative concerning community-based forest management and use is local knowledge and skills appropriate to the material uses of the forest. These often imply sustainable management of multiple products that are managed by the organization of permeable boundaries in space and time (see Cell 5A in Figure 3.2). This means that specified users (e.g. women only, charcoal burners, artisanal bamboo workers and traditional medicine practitioners) can cross into areas of forest at certain times for certain products, but are banned from doing so at other times (see Cell 5B in Figure 3.2). This type of boundary can be operated in a way that is highly sensitive to seasonality, rates of usage of different products at different times and harvesting techniques, and can be adapted to unforeseen circumstances such as drought (see Cell 5C in Figure 3.2; see Messerschmidt, 1986, for an account of the complexities of space–time organization of management in an area of the middle hills of Nepal). Centralized control of forests, on the other hand, encourages a simplified treatment of boundaries with exclusion of *all* users for *all* harvesting at *all* times.

Local governance institutions and secure tenurial rights over forest land

(Cells 6 and 7 in Figure 3.2)
An important (but, in practice, problematic) part of the popular forestry narrative is the crucial role of local governance. There are two aspects to this part of the narrative. The first concerns the long-established institutions discussed above. These have had a variety of property rights

attached, often to a local 'feudal' landlord, sometimes to the state, which had certain tax-raising rights. 'Participation' and, more explicitly, joint forest management, take as their starting point a completely different set of tenurial rights. CBNRM institutions just managed the forest (even if it may have been inegalitarian) – they did not 'participate', they just *did* it. Therefore, the linkage between CBNRM and long-established self-management, on the one hand, and participation and JFM, on the other, has to negotiate the historical fact of long-term encroachment and dispossession. The key element of this process is the abrogation of tenurial rights, both legal and illegal (see Cell 7 in Figure 3.2), as Chapter 1 has discussed in detail.

Decentralization of local government shares some, but far from all, of its underlying assumptions with PFM. Decentralization of governance during the 1990s in India and Nepal sought to build more robust and representative governance structures, introducing *panchayati raj* in India and the village development committees (VDCs) and district development committees (DDCs) in Nepal for the planning and coordination of service delivery structures. However, as the discussion above has shown, the movement towards local self-governance has also brought problems for PFM of accountability, coordination and rights of revenue collection and overall management.

The long history of decentralization has been picked up by IFIs in different forms, particularly in Africa, but also elsewhere. In general terms, the advantages of decentralization, not only of forest providers but also of all other service providers (e.g. public health, water and sanitation) are co-opted by the case for participatory forest management. While these advantages are more general and more widely applicable to other sectors than those deriving from CBNRM, they have, nonetheless, been used as theoretical justification for local institutions to assume an increasing role.

There are budgetary and strategic motives for decentralized governance, of which the passing of the burden of management and policing from official to local institutions is the most important. Also, paradoxically, decentralization may bring surveillance and control by the centre *closer* to local people and their forests. Devolving 'funds, functions and functionaries' and discretionary powers to local government may increase democratic redress or, on the other hand, extend the reach of the state. Ribot (2002) explains:

> Political *or* democratic decentralization *occurs when powers or resources are transferred to authorities representative of and downwardly accountable to local populations. Democratic decentralization aims to increase popular participation in local decision-making. Democratic decentralization is an institutionalized form of the participatory approach.*

Responses of the state forest administrations

The account of the classic state narrative of forest policy illustrates its consistency and power; but the popular narrative has clearly been increasingly well established and publicized among a growing national and international network, at least in rhetorical terms. To answer the challenges, the forest administrations, both institutionally and as individuals, have deployed various direct counterarguments, such as:

- PFM poses radical challenges and the necessary long-term transition cannot be rushed.
- Participation would result in more problems than it would solve, and does not lead to our well-established goals. Indeed, it leads away from them.
- The forest administration does not have (and *should not have*) the training to do social engineering and negotiation, rather than actual technical management of forests based on proper scientific principles.

- The forest administration has neither the resources nor the capacity to engage in the extensive and time-consuming negotiations necessary for PFM in many remote locations.
- If the resources made available for PFM were accessible to traditional forest management, the results would be much better.
- The forest service needs a strong, well-organized and centralized bureaucracy in order to deliver its historic mission, and PFM dissipates these essential characteristics.

There are many instances where these counterarguments are very persuasive; but essentially they present the deep-seated structural and discursive challenges presented by PFM. The second way in which the forest administrations have responded is to finesse the contradictions between PFM and the state-centred, classic model, rather than to argue and confront directly. As Chapter 2 has shown, forest administrations are facing increasingly powerful international coalitions, strong political responses from both civil society and within government (including members of forest departments themselves) and, most significantly, from the resistance of local forest users. To accede to these demands would require fundamental transformations away from a regulatory body and towards one more akin to a farmer-centred agricultural extension service. PFM requires such a radical transformation in the forest services of India (and, to a lesser extent, Nepal) because every aspect of the state narrative has a corresponding set of precedents and current practices. PFM is contradictory to most aspects of the state narrative and to corresponding practices.

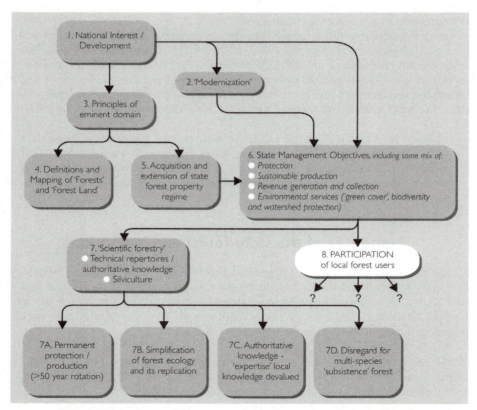

Figure 3.3 State forestry narratives after the 1980s: The participatory model

Source: Piers Blaikie and Oliver Springate-Baginski (original material for this book)

The main reactive strategy of both Indian and Nepalese forest administrations in reconciling their agendas with the popular narrative has been to make rhetorical adjustments to their narratives. 'Participation' is added to statements of intent, although the extent to which it is linked in a logical or practical sense tends to be much more limited since it remains at odds with many other aspects of the narrative. Figure 3.3 illustrates how 'participation' is disconnected in a discursive sense. The figure is simply a reproduction of Figure 2.1, with the isolated addition of 'participation' (Cell 8 in Figure 3.3). It may be an interesting exercise for the reader to attempt to make logically consistent links between the classic narrative (mapped in Figure 3.1) and PFM. Each attempt to make such linkages is beset with contradictions, both discursive (discussed in this chapter) and structural (discussed in Chapter 2).

Thus, the forest administration of India has been pressured to introduce an inconsistent new component to its classic narrative, which seems to admit the need for a more participatory and flexible form of forest management, and, therefore, superficially concedes some ground to popular arguments. PFM, *in practice*, contradicts its own classic narrative and associated practices, and in this sense it is argued that JFM in India is in many ways a means by which concessions are made to the popular narrative and where real changes are set out and practised. However, the contradictions explained above have shaped the way in which JFM is practised today. How JFM is actually practised in the field (the right-hand side of policy process Figure 2.1 in Chapter 2) turns out to be *more consistent with the classic than the popular forest management model*. Hardly surprising! This deep contradiction explains the ambiguity and the rhetoric – but not the practice – of PFM. The reality is that the letter and not the spirit of PFM regulations is followed – and even this is done in a patchy and selective way. As a result, there is a great deal of confusion and difficulty for the forest administration at most levels. It is worth repeating that individuals have a degree of discretion and the latitude to move towards a more participatory practice, but that the dominant narrative, set of practices and peer judgement of an individual's performance make it difficult for the individual to become 'too participatory'. Under existing conditions of employment, PFM is often under-resourced, very hard work, not particularly good for promotion, risky and uncertain, demands technical and personal skills, and requires procedures that are confusing, non-existent or contradictory – not an attractive prospect for many forest department employees.

In Nepal, by contrast, where the wide-scale implementation of PFM as community forestry coincided with democratization of the polity in the early 1990s, much more fundamental reforms to the narrative have occurred, at least in hill areas where the 'classical' narrative was hardly implemented prior to this. Nonetheless, many of the contradictions discussed in the Indian case also apply (even if less strongly) in Nepal.

Participation: From narrative to practicalities

PFM means that local people take part in decisions and actions about forest management and have access to, and rights to collect, forest products. There are other issues concerning equity between local forest users and the relative shares of forest produce which the state and local people should have, and these are discussed in Part III. A list of the practicalities of PFM includes the following:

- *constitutional decisions*, such as deciding on organizational and political forms for the management of forests (deciding on the appropriate institution, boundaries, definitions of 'forest', legitimate members of a group and institutional rules).
- *designs concerning rights of use and ownership of forest products* involving the sharing of these between the forest administration and local groups, and between individuals within the groups themselves;

- *decisions about forest management*, such as objective-setting and management planning, including the application of local knowledge and contributing to the production of knowledge about forests and forest management (i.e. research in the same way that farming systems research moved from centralized state farms with farmers participating at the state farm to on-farm and farmer-to-farmer research; here one needs to distinguish between objectives being set by local people or their having to confine themselves to the non-negotiable objectives determined by the forest department); and
- *forest management activities*, such as planting, thinning, pruning, harvesting and protecting (the key is whether local people are empowered to do this on their own according to mutually agreed plans or merely as labour working under forest department supervision, as in the case of JFM);

One of the key results of all four aspects of participation listed above is their impact on livelihoods. One hypothesis is that the impact *should* be beneficial since a participatory process of agreeing and implementing a working plan by a broad coalition of forest users will reflect subsistence needs and the opportunity for small-scale sales of forest products by poorer people. But a fundamental problem with the practice of 'participation' is that it tends to disregard existing inequalities in the agrarian political economy. Our studies on gender and equity impacts revealed serious intra- (and inter-) village and intra- (and inter-) gender differences. These issues are highlighted in Part II.

Conclusions

- The state forestry narrative is coherent, powerful, multi-stranded and conforms largely to its existing practice, but is contradictory in most of the aspects important to PFM.
- The popular discourse, particularly PFM, presents significant challenges for forest administrations, resulting in various responses, from partial, rhetorical adoption of PFM and 'riding out' the difficulties, selective adoption, (picking and choosing; diluting; target-chasing to the letter, but not in spirit; symbolic reform and changed practice; but mostly 'business as usual' and foot-dragging), or vociferous resistance to 'repel and roll back' the challenges. Examples of all these strategies are illustrated throughout this book.
- There are a number of points of negotiation in discourse and in practice where adaptations are possible so that most of the actors involved can feel that they have won significant ground. But other aspects are more structural in nature, such as referring to or dealing with resources, training and the production of forest knowledge, which will have to be tackled in the longer term through a wide range of strategies and alternative routes that may lead round, rather than through, forest administrations.

References

Adams, W. M. and Hulme, D. (2001) 'If community conservation is the answer in Africa, what is the question?', *Oryx*, vol 35, no 3, pp193–200

Agrawal, A. (2001) 'Common property institutions and sustainable governance of resources', *World Development*, vol 29, no 10, pp1649–1672

Agrawal, A. and Gibson, C. (2001) *Communities and the Environment: Ethnicity, Gender, and the State in Community-Based Conservation*, New Brunswick, New Jersey and London, Rutgers University Press

AISLUS (All India Soil and Land Use Survey) (1997) *Information Digest of All India Soil and Land Use Survey*, New Delhi, AISLUS

Apthorpe, R. and Gasper, D. (1996) *Arguing Development Policy: Frames and Discourses*, London, Frank Cass

Baden-Powell, B. H. (1882) *A Manual of Jurisprudence for Forest Officers*, Calcutta, Government Press

Baland, J. M. and Platteau, J. P. (1996) *Halting Degradation of Natural Resources: Is there a Role for Rural Communities?*, Rome, Food and Agriculture Organization, and Oxford, Oxford University Press

Banerji, A. (1996) *Joint Forest Management: The Haryana Experience*, Environment and Development Book Series, Ahmedabad, Centre for Environment Education

Bardhan, P. (1984) *The Political Economy of Development in India*, New Delhi, Oxford University Press

Batterbury, S., Forsyth, T. and Thompson, K. (1997) 'Environmental transformation in developing countries: Hybrid knowledge and democratic policy', *Geographical Journal*, vol 163, no 2, pp26–132

Berkes, F. (ed) (1989) *Common Property Resources: Ecology and Community-Based Sustainable Development*, London, Belhaven

Blaikie, P. M. and Brookfield, H. C. (1987) *Land Degradation and Society*, Methuen, London.

Blaikie, P. M. and Muldavin, J. (2004) 'Policy as warrant: Environment and development in the Himalayan region – working papers', *East–West Center*, no 59, April, Manoa, University of Hawaii

Brandis, D. (1883) *Suggestions Regarding Forest Administration in Madras Presidency*, Madras, Government Press, Madras

Brandis, D. (1897, reprinted 1994) *Forestry in India*, Dehradun, Natraj Publishers, reprint by WWF India

Bromley, D. W. (ed) (1992) *Making the Commons Work: Theory, Practice and Policy*, San Francisco, Institute for Contemporary Studies Press

Brown, D. (1999) *Principles and Practice of Forest Co-Management: Evidence from West-Central Africa*, EU Tropical Forestry Programme Papers, London, ODI

Bryant, R. L. (1997) *The Political Ecology of Forestry in Burma 1824–1994*, London, Hurst and Company

Campbell, J. Y. (1995) 'Evolving forest management systems to people's needs', in Roy, S. B. (ed) *Enabling Environment for Joint Forest Management*, Delhi, Inter India Publications

Clapp, R. A. (1995) 'Creating competitive advantage: Forest policy as industrial policy in Chile', *Economic Geography*, vol 71, no 3, pp273–296

Clegg, S. R., Hardy, C. and Nord, W. R. (eds) (1996) *Handbook of Organization Studies*, London, Sage Publications

Commander, S. (1986) *Managing Indian Forests: A Case for the Reform of Property Rights*, Network Paper 3, October, ODI Social Forestry Network, London, ODI

Court, J. and Young, J. (2004) *Research and Policy in Development Programme (RAPID) Briefing*, Paper No 1, October, London, ODI

DFID (UK Department for International Development) (2003) *Promoting Institutional and Organizational Development: A Sourcebook of Tools and. Techniques*, London, DFID

Farooqui, A. (1997) *Colonial Forest Policy in Uttarakhand: 1890–1982*, New Delhi, Kitab Publishing House

Forsyth, T. (1998) 'Mountain myths revisited: Integrating natural and social environmental science', *Mountain Research and Development*, vol 18, no 2, pp126–139

Forsyth, T. (2003) *Critical Political Ecology: The Politics of Environmental Science*, London, Routledge

FSI (Forest Survey of India) (1999) *State of the Forest Report*, Dehra Dun, FSI

FSI (2005) *State of the Forest Report 2003*, Dehra Dun, FSI

Gilmour, D. A. and Fisher, R. J. (1991) *Villagers, Forests and Foresters: The Philosophy, Process and Practice of Community Forestry in Nepal*, Kathmandu, Sahayogi Printing Press

GoWB (Government of West Bengal) (2001) *State Forest Report*, West Bengal, GoWB

Guha, R. (1983) 'Forestry in British and post-British India: A historical analysis in two parts', *Economic and Political Weekly*, vol 18, no 45/vol 18, no 46, pp1882–1947

Guha, R. (1989) *The Unquiet Woods: Ecological Change and Peasant Rebellion in the Himalaya*, Berkeley, University of California Press

Hajer, M. A. (1995) *The Politics of Environmental Discourse: Ecological Modernization and the Policy Process*, Oxford, Clarendon Press

Hobley, M. and Malla, Y. B. (1996) 'From the forests to forestry – the three ages of forestry in Nepal: Privatization, nationalization, and populism' in Hobley, M. (ed) *Participatory Forestry: The Process of Change in India and Nepal*, London, ODI, pp65–82

Joshi, A. L. (1993) 'Effects on administration of changed forest policies in Nepal', in Warner, K. and Wood, H. (eds) *Policy and Legislation in Community Forestry: Proceedings of a Workshop*, Bangkok, Regional Community Forestry Training Centre (ECOFTC)

Keeley, J. and Scoones, I. (1999) *Understanding Environmental Policy Processes: A Review*, IDS Working Paper 89, Brighton, Institute of Development Studies, University of Sussex

Kitching, G. (1982) *Development and Underdevelopment in Historical Perspective: Populism, Nationalism and Industrialisation*, London, Methuen

Kothari, A., Pathak, N., Anuadha, R. V. and Taneja, B. (1998) *Communities and Conservation: Natural Resource Management in South and Central Asia*, New Delhi, Sage Publications

Leach, M., Mearns, R. and Scoones, I. (1997) *Environmental Entitlements: A Framework for Understanding the Institutional Dynamics of Environmental Change*, IDS Discussion Paper 359, Brighton, Institute of Development Studies, University of Sussex

Long, N. and van der Ploeg, J. D. (1989) 'Demythologizing planned intervention: An actor perspective', *Wageningen Studies in Sociology*, vol 29, no 3/4, pp226–249

Mehta, L., Leach, M., Newell, P., Scoones, I., Sivaramakrishnan, K. and Way, S. (1999) *Exploring Understandings of Institutions and Uncertainty: New Directions in Natural Resource Management*, IDS Discussion Paper 372, Brighton, Institute of Development Studies, University of Sussex

Messerschmidt, D. A. (1986) 'People and resources in Nepal: Customary resource management systems of the Upper Kali Gandaki', in *Proceedings of the Conference on Common Property Resource Management*, 21–26 April 1985, Washington, DC, National Academy Press, pp455-480

Narain, S. and Agarwal, A. (1989) *Towards Green Villages: A Strategy for Environmentally Sound and Participatory Rural Development*, Delhi, Centre for Science and Environment

NPC (National Planning Commission) (2003) *The Tenth Five Year Plan*, Kathmandu, NPC

Ostrom, E. (1990) *Governing the Commons: The Evolution of Institutions for Collective Action*, Cambridge, Cambridge University Press

Pathak, A. (1994) *Contested Domains: The State, Peasants and Forests in Contemporary India*, New Delhi, Sage Publications

Punjab Erosion Committee (1931) Report, Punjab Erosion Committee, Government Press, Lahore, India

Rackham, O. (1990) *Trees and Woodland in the British Landscape*, (London, J. M. Dent and Sons, cited in Rietbergen, S. (2001) 'The history and impact of forest management', in Evans, J. (ed) *The Forests Handbook*, Abingdon, Blackwell Scientific

Ramakrishan, P. S and Misra, B. K. (1981) 'Population dynamics of *Eupatorium adenophorum spreng* during secondary succession after *jhum* and burn cultivation in north-eastern India', *Weed Resources*, vol 22, pp77–84

Ray, P. K. (1964) 'Kalimpong Forest Division: Past, present and future' in GoWB (1964) *Centenary Commemoration Volume*, Calcutta, Planning and Statistical Cell, West Bengal Forest Department

Regmi, M. C. (1978) *Land Tenure and Taxation in Nepal*, Kathmandu, Ratna Pustak Bhandar

Ribot, J. (2002) *Democratic Decentralization of Natural Resources: Institutionalizing Popular Participation*, Washington, DC, World Resources Institute

Roe, E. (1994) *Narrative Policy Analysis: Theory and Practice*, Durham, US, Duke University Press

Saxena, N. C. (1994) *India's Eucalyptus Craze: The God that Failed*, New Delhi, Sage Publications

Scott, J. (1977) *Moral Economy of the Peasant: Rebellion and Subsistence in Southeast Asia*, New Haven, Yale University Press

Scott, J. (1985) *Weapons of the Weak: Everyday Forms of Peasant Resistance*, New Haven, Yale University Press

Sen, G. (1992) *Indigenous Vision: Peoples of India, Attitudes to the Environment*, New Delhi, Sage Publications

Shankland, A. (2000) *Analysing Policy for Sustainable Livelihoods*, IDS Research Report 49, Brighton, Institute of Development Studies, University of Sussex

Shrestha, N. K. (2000) 'Turning back the clock: Where is community forestry heading in Nepal?', *Asia-Pacific Community Forestry Newsletter*, vol 13, no 2, December, pp53–55

Sutton, R. (1999) *The Policy Process: An Overview*, Working Paper No 118, London, ODI

Tucker, R. P. (1984) 'The historical roots of social forestry in the Kumaon Himalayas', *Journal of Developing Areas*, vol 13, no 3, pp341–356

Van Den Hoven, J. and Shrestha, S. (1998) *Distant Users Survey – Report*, Lahan, Nepal, Churia Forest Development Project

Wade, R. (1988) 'The management of irrigation systems: How to evoke trust and avoid prisoners' dilemma', *World Development*, vol 16, no 4, pp489–500

Walzer, M. (1994) *Thick and Thin: Moral Argument at Home and Abroad*, London, University of Notre Dame Press

World Bank (1992) *World Development Report 1992: Development and the Environment*, Oxford, Oxford University Press

World Bank (2005) *Preliminary Analysis Based on Nepal Living Standard Survey Report 2004*, Kathmandu, National Seminar, 11–12 May 2005

4

Understanding the Diversity of Participatory Forest Management Livelihood and Poverty Impacts

Oliver Springate-Baginski and Piers Blaikie

Livelihoods and the popular narrative

Forest resources continue to form a significant part of many local people's livelihoods in South Asia. However, livelihood linkages with forest land have widely been restricted or curtailed by state forest administrations (see Chapter 1). Thus, their resumption and protection is a central argument of the popular narrative, as Chapter 3 has indicated. If a substantial proportion of agrarian society relies upon forest land to a significant degree, and that reliance is compromised by the forest administration's policies and practices, then the argument that they should be involved in the management of that forest land is persuasive. Second, if people can be shown to be able to use the forest in a sustainable way through collective action, the state forestry narrative has to counter this claim if it is to maintain its credibility. It would need to make the case that its policies have a more pressing goal (such as green cover, wildlife/biodiversity conservation, large-scale plantations, the promotion of commercial timber over other forest products) that should be prioritized 'in the national interest'. Furthermore, it would need to demonstrate that forest policy is a zero-sum game in the sense that people's participation in forest management is gained exactly to the extent that other goals of the forest administration are lost. However, if livelihood links between local people and the forest are *not* significant, and the severance of rights of access by the state or via assertion of private property rights has *not* led to injustice, hardship and poverty, then the argument that the state should manage the forest according to the 'national interest' is carried.

This chapter sets out the framework for identifying the possible impacts of forest policy change towards either a more conservative and exclusionary direction or a more popular and participatory direction. This book has participatory (PFM) as its main focus, and therefore most attention is paid to the impacts of joint forest management (JFM) in India and community forestry (CF) in Nepal. However, it is useful to pay attention to aspects of forest policy that go against participation. The historical aspects have already been discussed in Chapters 1 and 2. Nevertheless, this book claims that there are aspects of current PFM policy that also limit participation in forest management. In order to identify these, it is necessary to have a more detailed model of livelihood use of forest lands. The points of engagement between policy change and the livelihood process are not only concerned with access to material forest products in a direct manner, although this is most important. Any policy works through an existing political ecology on the ground. Rural elites and local forest administration staff can interpret forest policy in particular ways. Some treat any change in forest policy as an entrepreneurial opportunity. There are many significant aspects of forest policy changes, including not only altered patterns of access to forest products, but also wage labour opportunities under JFM, changed marketing arrangements for non-timber forest products (NTFPs) and the creation of funds managed by user groups. The effectiveness of collective action can also become important in altering the pattern of livelihood activities.

The framework outlined in this chapter provides the basis for the empirical research in Part II. A brief explanation of the framework is provided, followed by the potential impacts that a change in forest policy may have on livelihoods. These impacts are set up here as hypotheses to be tested in the empirical research in Part II.

Livelihood systems and forest use in South Asia

Livelihood analysis became popular during the 1990s as a method of analysing local households' socio-economic circumstances. The basic concept of livelihood analysis is that households draw down and allocate a range of collective and private assets for use in activities that generate income streams for consumption, inputs and saving or investment. Carney (1998) provides a definition of livelihood based on the work of Chambers and Conway (1992), as follows:

> *A livelihood comprises the capabilities, assets (including both material and social resources) and activities required for a means of living. A livelihood is sustainable when it can cope with and recover from stresses and shocks and maintain or enhance its capabilities and assets both now and in the future, while not undermining the natural resource base.* (Carney, 1998, p4)

There is a wide variety of local livelihood–forest links across the different political ecologies of South Asia, reflecting the diverse range of socio-economic and ecological circumstances. Households use forests for a number of products and services: domestic needs such as fuelwood and construction timber; agricultural inputs; grazing and fodder; water; artisanal inputs; NTFP collection; processing and sale; and labouring opportunities.

The livelihood importance of forests and trees is closely interlinked with their cultural significance. The cultural importance of forests, trees and tree products has been historically emphasized in extensive folklore myths and spiritual and religious practices across South Asia. Authors such as Croll have emphasized the importance of the cultural imagination of tribal groups in the creative management and use of forests, as well as other environmental resources (Croll and Parkin, 1992). Sacred groves are perhaps the most evident manifestation of how cultural practices intersect with livelihood-related forest resource management. For millennia, biodiversity has supported the livelihoods and life of the people of India, shaping a diversity of cultures in which respect for nature and its myriad life forms has enjoyed a central place. Animals and plants have been revered – often worshipped – and many forests, rivers, mountains and lakes have been seen as abodes of the gods. The tradition of protecting patches of forests, dedicated to deities and/or ancestral spirits, as sacred groves by many Indian communities means that many of the sacred groves still provide a safe refuge to several endangered and threatened species of flora and fauna (Malhotra et al, 2001). They are also a nursery and storehouse of ayurvedic, tribal and folk medicine, and help in soil conservation and in nutrient cycling. Many of the sacred groves harbour water resources in the form of springs and ponds, which act as recharge for aquifers. However, threats from commercial forestry, infrastructure projects such as dams, railroads, highways, as well as cultural and economic changes in the communities themselves, have led to the weakening and destruction of these ancient practices.

> Banashankari Jatra *near the famous Badami town is a fair around the celebrated temple of Banashankari, the Goddess of the Forest. For three days the entire town turns green with hundreds of varieties of leaves and flowers decorating the town and the temple.* (Satheesh, 2000; Kalpavriksh, 2003, pp37–43)

Most rural populations, as well as many urban and peri-urban populations, depend to some extent on the use of forest products and services. In India, it has been estimated that of a total population of over 1 billion, an estimated 147 million villagers live in and around forests (FSI, 2000) and there are another 275 million for whom forests constitute an important source of livelihood (Bajaj, 2001). Gathering of fuelwood, fodder and NTFPs is an important subsistence and economic activity for poor women, and about 60 to 70 per cent of the gatherers are women (Gera, 2001; Kalpavriksh, 2003, p55). A slightly lower estimate has been suggested by the World Bank. By its assessment, over 100 million people are directly dependent on forests for their livelihoods, and in India a further 175 million are significantly dependent (World Bank, 2006). Although there are no estimates for Nepal, the proportion of the population that is forest-dependent is likely to be higher, with almost all of the 85.8 per cent of the population which is rural (Central Bureau of Statistics, 2003) depending to a great extent on forests for essential products – fuelwood, grazing and fodder, house timber and poles, and so on. Peri-urban areas also depend heavily on forests, particularly for fuel and timber, with an estimated 90 per cent of Nepal's fuel needs being supplied from forests (Central Bureau of Statistics, 2003). On a worldwide scale, the World Bank (2001, p14) has estimated that:

> *More than 1.6 billion people depend to varying degrees on forests for their livelihoods. About 60 million indigenous people are almost wholly dependent on forests. Some 350 million people who live within or adjacent to dense forest depend on them to a high degree for subsistence and income.*

These estimates must be treated with circumspection, however, as robust statistical sources to support them do not exist (Byron and Arnold, 1999; Angelsen and Wunder, 2003). The definition of 'dependence' is so complex and open to interpretation that the net can be cast as wide or as narrow as the user wishes. Although use of forests usually forms only one among a bundle of a household's livelihood activities, some households – for instance, shifting cultivators depending on forest land for subsistence food production – may have a very high level of dependence on forests. Other households such as settled agriculturalists who may be depending on forests for complementary or supplementary grazing, domestic inputs and, perhaps, NTFP collection for sale may have a more moderate level of forest dependence. Finally, some households may only have a low level of forest dependence; for instance, wealthy households may depend on the forest only for environmental services such as hydrological moderation to ensure that irrigation water continues in the hot season.

In order to clarify the definition of 'dependence', Table 4.1 (based on Angelsen and Wunder, 2003) categorizes the types of forest dependence of different groups.

For most rural households, forests provide basic domestic needs. The primary *fuel* for cooking and heating continues to be fuelwood, and as most households lack sufficient on-farm trees, this is likely to be sourced from local forests. For most urban and peri-urban areas, purchased fuelwood has been the prevalent cooking and heating fuel until very recently, although this situation is rapidly changing in South Asia with the increasing availability of low-cost (often subsidized) liquid petroleum gas for domestic use. Continuing

Table 4.1 *Importance of different forest benefits to different groups*

User groups	Types of economic benefits			
	Agricultural land and nutrients	Non-timber forest products (NTFPs)	Timber	Onsite ecological services
1 Forest dwellers				
i Hunters and gatherers	Minor benefits	Main benefit	Supplementary if transport access exists	Variable
ii Shifting cultivators	Main benefit	Important supplement	As above	Variable
2 Farmers living adjacent to forests				
i Smallholders	Major 'land reserve'	Supplementary	Supplementary if transport access exists	Variable
ii Landless	Not important	Important supplement	As above	Variable
3 Commercial users				
i Artisans, traders and small entrepreneurs	None	Important	Important	None
ii Employees in forest industries	None	Supplementary	Main benefit	None
4 Consumers of forest products				
i Urban poor and others	None	Some	Variable	None

Source: adapted from Angelsen and Wunder (2003)

urban fuelwood demand provides a ready market for rural fuelwood collectors. Forest and tree products are also essential to household maintenance in terms of providing timber and poles for house construction, and thatch for roofing and fencing.

Agriculture has historically been the mainstay of the rural economy in South Asia, contributing to households' food self-sufficiency and, for more productive landholdings, surplus for trade. Land generally remains the most important productive household asset, and access to it is a fundamental determinant of the economic status of most rural households. Agricultural land is generally cleared and privatized, and in hill areas, in particular, the boundary between forests and agriculture remains a fluctuating one, depending on labour availability and farming techniques. In many tribal areas, long-rotation forest fallows cultivation persists, although with uncertainty over tenure the rotation periods are diminishing in many places and settled agriculture is taking their place. Land-poor cultivators may also clear patches of forest for cultivating particular cash crops such as turmeric. Forests play a particularly important role in agriculture in terms of nutrient cycling (i.e. through composting of leaf litter, particularly in hill areas where availability of arable land is limited). In hill areas, nutrient cycling from forests to agricultural land is particularly important so that maintaining agricultural land productivity depends on much more extensive forest land for inputs (see Blaikie and Coppard, 1998, for an example in Nepal).

Keeping livestock is the second most prevalent rural livelihood activity after agriculture, and represents a major asset for many households. Livestock complements agriculture through providing draught power and manure, and forest lands typically supply grazing, cut fodder (grass and leaf) and leaf litter for animal bedding. Grazing practices can involve tran-shumant pastoralism (ranging between hill and plains areas, and between seasonally dry and other areas) or non-migrant grazing of livestock. Livestock are kept by landless, land-poor and land-rich households alike, although the livestock composition differs: landed house-holds with access to on-farm fodder generally keep larger herds of large ruminants (cows, bullocks and buffaloes), often including a pair of bullocks for draught power. Trends in keeping large ruminants are in flux, particularly in view of the increasing opportunity costs for labour (e.g. school-age children are less available), but also due to changes in market conditions:

> *Today, non-food functions of livestock are generally in decline and are being replaced by cheaper and more convenient substitutes. At the same time, the asset, petty cash and insurance functions of livestock are being replaced by financial institutions as even remote rural areas enter the monetary economy. Except for some parts of South Asia, the animal as draught power is declining as more farm-ers mechanize, partly attracted by government subsidies. Manure continues to be important in mixed farming; but its role in overall nutrient supply is diminish-ing because of the competitive price and ease of management of inorganic fertilizer. The same applies to animal fibres: although the demand for natural fibres is still high, and in many places even increasing, there are a growing number of synthetic substitutes for wool and leather.* (FAO, 1998)

The trend has been towards intensification: keeping smaller herds, increasingly stall-fed and cross-bred with non-local varieties for increased milk production. However, poorer house-holds with much more limited access to on-farm fodder supplies tend to keep flocks of smaller ruminants, particularly goats, which are more tolerant of browsing on lower-quality fodder.

Artisanal production, a major occupation of landless groups in South Asia, often involves using products from the forest. For instance, blacksmiths in Nepal producing metal farm implements and utensils require charcoal produced in the forest, as well as wood for utensil handles, the manufacture of bamboo screens and furniture, and silk production, as included in the case studies in Part II.

The collection, processing and sale of NTFPs, including medicinal herbs, forest fruits and foods, honey, leaves for making plates, vines for ropes and so on, all offer supplemen-tary incomes and nutrition, particularly to the poorest households. These activities offer income opportunities in the lean season when other possibilities are limited and the oppor-tunity cost of labour is low. They also offer safety nets in times of stress. In eastern India, for instance, collection of tendu leaf for local cigarette-making and sal and saliali leaf for plate-making are major village activities, and provide a significant contribution to the household incomes of poorer groups, as we shall see in Part II.

Labouring can be a major or supplementary part of many households' livelihoods, especially poorer ones. This may involve agricultural labouring for landowners, off-farm labour, both local and distant, and forest-based labouring for forest departments or corporations.

In the context of the South Asian climate, *seasonality* of forest use is a critical dimension. During the months of the hot season in April, May and early June, there are very few agri-cultural labour opportunities. For households without food reserves, it is a time when safety nets are critical. Collection and sale of forest products, as well as wage labour opportunities, can mean the difference between staying in the village or being obliged to migrate in search of work.

The *intra-household labour allocation* patterns across South Asia commonly involve women performing most forest-related activities, such as fuelwood and fodder collection, NTFP collection and processing, and supervision of grazing. Children are also involved in grazing supervision. Men generally perform most agriculture-related tasks and labouring. Men also tend to dominate local committees and decision-making processes, although this has begun to change in recent years with a dramatic growth in grassroots women-only self-help groups (SHGs), saving and credit groups, and women's groups (*Mahila Mandals*), as well as reservations for women on local government committees. Gender issues are particularly relevant when we consider who is influencing forest management decisions: although women are most directly affected by these decisions in terms of impact on their workload and labour productivity (especially the poorest), they may only have a minor influence on them and important decisions may continue to be controlled by men (Agrawal, 2001; Jackson and Chattopadhyay, 2001).

A general observed worldwide pattern in recent years has been a *diversification* of household livelihood activities, particularly towards cash income generation (Ellis, 2000). A comparison of livelihoods over a period of 20 years in Nepal (Blaikie et al, 2002) indicates that there has been remarkably little change in overall incomes and in the differentiation between rich and poor. What has changed is the reliance on non-agricultural income, such as cash from migrant labour and off-farm cash income for all wealth groups of households, which have allowed the substitution of purchased non-forest-derived agricultural and household consumption items such as kerosene, tin roofs and chemical fertilizer. However, the continuing importance of the forest in household subsistence and food production remains paramount, especially for the poor, who have access to more menial and badly paid off-farm income opportunities, limiting the extent to which they could substitute purchased and manufactured inputs for forest products. In turn, this means that they continue to rely on the forest much more than their wealthier neighbours.

Tribal people in forest-fringe areas often have a major dependence on forests, as forest products and flows generally play important roles in their traditional material culture. This includes cultivation in forest land and collection of forest products for direct subsistence use, processing and sale.

Poverty and forest policy

These forest-livelihood links form a more significant contribution to resource-poor households, which have fewer private assets, especially private land. *Landless households* typically depend on a combination of artisanal production, labouring, and forest product collection, processing and sale. This may include fuelwood collection. For *tribal groups*, shifting cultivation in forest areas has been a common practice, often bringing them into conflict with the forest administration:

> *High forest dependence and poverty reflect that other employment options that offer higher returns are not accessible to the poor.* (Angelsen and Wunder, 2003)

For this reason, many international donors, as well as South Asian governments, have focused poverty alleviation policies on forest-adjacent populations – for instance, in both India and Nepal's tenth Five-Year Plans:

> *There is growing impatience in the country at the fact that a large number of our people continue to live in abject poverty … the mandated reductions in the poverty rate … during the Tenth [and Eleventh] Plans … will still leave more than 11 per cent of the population … below the poverty line in 2012. Every*

effort, therefore, needs to be made to reduce the poverty rate even faster. (National Planning Commission, 2002, p7)

The overriding objective of development efforts in Nepal is poverty alleviation. In spite of noticeable progress achieved over the past decade, there is still widespread poverty. The Tenth Plan['s] … sole objective is to achieve a remarkable and sustainable reduction in the poverty level in Nepal. (National Planning Commission, 2003, p35)

Donors have also adopted forest-related policy alleviation measures. For instance, the World Bank's *Revised Forest Strategy* states that 'the strategy must give priority to poverty reduction' (World Bank, 2001).

As discussed above, although the absolute figure is uncertain, it is clear that many poor rural households are currently particularly dependent on forests; therefore, increasing the remuneration for these activities will lead to a rise in their income. Over the longer term, increased income may allow people to reach a position where they can move on to more remunerative livelihood activities. The policy implication would be that forest-dependent households should have their forest-based livelihood activities made more secure and remunerative in order to help them move out of poverty. There is some discussion in the literature as to whether forest-based livelihood activities may, in fact, be 'poverty traps', and some foresters suggest that because this is the case these activities should be discouraged (Angelsen and Wunder, 2003). However, where people lack alternatives, such policies only make poor households more insecure. Forest-based livelihood activities are engaged in because they are the most attractive available labour allocation.

Implementing participatory forest management (PFM) at the local level

In this book we consider a range of different implementation approaches to PFM. In the Nepal hills, community forestry has involved extensive handover of forest management authority to community forest user groups (CFUGs). In Nepal's *tarai*, community forestry has also proceeded, albeit at a much more constrained level. In both cases, donor support has been critical in promoting the transfer of management authority, although direct support to local groups has been very limited. In West Bengal, the JFM model emerged from a crisis in the legitimacy of the forest administration, and after innovative experimentation by forest department staff and local people, it was 'scaled up' with World Bank support. However, it has involved local people primarily in a protection rather than a planning role.

In Orissa, local forest-dependent people recognized the need to organize themselves in order to protect forests from the 1950s onwards, leading to the emergence of independent self-initiated forest protection groups. The forest administration only began to implement JFM in the late 1980s. Finally, in Andhra Pradesh, although there had been a history of state-supported forest *panchayats* pre-independence, these were suppressed during the 1950s, and it was only by the early 1990s that the forest administration began again to take an interest in PFM, although this was not scaled up until donor support became available in the mid 1990s.

Each of the case studies in Part II shows how forest control and management has historically been contested between local and state interests, each seeking to assert their priorities on forest land. The implementation of PFM by no means resolves this conflict, but rather presents new arenas and opportunities for different groups of people. Implementation involves a number of stages that may lead to the assertion of the priorities of either group. The initial precondition is a changed attitude from the forest administration: an acceptance by the forest department to work with, rather than against, local people. Subsequently, the implementation of PFM may involve a number of steps, each of which can affect local people's livelihoods in a range of ways. The 13 main steps involved are as follows:

1   changes to local forest management institutional arrangements;
2   local deliberation over forest management planning and decision-making;
3   local forest management practices;
4   wage labour opportunities;
5   changes to the condition of the forest resource;
6   changes to the levels and security of access and entitlements;
7   changes to the availability of livelihood-relevant products;
8   the labour productivity of forest users;
9   forest product-marketing conditions;
10 revenue-sharing;
11 community funds and local development works;
12 long-term sustainability of forest-livelihood links; and
13 long-term political empowerment.

Changes to local forest management institutional arrangements

PFM implementation may lead to the strengthening of existing institutional arrangements for communities' sustainable forest management, or, if deemed necessary, to the development of new institutions representing all 'legitimate' local forest users. At best, these institutions will have independent legal status and legal endorsement from the state and will receive sympathetic technical and facilitation support as needed. Control of decisions would be democratic in nature and driven by the local community. The institutions may give themselves, or be given, responsibilities for forest management planning and implementation.

On the other hand, implementation of PFM may lead to the replacement of effectively functioning customary institutions with perfunctory institutions (what might be called 'company unions') under the control of the forest administration, serving forest administration objectives rather than those of the local people. A further risk is that important sections of livelihood forest users may be excluded or their membership made contingent on prohibitive fees.

Local deliberation over forest management planning and other decision-making

These local PFM institutions may be empowered to conduct inclusive village-level forest management planning according to local people's livelihood needs. However, it is often the case that local people do not have significant influence in management planning; therefore, forest department priorities for timber production may be imposed on village land. Village micro-plans may be irrelevant to forest management decisions taken in divisional working plans. These can negate the prior livelihood use of forest land, particularly affecting the poorest (e.g. goat grazing and NTFP collection).

Local forest management practices

PFM can lead to the involvement of local people in sustainable forest management according to their own priorities. This is likely to involve protection of the forest area against outside use (exclusion of illicit cutting and organized timber mafias), perhaps planting and silvicultural measures, and regulation of forest use and forest product extraction. The downside may be that the forest department or local elites may be in a position to control this process according to their own priorities (i.e. in terms of species mix and forest use), which may lead to severe restrictions on local people's forest uses. Local people may be involved only as wage labourers or even be obliged to act as 'voluntary' labourers. Further, PFM may not address the politically difficult problems of illicit extraction by organized and politically powerful timber mafias.

Wage labour opportunities

Where PFM is implemented as a state scheme – and particularly where there is donor support – funds may be allocated to support community forest management activities.

Changes to the condition of the forest resource

Degraded forest areas may be regenerated through PFM and thereby become more productive for livelihood needs. On the other hand, if management provisions reflect the forest administrations' traditional timber orientation and do not consider local needs, there is a risk that village forest resources may be converted into plantations of non-livelihood-relevant species.

Changes to the levels and security of access and entitlements

In order for local people to derive livelihood benefits from forests they require secure entitlements. The implementation of PFM may lead to improved tenure security and access to the use of forests. For instance, in Nepal's community forestry programme, local people became the permanent *de jure* managers of the local forest through a 'handover' process, and the general view was that it was 'their' forest. This identification with the forest has been a major motivating factor seen widely across PFM areas, leading to increased vigilance in excluding outside forest users.

However, there are risks that entitlements may not become comprehensive or secure through PFM, and that they may even be reduced, particularly for more forest-dependent groups. This can happen in a number of ways. The forest administration may implement PFM in a manner that might create 'participatory exclusion' by setting one group against another, as in Andhra Pradesh, where non-grazer households in some villages have been formed into *Vana Samarakshyan Samiti* (forest protection committee, or VSS) groups to restrict the grazing of neighbouring households in forests. In this way, the use of forests for products and services critical to local people's livelihoods can be delegitimated. Enforcement against local use may have only been partially effective prior to PFM; but the effectiveness of enforcement may have increased through local institutions (e.g. for fuelwood collection in some areas of the Nepal *tarai*). Heavy fines may be imposed on the poorest and most desperate forest users.

Changes to the availability of livelihood-relevant products

If the forest condition improves, an increased and sustainable level of product and benefit flows may become available for livelihood use. In better PFM scenarios, this may be distributed equitably and thereby promote poverty alleviation. But there is a risk that if forest department or local elites' priorities are imposed against the wishes of local people, the forest ecology may be transformed such that livelihood products are no longer available.

The labour productivity of forest users

If the forest resource and access to it improves, it is likely that less time will be required to collect each unit of forest product, and it may be that increased quantities become available. On the other hand, if livelihood-related products diminish, this may lead to increased time to collect them or oblige collectors to seek alternative sources for products that are no longer accessible.

Forest product marketing conditions

If the organization of forest users leads to collective action over forest product marketing, there is a chance that improved marketing arrangements can be achieved. Value-added processing opportunities may emerge – such as leaf plate-making with machinery. More remunerative prices for forest products may also be achieved, such as from a pine-resin marketing CFUG network in the Kosi hills of Nepal. This may lead to increased forest-related and other income, higher employment and further enterprise opportunities. On the other hand, changes in forest management can lead to a restriction or delegitimizing of livelihood-oriented forest product marketing (e.g. bamboo collection by landless groups in Orissa and fuelwood collectors in the Nepal hills).

Revenue-sharing

Where high-value forest products, particularly timber, are marketed in PFM areas, substantial funds may be generated. Subject to the initial agreement, these funds may be distributed either to a community fund or directly to households.

Community funds and local development works

Funds may be generated through external support or local fund generation to reduce vulnerability (micro-credit, emergency distress payments, etc) and for community infrastructure development and labour creation (community hall, road and path building, water supply, etc).

On the other hand, funds may be distributed inequitably (as a result of which the greatest burden falls on the poor) and allocated according to external or elite priorities, thus benefiting rich groups most (e.g. schools and electrification are most accessible to elite homes).

Long-term sustainability of forest-livelihood links

Where self-sustaining institutional arrangements have been created they can lead to sustainable forest-livelihood links over the long term. This particularly depends on whether the local institution has lasting management authority that is not linked to fund disbursement. In the hills of Nepal, many CFUGs have continued annual management planning and product distribution activities despite a withdrawal of project or forest administration support over the recent conflict period. However, many local PFM institutions in India are meant to depend on the continuation of scheme-based funding, and inevitably stagnate and collapse after the support ends.

Long-term political empowerment

Greater ability to manage local development processes, and to negotiate with or stand up to the state, not just in forestry issues but in many other matters, is certainly a major outcome of self-initiated forest conservation/management processes and of many CFUGs in Nepal. In the hill areas of Nepal, decision-making in community forestry has been claimed to serve as training in local democratization since the resumption of democracy in 1990.

However, where PFM has been introduced as a state scheme involving cash disbursements for wage labour for local people to work on forest department plans, particularly where *de facto* control of forest use is taken away from local people (as has happened in the tribal hill areas of Andhra Pradesh), disempowerment is the more likely outcome.

In order to understand how these implementation *outputs* and their *outcomes* lead to livelihood *impacts*, we need a detailed model for livelihood analysis.

Understanding the impacts of PFM implementation on livelihoods

The introduction of PFM, through either state schemes or self-initiatives, typically affects the livelihoods of forest-using households in a variety of direct and indirect ways over time, often involving complex feedback loops. There is major potential for improving local livelihoods through PFM by increasing the productivity of the forest resource; improving entitlements to its use; improving labour productivity in terms of time taken to collect and value-added opportunities; and improving market relations. However, there is also the risk of conflicting outcomes. In order to assess the livelihood impacts of PFM implementation, we must consider three stages:

1   the *conditions prior* to intervention;
2   the PFM *intervention and 'gestation'* process (which may take several years), in particular:
    •   institutional development;
    •   changes to the condition of the forest resource; and
    •   changes in access and entitlements; and
3   the *post-intervention impacts*.

The pattern of distribution of the impacts of PFM also needs to be analysed in terms of why particular *areas and social groups* have been involved in PFM. PFM is implemented (and membership may be selected) for a number of different reasons, involving both supply ('push') and demand ('pull') factors. In India over the 1990s, JFM was formally targeted at so-called 'degraded forest lands', and it was not until the Ministry of Environment and Forests' (MoEF's) revised JFM guidelines (GoI–MoEF, 2000) that healthy forests were also recognized as legitimate areas for the scheme. In some states, such as Andhra Pradesh, the formation of groups has focused on poorer settlements, rather than all members of *panchayats*, including areas where shifting cultivation has been practised. In the Nepal hills, on the other hand, virtually all village-adjacent forests, regardless of their condition, have been handed to CFUGs, with their aim of including all forest-adjacent users.

Before the impact of interventions is considered, it is essential to take account of the *prior situation*. The pre-existing trends in institutional arrangements, forest use and forest condition will indicate whether levels of forest use have been sustainable or have resulted in deterioration of the forest condition. In many areas across India, JFM has mainly been implemented in what were *de facto* open-access areas that had become degraded. This is significantly different from self-initiated PFM areas, many of which are in 'good' forests on which local people depend and therefore seek to preserve.

The *time factor* is critical to observing PFM impacts. Due to the slow rate of growth of trees and the emphasis on exclusionary protection for regeneration, the initial implementation of PFM involves restrictions on the use of the forest. During this period, which may last for several years (or very much longer when long rotations are imposed by the forest department), forest users are forced to fulfil their needs in other ways. However, if the forest becomes more productive and access is secure, the positive impact on the poorest households' livelihoods may be greatest. Interim support or provision of alternatives can be very important in mitigating the negative impacts on poorer households during this 'stinting' period.

Understanding the way in which these outcomes filter through to specific livelihood impacts requires a detailed livelihood model. The 'five capitals' model (human, socio-political, financial, physical and natural) has commonly been used for analysing livelihoods, although there are a number of limitations to operating conventional livelihood models (Carney, 1998) for assessing the livelihood impacts of community forestry. In order to overcome these limitations, a revised and adapted livelihood model is used here for analysing the impact of PFM.

Household livelihoods can be analysed according to three major components: assets and entitlements, livelihood activities, and income and budgeting.

Household assets, entitlements and access to collective assets, as well as access to market opportunities

The conventional livelihood models commonly conflate household and collective assets, and where livelihood analysis relates to collective assets, it is essential to conceptually distinguish these from private assets. Thus, we explicitly separate them here. Additionally, conventional livelihood models tend to neglect social stratification and distributional/equity issues (relating to assets and benefit flows), which are critical to maintaining or alleviating poverty. Conventional models graphically represent an 'average' household; yet, important wealth-related differences between households exist, as do gender-related differences within households in their access to forests. Therefore, changes in forest policy affect households and their members differently, and these differences must be explicitly considered. Hence, 'poor', 'middle' and 'rich' households are graphically differentiated, as are the different households' access to the collective assets. These stereotypical labels have many different and interlocked dimensions, which are elaborated on in Part II.

Livelihood conditions are affected by wider external institutional and environmental contexts at district, regional, national and international levels (see also the discussion of the policy process in Chapter 2). Conventional and economistic livelihood models can de-emphasize the socio-political context. All households in an area bounded by a JFM or a CFUG are part of a wider context at international, national, state or regional (in the Nepal hills/*tarai*), and district levels. This has been explored in Chapter 2 and is reproduced in summary form in the top left corner of Figure 4.1. Thus, the changes in forest policy shaped by processes at the international, national, state/regional and district levels impact on the livelihoods of different groups of rural people at the local level. The *local* socio-political context is also critical, and in the model this is emphasized under 'local collective assets', although, of course, this is recognized to be something of an oversimplification.

Livelihood activities

On the basis of assets and entitlements, opportunities and market conditions, households make strategic decisions about how to invest their time and capital in a portfolio of activities. Householders select livelihood strategies (i.e. allocation of household time and resources to different livelihood activities) according to their household's assets and entitlement to collective assets in the context of market conditions and opportunities. It is critical to consider market conditions, including exchange relations and opportunities. As discussed above, livelihood diversification across developing countries is a common pattern employed to spread risk and maximize earning opportunities in an increasingly unpredictable, inter-linked and cash-based economy. Improving forest users' livelihood strategies requires improving the security and remuneration of existing activities, as well as facilitating the development of new options. On the other hand, reducing the security of existing strategies leads to 'pauperization' – making households vulnerable, desperate and dependent on less remunerative options. Livelihood activities are made up of domestic 'reproduction' activities, 'production' activities (e.g. agriculture, livestock keeping, labouring, etc), community services and expenditure on inputs.

Income

Income is generated as products, services and cash, which the household then decides how to allocate. Some part of the income will be spent on inputs, some is consumed. Any surplus may be reinvested into assets. If there is a deficit, the poor may need to borrow money to subsist. Over time, a consistent deficit can lead to indebtedness and a further decline in assets.

The actual impacts on livelihoods of implementing PFM may be analysed according to the following pattern – relating to the numbers in Figure 4.1.

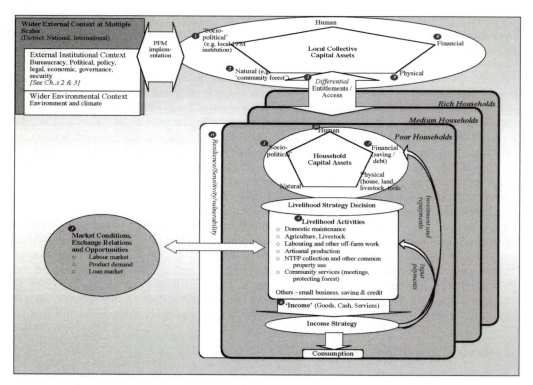

Figure 4.1 Livelihoods model showing the potential impacts of participatory forest management (PFM)

Source: Oliver Springate-Baginski and Piers Blaikie (original material for this book)

Table 4.2 *Potential impacts on local livelihoods of implementing participatory forest management (PFM)*

Aspect of PFM implementation	Possible pro-local people outcomes	Possible anti-local people outcomes	Household livelihood impact
Local forest management institution changed	Communities make institutional choices according to their preferences. Strengthening of existing or developing new representative institutional arrangements for communities' sustainable forest management, with independent legal status. Improved transparency.	Imposed replacement of functioning local institution/customary practices with perfunctory 'company union' serving forest department objectives and under its control. Exclusion of important sections of livelihood forest users. No transparency.	Socio-political capital affected ❶
Local planning and decision-making	Inclusive village-level forest management planning according to local people's livelihood needs – incorporating different groups' views (male/female, poor, occupational groups).	Forest department priorities for timber production imposed on village land, negating livelihood use, particularly of the poorest (e.g. goat grazing, non-timber forest product (NTFP) collection). High transaction costs and time input.	Socio-political capital affected ❶
Local forest management	Effective participatory implementation with regulated extraction. Illicit cutting and organized timber mafias excluded.	Forest department or local elites control the process. Local people involved as labourers or required to commit substantial time *gratis*.	Differential entitlement/access affected ❸
Wage labour opportunities	Additional income opportunities for several months of the year.	Short-term palliatives offered in lieu of access to forest land for normal livelihood use, such as shifting cultivation or grazing.	Livelihood activities affected ❹
Changes in forest resource condition	Degraded forests regenerated. Forest resources made more productive for livelihood needs.	Village forest resources converted to plantations of non-livelihood relevant species.	Natural capital condition changed ❷
Level and security of access and entitlements	Improved tenure security and access to use of forests.	Reduced entitlements of the poor to forest products critical to their livelihoods. Reduced security of tenure and/or reduced access to forests. Enforcement against local use increased through local institution.	Differential entitlement/access affected ❸

Table 4.2 *continued*

Aspect of PFM implementation	Possible pro-local people outcomes	Possible anti-local people outcomes	Household livelihood impact
Availability of livelihood-relevant products	Increased and sustainable level of product and benefit flows, distributed equitably and promoting poverty alleviation.	Forest ecology transformed so that livelihood products are no longer available.	Livelihood activities affected ❹ Income affected ❻
Labour productivity of forest users	Less time required to collect sufficient quantities.	Increased time to collect and or obliged to seek alternative sources for products that are no longer accessible.	Livelihood activities affected ❹
Forest-product marketing conditions	More remunerative prices for forest products and increased value-added processing opportunities. Increased employment and enterprise opportunities in forest-related and other income-producing activities.	Livelihood-oriented forest-product marketing restricted or delegitimated (e.g. fuelwood).	Market conditions affected ❺
Revenue-sharing from forest product harvest	Regular significant income to households from marketing of timber from PFM forests.	Forest administration timber revenue-sharing agreement not honoured; basis and calculations non-transparent; amount insignificant and offset into the distant future.	Income affected ❻
Community funds and local development works	Funds generated through external support or local fund generation allocated to reduce vulnerability (micro-credit, emergency distress payments, etc) and for community infrastructure development and labour creation.	Funds generated inequitably and greatest burden falls on the poor. Funds allocated according to external or elite priorities and benefiting rich groups most (e.g. schools and electrification are most accessible to elite homes).	Collective financial capital affected ❽ Physical capital assets affected ❾
Long-term sustainability of forest–livelihood links	Self-sustaining institutional arrangements leading to sustainable forest–livelihood links.	Funding- or scheme-dependent arrangements collapse after support ends.	Resilience affected ⓫
Long-term political empowerment	Greater ability to negotiate with or stand up to the state, not just in forestry issues but also in many other matters	Increased dependency on state (i.e. cash disbursements and wage labour.)	Political capital affected ❶

❶ *Local collective socio-political and cultural capital*

In most forest areas across South Asia there has been some history of local political and technical experience of managing forests, either informally according to customary norms or with formal institutional arrangements. Collective cultural assets – for instance, tribal self-identity, cultural practices relating to decision-making and material culture relating to forest resource management and use – are crucial here. These have generally been disrupted by the assertion of state forest control over the 19th and 20th centuries (see Chapter 1). Some still persist, sacred groves being perhaps the best-documented example (Gadgil and Vartak, 1981; Deb, 2006).

Where PFM is designed and implemented by the state, the initial intervention usually involves the creation of new local institutional arrangements. The cases of the hills of Nepal (Chapter 5) and Orissa (Chapter 8), and elsewhere in India, mentioned throughout Part I all give some illustration of their importance, and also of the problems which PFM generates when state-designed institutions clash with pre-existing forest management institutions.

Other local institutions, such as the village development committees (VDCs) in Nepal and the *panchayats/gram sabhas* in India, may also offer socio-political capital for forest management and livelihoods in the form of existing rules and norms that reduce uncertainty and transaction costs. On the other hand, conflicts over authority may emerge, particularly under conditions of village factionalism (see Wade, 1988, for a discussion of local 'corporate' behaviour in Andhra Pradesh).

The participation of households and individuals in local institutions, enhanced knowledge of rights and duties involved in securing a livelihood, and any enhanced benefits from collective action all involve drawing on socio-political capital. There may also be crucial networks and contacts that are necessary to access other capital. These are very difficult to quantify, and evidence will be anecdotal. New collective socio-political 'capital' may or may not represent an improvement for local households, depending on the specific institutional arrangements and how these are linked with pre-existing institutions. However, the disruption of pre-existing structures may offer an opportunity for poorer households to improve their position.

Household 'socio-political' and cultural capital

This refers to the set of social relationships on which people can draw in order to expand livelihood options. These include kinship, friendship, patron–client relations, reciprocal arrangements, membership of formal groups, and membership of, or informal access to, organizations that provide loans, grants and other forms of insurance. Improved household socio-political capital may occur through the emergence of participatory local institutions, particularly if these are inclusive of women and poorer groups.

❷ Collective natural capital

The main objective of the forest administration has been to involve local people in improving the condition of forests. Whether this benefits local livelihoods has generally been seen as a secondary consideration by forest departments, and – at most – a potential incentive for getting local people to participate in forest protection.

Household natural capital can also be important, especially for wealthier households with greater private lands that feature forests and gardens. However, PFM does not tend to have a significant impact on these lands.

Where policy change affects existing natural capital, this is an outcome of political history. The quality and extent of existing forests and their structural and species composition shapes the way in which policy itself is specified locally. For example, in India, JFM has largely been extended to 'degraded' forests only. Therefore, the natural capital that can be drawn on under any new policy will tend to be less in terms of diversity, quantity and commercial value than from non-degraded forest (although this is very different in many self-initiated processes, where existing natural capital might be quite rich). The key impact of policy is the way in which it induces changes in human behaviour. This, in turn, alters the use and management of natural resources, which shapes the distribution and flow of natural resources (forest products, soil fertility, and available surface water and groundwater) to people and livestock. Many other factors not connected in any way with the policy process may well be dominant driving forces of environmental change, and it will be difficult to isolate the impact of forest policy change on the environment.

Turning an open-access resource into a common pool resource under regulated use is likely to lead to improved resource conditions and, therefore, productivity. Restricted access may successfully exclude outside opportunistic extraction – for example, from commercial and petty commodity timber cutters – but can, of course, also have negative implications for the livelihoods of customary local users if they are excluded from the group altogether, or if group decisions are biased against the interests of particular users. Examples include disallowing the collection of certain products (charcoal, fuelwood, grazing, etc) and the introduction of an interim 'stinting' period for regeneration of degraded land, which may affect poorer people, who often rely on the forest to a greater extent than the less poor.

Improved environmental services such as enhanced water retention and hydrological regime, reduced soil erosion and run-off, and increased agricultural productivity may also ensue from improved natural capital. This is difficult to demonstrate and measure, and in the literature there has been much controversy and uncertainty remains. The issue does, however, receive substantial support from local people, who have benefited from good forest regeneration and protection, and from scientific studies on the importance of trees in watershed protection. It is also often the motivation for community action, especially in self-initiated PFM.

❸ *Differential entitlements and access*

Changes in local collective assets do not automatically translate into livelihood changes for local households; rather, the distribution of the relative costs and benefits is generally spread differentially according to the mode of implenting the policy and the relative wealth and forest dependency of the different households. Whether implementation of PFM is pro-rich or pro-poor critically depends on this distributional aspect, and it is a common experience that local forest management institutions replicate the inequities of the local society. Committees are typically dominated by local elites, whose decisions often favour their own interests in various ways, rather than those of the poorest. As we shall see in Part II, differential impacts can be achieved through conditions of membership, forest closure, enforcement of rules, conditions for the distribution of products and so on. Even the requirement of attending meetings can be a serious constraint to poorer households who are obliged to engage in daily wage labour in order to provide enough income to buy food.

Changed 'entitlements' to forest through tenure reform can legitimate existing use practices and may transform relationships with forest department field staff. The formation of PFM groups may legitimate use practices and reduce coercive/intimidating behaviour from forest department field staff. But, marginalized and politically weak groups which had enjoyed informal access before tenure reform may be excluded in new arrangements, and livelihood practices and tenure may also be threatened in different ways – for instance, PFM groups in Paderu and the outlawing of *jhum/podu* (shifting cultivation). The implementation of PFM doesn't normally improve household capital assets directly, although it may do so through longer-term feedback loops. The main impact on the household may be changed options for livelihood activities.

❹ *Livelihood activities*

If the availability of products in the forest increases *and* if households have the entitlement to collect those products, then livelihood activities may become more productive and/or remunerative.

A range of forest-related livelihood activities may be affected by PFM. Fuelwood collection and other aspects of household maintenance typically require long hours, and this is

often culturally seen as the responsibility of women. Reducing the time taken to collect sufficient fuelwood can improve women's welfare in terms of freeing up time for other tasks and
remunerative work.

The availability of fodder and access to grazing land can be critically affected by PFM,
both positively and negatively. Regeneration of forest areas often initially involves closure of
an area for grazing and extraction, and these strictures frequently weigh most heavily on the
poor, who have the fewest alternative options.

Wage labour opportunities are particularly important for rural households without self-
sufficiency in food production. If there are inadequate local labour opportunities available,
households may be forced into seasonal out-migration in search of work. PFM programmes
sometimes offer labour opportunities, particularly when they are funded by donors – for
example, the World Bank, which supported the Andhra Pradesh joint forest management
and community forestry projects (World Bank, 1994, 2002).

One of the main determining factors here is entitlement. Only if the household has
secure entitlements will it benefit sustainably from any changes. Entitlements do not always
improve with PFM, and even when they do their security is rarely established. Artisans such
as blacksmiths, woodcarvers and builders, as well as charcoal makers, fuelwood sellers and
collectors of NTFPs, can be particularly dependent on access to forests for their livelihoods,
and are negatively affected if access is restricted.

There are likely to be significant costs associated with membership of local institutions
in terms of the time required to participate in meetings and activities, as well as financial
costs such as membership fees, charges for product extraction, donations and fines
(Adhikari, 2002).

Poorer households are typically exposed to the risk of seasonal consumption-based debt.
The lean season in summer is the period when deficits are most likely to occur. Interlinked
loans and NTFP sales to traders constitute a common pattern in Orissa, for instance, where
loans are given for consumption use in the lean season on the assurance that NTFPs gathered in the monsoon season will be traded to pay off the loan, often at depressed prices.
Increased incomes may lead to reduced indebtedness and may lower the need for distress
sale of produce.

❺ *Market conditions*

Wage labour opportunities in forest management works (e.g. clearance, planting, and soil
and water conservation) may often be available in PFM schemes, but generally depend on
state support and, especially, on donor financing These may offer additional income opportunities for some months of the year, as in the case of the World Bank-supported Andhra
Pradesh JFM and CF projects. However, these can be used as short-term palliatives by forest
administrations, offered in lieu of access to forest land for normal livelihood use, such as
shifting cultivation or grazing. Upon cessation of wage labour, many households return to
now illegal farming practices, such as shifting cultivation and extraction of forest products
forbidden by the management plan.

❻ *Income (goods, cash and services)*

If PFM has a positive impact on households' livelihoods, this is likely to be reflected in
increased income in terms of benefit flows in cash and kind. This may involve increased
forest products for own use or for marketing with the resultant cash income. On the other
hand, exclusion from important aspects of forest-related livelihood activities may lead to
reduced income – for instance, through exclusion from access to forest land for shifting
cultivation (see the case study on areas of Andhra Pradesh in Chapter 8).

❼ *Household financial capital (savings/debt level)*

A household's level of savings/debts is a critical indicator of its well-being. If income increases, any surplus may be saved, providing future options for expenditure on emergencies or investments. On the other hand, if incomes decline, the need for ready cash to cover consumption costs may lead to indebtedness or the disinvestment of essential productive assets, such as land. Membership of savings and credit groups has become a frequent phenomenon across poorer areas.

❽ *Collective financial capital*

The generation of collective financial capital may be both an outcome of changes in forest policy (in terms of the establishment of savings and credit 'self-help' groups and micro-credit schemes financed by forest produce incomes) and an impact (in terms of altered access to assets on the part of households). Interest rates in rural areas of South Asia have traditionally been exploitatively high and dependent on collateral such as land. Additionally, interlinked debt and NTFP sales have been a common phenomenon, particularly in tribal areas of India, where, as mentioned above, poor households can often be caught in tied loan arrangements, borrowing at high interest rates in the hot season against supplies of NTFPs to be supplied to the lender, typically a local trader, in the monsoon season. Local collective financial assets generated through community institutions or project interventions can provide a less exploitative alternative. The PFM institution fund may also be lent out to members and be invested in milk cattle, buffalo, mechanized hand tillers and chaff-cutting machines to improve livelihoods. There are some very optimistic examples from Nepal where local CFUGs have mobilized their funds for micro-credit (setting aside part of their funds for lending to poorer households on preferential terms), although it is a more general experience that micro-credit is not a priority of those on CFUG committees.

❾ *Collective physical capital*

Although this may not seem to follow logically from programmes or schemes to improve forests, works to improve physical infrastructure, such as roads, village meeting halls or water supplies, are often involved in PFM implementation, particularly in Nepal. This may be in terms of so-called 'entry-point activities' as part of JFM schemes, where the Indian forest departments seek to disburse funds for community development, or part of a community's own decisions, especially in many self-initiated PFM examples. It could also be a result of the political empowerment that PFM brings, with spin-offs in the community's ability to negotiate with rural development line departments or directly with state governments. Second, village infrastructural development in Nepal has often occurred over the longer term after CFUGs have generated collective funds and used them accordingly.

Physical infrastructure (rural roads, suspension bridges, health centres, processing facilities) may have considerable, although often unintended impacts on livelihoods. The existence of collective physical capital is not usually directly affected by forest policy, but it can affect its impact. Road provision can provide closer supervision, assistance and control by the forest administration. Roads can also provide better physical access to markets that may be exploited both legally and by forest mafia.

❿ *Household human capital*

Successful forest policy may have a wide impact on the development of so-called 'human capital' and, hence, on the capabilities of individuals to secure their own well-being, although indicators for this impact are elusive and multiplicative. In turn, availability of human capital – for example, of relevant traditional knowledge and practices – is vital to PFM. For instance, new organizational roles for women on committees to oversee collective action, such as new savings groups or adult literacy classes, may be established. The development of human capital may also be linked to the freeing-up of time by providing more accessible drinking water or labour-saving technologies in agriculture, which, in turn, allow disadvantaged groups to spend time on accumulating skills, confidence and networks. The acquisition of new 'development knowledge' through literacy, the radio and personal networks is also important. Such knowledge can be developed through forest policies that allow forest products to be marketed by local institutions, or at least a major part of the proceeds to be retained by them.

⓫ *Resilience, sensitivity and vulnerability*

Resilience against shocks of drought and crop failure, civil disturbances, natural disasters, landslides, house fires, and so on can also be affected by forest policy. Improving local forests and access to them can increase local people's livelihood resilience in a number of ways. If options for lower-productivity remunerative activities, such as NTFP collection and processing, become available, these can offer opportunities in times of stress, especially for the poor, who lack alternatives. Grants of trees or cash by local institutions to reconstruct houses, or support for social ceremonies, such as a loan of utensils (which have become a frequent practice in the Nepal hills), can mitigate the negative impact of major financial burdens. However, if these options are not available as a result of changing forest composition and/or impeded access, the poor may become not only poorer, but even more vulnerable to shocks. Again, long-term political empowerment may also be a crucial gain.

This discussion illustrates the diverse ways (both direct and indirect) in which PFM can potentially affect livelihoods at the individual and household level, as well as at the collective and institutional level. While the primary impact of PFM on the ground remains the changing access to, and production of, forest land, the discussion has drawn attention to the many other policy 'impact points' that produce changes in the supply and use of different livelihood capitals.

References

Adhikari, B. (2002) 'Household characteristics and common property forest use: Complementarities and contradictions', *Journal of Forestry and Livelihoods*, Kathmandu, Forest Action, vol 2, no 1, pp3–14

Agrawal, B. (2001) 'Participatory exclusions, community forestry, and gender analysis for South Asia and a conceptual framework', *World Development*, vol 29, no 10, pp1623–1648

Angelsen, A. and Wunder, S. (2003) *Exploring the Forest-Poverty Link*, Bogor, Centre for International Forestry Research

Bajaj, M. (2001) 'The impact of globalization on the forestry sector in India with special reference to women's employment', Paper commissioned by the Study Group on Women Workers and Child Labour, National Commission on Labour, New Delhi, Government of India

Baumann, P. (2000) *Sustainable Livelihoods and Political Capital: Arguments and Evidence from Decentralisation and Natural Resource Management in India*, London, Overseas Development Institute

Bebbington, A. (1999) 'Capitals and capabilities: A framework for analyzing peasant viability, rural livelihoods and poverty', *World Development*, vol 27, no 12, pp2021–2044

Blaikie, P. M., Cameron, J. and Seddon, D. (2002) 'Understanding twenty years of change in west-central Nepal: Continuity and change in lives and ideas', *World Development*, vol 30, no 7, pp1255–1270

Blaikie, P. M. and Coppard, D. (1998) 'Environmental change and livelihood diversification in Nepal: Where is the problem?', *Himalayan Research Bulletin*, vol XVIII, no 2, pp28–39

Byron, R. N. and Arnold, J. E. M. (1999) 'What futures for the people of the tropical forests?', *World Development*, vol 27, no 5, pp789–805

Carney, D. (ed) (1998) *Sustainable Rural Livelihoods: What Contribution Can We Make?*, London, DFID

Central Bureau of Statistics (2003) *Statistical Yearbook of Nepal*, Kathmandu, Nepal Government Planning Commission

Chambers, R. and Conway, G. (1992) *Sustainable Rural Livelihoods: Practical Concepts for the 21st Century*, IDS Discussion Paper 276, University of Sussex, Institute of Development Studies

Croll, E. and Parkin, D. (eds) (1992) *Bush Base: Forest Farm: Culture, Environment and Development*, London, Routledge

Deb, D. (2006) 'Sacred ecosystems of West Bengal', in Ghosh, A. K. (ed) *Status of Environment in West Bengal: A Citizens' Report*, Kolkata, ENDEV

Ellis, F. (2000) *Rural Livelihoods and Diversity in Developing Countries*, Oxford, Oxford University Press

FAO (United Nations Food and Agriculture Organization) (1998) 'Livestock issues in Asia', in *Agriculture 21*, Rome, FAO

Fisher, R. J. (2000) 'Poverty alleviation and forests: Experiences from Asia', Paper prepared for a Workshop on Forest Eco-spaces, Biodiversity and Environmental Security, Amman, 5 October, IUCN Conservation Congress 2000, Regional Community Forestry Training Centre for Asia and the Pacific (RECOFTC)

Forest Action (2003) *A Survey of Priority Problems of the Forest and Tree Dependent Poor People in Nepal*, Nepal, Forest Action and DFID

FSI (2000) *State of Forest Report 1999*, Forest Survey of India, Dehra Dun

Gadgil, M. and Vartak, V. D. (1981) 'Sacred groves in Maharashtra: An inventory', in Jain, S. K. (ed) *Glimpses of Indian Ethnobotany*, New Delhi and Oxford, IBH Publishers, pp279–294

Gera, P. (2001) 'Background paper on women's role and contribution to forest based livelihoods', Paper prepared for Human Development Resource Centre, New York, United Nations Development Programme

Gilmour, D., Malla, M. and Nurse, M. (2004) *Linkages between Community Forestry and Poverty*, Bangkok, RECOFTC

GoI–MoEF (Government of India–Ministry of Environment and Forests) (2000) *Strengthening of Joint Forest Management Programme: Guidelines*, New Delhi, MoEFF

Guhathakurta, P. and Roy, S. (2000) *JFM in West Bengal: A Critique*, New Delhi, WWF-India, pp60–63, pp65–73

Humagain, K. H. (2003) 'Gender dynamics and equity in CPR management: A case study of Baidol Pakha community FUG', in Timisina, N. P. and Ojha, H. R. (eds) *Case Studies on Equity and Poverty in the Management of Common Property Resources in Nepal: Proceedings of Workshop on CPR and Equity: Exploring Lessons from Nepal*, Jawalakhel, Forest Action

Jackson, C. and Chattopadhyay, M. (2001) 'Identities and livelihoods: Gender, ethnicity, and nature in a south Bihar village', in Agrawal, A. and Sivaramakrishnan, K. (eds) (2001) *Social Nature: Resources, Representations and Rule in India*, New Delhi, Oxford University Press, pp147–169

Kalpavriksh (2003) *Draft India National Biodiversity Strategy and Action Plan*, Pune, Kalpavriksh

Kanel, K. R. and Niraula, D. R. (2004) 'Can rural livelihoods be improved in Nepal through community forestry?', *Banko Janakari*, vol 14, no 1, pp19–26

Koos, N. (2000) *Environments and Livelihoods: Strategies for Sustainability*, Oxford, Oxfam Academic

Kunwar, P. (2003) 'Grazing management practices in upper Mustang', in Timisina, N. P. and Ojha, H. R. (eds) *Case Studies on Equity and Poverty in the Management of Common Property Resources in Nepal: Proceedings of Workshop on CPR and Equity: Exploring Lessons from Nepal*, Jawalakhel, Forest Action

Malhotra, K. C., Yogesh, G. and Ketaki, D. (2001) *Sacred Groves Of India: An Annotated Bibliography*, New Delhi, Indian National Science Academy and Development Alliance

Mukherjee, N. (2002) 'Forest protection committees of West Bengal: Measuring social capital in joint forest management', *Economic and Political Weekly*, Mumbai, Sameeksha Trust, 20 July, pp2994–2997

Perlis, A. and Warner, K. (2000) *Forests, Food Security and Sustainable Livelihoods*, Rome, FAO

National Planning Commission (2002) *Tenth Five-Year Plan (2002–2007)*, New Delhi, Government of India

National Planning Commission (2003) *Tenth Five-Year Plan (2002–2007)*, Kathmandu, National Planning Commission, Government of Nepal

Pokharel, B. and Nurse, M. (2004) 'Forests and people's livelihoods: Benefiting the poor from community forestry', *Journal of Forestry and Livelihoods*, Kathmandu, Forest Action, vol 4, no 1, pp19–29

Rangachari, C. S. and Mukherji, S. D. (2000) 'Old shoots, new shoots: A study of JFM in Andhra

Pradesh, India', in *Production Systems and Resource Use in India: Part III*, New Delhi, Winrock International-Ford Foundation, pp20–52

Satheesh, P. V. (2000) *Biodiversity Festivals*, Note for Technical and Policy Core Group (TPCF) Members, New Delhi, National Biodiversity Strategy and Action Plan

Satyal Pravat, P. (2004) *Community Profile Report: Forestry Sector in Nepal*, Cambridge, Forests Monitor, pp8–12

Saxena, N. C. (2003) *Livelihood Diversification and Non-Timber Forest Products in Orissa: Wider Lessons on the Scope for Policy Change?*, London, Overseas Development Institute

Solesbury, W. (2003) *Sustainable Livelihoods: A Case Study of the Evolution of DFID Policy*, London, Overseas Development Institute

Timisina, N. P. and Ojha, H. R. (2003) 'Case studies on equity and poverty in the management of common property resources in Nepal', in *Proceedings of a Workshop on CPR and Equity: Exploring Lessons from Nepal*, Kathmandu, Forest Action, pp32–40, 77–84

Wade, R. (1988) *Village Republics: Economic Conditions for Collective Action in South India*, Cambridge, Cambridge University Press

World Bank (1994) *C2573: Andhra Pradesh Forestry Project Agreement*, Washington, DC, World Bank

World Bank (2001) *A Revised Forest Strategy for the World Bank Group*, Washington, DC, World Bank

World Bank (2002) *C3692: Andhra Pradesh Community Forest Management Project – Project Agreement*, Washington, DC, World Bank

World Bank (2006) *India: Unlocking Opportunities for Forest Dependent People*, Delhi, Oxford University Press

Zerner, C. (2000) *People, Plants and Justice: The Politics of Nature Conservation*, New York, Columbia University Press

Part II

Participatory Forest Management: Reality in the Field

This part of the book examines the situation in three particularly important Indian states (Orissa, West Bengal and Andhra Pradesh) and two regions in Nepal (the mid hills and the *tarai*).

Nepal

Nepal is a relatively small state in a geopolitically sensitive location, with extensive poverty. Nepal has been undergoing dramatic political turmoil over recent years. Nepal has, for decades, been favoured with substantial donor support.

Nepal's hills

Participatory forest management (PFM) in the form of community forestry began to be implemented here from the late 1970s, with considerable donor support. With substantial progress already achieved by the time of democratization in 1990, policy reform quickly led to major scaling-up of forest handover to community forest user groups (CFUGs).

Nepal's *tarai*

Nepal's plains area presents an entirely different management challenge from the hills. The extensive high-value sal forests have been under clearance for agricultural land since the malaria eradication programme of the 1950s and 1960s, and the rapidly increasing population depend on the receding forests for fuelwood and construction timber. Despite handover of much of the remaining forest to local communities under the auspices of community forestry (CF), there is disagreement over the appropriate institutional arrangements to ensure the sustainable management of the forests and the equitable distribution of its products.

India

In contrast to Nepal, India is a large and powerful country. Within India, states' widely differing local circumstances intersect with national forestry processes and structures, particularly since forestry remains a concurrent subject under the constitution, with different aspects coming under the purview of either the central government or the state governments. The three states chosen for study were selected on the basis of their high levels of poverty and high levels of forest-dependent poor households, as well as to reflect a range of different patterns of PFM: local self-initiated, administration initiated and donor promoted.

West Bengal

West Bengal has extensive historically contained forest areas and a high tribal population. It was here that the initial experiments which led to joint forest management (JFM) began during the 1970s by forest department staff, with the cooperation of local people. Hence, it is here that livelihood impacts have been at play the longest. The World Bank has been involved in supporting institutional change programmes for the forest department.

Orissa

Orissa is among the poorest states in India, with very weak governance, a high proportion of tribal population and extensive forested areas. For many decades, there has been a widespread movement of self-initiated forest protection groups, where communities depending for their livelihoods upon forests have been actively protecting them. During the 1990s, the forest department embarked on a JFM programme, which has frequently involved transforming these self-initiated groups, a move that has not always been popular.

Andhra Pradesh

This state features a wide variety of local ecological conditions, from the hill forest tribal belt across the north, the central Telangana region around Hyderabad, the well-irrigated coastal Andhra region, and the arid rain-fed Rayalaseema region to the south. Forest-dependent tribal populations are distributed across the state, and relationships with the forest department have been strained for decades, particularly in tribal areas. Over recent years, the World Bank-supported joint forest management projects have attempted to reform this.

The following regional chapters (Chapters 5 to 9) are composed of three main parts:

1   a review of the forest policy framework and PFM implementation scenario;
2   an assessment of the outcomes and impacts of PFM implementation based on primary study data; and
3   an analysis of the policy process – explaining impact patterns and exploring focal issues.

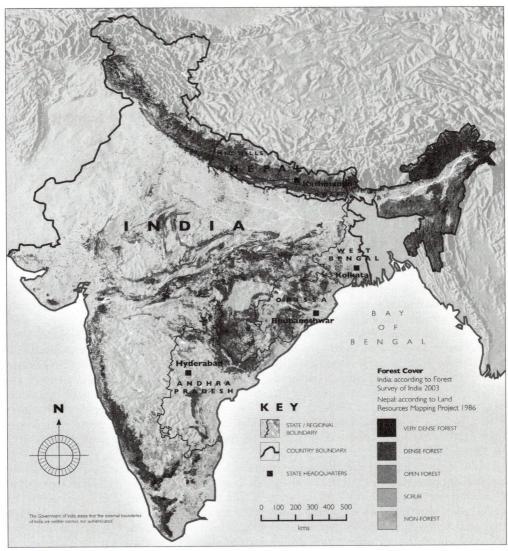

Map 1 Study regions of South Asia

Source: Jonathan Cate (original material for this book)

5

Community Forestry in the Nepal Hills: Practices and Livelihood Impacts

Om Prakash Dev and Jagannath Adhikari

Synopsis

This chapter examines the pioneering experience of community forestry (CF) implementation in the ecologically and socially diverse middle hills of Nepal. A range of traditional forms of customary local forest management historically existed in the hills, although, since the 1950s, the modernizing state had attempted to replace these with 'scientific' forest management. In response to the resultant problems of *de facto* open access, deforestation and adverse impacts on local people's livelihoods, the 1989 Master Plan for the Forestry Sector and the 1993 Forest Act (facilitated by the resumption of democracy in 1990 and donor support), promoted community forestry. This chapter considers the field experience of community forestry by looking at four diverse districts across Nepal – Dhankuta, Kavreplanchok, Kaski and Dadeldhura – and focuses on 14 community forest user groups (or CFUGs). Data was collected through extensive field study over the three years of 2003 to 2005.

Forest cover has generally improved under community forestry (having recovered from a prior degraded state in 11 of the 14 study CFUGs), although in one CFUG, the tree species promoted were not useful to local forest users. The quantity of annual fuelwood and timber collected has increased for all households, except for the very poor, whose fuelwood collection has, worryingly, actually marginally declined. In terms of fodder, grass and leaf litter, although grazing opportunities have declined for all due to forest closure, the middle wealth-rank households are receiving increased benefits in terms of fodder collection compared to pre-participatory forest management (PFM). Overall, 'very poor' households have lost out in terms of forest product collection since PFM. There has, furthermore, been little significant improvement in the generation of community assets as most hill CFUGs have only generated relatively limited financial resources.

Although there are regular elections, meetings and recording of discussions and decisions in almost all CFUGs, indicating systematic institutional functioning, the decision-making in CFUGs was found to be generally in the hands of village elites, and the participation of women was below quota levels. A lack of information reaching local women and poor and illiterate people has created barriers to their taking decision-making roles, due to caste, gender and class division and segregation. Because of entrenched social and cultural norms emanating from the existing social structure, the real advantages have not gone to the neediest people, who also lack the capacity to influence decision-making in their favour.

Participatory forest management (PFM) policy and context

Prior to 1950, when the country was largely closed to the outside world, traditional or indigenous forest management practices differed from place to place according to the different ecological and cultural contexts: *Talukdar* (village headman responsible for the management of a forest); *kipat* (a type of community ownership of land particularly common in east Nepal); *Riti-Thiti* (a traditional way of managing forest, pasture and *khoria* (swidden) land in Gurung villages); *Mukhiya/Jimmawal* (a system under which village revenue collectors are responsible for the management of local forest); and *Mana-Pathi* (users pay a specified amount of grain annually for the protection of forest or common land by a locally appointed forest guard) (Regmi, 1978; Fisher, 1989; Bartlett and Malla, 1992; Hobley, 1996, p87; Gurung, 1997). After the end of the Rana dynasty in 1950, the country was opened to the outside world and embarked on a path of modernization and development. Part of this process involved the nationalization of forests in 1957. This policy was partly introduced to reclaim the forests, a major asset, especially in the *tarai*, which had been distributed to ruling elites as *birta* (land grant with tax exemption) and *jagir* (land grant with tax), and partly to allow the implementation of scientific/technocratic methods to increase timber and revenues for the state. But by the early 1970s it had become apparent that this policy, because it had disregarded customary local rights and traditional institutions that had been conserving and utilizing forests, was leading to an open-access situation. This deforestation was adversely affecting people's livelihoods, particularly the poorer hill people who lacked land on which to grow their own trees. The government forest administration had not been able to develop an effective forest management system that regulated forest use; but many local people had lost their feeling of ownership. Despite this, indigenous practices based on collective use rights persisted in some local communities, especially in the hills and mountains (Messerschmidt, 1993).

Box 5.1 The origins of community forestry in Nepal

Om Prakash Dev

Many organizations take credit for the initiation of the community forestry programme in Nepal. However, those involved often comment that it is, in fact, local villagers who should really receive the credit. For instance, Laxman Dong (a *Tamang* ethnic group farmer and village leader) and the inhabitants of remote Banskhark village in Sindhupalchok district had many forest management innovations that they were able to share with foresters and other development practitioners during the 1970s and 1980s. For example, there were various customary rules around extraction and planting in the forest (e.g. five trees must be planted in communal land for any birth or death). There was also a specific location for making funeral pyres to reduce fuelwood use, and an established seedling nursery. In this way, customary local practices and more recent adaptations became the basis for the state community forestry programme.

The recognition of the functions of these local management systems and the failure of centralized control, policing and management led to the formulation of a new Forest Policy in 1976 (and Forest Rules in 1978), which initiated the sharing of forest management responsibility with local communities. However, it failed to recognize the legal rights of local communities in the forests that they were managing. Subsequently, the 1989 Master Plan for the Forestry Sector emphasized the role of community management to conserve forests and meet rural people's basic forest product needs. The democratic government formed after the 1990 popular revolt fully approved this strategy and introduced a range of legislation strengthening community control of local resources (e.g. 1993 Forest Act and 1995 Forest Regulations). These developments led to the rapid implementation of CF in the hills. Donors, for whom environmental conservation was a priority, saw CF as a way of addressing the problem of deforestation and, at the same time, of improving local household livelihoods. Donors therefore provided a financial incentive for the government to develop these policies further and to speed up the implementation process. Partly due to the government's responsiveness, donor support in the forestry sector rapidly increased for the implementation of CF. This involved 'handing over' to local people the management responsibility for forests adjacent to settlement as 'community forests'. Local people were formed into community forest user groups on the basis of their actual use of the forest in question. Later, the leasehold forestry (LHF) model also emerged, which sought to lease degraded forest or barren land to poorer households. LHF was introduced to provide positive discrimination towards the welfare of the poor, although its implementation has been extremely slow, partly due to government priority on CF and also to the fact that most of the prospective forest areas had already been handed over as community forests.

From 1993 to 2005, most of the forest areas in the middle hills had been handed over to communities. By early 2006, there were more than 14,000 CFUGs in the country managing 1,184,824ha of forest involving an estimated 1,633,408 households (although double counting is a problem here as many households are members of more than one CFUG). Of these, the vast majority were in the hills: about 13,000 CFUGs, managing 1,017,090ha of forest, involving an estimated 1,352,299 households (DoF, 2006).

Community forestry has generally been taken as a panacea for both forest conservation and poverty reduction. For example, the tenth Five-Year Plan (2002–2007), and the poverty reduction strategy paper (PRSP) emphasize that CF may be one way to reduce poverty. However, there is only limited understanding of how exactly CF has contributed to household livelihoods (e.g. to food security and basic needs, contribution to livelihood systems and the general well-being of household members). Some studies have claimed that the poor and marginal people are experiencing significant discrimination in CF (see, for example, Graner, 1997; Sharma, 2002, Timsina and Paudel, 2003). These studies have highlighted various injustices, including the poor and the marginalized being given a greater share of the work of forest conservation without commensurate access to benefits. These studies have primarily focused on the impact of CF without extensive reference to policy or implementation processes; nor have they explored the mechanisms or processes that lead to the inequitable sharing of benefits and costs. This chapter aims to provide a detailed understanding of how the CF policy and implementation processes have influenced the livelihoods of different groups of people, particularly the poor.

Socio-economic and political profile of Nepal's hills

Until very recently, the hills were the centre of Nepal's national life. Due to their mild climate, the availability of agricultural land and trade routes between lowlands and highlands passing through the hills, the historical locus of political control was in the hills, and the national identity was closely linked with hill culture. However, with the eradication of malaria in the *tarai* during the 1950s, the balance shifted, and due to the relative inaccessibility of the hills many people migrated to the plains. Today hills and *tarai* have more or less equal populations – each with about 46 per cent of Nepal's population. The mountains, which lie north of the hills, contain less than 8 per cent of the population and are even more rugged and inaccessible than the hills.

Agriculture and allied activities, such as forestry and animal husbandry, have been the primary sources of the population's livelihood, even though cash income from non-farm employment (particularly remittances) is growing and livelihoods are typically composed of multiple activities. Most of the hills are characterized by inaccessibility to motorized traffic, and some areas are as much as a week's walk from the nearest district headquarters. There are highly diverse and often fragile agro-ecological systems, with a heterogeneous social (caste and ethnic groups) composition. The hierarchical nature of caste, still seen in social practice, means that a rough correlation is seen in ascribed caste status and control over resources and access to education. Higher status castes such as Brahmins and Chettris, and certain wealthier families of ethnic groups, or *Janajati*, generally control more resources and political power and are more educated than poorer groups and Dalits. Out-migration has, over recent decades, led to a shortage of on-farm labour and a gradual decline in investment in farming and the management of communal natural resources. In areas of high out-migration (e.g. west-central Nepal), families have started to abandon the fragile and relatively unproductive land (Adhikari, 1996). There is also a gradual feminization of farming and forest use as women assume the greater part of these activities (Adhikari, 1996) due to male migration. There is wide variation in these socio-economic variables across the region (from west to east), as well as between ridge and valley bottoms, between different elevations and between socio-cultural groups. For example, seasonal migration is extremely high in the far and mid-west regions, and temporary migration is high in the central region. The injection of remittances is prominent in the west central region and in the hills. Valley bottoms are locations of higher agricultural productivity and cash crops are prevalent, whereas hill farms are more integrated with forests and depend on them for nutrient cycling. These changes and diversities have implications for who is able or needs to participate in community forestry decision-making. They also influence the structure of the community, which has become less stable due to internal pressures (population growth, migration and decline in household resources) and external pressures (greater dependence on markets and a need for more cash income) (Blaikie et al, 2002). Overall, people managed their resources in the context of local cooperation in the past; but due to less stable communities and out-migration, they now have much more limited time for collective activities including community forestry.

People's dependency on outside income (migration) and non-farm income (labouring) is growing (Adhikari, 1996; Blaikie and Coppard, 1998; Seddon et al, 1998). The injection of cash income in the form of remittances is said to be one of the main causes of poverty reduction in Nepal, which witnessed a fall of 11 per cent in poverty (from 42 per cent in 1995 to 1996, to 31 per cent in 2003 to 2004) (CBS, 2005). But, again, the poorest 20 per cent of households are not generally able to access remittance-producing employment, which means that their dependency on local resources remains central to their livelihoods. The gap between wealthier and poorer households is growing in Nepal, and this inequality is generally viewed as harmful to the management of common forest resources as these groups' interests in local production and the use of the forest diverge.

Another major change in Nepal has been the emergence of armed conflict between the Communist Party of Nepal (Maoist) and the government since 1996, which has been particularly pronounced in the mid-west and far west hills. The conflict, in abeyance for the time being, has laid a serious burden on the progress of CF due to the displacement of wealthier and younger members of the community able to move out of the affected areas, a general decline in production, and threats to CFUGs from rebels and government army alike. The growing male-specific migration, now exacerbated by the conflict, means that even some executive members of CFUGs have left the villages. Thus, the responsibility of managing resources, including farming and forests, is increasingly placed on women. This feminization of resource management has some important implications for PFM.

The government has, in recent years, become concerned that Maoist groups active

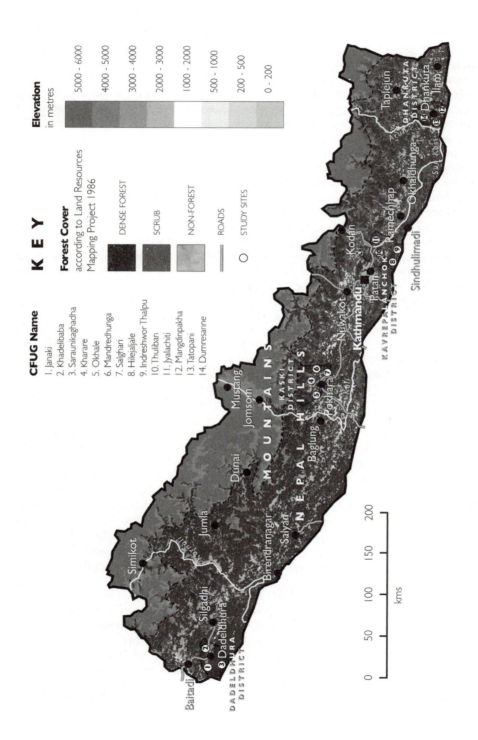

Map 2 Mid-hills of Nepal showing forest cover and study areas

Source: Jonathan Cate (original material for this book)

across Nepal might acquire financial resources via 'donations' or extortion from CFUGs, and has therefore often blocked the bank accounts of selected CFUGs. In 2004, on the charge that they had been making donations to Maoists or otherwise misusing their funds, the government seized the bank accounts of all CFUGs in Kavreplanchok district. Maoists were, until very recently, charging CFUGs, people collecting non-timber forest products (NTFPs) and business people 10 per cent 'war taxes' on sales of 'low-quality timber', 20 per cent on sal timber and 25 per cent on *Acacia catechu* spp. The Maoists had imposed four operating conditions on CFUGs: first, they required them to register with their (alternative) government; second, they required the formation of a new committee in the presence of their own representative; third, they required the forest to be given a Maoist 'martyr's' name; and, lastly, they required the aforementioned taxation of the CFUG's income. They have also destroyed the infrastructure of some forest-based industries that did not comply with their tax demands (Bhatt, 2005, p30). Maoists have also destroyed a majority of the district forest officer (DFO) field offices in all four districts of this study, and all district forest offices in Dhankuta and Dadeldhura.

The implementation of CF in Nepal has been seriously affected by the advance and retreat of the broader democratic environment. The resumption of democracy in 1990 was favourable for this recognition of the rights of users to their local resources. But since 2003, and particularly after 1 February 2005, when the King dissolved democracy, there have been severe problems in implementing PFM, particularly in field support due to the gradual withdrawal of donor project-supported field activities. Additionally, forest users are restricted in networking and advocating for their rights; therefore, it has become difficult to implement any rights-based empowerment programmes.

Overview of forests

Of Nepal's total land area, 29 per cent is classified as forest and an additional 10.6 per cent as shrub and/or degraded forest. The hill and mountain forests account for about 53 per cent of the country's total forest area (approximately 2.1 million hectares) (MoPE, 2001). The type of forest in the hills varies across the ecological belts from south to north and east to west due to climatic differences. There is, however, extreme diversity of species within the different general forest types (such as tropical mix hardwood, sal (*Shorea robusta*) and Katus Chilaune (*Schima-castonepsis*), Chir pine (*Pinus roxburghii*) and mixed oak and rhododendron), and in each of these regions numerous other tree species are also prevalent, along with a complex range of other plants and herbs, offering numerous forest products, including various medicinal and aromatic plants (MAPs). These forests are well integrated with farming systems, providing many direct and indirect inputs. A number of studies have indicated that community forest management (CFM) has reversed prior deforestation. For example, Branney and Yadav (1998), based on a survey of community forests in the Koshi hills region of eastern Nepal, found that:

> The overall indications are that [community] forest condition is improving, particularly in relation to the number and growth of young stems, which, if present trends continue, will serve to regenerate the forest.

Our evidence from a wide range of sites across the hills confirms this assertion. The growth of young stems has increased the forest base and the possibility of a sustainable supply of forest products at all 14 study sites.

The role of forests in agrarian livelihoods in the middle hills

The hill farming system, still the primary source of livelihood security for most rural households, is interlinked with forests and livestock. Forests provide various essential inputs to farming systems (see Figure 5.1).

Farming alone is rarely sufficient these days to meet households' livelihood requirements, mainly because of small and declining landholdings (0.7ha per household in the hills; CBS, 2003) and low productivity. Therefore, most households have diversified their livelihood sources (Adhikari, 1996; Blaikie and Coppard, 1998), and forests have been helping in this diversification in various ways. Additionally, out-migration, the penetration of the cash-based economy and labour shortages have all brought about the substitution of forest products by purchased inputs – for example, kerosene and liquid petroleum gas (LPG) in urban areas for fuelwood, chemical fertilizer for fodder and compost, tin roofing for roofing grass, and steel for timber around towns and cities. The practice of keeping a small dairy for cash income has spread among smaller landholders and marginal farmers around towns and cities (Tulachan and Neupane, 2001). Since they have only small areas of land on which to produce animal feed and fodder, they depend on the forest for these inputs.

Because the forests are so closely integrated with farming systems, it is difficult to estimate the equivalent monetary value of the energy that flows to households through the process described above. But given that households' dependency on local income and production has declined, as is evident in the declining food self-sufficiency (Adhikari and Bohle, 1999), it is likely that the contribution of forest to total income has also been declining. Because of labour shortages for the collection of forest products, especially NTFPs such as fodder, leaf litter and other edible and non-edible products, income from forests has declined. This decline might have taken place anyway, but at differing levels for different wealth groups. For the very poor, who still depend primarily on local opportunities, dependency on forests remains high. A study conducted in a mixed-village in west-central Nepal in 1989/1990 revealed that forests contributed 1232 Nepali rupees (6.8 per cent of the total to average household income). The poorest group (Dalits) had derived 2189 rupees (17.6 per cent of their income) from forest production, while wealthier groups' (e.g. Brahmin and Chettri caste groups) derived share of total forest income was estimated at 4.3 per cent and 4.7 per cent, respectively (or 876 and 1080 Nepali rupees) (Adhikari, 1996, p210). Since Dalits are effectively landless, they depend on household enterprises such as caste-based work, repairing and producing metal tools and utensils, bamboo baskets, mats and storage tanks. Raw materials such as charcoal and different species of bamboo are obtained from different types of forest, sometimes requiring more than two days' walk. Dalits and poorer groups also collect a wide range of wild vegetables (*ningalo* shoots, *nieuro*, nettles and mushrooms*)* and fruits, such as *aalcha* (*Adhatoda vasica*); wild cherry (*Castanopsis indica*); walnut (*Juglans regia*); guava; mango (*Mangifera indica*); timur (*Zanthoxylum armatum*); bel (*Aegle marmelos*); lapsi (*Spondias axillaries*), kafal (*Myrica esculenta*); bair (*Zizyphus mauritiana*); and amla (*Emblica officinalis*) for their own consumption and for sale in local markets.

Forests also contribute to human well-being through their environmental functions and services. In the past, they were considered important in checking landslides and flooding on the plains, and this view was reflected in conservative forest policies and plans. During the late 1970s and 1980s, the Theory of Himalayan Environmental Degradation (THED) (e.g. Eckholm, 1976) had a major influence on increasing financial support for forest protection in Nepal (Ives and Messerli, 1989). However, there remains inadequate understanding of environmental functions of forests in controlling such phenomena as landslides and erosion (Ives, 1987; Ives and Messerli, 1989).

Since forest types in the hills vary by altitude and local climate, they relate to livelihoods in different ways across the different zones. Although there may be both broadleaf and

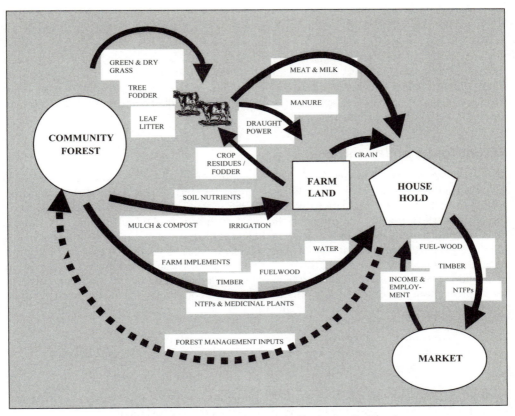

Figure 5.1 The role of forests in mid-hills agrarian livelihoods

Source: Om Prakash Dev and Jaganneth Adhikari

coniferous forests accessible to hill households across altitudes they usually prefer broadleaved tree species due to their multiple uses. The wide variation in forest type makes it very difficult to estimate the contribution of forestry to the country's gross domestic product (GDP), based on extrapolating from case studies (NPC, 1997).

Although hill forests are relatively rich in NTFPs such as medicinal and aromatic plants, current forest management approaches have largely been limited within the narrow sphere of fuelwood and timber production. Community-led forest product marketing is, as yet, very rarely included in operational plans or supported by service providers.

Most of the country's MAPs are mainly traded to India and, to a lesser extent, China. Due to the very limited markets, processing facilities and support, local collectors are obliged to sell these materials in their raw state, sometimes at rates as low as the labour cost incurred in collection. Edwards (1996) claims that 10,000 to 15,000 metric tonnes of herbs from more than 100 species are traded annually from the middle hills and high mountains, although the primary collectors get only a small percentage of the final price paid at wholesale markets. Even senior bureaucrats blame the confusing current policy environment for the lack of effective and remunerative marketing of MAPs, leading to their low contribution to poverty reduction (e.g. Kanel, 2000). The centralized and non-transparent policy environment (e.g. regarding NTFP collection, processing and marketing licensing) has undoubtedly provided lucrative opportunities for corrupt practice to those in positions of power. On the other hand, CFUGs are often unaware that they have the legal right to regulate extraction of NTFPS in their forests. Major national-level Nepali NTFP traders are themselves only small players in international markets, and complain of the monopoly of international traders and open-access over-extraction in the country, leading to declining product availability and threatening the sustainability of their business (J. T. Thapa, pers comm on Nepali Herbs Enterprises, 2003).

Recent forest policy and issues

The legal framework for managing forests is determined by Nepal's 1990 National Constitution, as well as by the 1976 Forest Policy, the 1993 Forest Act and the associated 1995 Forest Rules and Regulations and the 1995 Community Forestry Guidelines. Even though the 1993 Forest Act provides the legal authority to DFOs to hand over forest management to user groups and considers CFUGs as independent local institutions, it maintains the primary management objective as 'environmental protection' – that is, protection of forests and tree species existing there, control of landslides and erosion, and conservation of wildlife. The category of livelihood benefits is treated as secondary to, and derivative of, forest protection. For example, the operational plan, a condition for handover of the forest to a CFUG, is defined in the act and regulations as:

> ... a document ... prepared in relation to development of forest by maintaining the environmental equilibrium, preservation and utilization of forest products, its sales and distributions. (1993 Forest Act, Section 2(d))

Even in the 2001 Community Forestry Development Guidelines, it is clearly specified that supervision and control over the forest is to be managed by government officials. For example, it states that CFUGs should work in accordance with the technical advice, recommendations and directives given periodically by district forest office staff, who are, in general, sensitive to the issue of deforestation and have the authority to withdraw the forest if they perceive this happening. Because of such provisions or conditions, users are directly or indirectly obliged to protect the forest as the overriding management objective. The forest administration seems to find conservation and livelihood objectives mutually incompatible, whereas donors, non-governmental organizations (NGOs) and other stakeholders such as government bodies challenge this presumption and consider the community forestry policy as a means to improving hill people's livelihoods *and* to improving forest condition.

The context and disposition of the various actors involved in CF at different levels vary widely. Staff from the DoF, from NGOs, from bilateral projects and from local CFUGs often have contrasting opinions regarding CF, especially on how it should be implemented. Within the forest bureaucracy, personnel following the narrative of orthodox technical forest management (involving forest protection, yield regulation and timber production) still hold the most senior positions. Their ideas and views therefore have more influence on field practice. A new type of professional trained in CF (mainly in Nepal itself) has also been entering the bureaucracy, or has initiated innovations in PFM. For example, NGOs and the Federation of Community Forest Users, Nepal (FECOFUN) have prepared much extension material related to forest policy, governance and the CF process, and in 2005, people working in donor-supported projects, NGOs and FECOFUN established the Kathmandu Forestry College (KFC). Even so, they still lack capacity to influence the community forestry implementation process (as seen in a few truly participatory CF projects, such as FECOFUN's Ford Foundation-funded Women's Empowerment Programme).

The district forest office and its staff are given the primary responsibility for protecting the forest. There is no mention of poverty reduction or of rural development in their job descriptions. Their major concern is to reverse deforestation and to regulate forest product extraction, for which end they impose conservation-oriented policies and practices on CFUGs. This, in essence, reflects the original goal of the DoF prior to the advent of community forestry: to control people's use of the forest and to technically manage the forest.

Despite the fact that the DoF has not been able to control people's use of forest resources or to manage it properly with local participation, the implementation of the 1993 Forest Act caused a large section of bureaucrats to feel disempowered. The government also felt that it was losing an important source of revenue. It was particularly concerned about its inability to meet general administrative costs with forest revenue, which, indeed, it has not been able to do since 1994/1995. Prior to 1993, especially before 1990, the forestry sector had generated a significant surplus for investment in development (Tiwari, 2002, pp170–171). Realizing this, the government wished to increase both its control of CF and its revenue, especially in *tarai* community forests (see Chapter 6). To this end, it increased the tax on sales of timber from time to time. But with pressure from CFUGs, particularly FECO-FUN, hill community forests have been exempted, and the rate was dropped to 15 per cent for timber from *Shorea robusta* and *Acacia catechu* species from *tarai* community forests only. The 1998 Local Self-Governance Act has further complicated the taxation of CF. According to this act, village development committees (VDCs) and district development committees (DDCs) have the authority to raise taxes from community forest products.

PFM implementation processes and practices

Initially, after the enactment of the 1993 Forest Act, the DFO and project field staff formed groups and wrote operational plans, with processes developed by the projects. The standard CF guidelines that were created with the assistance of donors have been revised by the DoF to bring uniformity to CF processes throughout the country. However, the basic problem with these guidelines is that they assume that there is no previous system of forest management and impose a uniform system, although as discussed above, various systems of indigenous forest and resource management were still operational in the hills. The current system has therefore replaced some effective and localized traditional practices (Tiwari, 1996), and gives low recognition to the diversity of forests and multiple livelihood needs of the people.

CFUGs are legally independent organizations. They have to prepare their constitution and operational plan, with the possible assistance of district forest office staff, and there are detailed government guidelines mandatory for the DFO to follow in relation to investigation, negotiation, planning and monitoring. In order to follow CF guidelines, district forest office staff adhere to a number of tasks, including determination of the CF area, development of a map, creating a constitution and an operational plan, registering the group as a CFUG and issuing a certificate for the handover. These steps increase their involvement and thus provide field staff with enormous authority to influence and even dominate CFUGs decisions. Moreover, the district forest officer has the authority to withdraw forests from the control of a CFUG committee (although they are obliged to return them or appeal to the regional director of the Ministry of Forests and Soil Conservation (MoFSC) within 35 days). Therefore, CFUGs are always under the DFO's influence.

In order to standardize forest management and forest product yield regulation across the country, the DoF recently also introduced the 2005 Forestry Inventory and Yield Regulation Guidelines, which focus on technical issues, such as forest inventory methods, growing stock and yield calculation (primarily to regulate annual harvesting of forest product, especially timber and fuelwood). Furthermore, since 2000, the government has required CFUGs to prepare operational plans using up-to-date forest inventories. This presents difficulties for CFUGs since they do not have the necessary expertise or procedural training, and must depend on district forest office staff or external service providers for forest technical services.

Although it is not mandatory, 4 of the 14 study CFUGs have found it useful to prepare an annual action plan independently of external service providers, based on their own needs and available resources. A few other CFUGs are preparing such plans on a more *ad hoc* basis, although most were found to be unaware of such planning processes and their value. According to the views of the study CFUGs, the annual plans are more practical, needs based and effective than the formal and legal operational plans, which are largely prepared by district forest office staff to satisfy their legal and forest conservation objectives. The district forest sector annual plans also do not match the needs of the CFUGs, but are based on district forest office staff capacity, their technical knowledge and the available funds from donor agencies.

Land tenure

The current PFM regime gives some proprietary rights to the community (Bhattarai and Khanal, 2005, p57), although the ownership of the CF land lies with the DoF. Local users have the right to use the forest under a number of conditions, such as when they are 'not destroying the environment'. CFUGs are not completely confident about the security of their rights since the government revises its policies from time to time. Bhattarai and Khanal (2005, p115) list six changes in government policy through orders and directives during 2000 to 2004: all related to enhancing government revenue through increasing taxation on CFs, and there were also a few cases of 'squatter settlement committees' and of government security forces appropriating CFs for the establishment of a military camp or for resettlement.

Until the resumption of democracy in 1990, the Government of Nepal had total power to acquire all types of land. Even private property could be acquired with or without compensation. As a result, Nepal devoted as much as 18 per cent of its land to protected areas, evacuating people from their land (even ancestral land) without adequate compensation. However, during the 1990 democratic constitution, this provision was removed, and now the government cannot acquire private land without paying due compensation. But as CF land is government owned, the government can still withdraw its control from communities without compensation for the trees or the investments that communities have made in protecting the forest.

Forest users' tenurial insecurity, stemming from inconsistent forest policy, is undermining communities' ability to invest extra work and long-term planning in CF. During Nepal's conflict period (1997 to 2006 – in abeyance for the time being), the government, especially its army, took over forest land in many CFs (for settlements, roads and military barracks, etc) without consultation or without compensating communities. This has discouraged many CFUGs from working actively to protect the forest. The government, including the DoF, considers all forest land, including CF land, still as their property and feels free to expropriate forest lands from CFUGs.

There are communities in Nepal who still live in forests or on other land not under their legal control. For example, the *Chepangs* and *Raute* of Chitwan, Dhading and Surkhet districts, who live in or near national forests, carry out shifting cultivation and other activities, but lack legal title to the land they use. Some of their lands have already been given to other communities under CF. The problem here is that ownership of land has been customarily defined, and these people need land not only for their livelihoods, but also for other customary purposes, such as cemeteries, marriage rituals and so on.

Spatial and temporal variation: Direction of policy

The rigid nature of operational plans, with their fixed contents (including prescribed annual yields of forest products, plans for specified tree plantation, penalties and list of prohibited activities), means that there is little scope to take into account spatial variations of local ecologies, livelihoods and ongoing forest management practices. Currently, the valleys, high-altitude dense and inaccessible forest, and the ecological conditions in the eastern and western hills receive uniform treatment. But realization of this wide variability is growing. Given the fact that some flexibility is afforded the district forest office and local community, spatial variation can be taken into account in managing the forest. Nevertheless, community forestry and forestry inventory guidelines provide a rigid and uniform framework that must be followed. These guidelines are not equally applicable in all ecological belts because of varying population densities and resource endowments.

Forest policies have also been influenced by donors' interests, experiences and priorities. At present, donors are showing commitment to PFM, along with poverty reduction, environmental protection, good governance, social inclusion and development. PFM is also seen as a way of increasing the access of poor people to resources and income in the absence of more radical agendas such as land reform. There are also variations in donors' priorities and modalities of implementation. For example, some projects had been supporting the capacity development of the DoF to implement CF. However, the interventions have had limited field impact because field staff were not mobilized and grassroots NGOs were not involved in the project. On the other hand, other projects focus on livelihoods from CF and emphasize mobilizing field support capacity to CFUGs through the involvement of grass-roots NGOs.

The role of donors has been important in expediting the forest handover process. From early in the 1990s, they emphasized the rapid geographical spread of community forestry. This 'quantity over quality' approach has led to the successful rapid scaling-up of CF, although it may have inevitably diverted attention from innovative ways of implementing the handover process and post-formation support. Donors' approaches have primarily focused on training district rangers and forest guards in CF extension. Initially, donors were concerned with achieving environmental benefits: increasing forest cover so that problems of erosion, land degradation and downstream flooding could be mitigated. These agendas were influenced by the Himalayan Environment Degradation Theory of the 1970s and 1980s, and only since the late 1990s have the donors become concerned with livelihoods, poverty reduction, social development (women, minorities and Dalits) and local governance. They have employed NGOs to promote these issues in a few cases; but this was phased out after the disruption of democracy by Maoists; the overthrow of democratic rights by the king; and consequent withdrawal of donor assistance; NGO involvement in the 'community forestry for livelihoods and poverty alleviation' agenda is now expected to increase with the resumption of the democratic process in 2006.

Research methodology

The methodology devised for this research was conducted in four hill districts: Dhankuta, Kavrepalanchok (Kavre), Kaski and Dandeldhura. These districts were selected purposively to capture the diversity in implementation approaches adopted by donor-supported forestry projects to cover both accessible and remote districts, and to include districts from a range of different development regions. Research was carried out at three levels: household, community and district. The four study districts are shown in Map 2.

Communities were selected randomly within the districts. First, range posts were chosen randomly, and very roughly 5 per cent of the CFUGs within these were also selected at random (resulting in three CFUGs being selected in two of the districts and four CFUGs in the other two). In each CFUG, the wealth status of each user household was determined independently using a participatory rural appraisal (PRA) exercise (three local informants were asked and their ranking averaged out). Twenty per cent of households were then randomly selected from each wealth category. Information was gathered to assess the situation before CF implementation. Considerable time was taken in the study villages and different approaches were used and different data collected to triangulate findings, including interviews and focus groups of men and women of different hamlets. The characteristics of the communities studied are presented in Table 5.1.

Table 5.1 *Characteristics of the community forest user group (CFUG) studied*

District	CFUG name	Forest area (ha)	Main forest type	Total house-holds	Community forest (CF) area per household	Number of ethnic/ caste groups	Men/ women in committee	Fund in CFUG at present (rupees)
Dadeldhura	Janaki	318.00	Sal	87	3.7	5	9/2	62,400
	Khadeli Baba	91.25	Pine	53	1.7	5	6/1	63,133
	Saraunikaghada	13.29	Oak	40	0.3	3	6/5	12,000
Kaski	Khahare	5.58	Katus Chilaune	17	0.3	1	5/4	5000
	Okhale	35.00	Katus Chilaune	205	0.2	8	10/1	7000
	Mandredhunga	14.00	Katus Chilaune	75	0.2	5	7/4	6000
	Salghari	122.11	Sal	47	2.6	4	9/4	49,912
Kavre	Hile Jaljale	94.89	Pine	223	0.4	4	11/0	20,000
	Indreshwor Tharpu	57.45	Katus Chilaune	254	0.2	7	8/5	49,286
	Thuli ban	63.21	Pine	337	0.2	7	8/5	4,000
	Jyalachitti	25.92	Pine	232	0.1	4	9/4	23,480
Dhankuta	Mangdin Pakha	21.01	Pine	35	0.6	1	5/4	3065
	Tatopani	148.69	Sal	140	1.1	7	9/4	3086
	Dumresanne	128.20	Pine	154	0.8	7	9/4	15,000

All of the communities selected for the study were mixed in terms of social composition and wealth, a usual feature of hill villages. Most had been implementing CF for about ten years. The major occupation of all households was farming, but they were engaged in multiple activities: livestock rearing, food crop production, cash crop production and home enterprises. A significant proportion of households were also engaging in seasonal or temporary migration for wage labour. Apart from CF, these communities also had many other ongoing local development programmes, and many forest users were members of several other development groups. Informal money-lending practices are prevalent.

Households were categorized into four groups: rich, medium-rich, poor and very poor, based on a participatory wealth-ranking exercise in which local key informants developed their own wealth status criteria that included: landownership, food self-sufficiency, sources of cash income and condition of housing. Of the total 296 households in 14 CFUGs studied, 16 per cent were rich, 37 per cent medium-rich farmers, 31 per cent poor and 16 per cent very poor. Generally, the higher-status castes in the Hindu caste hierarchy and *Janajatis* (those belonging to non-Hindu groups, such as the Gurungs, Magars or Rais) were in the rich and medium groups. A large proportion of very poor households were *Dalits* (low-status caste Hindus). The medium-rich households were predominantly farming households utilizing their limited land resources through intensive labour input, seeking to maximize resource productivity through innovation. The powerful politically active persons had generally left the village at the time of the study because of the conflict, staying in the *tarai*, towns and cities. But despite the conflict, the CFUGs were still functioning, coping with various constraints imposed by the 'two government' system (i.e. the Maoists and the official government). CFUGs were expected to obey orders from both sides, and many had been donating money and food to the rebels and enduring the concomitant harassment by the security forces. The functioning of CFUGs had also slightly changed during the conflict: with government agencies unable to reach the villages, CFUGs had seen no government staff, but were interacting with rebels constantly. They were found to be incorporating some of the rebels' demands, such as donating some 'tax' from the profits from sales of forest products and showing the rebels their records.

Outcomes and impacts

Processes of local forest management institution

Prior to CF, all of the CFUGs studied had followed traditional forest management systems under which village chiefs were authorized to look after the forest by feudal elites. With the implementation of CF, the district forest office staff persuaded the forest users to form CFUGs and to apply for handover of adjacent forest as CF. Initially, the DoF field staff made contact only with the previous decision-makers or the village elites. Poorer and marginal groups such as Dalits received little information about the new system and were thus often left out from membership. This was clearly seen in Kaski district, where after the formation of CFUGs, the Dalit groups had to petition the DFO for inclusion in the existing CFUGs in their village. After protesting, about 60 per cent were included by paying an amount of money for membership; but for some, the fees were intended to be prohibitive. For example, one excluded Dalit family refused to pay the requested 14,000 rupees for membership.

Indirect exclusion was also created due to the establishment of rigid forest boundaries and the related rights of members and non-members, particularly in the context of hill forest use, where the availability of different forest products from across a range of altitudes is often important to livelihoods. For example, in Lahachok village (Kaski district), during the past people had access to different types of forests at altitudes ranging from 900m to 3000m, and therefore could gather a range of products useful for their diversified livelihood strategies. They would bring from the high hills herbs and wild vegetables, such as the shoots, leaves and roots of *nigalo* (*Arundinaria* spp) for making bamboo mats and baskets, the best varieties being available at higher altitudes. Today the forest has been allocated to different users, with the high-altitude forest allotted to villages close by and the previous users from the lowlands no longer considered members so that they are barred. These restrictions adversely affect the livelihoods of the resource-poor, in particular, who depend most on common property. In another example, for the people of the mountainous Karnali region, the transhumance system of raising animals was combined with trade, prior to CF. In winter, they used to bring their animals to the lowlands to graze in forests or fallow land, paying a forestland grazing charge to the district forest office. But as forests in the hills were converted to CFs, grazing was prohibited to the previous users, now non-members. It has been widely reported that transhumance, already drastically reduced after Tibet was annexed by China, has declined further due to CF. This decline has created food insecurity in the Karnali zone (Adhikari, 2003). The other tribal groups, such as the *Raute*, who roam the forest and live by making wooden utensils and exchanging them for food grains, consider CF their biggest threat because it impedes their mobile life.

Just after the 1993 Forest Act came into force, there was an initial rush to increase the number of CFUGs. From 1993/1994 to 1999/2000, there was a rapid increase in the number of CFUGs in all of the districts studied, with as many as 40 to 60 CFUGs formed per district per year, with slight variation from district to district, mainly due to differences in donor support. After this period, there has been stabilization or slow growth, with less than ten CFUGs being formed per year in each district.

CFUG internal institutional management and participation

The extent of CFUG members' participation and influence in decision-making was found to vary according to wealth status and the nature of the task under consideration. For example, participation is high when decisions are made on forest product prices and fund mobilization. CFUG decisions are taken either in users' assemblies or in committee meetings. Most users were aware that decisions are actually taken by committees; but they were unaware of participatory decision-making concepts. However, in 12 out of the 14 study CFUGs, it was clear that the chairperson, village elite and/or secretary dominated the decision-making process, and women and Dalits' participation was negligible. The CFUG executive committees contain members who are influential in village politics who may come from a wealthy and educated background, as shown in Table 5.2. The representation of poor households is only at 'member level', and that, too, at only 10 per cent. Women's membership in CFUG executive committees is at only 20 per cent, less than the 33 per cent standard government quota for most development activities, including forestry.

Table 5.2 *Background of executive members of the 14 CFUGs studied*

Executive members	Wealth status (% members)			Gender (%)		Village development committee (VDC) leaders (%, present or past)	Education (% members)		
	Rich	*Medium*	*Poor*	*Male*	*Female*		*Educated*	*Literate*	*Illiterate*
Chairpersons	55	45	0	100	0	65	55	45	0
Secretaries	35	65	0	100	0	5	100	0	0
Vice-secretaries	20	80	0	100	0	0	100	0	0
Treasurers	85	15	0	100	0	25	20	65	15
Members	35	55	10	80	20	20	45	20	35
Advisors	75	25	0	100	0	15	35	65	0

Source: 2004 field study

Networking and linkage of forest user groups

In all of the districts, CFUGs had developed both formal and informal networks with neighbouring CFUGs. Of the 14 study CFUGs, 8 were registered with FECOFUN, which has formed district and forest department range-post levels and national-level working committees in more than 50 districts, with the support of donor agencies. In Kavrepalanchok district, three of the four study CFUGs had networked with independent CFUG coordination committees at both VDC (the lowest administrative level of government) and forest department range-post levels. The objectives of developing these relationships should be to resolve conflicts, develop bargaining power with the local governments and district forest office and to exchange plant resources (e.g. seedlings and saplings), as well as to initiate social networking. However, even though FECOFUN has developed in all districts, due to lack of these sorts of benefits there seems to be a reluctance to renew group membership (five out of eight CFUGs had not done so). The main reasons given for this included lack of support and limited benefits from membership, lack of awareness of FECOFUN's role and responsibilities, unaffordable renewal fee, and political differences (e.g. the Mangdinpakha CFUG committee had a different political affiliation from that of the majority of the FECO-FUN district committee members, who were from the one of the Communist parties.

Forest user organization and cohesion

CFUGs have provided new spaces for villagers to come together and discuss issues that are not only related to the forest, but also to the village as a whole. This seems to have increased interaction in villages, although, again, only those who can afford the time participate in the meetings, and non-members and poorer people are left out.

CFUG meetings often address disputes and conflicts, including differential access to forests between hamlets, lack of accountability, lack of equitable benefit distribution and so on. These deliberations have, to some extent, increased caste consciousness and awareness of discrimination, and this is especially evident in Dadeldhura district. CFUGs lack the cohesion of ritual, religious or cultural underpinning of pre-existing groups, and although they may have some traditional roots in village forest use practices, they have been formed through external intervention and function separately from other local development groups under the guidance of the district forest office and its field staff. Because of these factors, CFUGs tend to lack common values and norms to harmonize and sustain them, at least in the initial period after formation. The extent of heterogeneity varied in the CFUGs studied; few were homogeneous in terms of ethnic or caste composition, and whereas some heterogeneous groups were found to be effective because they followed traditions of community development work, some of the more homogeneous CFUGs studied found it difficult to punish offenders because of the close social relationships.

Conflict management: Types and capacity

None of the 14 CFUGs were completely conflict free, and the nature and severity of conflicts varied. In seven CFUGs, conflicts were minor, arising from issues such as the dictatorial working style of the chairperson and executive committee, decisions over fuelwood harvesting and fodder/grass collection, fundraising, and allocation of opportunities for participating in training and study tours organized by the district forest office or by donor-supported NGOs. Such conflicts did not seriously hamper the work of the CFUGs. Rather, they encouraged resolution and innovation in a healthy and amicable manner, although personal rivalry between elite members and those who stood for the position of chairperson often developed into factionalism and led to conflict. Overall, however, these conflicts did not seem to lead to improved working practice or bring changes in favour of marginalized groups.

The cause of conflict in three CFUGs was the exclusion or inclusion of members. In the Thuliban CFUG, this conflict related to the elite members, who were immigrants who sought to bring in new households from the nearby town and bazaar areas against the wishes of the general body. In the Kaski and Dadeldhura CFUGs, neighbouring villagers who had been users of the forest, but who were excluded from membership, continued to collect products. In five CFUGs, the perceived unfair distribution of benefits was causing conflict. These conflicts did not seriously affect the CFUGs, which were still functioning normally. Forest boundary conflicts that are common in all districts emerged from poor forest handovers (e.g. unnegotiated or unclear forest boundaries, or their overlap with other community forests and private lands). In four sites in Kaski and Dadeldhura, seasonal forest product collection from high- and low-altitude forests had been stopped due to the handover of forests to neighbouring communities without incorporating the traditionally seasonal users from mid altitudes. Another issue causing conflict has been the illegal pine resin collection by resin companies in Dadeldhura. Resin collection was not specified in the operational plan and, thus, charges were not paid by the resin company to the CFUG, whereas in Dhankuta the resin contractors pay the CFUGs to collect in their community forest.

Decision-making

Community forest policy (the 1993 Forest Act, the 1995 Forest Regulations and the Community Forestry Guidelines) has provided rules for local decision-making processes, including elections, minutes of decisions, record-keeping and accounting, discussion in meetings, and so on. During the past, it was elite controlled; one village chief decided on these matters and there was no transparency. However, although in CF there is a formal process, decision-making positions are often taken by the previous village rulers. The control of decision-making mechanisms by elites and the continuation of the traditional elite-biased working practices in CFUGs mean that poorer and marginalized people have little voice or influence to change policies and rules in their favour.

There are two main problems with decision-making practices in most CFUGs. The first is that because the CFUG institution is often weak, the same decisions are taken in consecutive meetings. For example, CFUGs in Kavreplanchok district had repeated a decision about their forest inventory for some years running, but were not able to implement it because of a lack of proper knowledge of inventories and an absence of support from service providers (the district forest office and NGOs). Lack of proper micro-planning and lack of support from service providers (the district forest office and the project) and local government were the main constraints in not implementing decisions. Another reason is non-cooperation from CFUG members. CFUGs also lacked an internal monitoring system to check on the implementation of decisions. Many CFUGs do not have formal handover from previous committee members, resulting in an information gap between new and old committees.

A second issue, seen in the study sites, was that decisions which needed external assistance were only partially implemented. In Dadeldhura and Dhankuta, decisions relating to enterprise development have not been implemented for a long time. Coordinating support from service providers has been a major constraint for CFUGs.

Impact on forests

Since forest use is directly linked with hill livelihood systems, the increase or decrease in its quality and quantity have implications for livelihoods in various ways.

A wide range of silvicultural forestry operations has been undertaken in all study CFs since the formation of CFUGs, including selective tree felling, pruning, plantation and coppicing. The main strategy is to protect natural forests, rather than to establish new plantations. But in 6 out of 14 CFs, the major tree species were not those that the CFUG preferred to grow. Five of the CFUGs wanted to convert Chir pine (*Pinus roxburghii*) forest to broad-leaved species that are more useful to them than pine, and the Khahare CFUG in Kaski did not want the broadleaved species that grew naturally in their forest. These six CFUGs wished to change the composition of their forests but were unable to do so due to the lack of support and technical advice. Four CFUGs recorded their investment in forest management: households have contributed one week's labour each year for the past six years.

Comparing forest condition before and after CF is difficult in the absence of baseline data; but according to CFUG members interviewed, forest condition in terms of regeneration and plant density has improved significantly in 10 of the 14 sites studied, and moderately in a further 2 sites. At three of these sites water sources have noticeably increased, rock slides have reduced and consequent damage to houses and farms has significantly lessened due to dense forest on the hills (in study sites of Kaski and Dadeldhura districts). CFUG members considered that the forest had improved because of the increased regulation of forest product harvesting; the reduced illegal harvest and theft due to protection of the forest from grazing, encroachment or fire; tree plantation; and the changed behaviour of forest users concerning forest protection due to greater knowledge. In the Tatopani and Dumresanne CFUGs of Dhankuta district, no significant change in forest condition has been perceived. In Tatopani, the forests are large and are used by local people as before; in Dumresanne, the forest is managed for resin tapping and site quality has not improved, so people have not perceived a significant change in forest condition.

Forest survey results

A forest inventory of tree, shrub and herb species was carried out in the CFs studied, using random sampling procedure to select plots. The volume of trees was calculated (using a 'form factor' of 0.60), and the mean annual increment (MAI) was calculated using the average age of plants as estimated by local people. Therefore, although the volume and MAI are approximate, they provide an idea to enable comparison across CFs.

The qualitative assessment of local people discussed above shows that growing stock in the forests has significantly increased since the formation of CFs. The quantitative data obtained through survey show that forests have significantly regenerated on all sites, although at a few sites, such as Khahare CF in Kaski, regeneration is profuse, but not of those species preferred by local people.

Table 5.3 *Forest inventory results: Growing stock*

District	Forest user group (FUG)	Seedlings (<4cm diameter) (Number/ha)			Saplings (4–9.9 cm diameter) (Number/ha)			Pole/tree (>10 cm diameter) (Number/ha)		
		Good >5000	Average 2000–5000	Poor <2000	Good >2000	Average 800–2000	Poor <800	Good >300	Average 150–300	Poor <150
Dadeldhura	Janaki			1867			759	416		
	Khadeli Baba			1933			600	451		
	Saraunikaghada	83,000			3988				290	
Kaski	Khahare	15,433					316	475		
	Okhale			1940		1705		735		
	Mandredhunga		4065				749	623		
	Salghari		3435			1685		422		
Kavre	Hile Jaljale	29,013					360		206	
	Indreshwor Tharpu	10,911			2049			956		
	Thuli ban		3195				623			
	Jyalachitti		3090				286			144
Dhankuta	Mangdin Pakha		2495			1729		474		
	Tatopani			1446		1750				137
	Dumresanne		3723			1582			299	

Source: 2005 field study

As shown in Table 5.3, only four CFs had poor regeneration, which indicates that forests are regenerating with proper protection by the CFUGs. If this trend continues, there will be more dense forests with adequate regeneration, saplings and trees. The growing stock and mean annual increment are higher in western Nepal than in the east. There could be several reasons for this: the western districts receive heavier rainfall (especially in Kaski district), promoting rapid regeneration, and economic conditions are better due to the high injection of remittances, leading to less pressure on the forest.

Under the 2005 Forest Inventory and Yield Regulation Guidelines, DFOs require harvesting to be under 50 per cent of the estimated mean annual increment. However, the estimation of the mean annual increment is necessarily very approximate for natural forests that have different-aged mixed crops, and since the livelihood needs of CFUGs are diverse, the CFUGs rarely extract as much as the prescribed annual allowable timber cuts, and forests are generally underutilized. If conditions remain the same, the forests will become overstocked and have an even-aged structure that will not be able to provide the sort of sustained yield which could be obtained from a mixed-age forest.

Livelihood impacts

A household may be affected by PFM in two main ways. It may directly benefit from, or be deprived by, changes to forest use; second, it may also be affected through the community level, depending on its access to community resources and development. These aspects are addressed in sequence.

Impacts at the household level

Forest product collection: Availability and collection time

Households have, on average, experienced a marginal increase in the availability of various products after PFM, except for grazing access, which has become restricted. However, not all households have experienced the same change. Table 5.4 shows that rich households have experienced a slight decline in fodder and green grass after PFM, but have obtained more of other products. Medium households have experienced an increase in the availability of all forest products. Poor households found an increase in the availability of all products except dry grass. The very poor households seem to be most adversely affected by PFM, with reduced access to timber, fuelwood, fodder, dry grass and leaf litter.

In terms of equity in product distribution, it seems that this has been maintained only in the case of fuelwood in some sites. There is a general trend of decline in the availability of all other products as one moves from the wealthier to the very poor classes. A significant disparity is seen in the case of timber. Timber prices fixed by CFUGs for internal distribution are prohibitively high for poor households. Wealthier and medium-rich households take an extremely large amount of timber compared to the poor and very poor. Medium-rich households collect more fodder, green grass and leaf litter (see Table 5.4).

In all CFUGs studied, households were found to be saving time on collecting forest products since CF was introduced. Households had previously to spend more than 4.5 hours a day, on average, to collect a *bhari* (a backload of approximately 30kg to 40kg in weight) of fuelwood over three months of the year prior to CF, whereas it now takes only three hours a week when the forest is opened for forest product collection by the CFUG (generally in winter). The time needed to collect a *bhari* of fuelwood has drastically reduced in Kaski and Kavreplanchok, whereas in Dhankuta it has decreased by a smaller margin. This has benefited women most since it is they who perform this work, and at a time when the out-migration of young men is increasing, this time saving has been important for households.

Table 5.4 *Average amount of forest products taken by households in a year and the changes in the availability after participatory forest management (PFM) (in comparison to pre-PFM situation)*

Forest products	Rich (n = 46)		Medium rich (n = 119)		Poor (n = 83)		Very poor (n = 46)	
	Amount taken	*Change*	*Amount taken*	*Change*	*Amount taken*	*Change*	*Amount taken*	*Change*
Fuelwood (35kg backload)	43	+4.3 (+10%)	42	+2.5 (+6%)	44	+0.9 (+2%)	36	−0.4 (−1%)
Timber (cubic feet)	23	+2.3 (+10%)	21	+7.3 (+35%)	3	+0.1 (+4%)	0.16	+0.01 (+7%)
Fodder (backload)	64	−9.0 (−15%)	71	Same	52	+2.6 (+5%)	51	−8.7 (−17%)
Green grass (backload)	132	−26.0 (−18%)	188	+47.0 (+25%)	184	+18.4 (+10%)	66	+6.7 (+10%)
Dry grass (backload)	140	+5.6 (4%)	169	+15.2 (+9%)	123	−8.6 (+10%)	46	−4.6 (−10%)
Leaf litter (backload)	16	+0.6 (4%)	211	+10.5 (+5%)	134	+14.7 (+11%)	55	−2.7 (−5%)
Grazing facilities		Declined		Declined		Declined		Declined

Source: 2004–2005 field study

The experience of saving time is not equal in all districts or in all wealth groups. For example, in Dhankuta, there has been some saving in fuelwood collection time; but this is not significant since forest users collect fuelwood from the same forests as before CF. In other districts, there has been significant reduction in fuelwood collection time because the availability of fuelwood has increased in their CFs through regulated access. Very poor people in Dhankuta and Dadeldhura districts take longer to collect fuelwood because they still rely on national forests – they generally cannot afford the charges claimed by the CFUGs, of which they are members. The poorer groups collect dry branches and twigs that are free of charge and can be gathered at any time of the year, so their access to quality fuelwood is very limited in CF. Therefore, for the very poor there have been no time savings. In Janaki CFUG (Dadeldhura district) and in Tatopani and Dumresanne CFUGs (Dhankuta district), collection times for fuelwood increased because the CFs were totally conserved for the growth of tree stock or for resin tapping in order to generate funds for the CFUGs. Users in these CFUGs have to collect fuelwood and other forest products from nearby national forests or find alternatives.

Cash income from community forestry

Cash income from forests was found to have declined after CF. The members of the 14 CFUGs had earned, in total, approximately 0.8 million rupees in the year before CF; but during the time of study (after 10 to 13 years of CF), the annual cash income obtained by the same groups had declined to only 0.4 million rupees. Before CF, the main sources of household cash income had been work as forest watchers or tree nursery technicians, wage labour, collection and sale of fuelwood/timber/fodder and NTFPs, and supply of charcoal to blacksmiths. Some traditional cash income sources, such as blacksmith work and sales of fuelwood and forest fruit, had also declined as the resources were increasingly controlled by the CFUG executive committees and by district forest offices through operational plans. For example, in a CFUG in Kaski district, poorer households used to collect small bamboo from the forest; but this is now community regulated and they have to pay a charge to collect more than for their immediate needs.

On the other hand, mobilization of CF funds has generated new opportunities, such as employment for new school teachers and local farm-based income-generating activities (e.g. vegetable growing and beekeeping). The funding for these types of employment and income generation activities are substantial where there is donor project support for PFM, although this is generally dropped after the phase-out of the PFM project.

Cash income generation from CFs was pronounced in medium-rich households, where it has increased from the pre-community forest situation due to their abilities, qualifications and resources. In the studied villages, there has been some decline in the activity of black-smiths, which is a general pattern in Nepal since the influx of cheap, industrially produced iron tools from India with the development of road access. There has been a small increase in the cash income of families involved in agro-forestry activities within CF.

Impact at the community level and access of households

Development of community infrastructure
Fifty-six per cent of respondent households felt that the impact on their community infra-structure has been 'very good', 7 per cent that it has been 'moderate', and the remaining 37 per cent were unaware of improvements through CFUG funds. The main infrastructure improvements included the school, village trail, irrigation channels, drinking water, health facility, temple, *Chautaro* (resting place with a tree) and community building. Table 5.5 shows that there has been some improvement and that development, such as electricity and irrigation, was entirely for the benefit of wealthier groups. Considering that the development mentioned in Table 5.5 has taken place over a decade in 14 CFUGs, it cannot be considered a major instigator of change in absolute terms. But, since this infrastructure did not exist before CF, even a small change or improvement might make a good impression on respondents. As a result, a majority of respondents might have taken the change positively.

Table 5.5 *Improvement of community infrastructure after community forestry (CF) (1993/1996 to 2004)*

Infrastructure	Number of study CFUGs	Quantity	Contribution of CFUGs		Main beneficiaries
			Nepalese rupees	% of cost	
Village trail	8	45km	226,000	50	All
Temple	1	One	65,000	85	All
School support	9	Nine schools	527,001	25	Wealthy and some poor
Electricity	1	One village	300,000	30	Wealthy
Water supply	5	Five projects	214,000	35	All
Health facility	1	One building	310,00	20	All
Improvement of *kulo* (irrigation channel)	5	20km	200,000	35	Wealthy

Source: 2004–2005 field study

Community funds generated and invested after PFM
The internal fund generation capacity of hill CFs is very low. The CFUGs studied were able to raise an average of 5876 rupees per CFUG annually over 10 to 13 years. At less than

US$100, this is quite a meagre level. Of the various sources of this income, external support contributed about 70 per cent of the total. This small resource has been used for community priorities, mainly infrastructure such as school maintenance, small irrigation, trails, etc. All CFUGs have bank accounts under the joint signatures of the chairperson and treasurer. General users are not generally aware of their CFUG's fund status; but literate users questioned fund availability and expenses at meetings.

CFUGs have also given loans from their funds. But, again, the amount is small – only 86,500 rupees by 5 (36 per cent) of the 14 study CFUGs over 10 to 13 years. A breakdown of loan disbursement by purpose and wealth rank is presented in Table 5.6, which shows that most of the loans have gone to poorer families. However, again it must be stressed that the amount (17,400 rupees per CFUG in the past 11 years) is small.

Table 5.6 *Loans given to forest users from CFUG funds after community forestry (1993/1996 to 2004)*

Items	Total loan		Rich (n = 46) (%)	Medium (n = 119) (%)	Poor and v. poor (n = 129) (%)
	Number	*Amount (Rs)*			
Goats	20	10,000	0	0	100
Bee-keeping	4	15,000	50	50	0
Forest-based cash crop cultivation loan	20	40,500	0	30	70
Non-timber forest product (NTFP) cultivation	7	12,000	0	100	0
Support to vulnerable groups in sickness	2	9000	0	0	100

Source: 2004 field study

Livestock raising and income

The reduction of per household livestock population (see Table 5.7) in the Nepal hills is a general long-term trend (see, for instance, Blaikie et al, 1980). However, this study shows that changes in forest management are also contributing to this trend, albeit differentially for different households. Overall, there is a general decrease in large livestock in the hills due to lack of labour, markets and fodder availability. In rich households, all types of livestock, particularly cows, have declined to the greatest extent, while medium-rich households have only slightly fewer livestock and buffalo have slightly increased. Poor households have increased their numbers of pigs, but reduced their goat holdings. The slight increase in buffalo numbers was found to be due to the availability of loans and the presence of dairy facilities and markets along the highway, but not as a result of CF policy.

The change in livestock numbers is not only a result of CF; but CF has led to further reduction in grazing land and reduced access to fodder and green grass. In response, the wealthier households have changed to stall feeding so-called 'improved breeds' of cows and buffaloes. It is only the medium-rich households that still heavily depend on their small dairies for cash income, and the fodder and grass produced on their land and in CFs can still support substantial levels of livestock.

The impact of restrictions on forest grazing and the decline in access to fodder and green grass from the forest is most severe among the poor and very poor households. As a result, goat holdings in poor households have also decreased, although they have been able to

Table 5.7 *Changes in livestock numbers (per household) as a result of community forestry*

| Livestock type | Wealth category of respondent households | | | | | | | | | | | |
| | Rich (n = 46) | | | Medium rich (n = 119) | | | Poor (n = 83) | | | Very poor (n = 46) | | |
	Now No. per house-hold	Before CF No. per house-hold	Change % per house-hold	Now No. per house-hold	Before CF No. per house-hold	Change % per house-hold	Now No. per house-hold	Before CF No. per house-hold	Change % per house-hold	Now No. per house-hold	Before CF No. per house-hold	Change % per house-hold
Cow	1.22	2.59	−52.90	1.38	1.54	−10.39	1.08	1.42	−23.94	0.67	1.2	−44.17
Buffalo	1.46	1.89	−22.75	1.13	1.06	6.60	0.67	0.90	−25.56	0.39	0.54	−27.78
Oxen	0.91	1.41	−35.46	0.87	1.02	−14.71	0.58	0.64	−9.38	0.63	0.61	3.28
Goat	3.54	5.72	−38.11	1.03	1.18	−12.71	1.40	1.95	−28.21	1.67	2.78	−39.93
Pig	0.19	0.28	−32.14	0.27	0.08	237.50	0.31	0.18	72.22	0.5	0.24	108.33

Source: 2004–2005 field study

maintain some of their goats mainly due to access to micro-credit, not only from CFUGs but also from other development projects.

Direct employment in community forestry activities

Community forestry has not brought any significant improvement in employment opportunities. A very few households (10.5 per cent of CFUG members, mainly in the medium-rich and poorer groups) have benefited from forest watcher jobs and occasional direct employment for a few months per year in jobs such as resin tapping, timber harvesting, NTFP collection, pruning and implementing social development activities (e.g. improvements to the village trail, drinking water and temple).

Ownership of physical assets and access to community infrastructure

Apart from community-based small infrastructure development, CFUGs have also supported individual households in developing physical assets (such as toilets, bio-gas and animal sheds) through projects funded by PFM and international funding institutions (IFIs) and executed by NGOs. They were not implemented in all CFUGs. Only active CFUGs in contact with projects and donors have been able to bring these activities to their villages and implement them – for instance, the sanitation awareness campaigns of a few active CFUGs in Kaski and Kavreplanchok districts may have contributed to users constructing toilets.

Benefits from community development activities

Households have generally benefited from community development activities initiated by CFUGs, although, again, their total impact is low. The benefits are low-cost community services such as support for emergency health treatment, loans, organization of literacy classes and providing timber during natural calamities. All CFUGs have purchased utensils to lend out for social events and a stretcher for community use. With a few exceptions, medium-rich households seem to derive the most benefit from these community services.

Environmental benefits and cost

Forest users had experienced some environmental benefits, as well as hazards and environmental costs, after CF was implemented. For example, two CFUGs reported fewer landslides and rock falls. In one location in Kaski, there has been more well water available for a longer duration in the hot season, although in other CFUGs no such impact was noticed. Six CFUGs reported an increase in the productivity of their land due to the flow of forest humus to agricultural land. In Kaski and Dadeldhura districts, people collect leaf litter for livestock bedding and fodder from CFs and then compost them with cow dung.

Surprisingly, perhaps, adverse environmental impacts were also noticed. Four CFUGs complained about the effects of growing trees casting too much shade over crops on the land of some of their members. The depredation of wildlife was more widespread. Nine CFUGs complained that their crops are destroyed by wildlife (e.g. by monkeys, bears and cheetahs), which has increased after CF because of improved habitat and greater restrictions on killing wildlife. These restrictions have been clearly specified in CFUG regulations as the DoF is considered the main authority to protect wildlife outside protected areas. Hence, the DoF includes restrictions concerning wildlife in CFUG operational plans.

Change in vulnerability and well-being

The vulnerability of the poor at times of crisis has increased with CF, and there has been a decline in their coping abilities. Before CF, the forest acted as a cushion for the poor, who collected wild fruits, herbs, vegetables and root crops for support and nutrition in lean times. However, the forest condition was widely in decline. Even though forests have generally recovered with CF and resource stocks have increased, the restrictions imposed (in terms of collection time and the nature of products to be collected) have increased vulnerability. Even if collection of these products is allowed in operational plans, there is usually a fixed charge that the poor cannot afford. There has also been increased uncertainty regarding fodder availability in the dry season.

On the other hand, in a few cases CF was found to have reduced vulnerability by providing emergency support, such as timber after natural calamities (e.g. house fires, landslides and floods), and fuelwood for festivals, although this support was not only targeted at poor households. Cases of CFUGs providing funds for emergency medical treatment were also reported. However, on balance, these benefits were inconsiderable compared to what was lost.

Governance and the policy processes: Explaining the livelihood impacts

Having reviewed the impacts of implementing CF policy in the Nepal hills, we now consider why they have turned out as they have.

Policy and governance processes

Policy-making in the forest sector has historically been influenced by the political power of the prevailing regimes due to the high value of both the timber and the land as major national assets. This was strongly prevalent during autocratic periods, including the Rana and *panchayat* regimes, and in the recent period from February 2005 to mid 2006, during which the king dissolved the democratic government.

Since the resumption of democracy in 1990, a number of new actors have become keen to influence forest policy formation, particulary democratic representatives, civil society and donors. However, the formal policy-making process still lacks substantial stakeholder participation since it still lacks a consultative and inclusive process. Since the resumption of democracy, the strength of civil societies has increased and these have acted as lobbying bodies (e.g. forest user networks and civil society). However, in many cases, stakeholders are ignored altogether during decision-making. For example, since 2000, the government has increased taxes on income from surplus timber sales from CF without wider consultation.

This practice grew during the period of February 2005 to mid 2006, during which time the king dissolved democratic government. For example, in 2005, in the absence of parliament, regulations presented as 'ordinances' to privatize 'protected areas' (except the Annapurna Conservation Area, a protected area given solely to the king-controlled King Mahendra Trust for Nature Conservation, or KMTNC) were made without consulting the people, as well as a section of the forest bureaucracy itself (Adhikari and Ghimire, 2006, p4). However, since Jan Andolan II, or the People's Movement of April 2006, these ordinances have been scrapped by the present popular coalition government.

There has also been heavy donor influence on policies. The alarmist Theory of Himalayan Environmental Degradation, which forecast ecological doom in the country, led to strong donor support for forestry, with an emphasis on reforestation. The forestry sector's Master Plan, with its focus on CF, was also donor-supported. IFI projects have funded higher education to most DoF personnel and have used the experience gained in field PFM implementation to influence policy at the centre (for example, in 1993, the Nepal–Australia Forestry Project helped to develop CF guidelines based on the ideas and processes of grass-roots-level CF implementation in the project districts). However, due to the insurgency and the political crisis from 1996 onwards, IFIs have not been as successful in influencing policy as during the early 1990s.

District-level actors such as the district forest office and donor-supported PFM projects must remain within the policy and legal framework set at the national level, but have a degree of latitude to make their own strategies for implementing PFM. District development committees (DDCs) and village development committees (VDCs) can also frame their own terms for implementing PFM, as facilitated by the 1998 Local Self-Governance Act. In practice, however, there is only limited variation evident from one district to another.

The CF policy was not explicitly disseminated at the local level; therefore, local people, especially marginal groups, have limited knowledge about the policy or its provisions, unlike the elites, the educated and those with access to information who understood the changes more clearly. In many places, this led to the exclusion of marginal groups. Although, in the past, these groups had access to forests through local social relations, the new system created two legally distinct categories of people: CFUG members in power and the powerless. The powerless are largely defined as those individuals who had little political influence and were therefore excluded, leaving them worse off than before CF.

Current policy process

With the 1993 Forest Act governing CFUGs already in place, new policy guidelines in the form of circulars and directives are frequently issued to revise or refine it. In the absence of debate, these have increased the control of the state, including the district forest office. Most CFUGs are not informed about the contents of the circulars and directives and so do not come to know about them for some time, if at all. Recent policy emphasis on poverty reduction, also through PFM, was introduced after the ninth Five-Year Plan (1997–2002) and was especially emphasized in the tenth Five-Year Plan (2002–2007). But this national-level priority has hardly been translated into district- and local-level plans, particularly under the prevailing conflict situation, and, therefore, the impact on poverty has remained low.

Implementation processes

CFUG regulations have imposed a rigid and standard format for implementing CF across the hills, even though the forestry ecology and livelihood context differ from one location to another. There are some ambiguities in the rules and regulations that give potential flexibility to district forest offices. In general, however, the district forest office and its staff do not take risks in developing innovative implementation approaches in terms of social development activities since this might invite questioning from their seniors and therefore compromise their career prospects.

There is a set system of implementation included in the CF guidelines. The district forest office asserts itself through advice and through the drafting and/or approval process for the operational plan. Since this plan is also a technical forestry document, district forest office staff have greater say in composing its contents.

Donor support has been useful for DFOs' capacity-building, enabling and prompting them to visit and guide CFUGs, and thereby increasing their exposure to the field situation. However, the district forest office, with the prompting of donors, pursued the formation of CFUGs in a rapid and often hasty manner, without proper emphasis on the quality of the process, inclusion or meeting the challenges imposed by social structure and dynamics. Donor projects have also supported NGOs and networks, such as FECOFUN, in social mobilization only to a limited extent.

Most district forest office staff, in interviews, considered the new provisions for forest inventory, lack of support for operational plan revisions and helping CFUGs to tap livelihood opportunities as problems in the implementation process. There are now too many groups for the district forest office to be able to attend to their support needs alone. The district forest office was found to lack adequate human resources to cope with the new challenges brought about by PFM. Staff did not have enough incentive (e.g. in terms of a field allowance) for field-based work.

Local institutional outcomes

The creation of thousands of grassroots-level CFUG institutions has been a major achievement. The regular election system for selecting executive committees by users provides continuity to the administration of CFUGs. These executive committees are controlled by village elites, leaders and wealthier persons, and there are persistent barriers to poor and illiterate people standing for positions in executive committees. The prevailing view in villages is still that educated, confident people should fill the responsible CFUG positions since they are more competent to talk to officers and outsiders. Literacy and accounting skills are also considered important as one has to read rules and regulations, as well as letters from the forest office, take down minutes, and maintain accounts and records. Moreover, poorer people and Dalits need to work every day as wage labourers and hardly have time to attend, let alone run, meetings. Their participation in meetings was found to be generally low, and even the general assembly lacks the presence and participation of low-caste and poor people.

The working culture of CFUGs is similar to that of the other more traditional village institutions. There is an enduring lack of communication across caste and class within CFUGs. And in an unequal socio-economic context, CF has not yet benefited poorer groups, but has often provided hidden subsidies to the well-off, who profit from forest products at subsidized rates – particularly from the CF support provided by government and donor agencies.

Some CFUGs have initiated new activities to address these problems and to improve their inclusive decision-making and equitable benefit-sharing, both with and without donor project assistance. But these activities are confined to limited geographical areas and there is still a lack of mainstreaming and continuity. Similarly, PFM projects have initiated innovative activities in this area; but these have not continued after the withdrawal of IFI assistance. For example, in Dadeldhura and Kaski districts, the Danish-funded Natural Resource Management Sector Assistance Programme (NARMSAP), and in Kavreplanchok, the Nepal–Australia Community Resource Management and Livelihood Project (NACRMLP), both worked on CFUG governance issues; but after their withdrawal most of the best practices discontinued due to lack of support from the district forest office, NGOs and other service providers. This led to a lack of continuity of innovations in PFM institutional development. District-level implementation and coordination committees and groups created by donors (e.g. district project CF support offices) also generally exist only for the duration of the project and lack permanent institutionalization.

Tenure issues

Tenurial issues have affected PFM outcomes in the hills. These include legal government ownership of CF land and its practice of confiscating it (particularly the army taking over some community forests), the lack of recognition for indigenous peoples' historic 'community lands', and the frequent practice of CF legalizing the rights of those included as members, but denying other traditional users' rights as non-members. Because of these problems, forest users are still not confident about whether the government will allow them to continue with CF. The indigenous peoples, who have long held forest land as community property without formal legal rights (e.g. deed papers), resent having lost their resources.

Major policy actors, and their changing characteristics and influence

At the national level, the MoFSC and DoF show an interest in CF primarily as a vehicle for the protection of the forest. Focus on poverty reduction, social inclusion and livelihood improvement objectives in the forest sector has been promoted by donors, and is reflected in the employment of social scientists, women and people from marginalized sections of the society in donor-supported projects if not in regular DoF programmes. FECOFUN, NGOs and civil society are generally in favour of CF and have resisted government policies, such as increasing revenue generation from CF and strengthening of the power of district forest offices over CFUGs. But, again, their emphasis has been on the 'community' rather than on poor, marginalized and disadvantaged people.

DoF personnel at all levels have now come to recognize that local people in the hills can manage the forest by themselves (although the DoF still wishes to retain ultimate control). A new cohort of foresters is being positively trained for CF, and the teaching of social, institutional and forestry knowledge (informed by the experience gained from PFM implementation) has begun at the national forestry schools in Hetauda and Pokhara. To promote this type of training, donors have provided financial support of different types (mainly fellowships) for higher education and on-the-job training. However, despite this knowledge, the government still seems not to have developed specific mechanisms for improved technical forestry suitable for PFM, inclusive decision-making and equitable benefit-sharing in CFUGs.

At the field level, the key actors are the CFUGs, NGOs and community-based organizations (CBOs), who interact with VDCs and other 'user groups' formed by different sectoral support services of the Government of Nepal (e.g. agriculture, livestock, women's development and irrigation services), as well as by NGOs and big international non-governmental organizations (BINGOs). Because of a recent emphasis on the 'group approach', a large number of groups have been formed in the villages by these agencies. It is not surprising to find a member of a CFUG belonging to as many as 12 other village groups. There are no mechanisms, as yet, to coordinate these groups; as a result, there are numerous duplications in activities. For example, most of the user groups carry out savings and credit, income generation activities and social development. This is a burden on poor people since they are expected to attend the monthly meetings of these different development groups, pay donations, levies and other charges, and provide free labour for village infrastructure improvement works.

Village development committees are the lowest level of local government and have much overlap with CFUGs in terms of charging taxes and managing livelihood-related development. VDCs and CFUGs are frequently now coming together in informal meetings to coordinate their respective work.

Donors' orientation towards CF has undergone a major change over the last decade, from being forest focused to viewing forestry more as a social arena and potential vehicle for poverty reduction. They employ social scientists to this end, and apply social inclusion to their own personnel and programmes. Despite this, they have not been able to reorient government forest policies for social development and poverty alleviation. For example, there are, as yet, no policies, guidelines or mechanisms in place within the forest sector for social inclusion, women's development, equity and justice, or employment generation.

The judiciary has shown a tendency to support local communities' rights to local resources. For instance, when the government increased the tax on CFUG timber sales to 40 per cent, the Supreme Court gave a verdict in a public interest litigation case that this should not be levied.

The politics of information and knowledge about PFM in Nepal's hills

Our study of CFUGs has indicated that there are several barriers to the flow of information to marginalized people (such as Dalits, women and illiterate people) regarding CF policy, DoF activities and CFUG decisions. Government staff generally contact village elites, on whom they depend for advice and information when planning forest development activities, and in this way developmental planning is often biased towards elite priorities, rather than towards the needs of the majority and marginal groups. Furthermore, government staff, as well as larger CFUGs, pass information through written notices, which makes access difficult for illiterate people (mainly women, Dalits and the poor). Overall, it is apparent that lack of access to information is a major cause of exclusion in CFUGs.

Despite great opportunities in decentralized forest management planning, government staff's technical knowledge has been gradually replacing local knowledge in forest management practices. Historically, local people have developed complex local knowledge systems and practices relating to forests (including identification, management, harvesting and utilization of plants). However, this body of knowledge is hardly acknowledged; rather, standardized forestry techniques are promoted through operational plans written primarily by DoF staff or with their advice and consent.

The government requirement that forest inventories are carried out by government staff, or a technical person recognized by the DoF, is another problem – it implies the supremacy of 'scientific' knowledge and the concern only for conservation. It is not that people do not know the relative health of the forest, how much can be harvested and what tree species are important for them. But what needs to be planted or not planted and how it should be managed are guided by government. This increases the control of 'experts', leading to the erosion of local knowledge and the obstruction of local experimentation and the development of new knowledge.

The necessary mechanisms for conducting research on people's forestry-related issues, and for disseminating or scaling up knowledge and practices gained through local people's experience, are lacking. Externally supported projects have helped here: they have published newsletters and books that cover some of the grassroots experiences of local people and their informal innovations. However, projects are constrained by a time horizon limited to their lifetime, and many of their initiatives do not continue beyond this.

Action to change policy and practices

The study CFUGs and the service providers identified a number of policy changes necessary to enable CFUGs to manage CFs sustainably and equitably. These include promotion of improved governance and transparency at all levels, promotion of participatory planning, and making service providers more responsive to the expressed needs of CFUGs.

Policy guidelines need to make inclusive decision-making and governance by CFUGs mandatory. Additionally, CFUGs need to play a role in district-level forest-sector planning in order that district support processes are based on the needs of forest users and reflect the pro-poor spirit of the tenth national Five-Year Plan.

The district-level CF monitoring system is currently focused on providing information for central authorities, such as the DoF, the MoFSC and the National Planning Commission (NPC) on the progress of service providers such as the district forest office and projects against annual targets. It does not focus on the development of the capacity of CFUGs, and information collected from the grassroots becomes diluted by the time it has been aggregated for the central authorities. A system of self-monitoring of CFUGs, in which they themselves develop criteria according to their needs, would be more valuable as a developmental process. These criteria could be incorporated within operational plans and CFUG constitutions. The procedure to create these documents also needs to be simplified so that users themselves can incorporate their knowledge and needs. This self-monitoring system would likely lead to increased complexity in the compilation of data at the regional or national levels; but with computerization, this increased detail could be a benefit rather than problem.

So-called 'public' or 'social' auditing is becoming a popular instrument for making transparent, and thereby improving, CFUG fund management and use. Public auditing and reporting by CFUGs themselves to a gathering of forest users was effective in terms of reviewing progress, communicating the budget and the expenses for different activities, finalizing the pressing needs of forest users and communities, and deciding on new CFUG plans. This is generally done at the user assembly, often held on a half-yearly or annual basis, and is far better than the unaffordable audit by registered auditors or government-employed accountants, whose reports have no meaning for the common villagers. The greatest current challenge is the lack of facilitation skill among executive members and service providers.

Main opportunities and constraints to achieving PFM

The development of the local-level institutions and enhancement of the forest resources provide a basis for the sustained management of resources and increased incomes of local people. There is also the potential to modify these institutions to benefit the poorest and marginalized communities if proper policies to include them in decision-making are set out. However, the existing unequal social and political structure that keeps the poor and marginalized powerless and disempowered remains a major constraint. Much of the resources of donor projects are focused in this area, although the majority of CFUGs and DoF staff at grassroots level still do not seem responsive to their needs and priorities. Without decision-making power and information, the poor and marginalized cannot make decisions or use their influence or voice to increase their benefits or to make policies appropriate to themselves.

Existing PFM policy trajectories and their impacts

The current pattern of PFM outcomes and impacts in the Nepal hills, as discussed in this chapter, has been due to the continued predominance of elite control of CFUGs and the leading role of the DoF. Because they have emphasized closure of forest and a conservative approach to forest management, forest cover has increased, but benefit flows have been constrained. There has been a slight increase in the flow of fuelwood and fodder; but timber has been emphasized as a priority product, rather than the diverse products and flows important to poor and disadvantaged households. Frequent exclusion of these groups from membership and benefits has led to hardship. Community funds have generally been allocated to pro-rich priorities, such as roads, electrification, temples and school development. *Equal* rights principles for all community members in the forest have meant that consideration of the poor's special needs and positive discrimination in their favour have not been achieved.

Nevertheless, there have been a growing number of experiments to increase the influence of poor/marginal groups in decision-making – for example, by forming '*tole*', or hamlet-level groups or committees, social mobilization and public auditing by trained facilitators funded by donor agencies. A few CFUGs themselves have initiated innovations that discriminate positively in favour of the poor and marginalized (such as by giving them proportionately more products or unused forestland for their cultivation of forest-based crops). Many service providers have later adopted this as a strategy for self-employment for the poorer groups; but the impacts of these methods and the possibilities for scaling up have not been well documented or promoted after donor support ended. The persistence of the attitude that the DoF works are not aimed at reducing poverty, but at the conservation of forests, still guides practices at various levels, especially at the grassroots level, and this has also led to the outcome of CF's minimal impact on poverty reduction. The further issue of legal duplication – as in the Local Self-Governance Act, which overrides the authority of CFUGs – has added to the lack of innovation and efficiency in CFUGs.

Unequal landownership, feudalism and unjust social and economic relationships are the major factors that allow discriminatory decision-making in favour of wealthier groups to continue. Even despite these factors, poverty alleviation is still a difficult proposition for hill-based CFs: there are very limited absolute levels of resources (monetary and forest based) to mobilize. Furthermore, hill forests close to settlements are rarely resource rich and therefore are mainly used for subsistence. They lack, for instance, the high-value NTFPs found mainly in the high mountains.

Finally, rural livelihoods in the hills have traditionally been *subsistence oriented* and local people have been slow to pick up on the many new enterprises. There are, however, a few examples of CF-based 'enterprise development', which suggests that service providers can play a key role in helping CFUGs to set up an enterprise. This was seen in Kavreplanchok district in the development of a wood-processing sawmill enterprise with financial support from IFIs (Singh, 2005). Nevertheless, many attempts at enterprise development have not been sustained and have terminated after the projects were phased out.

What is 'best' PFM practice under different (often hostile) conditions, and how can it be scaled up or replicated?

Best practices vary from place to place depending on the socio-ecological and political environment. The best practices, which can be replicated with suitable modification to suit local conditions, mainly relate to improving the role and benefit flows to the poorest and to promoting enterprise development:

1   *Closer collaboration between local governments and CFUGs.* This has been helpful in developing forestry-related enterprises. For instance, in Dhankuta the DDC, along with an NGO project and the CFUG, helped to set up an essential oil-processing plan for the CFUG. This is an example of an interactive and responsive approach by local government. In a few cases, VDCs have also provided matching funding for various activities, which is important in the hills, where CFUGs' resources have been low.

2   *Income generation and employment creation.* Allowing the poor to use resources that are not directly of interest to the elites can bring in extra income. For example, the elites do not challenge the production of briquettes from *banmara* weed (*Lantana* spp) or growing grasses for essential oil production by the poor, nor do they wish to share the income

from them as these activities are not sufficiently lucrative. The practice of providing the poor with forest-related work (e.g. NTFP collection in Dadeldhura and Dhankuta districts) has provided them with employment.

3   *Creation of separate representative committees of women and poor and marginalized people within CFUGs, and increasing their presence at executive meetings to present and make decisions on their agendas.* This best practice allows the agendas of the poor to come to the fore. For example, in one CF where a poor group was given the authority to invest group funds, it decided to use them as credit to support goat farming in poor households. This practice seems to balance the absence of marginalized people's voices on executive committees. In this way, the distribution of burden and benefits would also be equitable.

4   *Community–private partnerships within CFUGs.* In Kaski district, one CFUG member made a huge investment to develop stone-cutting using stone from the CF. He employed the poor members of the CFUG and was given this opportunity for investment, as the CFUG had no funds.

Conclusions

Community forestry policy, which evolved from the traditional forest management practices of local people for subsistence use, has been implemented across the middle hills of Nepal with the assistance of donor agencies. It has created opportunities through which local communities can ensure a sustained supply of subsistence forest products and can generate income. Opportunities vary across CFs depending on forest condition, type, size, availability of forest products, location and accessibility to the market. Most CFUGs have become viable institutions and have demonstrated their capacity to manage forests and generate small funds for rural development. In most CFs, the condition of forests has improved due to effective forest protection measures, such as regulating the collection of forest products, ensuring people's participation in plantation, thinning and pruning, and controlling forest fires. It has affected the livelihoods of people by regulating access to forest resources, and generally improving rural infrastructure and social cohesion to some extent. Despite its success, CF policy requires further development in order to overcome differences of objectives between local communities and the DoF. Forest users want a greater contribution to their livelihoods, rather than just improvements to the condition of forests.

Current CF policies and implementation strategies are generally inadequate to address the needs of the hill farming system, which remains an important source of livelihood security for most rural households. The DoF lacks strategies to ensure a sustained supply of fuelwood, herb and wild foods for household needs, as well as fodder, grass and grazing facilities to guarantee livestock and nutrient flows from the forest to agricultural land in order to maintain agricultural productivity. The absence of an equitable timber-sharing mechanism, for instance, means that timber supports for poorer households are generally absent, with the result that their house quality has not improved. Similarly, forest-based employment and income generation through timber-and NTFP-based enterprise development have yet to be developed by CFUGs. Hill forests are generally considered of lower value compared to those of the *tarai*, although CFs in the hills with easy road access have been involved in lucrative business through trading timber and NTFPs in Dadeldhura, Dhankuta and Kavreplanchok districts. CFUGs within each district should be individually categorized by district forest office staff based on the economic potential and specific support needs of each.

This study has revealed that there is potential for CF to contribute to household livelihood improvement and the empowerment of men and women in poorer groups, who are typically more dependent on forest resources than richer groups (although this potential is not uniform at all sites). Thus, CFs have only contributed to household livelihood strategies to a limited extent, and in some cases they have worsened the conditions of the poorest – usually disadvantaged groups such as the Dalits and poorer groups where forest-based income generation activities and access to forest products have been rationalized and controlled by CFUGs.

There are different areas of best practice in varying CFUGs; but how far these can be replicated in other CFUGs is difficult to ascertain. The best way would be to use an interactive approach to discover which interventions benefit the poor but would not be challenged by non-poor groups. Poor and other socially excluded groups such as women know what works for them, and if they are given the decision-making role and an opportunity to invest in activities that benefit them, they can do so with little external guidance. Therefore, locally appropriate methods, such as fixing their own quotas or forming their own sub-committee to present their agendas, for increasing the poor's participation in decision-making bodies are important. Women-only CFUGs can also be formed to encourage women's participation in CF decision-making and social activities.

The contribution of CF to household livelihoods depends to a large extent on the quality of the support structure and the 'enabling environment' provided by CF policy, local government and other stakeholders. So far, the National Forest Sector Policy and plans appear to have become diluted by the time they reach the district and village levels.

The periodic and annual plans of district forest offices tend to diverge from both CFUGs' objectives and the National Planning Commission's Five-Year Plans. Similarly, effective and comparable monitoring structures to measure the livelihood impacts of PFM implementation and to improve the policy accordingly do not yet exist. Greater participation of NGOs and the private sector in service delivery is needed; and the role of DoF field staff needs to be redefined to focus on livelihood-based forest management.

The role of DoF field staff has increased since the formation of the CFUGs; but due to their lack of capacity to support CFUGs in any way beyond technical forestry and legal matters, they are unable to meet the demand for support in the institutional development of CFUGs, forest-based livelihoods, equitable benefit-sharing and rural development. Therefore, the services available to CFUGs from the DoF are very limited and focus on preparing the operational plan and advice in forest development and management. CFUGs in remote parts of the country rarely get any support at all. Some donor-supported projects have contracted local NGOs and CBOs as new service providers to CFUGs. During project lifetimes, at least, NGOs (who also have only limited capacity) provide services to the CFUG in social mobilization, rural infrastructure development and income generation activities. Except for DoF field staff support in operational plan revision and forest management, most support services stop after the phasing-out of the project. This happens due to a lack of coordinated mechanism among service providers and government. An inclusive process of planning and service delivery among service providers and stakeholders such as CFUGs, the district forest office, VDCs, DDCs, CBOs, NGOs, the private sector and projects under the coordination of DDC (district government) and VDC (local government) is needed in order to meet CFUG demands, and to ensure that the CF programme continues to evolve.

References

Adhikari, J. (1996) *The Beginnings of Agrarian Change: A Case Study in Central Nepal*, Kathmandu, T. M. Publications

Adhikari, J. (2000) *Decisions for Survival, Farm Management Strategies in the Middle Hills of Nepal*, New Delhi, Adroit Publications

Adhikari, J. (2003) 'Food security in the context of globalization', in Chene, D. M. and Onta, P. (eds) *Social Science Thinking in Nepalese Context*, Kathmandu, Social Science Baha, pp217–249

Adhikari, J. and Bohle, H.-G. (1999) *Food Crisis in Nepal: How Farmers Cope*, New Delhi, Adroit Publications

Adhikari, J. and Ghimire, S. (2006) 'New ordinance is anti-people', *The Kathmandu Post*, Kathmandu, 25 January, p4

Balland, J. M., Platteau, J. P. and Bardhan, P. (2000) 'Irrigation and co-operation: An empirical analysis of 48 irrigation communities in south India', *Economic Development and Cultural Change*, vol 48, no 4, pp845–865

Bartlett, A. and Malla, Y. (1992) 'Local forest management and forest policy in Nepal', *Journal of World Forest Resource Management,* vol 6, no 2, pp99–116

Bhatt, B. (2005) 'More destruction than Conservation', *National Weekly,* vol 6, no 15, pp30–31

Bhattarai, A. M. and Khanal, D. R. (2005) *'Community, Forests and Law of Nepal: Present State and Challenges*, Kathmandu, Federation of Community Forest Users Nepal and Forum for Protection of Public Interest

Bingen, J. (2000) *Institutions and Sustainable Livelihoods*, Michigan, Michigan State University, www.livelihoods.org

Blaikie, P., Cameron, J. and Seddon, J. D. (1980) *Nepal in Crisis: Growth and Stagnation at the Periphery*, New Delhi, Oxford University Press

Blaikie, P., Cameron, J. and Seddon, D. (2002) 'Understanding 20 years of change in west-central Nepal: Continuity and change in lives and ideas', *World Development,* vol 30, no 7, pp1255–1270

Blaikie, P. and Coppard, D. (1998) 'Environmental change and livelihood diversification in Nepal: Where is the problem?', *Himalayan Research Bulletin,* vol 2, pp28–39

Branney, P. and Yadav, K. P. (1998) *Changes in Community Forest Condition and Management (1994–1998)*, Report G/NUKCFP/32, Kathmandu, NUKCFP

Brower, B. (1990) 'Range conservation and Sherpa livestock management in Khumbu, Nepal', *Mountain Research and Development*, vol 10, pp34–42

CBS (Central Bureau of Statistics) (2003) *Statistical Year Book of Nepal 2003*, Kathmandu, CBS

CBS (2005) *Poverty Situation in Nepal*, Kathmandu, CBS

Chapagain, D. P., Kanel, K. R. and Regmi, D. C. (1999) *Current Policy and Legal Context of the Forestry Sector with Reference to the Community Forestry Programme in Nepal*, Kathmandu, NUKCFP

Dayton-Johnson, D. (2001) *Peasants and Water: A Review Essay on the Economics of Locally-Managed Irrigation*, Halifax, Nova Scotia, Dalhousie University

DoF (Department of Forests) (2005a) *Community Forest User Group Data Base 2005*, Kathmandu, Community Forestry Division, DoF

DoF (2005b) *Forest Cover Change Analysis of the Tarai Districts (1990/1991–2000/2001)*, Kathmandu, His Majesty's Government, Ministry of Forest and Soil Conservation, DoF

DoF (2006) *Community Forest User Group Data Base 2005*, Kathmandu, Community Forestry Division, DoF

Eckholm, E. P. (1976) *Losing Ground: Environmental Stress and World Food Prospects*, New York, World Watch Institute

Edwards, D. M. (1996) *Non-Timber Forest Products from Nepal, Aspects of the Trade in Medicinal and Aromatic Plants*, Monograph No 1/96, Kathmandu, Forest Research and Survey Centre (FORESC), Ministry of Forests and Soil Conservation

Fisher, R. (1989) *Indigenous Systems of Common Property Management in Nepal*, Working paper no 18, East–West Centre, Honolulu, Environment and Policy Institute

Gilmour, D. (1989) 'Management of forests for local use in the hills of Nepal: Changing forest management paradigm', *Journal of World Forest Resource Management,* vol 4, no 2, pp93–110

Gilmour, D. and Fisher, R. (1991) *Villagers, Forests and Foresters: The Philosophy, Process and Practice of Community Forestry in Nepal*, Kathmandu, Sahayogi Press

Graner, E. (1997) *The Political Ecology of Community Forestry in Nepal*, Saarbrucken, Freiburg Studies in Development Geography, Verlage fur Entwicklungspolitik Saarbrucken Gmbh Auf der Adt D-66130

Gurung, R. (1997) *Forest Management by the Gurungs and Bahuns in the Annapurna Range: A Comparative Study*, Kathmandu, Department of Sociology and Anthropology, Tribhuwan University

HMGN (His Majesty's Government of Nepal) (1993) *Forest Act 1993*, Kathmandu, Law Books, Government Press

HMGN (1995) *Forest Regulations (Official Translation)*, Kathmandu, Ministry of Forests and Soil Conservation

Hobley, M. (ed) (1996) *Participatory Forestry: The Process of Change in India and Nepal*, London, ODI

Hobley, M. and Malla, Y. (1996) 'From forest to forestry: Three ages of forestry in Nepal – privatisation, nationalisation and populism', in Hobley, M. (ed) (1996) *Participatory Forestry, Process of Change in India and Nepal*, London, ODI

Ives, J. D. (1987) 'The Theory of Himalayan Environmental Degradation: Its validity and application challenged by recent research', *Mountain Research and Development,* vol 7, no 3, pp189–199

Ives, J. D. and Messerli, B. (1989) *The Himalayan Dilemma: Reconciling Development and Conservation*, New York, Routledge and the United Nations University

Johnston, R. and Libecap, G. (1982) 'Contracting problems and regulation: The case of the fishery', *American Economic Review,* vol 72, pp1005–1022

Joshi, A. L. (1996) 'Effects on administration of changed forest policies Nepal', in *Policy and Legislation in Community Forestry: Proceedings of a Workshop (27–29 January 1993)*, Bangkok, RECOFTC, pp103–113

Kandel, K. and Kandel, N. (2005) *Impact of Community Forestry in Poverty Reduction: Analysis of NLSS (2003–2004) Data*, Kathmandu, CBS

Kanel, K. R. (2000) 'Analysing policy for poverty alleviation: An example from non-timber forest sub-sector', *Banko-Jankari*, vol 10, no 2, pp3–8

Karna, A. (1998) *Critical Examination of Current Approaches to Participatory Community Forestry Planning in Nepal*, MSc thesis, Reading, UK, University of Reading

Messerschmidt, D. (1993) 'Linking indigenous knowledge to create co-management in community forest development policy', in *Policy and Legislation in Community Forestry: Proceedings of a workshop, 27–29 January 1993*, Bangkok, Ford Foundation/Winrock International/Australian International Development Assistance Bureau

MoPE (Ministry of Population and Environment) (2001) *State of the Environment Nepal (Agriculture and Forests)*, Kathmandu, HMG

National Planning Commission (2004) *District Development Profiles 2004*, Kathmandu, National Planning Commission (NPC) Nepal

NPC (National Planning Commission) (1997) *The Ninth Five-Year Plan (1997–2002)*, Kathmandu, National Planning Commission, Government of Nepal

NPC (2003) *The Tenth Five-Year Plan/Poverty Reduction Strategy Paper*, (2002–2007), Kathmandu, National Planning Commission, Government of Nepal

Ostrom, E. (1990) *Governing the Commons: The Evolution of Institutions for Collective Action*, Cambridge, Cambridge University Press

Ostrom, E., Schroeder, L. and Wynne, S. (1993) *Institutional Incentives and Sustainable Development: Infrastructure Policies in Perspective*, Boulder, CO, Westview Press

Pokharel, R. K. (2005) 'A local perspective on indicators of successful community forestry program: A case of Nepal's Kaski district', *Forestry*, Journal of Institute of Forestry, vol 13, pp29–34

Poteete, A. R. and Ostrum, E. (2004) 'In pursuit of comparable concept and data about collective action', in *Proceedings of the Fourth National Workshop on Community Forestry*, Kathmandu, DoF, pp385–398

Regmi, M. C. (1978) *Land Tenure and Taxation in Nepal*, Kathmandu, Ratna Pustak Bhandar

Seddon, D., Adhikari, J. and Gurung, G. (1998) 'Foreign labour migration and remittances economy of Nepal', *Himalayan Research Bulletin,* vol 2, pp3–10

Sharma, A. R. (2002) 'Community forestry development program', in Ministry of Forests and Soil Conservation (ed) *Our Forest*, Kathmandu, MoFSC, pp14–28

Singh, H. B. (2005) 'Choubas-Bhumle community saw mill: Empowering local people', in Patrick, B. D., Chris, B., Henrylito, D. T. and Miyuki, I. (eds) *In Search of Excellence: Examplary Forest Management in Asia and the Pacific*, Bangkok, FAO and RECOFTC

Soussan, J. G. Shrestha, B. K. and Uprety, L. P. (1995) *The Social Dynamics of Deforestation*, United Nations Research Institute for Social Development (UNRISD), London and New York, Parthenon

Springate-Baginski, O., Dev, O. P., Yadav, N. P., Keif, E. and Soussan, O. J. (2001) *Community Forestry in Nepal: Progress and Potentials*, London, DFID

Tang, S. Y. (1994) 'Institutions and performance in irrigation systems', in Ostrom, E., Gardner, R. and Walker, J. (eds) *Rules, Games and Common Pool Resources*, Ann Arbour, MI, University of Michigan Press

Timsina, N. and Paudel, N. S. (2003) 'State versus community: A confusing policy discourse in Nepal's forest management', *Journal of Forest and Livelihood,* vol 2, no 2, pp8–16

Tiwari, S. (1996) *Community Forestry in the Hills of Nepal: A Property Rights Approach to Resource Management*, MSc thesis, Edinburgh, Scotland, University of Edinburgh

Tiwari, K. B. (2002) *Historical Review of Nepal's Forest Management*, Kathmandu, Multi Graphic Press

Tulachan, P. M. and Neupane, A. (2001) *Livestock in Mixed Farming Systems of the Hindu Kush-Himalayas: Trends and Sustainability*, Kathmandu/Rome, ICIMOD and FAO

Varughese, G. and Ostrom, E. (2001) 'The contested role of heterogeneity in collective action: Some evidence from community forestry in Nepal', *World Development,* vol 29, no 5, pp747–765

Participatory Forest Management in the Nepalese *Tarai*: Policy, Practice and Impacts

Binod Bhatta, Akhileshwar L. Karna, Om Prakash Dev and
Oliver Springate-Baginski

Synopsis

The northern *tarai* region (also called *terai*) of Nepal was densely forested and sparsely populated until the malaria eradication programme took effect in the 1960s. Since then, much of it has gradually been cleared and converted into agricultural land for the resettlement of the hill population. The harvesting of extremely high-value sal (*Shorea robusta* spp) forests for revenue has also been a major factor in the deforestation. There has been no systematic management of forests, and timber mafias conduct large-scale operations in the *tarai*.

The problem of the destruction of forests was gradually recognized during the 1980s; but there has been very limited political will to address this problem since it conflicts with the patronage politics of land allocation to hill migrants. The *tarai* has also seen extensive oppression of its tribal populations, including *Tharu* forest dwellers, by armed Department of Forests (DoF) staff, which has not entirely ceased. Community forestry (CF) has been partially implemented in some *tarai* areas along lines similar to those used in hill forests. However, the ambivalent attitude of the DoF to handing over these forests has led to ongoing institutional problems, which, as mentioned in previous chapters, primarily involve the exclusion of distant forest users; elite control of benefit flows; lack of organized livelihood support from participatory forest management (PFM) for poorer groups; the deteriorating condition of government-managed forests; government allocation of forests for infrastructure development; and security forces' use of forests during the recent political conflict.

This chapter first considers the policy context, then assesses the implementation and impact of PFM in the *tarai*, and then looks at governance issues. Finally, it analyses the potentials and constraints of PFM in the *tarai*.

Policy and context in the *tarai*

Socio-economic profile of the *tarai* and politics

The Nepalese *tarai* forms the southern plain land of Nepal. Of the 75 districts in Nepal, 18 have 'true' *tarai* areas (i.e. plains only), although a further 6 'inner *tarai*' districts consist mainly of valleys in the Siwalik foothills. Altogether, the *tarai* covers about 23 per cent of the total land area of the country and accommodates 47 per cent of the country's population.

The *tarai* forests form a narrow belt to the north of the fertile Gangetic plains. The major tree species is sal (*Shorea robusta* spp), a very high-value timber, although there is remarkable species diversity in the forest. After the malaria eradication programme began in the late 1950s, migrants from the hills cleared substantial forest areas and created new settlements, bringing about significant changes in the *tarai*'s land system and agro-ecology. The indigenous peoples and densely populated areas of the *tarai* are now concentrated mostly in the southern belt far from forests.

Nepalese society is multi-ethnic, consisting of the descendants of three major classes of migrants from India, Tibet and Central Asia (Bista, 1991), as well as indigenous groups. Some of the earliest indigenous inhabitants of the *tarai* still live here: Tharu, Danuwar, Rajbanshi, Satar, Mushahar, Dushadh, Tatma, Khang, Halkhor, Khatwe, Dom, Mallah and Kewat, among others. The recent settlers from the hills took over the forest lands from these indigenous inhabitants, who were tricked into losing much of their land and property.

With its fertile land, both irrigated and with potential for irrigation, the *tarai* has been called 'the grain basket of Nepal', feeding not only the *tarai* people but also exporting grain to the hills and mountains. It has many forest-based industries (e.g. sawmills, medicinal plant processing and wood-based industries) and is also suited to horticultural production. The many food processing and manufacturing industries established here provide employment for a large number of people, including non-natives.

Economically, the *tarai* has been a relatively prosperous area compared to the hills and the mountains because, apart from its productive agriculture and other resources, it has good access to markets both within Nepal and in the northern states of India. There is a strong market linkage between Nepal and India in both legal and illegal trades, including in timber and non-timber forest products (NTFPs) through its 'leaky' open border. Despite this economic dynamism, the *tarai* also has a high and growing incidence of poverty that equals the overall national poverty index. At present, 41 per cent of the *tarai* population currently lives below the poverty line (CBS, 2002). Some 4.2 million indigenous peoples of the *tarai*, such as Tharus and Danuwars, as well as disadvantaged caste groups (Dom, Chamar, Dushadh, Mushahar, Mandal, Khang, Halkhor, Kumhar and Tatma), live in extreme poverty.

The inner *tarai*, covering the valleys of the Shivalik foothills, contrast with the *tarai* proper. The major occupants of the inner *tarai* are Tharu and Danuwar, among other native ethnic groups and migrants from the hills and mountains, usually settled in the central part of the inner *tarai* and surrounding forest areas. Such conditions have important implications for PFM, and so the *tarai* and inner *tarai* need to be examined and understood separately in relation to PFM.

Overview of forests today

The 24 districts of the *tarai* and inner *tarai* contain 2,039,257ha of forest (Adhikari, 2002), or 47 per cent of the total land area. The high value of *tarai* forest products and land, as well as its easy access to market, are strong incentives for illicit extraction and encroachment, resulting in forest degradation, clearance and conversion to other land uses. The *tarai* forest area was reported to be declining at an annual rate of 1.3 per cent between 1979 and 1991 (DFRS, 1999), and although this has slowed in the east, the western *tarai* forests continue to degrade due to migration from the hills, illegal timber trading, and allocation of forest land for other uses and initiatives, such as educational institutions, military barracks, resettlement schemes and infrastructure (e.g. roads and urban areas). In 2004, about 97,050 households had 'illegally' encroached on 70,256ha of national forest across the *tarai* region, which represents 3.45 per cent of the total forest land in the *tarai* and inner *tarai* (Adhikari, 2002).

The DoF has handed over 308,168ha (15 per cent of total forest land) of forest to 2157 community forest user groups (CFUGs) in the region (1200 in *tarai* and 967 in inner *tarai*), as illustrated in Table 6.1. A large proportion (1,401,723ha) of national forests remains under DoF management.

Table 6.1 *Land use and forest 'encroachment' recorded in 2004 in the* tarai

District	*Tarai/* inner *tarai* regions	Total area (ha)*	Forest land (ha)**	% of district classified as forest land	Forest 'encroach- ment' (ha)**	Number of CFUGs***	Total CF area (ha)***	% of forest area under CF
Udaypur	Inner *tarai*	206,300	124,242	60	7000	161	34,065	27
Sindhuli	Inner *tarai*	249,100	167,947	67	1089	223	36,711	22
Makwanpur	Inner *tarai*	242,600	163,413	67	1727	255	38,118	23
Chitwan	Inner *tarai*	221,800	128,500	58	2070	24	9293	7
Dang	Inner *tarai*	295,500	161,900	55	1797	373	66,250	41
Surkhet	Inner *tarai*	245,100	177,855	73	163	199	31,540	18
Jhapa	*Tarai*	160,600	15,000	9	1990	28	7685	51
Morang	*Tarai*	185,500	55,500	30	537	28	3260	6
Sunsari	*Tarai*	125,700	15,000	12	1752	12	180	1
Saptari	*Tarai*	124,100	27,200	22	31	105	10,087	37
Siraha	*Tarai*	118,800	21,800	18	1084	80	12,925	59
Dhanusha	*Tarai*	118,000	24,800	21	1038	29	8032	32
Mahottari	*Tarai*	100,200	21,900	22	565	53	1150	5
Sarlahi	*Tarai*	100,200	25,600	26	647	31	2940	11
Rautahat	*Tarai*	112,600	29,400	26	193	13	1050	4
Bara	*Tarai*	119,000	47,200	40	1340	13	1818	4
Parsa	*Tarai*	135,300	71,700	53	38	25	102	0
Nawalparasi	*Tarai*	216,200	100,100	46	5673	34	2638	3
Rupandehi	*Tarai*	136,000	29,600	22	7003	43	7698	26
Kapilvastu	*Tarai*	173,800	68,800	40	691	24	1455	2
Banke	*Tarai*	233,700	148,800	64	658	77	8367	6
Bardiya	*Tarai*	202,500	124,800	62	1419	163	7528	6
Kailali	*Tarai*	323,500	200,000	62	19,527	128	10,590	5
Kanchanpur	*Tarai*	161,000	88,200	55	9989	36	4686	5
Total		4,307,100	2,039,257	47	68,021	2,157	308,168	15

Source: *District Development Profile (2004); **Adhikari (2002); ***DoF (2006)

During the Rana rule, valuable timber and other forest resources were exported from the *tarai* to India (Mahat et al, 1986), and these were the state's main source of revenue. The DoF was established in 1942 to streamline the supply of high-value timber to India. When the Rana regime ended in 1951, about three-quarters of the *tarai* forests were under some form of private ownership, such as *birta* (a tax-exempt land grant to rulers' relatives and favoured elites) or job (*jagir*), granted during the regime to administrative, military or political elites. The 1957 Private Forest Nationalization Act and other related acts, such as the 1961 Birta Abolition Act, brought these forest resources under state ownership and ended the discretionary granting of large parcels of forests to elites. This shifted the responsibility of forest management onto the DoF, which had to cope with limited infrastructure and capacity, and contradictory policies that demanded forest land for uses such as resettlement and infra-structure development. Large areas of *tarai* forest were also cleared during political crises in the country due to the *de facto* open-access situation – for instance, during the 1980 refer-endum, during the people's movement in 1989 and, more recently, during the insurgency from the late 1990s to 2006.

The *tarai* forests have never been managed systematically or even according to basic sustainable forestry principles, despite the presence of enabling conditions such as roads, markets and productive land. Not until recently did the ruling political powers even see any benefit in introducing a sustainable management system. Even after the fall of the Rana regime, harvesting of trees was sought, rather than management of the regeneration process. Since land value in the *tarai* has always been very high, forest use has had to compete with other land uses and has gradually yielded to them – senior planners in the government have always considered *tarai* forest as potential land for settlement and agriculture. Forest management plans have been prepared in recent years, but never implemented, mainly due to the weak commitment of politicians and policy-makers, as well as social contestation over their contents. Even after the introduction of the 20-year Master Plan for Forestry Sector (MPFS) in 1989, and despite several management initiatives, such as preparing forest management plans, attempting large-scale plantation and experimenting with collaborative forest management (CollFM) on the part of the DoF to manage and enhance *tarai* forests, revenue from forests has declined rapidly as they have been cleared for agriculture, settle-ments, infrastructure and other purposes, and because the remaining forests are not being managed in a sustainable intensive manner. The *tarai* forests, if sustainably managed, could enhance the livelihoods of the poor and marginalized people living in and around them. Sustainable harvesting of timber is estimated to potentially yield around US$150 million annually (Hill, 1999).

Role of forests in agrarian livelihoods

Until the mid 1980s, the indigenous *tarai* communities used the forest for their livelihoods for the months of the year when they had no other employment, supplementing their food with various forest products (roots, tubers, leaves, climbers, flowers, fruits, hunted animals and birds) or collecting forest products (timber, firewood, medicinal plants and grasses) to sell in order to buy food. However, after the malaria eradication programme of the 1960s, migrants from the hills rapidly settled in the northern *tarai* close to forests, which were grad-ually cleared, their frontiers receding northwards. The indigenous inhabitants of the *tarai* lost their access to forests and became marginalized and vulnerable, so that they are now some of the poorest groups in the country. Some of these people continue to collect and sell forest products in urban areas and in neighbouring India, despite this being illegal, as they have very few other livelihood options.

The new settlers have massively increased the population of the *tarai* and changed its livelihood practices. There were over 11.2 million people (about 1.965 million households) in the *tarai* region in 2001, of which 76.3 per cent had agriculture as their main source of livelihood (representing 44.2 per cent of the total agriculture-dependent households of the country) (CBS, 2004).

The skewed distribution of agricultural land is one of the major causes of widespread poverty in the *tarai*: 72.8 per cent of the agricultural land is owned by 34.6 per cent of house-holds; 41.3 per cent of *tarai* households own less than 0.5ha of land; and 12.4 per cent are completely landless (CBS, 2004).

Forests and trees form an important part of the *tarai's* rural livelihood system, particu-larly in supporting subsistence farming where agriculture, livestock and forestry systems are interdependent (Gilmour and Fisher, 1991). Forests supply products and services such as

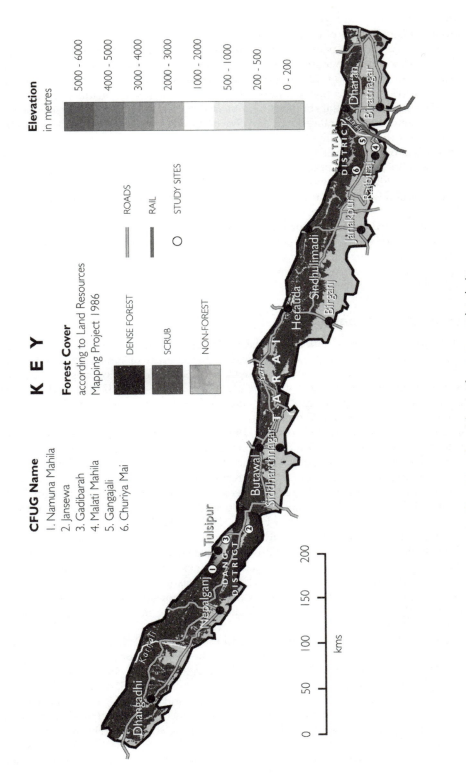

Map 3 Nepal *tarai* showing forest areas and study sites

Source: Jonathan Cate (original material for this book)

timber, firewood, fodder and grazing leaf, medicinal plants, and construction stones and sand, as well as maintain hydrological systems.

The middle class and rich people of the northern *tarai* also rely on forests for their everyday needs (e.g. firewood, timber and NTFPs). Those living along the northern forest fringes are closer to the forest, and so it plays a greater role in their livelihoods. They consequently consume more forest products than people settled far from the forest (Bhatta, 2002). As the forests recede, the people of the southern *tarai* meet their requirements from trees planted on their private land, agricultural residues and cow dung. They also buy such products from people who collect them (illegally) from the forest.

Livestock is another important component of most rural livelihood systems in the *tarai*, and most households, whether poor or rich, own some livestock, although the type, number and breeds vary. Herds are commonly put to graze in the forest for some seasons, as well as on fallow agricultural fields and on village communal land. Many well-off families keep large herds of livestock, although these tend to be of the less productive native breeds. Such people bring their herds to forests for grazing, especially during the dry season when they build them temporary sheds. Poorer people have fewer and smaller ruminants, and since they cannot afford to bring them to the forest to graze because it requires a large herd to become profitable, they engage a member of the family to move cattle around within communal land in the village.

The history of forest policy

During the unification of the country in the late 18th century, the government granted land to favoured civil and military elites and employees. Large areas of *tarai* forest, as well as degraded forest or wasteland, were given as grants and their deforestation was encouraged. Ex-servicemen were also encouraged to settle along the Nepal–India border and to clear forests in strategically important areas in the *tarai* and inner *tarai*, such as in Makwanpur district, for military camps to defend the country against invasion by the British–Indian army.

From the 19th to the early 20th century, the Ranas used the *tarai* forests to generate revenue for the state, as well as to fund their lavish expenses. In 1925, the then Rana ruler hired an experienced British forester, J. V. Collier, to 'supervise and improve' timber felling in the *tarai*. Collier produced a report in 1928 recommending 'extensive clearance of *tarai* forests for conversion into agriculture and settlements'. From 1927, the government Kath Mahal became the central office to manage the extraction of *tarai* timber for supply to India for revenue. It became the Department of Forests (DoF) in 1942.

By 1950, one-third of *tarai* forests had been granted as *birta*, three-quarters of which went to Rana families. Different systems of forest management, such as *talukdari* and *kipat*, operated in conjunction with *birta*. In the *talukdari* system, *talukdars* (local functionaries of the state) were responsible for revenue collection, whereas in the *kipat* system, *jimmawals* (local headmen) were responsible for collecting taxes and managing forest resources (Fisher, 1989; Bartlett and Malla, 1992; Hobley and Malla, 1996). The main aim was to maximize revenue from *tarai* forests.

After the overthrow of the Rana autocracy and the initiation of democracy in 1957, the government nationalized all private forests and made DoF the primary custodian and manager of forest land. However, it lacked the administrative and technical capabilities to assume this role effectively.

Over the 1960s, the 1961 Forest Conservation Act and the 1967 Forest Protection (Special Arrangements) Act were passed, increasing the power of the DoF and creating an atmosphere of antagonism between rural people and forestry staff. Several other acts, such as the 1959 Land Grant Abolition Act, the 1963 Birta Abolition Act, the 1964 Land Reform Act and the 1974 Pasture Land Nationalization Act, also increased the DoF's power (Hobley and Malla, 1996). Despite these powers, the small size and limited technical capacity of the DoF rendered it ineffective (Gilmour and Fisher, 1991). Yet, because of the revenue potential from cutting the accessible sal forests in the *tarai*, the DoF's capacity was almost entirely focused there.

Clearance of *tarai* forests rapidly accelerated during the 1960s and 1970s for two main reasons. During the late 1960s, as mentioned, there was a major malaria eradication programme in the *tarai*. Additionally, during the late 1960s and 1970s, the construction of the East–West Highway, aligned through extensive forest areas, rendered the now malaria-free forests vulnerable to settlement and exploitation. The attraction for migrants from the hills of converting *tarai* forest land to agricultural use and settlement rapidly increased, and the relocation of hill communities to the *tarai* (*paharisation,* as termed by Jha, 1993, quoted in Bhattarai et al, 2002) became a policy priority. The government resettlement programme under the Nepal Punarvas Company (Nepal Resettlement Company) encouraged the clearance of *tarai* forests for the distribution of land to hill people. Thereafter, many families began to migrate from the hills to the *tarai*, and the illegal encroachment of forests grew. The government also distributed forest land to ex-servicemen and people of Nepali origin returning from Burma for settlement until the end of the *panchayat* period. This internal migration, mostly from the hills, has been very substantial. Of the total population of Nepal, 22 per cent were internal migrants (CBS, 2000).

During this period, there was a boom in illegal forest cutting. With the increased population and expansion of urban areas, demand for both timber and firewood soared. The limited legal supply was not sufficient to fulfil even the basic demand. Since the DoF's policing and patrolling efforts were ineffective, there was furtive extraction of forest products on a massive scale by both poor rural people and well-organized forest mafias, often in nexus with forest staff and local political elites.

During this period, DoF staff, despite being trained in forestry, were implementing neither sustainable, nor pro-people forest management policy, but rather were acting as the main instrument of forest clearance for in-migration, often under political patronage, while at the same time enforcing ruthlessly coercive anti-people policing in the forests that they had not destroyed.

DoF staff also had the task of allotting harvesting plots and marking trees for felling by the Timber Corporation of Nepal (TCN), the Firewood Corporation (FWC) and other His Majesty's Government of Nepal (HMGN) boards and companies. The TCN was responsible for the harvest and sale of timber. Politicians often used the TCN for their own benefit, and many forest ministers recruited their relatives or political supporters into the corporation without following proper recruitment procedures or requiring relevant qualifications. Gyawali and Koponen (2004) write:

> Tarai *forests have been a lucrative source of benefits to the powers that be, whether the Ranas, the Pancha or the post-1990 Congress or Communist parties. Furthermore, timber smuggling networks have been historically well organized as a transboundary enterprise in the* tarai*, and in recent years these have been able to take advantage of lax, newly formed community user groups.*

Armed forest guards were recruited and trained on the assumption that local people must be stopped from entering the forests. People's participation in forest protection and management was not formally recognized or sought until 1976 when the National Forestry Plan recognized the role and value of their participation in forestry programmes. It laid the foundation for a comprehensive plan of action and included objectives such as protecting the natural environment, developing forests as an important economic resource by enhancing people's self-sufficiency, increasing state revenue from forest products, and soliciting the broader participation of people in forest conservation and management (HMGN, 1976). Following this plan, *panchayat* forest (PF) and *panchayat* protected forest (PPF) regulations were promulgated in 1978, promoting the involvement of local people in forest protection and management. Although the national forest management paradigm had begun to change, people's participation through PFs and PPFs was hardly sought in the *tarai* compared to the hills.

The World Bank-supported *Tarai* Community Forestry Development Project (TCFDP) was implemented from 1985 to 1994. This project was designed following the Indian social forestry model of diverting the fulfilment of rural people's need for tree products from forests to new common-land plantations of exotic fast-growing species. It involved the conversion of roadsides, canal banks and village common land into village woodlot plantations, requiring much effort from the local people to plant and tend. It employed people to raise and plant seedlings essentially as paid labourers, as well as promoting planting on private land. Although such efforts were successful in raising awareness, they did not promote any sense of 'ownership' of the forest on common land. Unfortunately, the project was as unsuccessful as the social forestry efforts in India.

During 1988 to 1989, a trade and transit embargo was imposed on Nepal by India in support of Nepal's democracy movement. During this period, a shortage of fossil fuel imports diverted the pressure for firewood to the remaining *tarai* forests, and large areas of forest were degraded or destroyed. The DoF began to seek people's participation in protecting the *tarai* forests – during the unrest, forest staff recognized that they were unable to fulfil their responsibilities alone. In certain *tarai* districts such as Rupandehi and Jhapa, forest protection committees (FPCs) were formed by district forest offices with the promise that the forests they were helping to protect would later be handed over to them as CFs. However, FPCs generally included elites, local leaders and other influential people, and virtually no members of poor or marginalized groups. This has had a significant bearing on the development of CFs in the *tarai* in these areas, where capture by the elite has become a major problem.

The concept of CFUGs was further consolidated with the Master Plan for Forestry Sector (HMGN, 1989). Emphasis was on 'developing and managing forest resources through active participation of individuals and communities to meet their basic needs'.

The long-term objectives of the MPFS were to meet the people's basic needs, protect land from degradation, conserve the ecosystem and genetic resources, and contribute to agricultural productivity and the national economy by creating income-generating and employment opportunities (HMGN, 1989). Despite this encouraging rhetoric, the MPFS remained silent on a specific strategy for the *tarai*, except that it emphasized 'forest development to be concentrated in the *tarai* and middle mountains where wood and fodder deficits are to be prevalent' (HMGN, 1989).

Before MPFS, donor interest in, and support for, *tarai* forestry, other than in World Bank projects, was limited. The government appeared content with the status quo since it allowed it full control over forests. However, after the MPFS, the government sought the support of donors, ostensibly to protect the *tarai*'s degrading forests. This was basically initiated by the bureaucracy in order to cope with the decreasing budget allocation of the DoF as the revenue from forests decreased.

During the early 1990s, after the restoration of democracy, there was a flood of donor projects and donor funds into the country, including for the management of *tarai* forests. In response to the government's request, two bilateral donors began to support forestry development activities in some *tarai* districts. In 1992, the German Agency for Technical Cooperation (GTZ) funded the Churia Forest Development Project (ChFDP), an integrated social forestry project, in three districts: Siraha, Saptari and Udaypur. The main focus of this project was on supporting the livelihoods of people living in or dependent on the Churia and *tarai* forests. The project adopted a three-pronged strategy, enhancing the productivity of forest resources, reducing local consumption of forest products and providing alternative employment opportunities. The first strategy promoted CF in government forests, and agro-forestry and the cultivation of forest crops on private farms. The second promoted more 'efficient' devices, such as improved cooking stoves and biogas units. The third strategy provided micro-credit for different income-generating activities (IGAs) for the people whose livelihood depended on the sale of forest products.

The Forest Management and Utilization Development Project (FMUDP) was established in 1991 with the assistance of the Finnish International Development Agency (FINNIDA). It initially focused on studying the regeneration potential and growth rate of forests in a few *tarai* districts. Later, based on its findings, it developed 'model' five-year forest management plans for *tarai* forests, known as Operational Forest Management Plans (OFMPs). Later, the Ministry of Forests and Soil Conservation (MoFSC) prepared OFMPs for 18 *tarai* districts, following the Bara district OFMP model and process. However, this planning process was fundamentally flawed. First, plans were developed without significant consultation with local stakeholders. Second, they classified the forest area by restricting the area of potential CFs, categorizing forests into three broad categories: putting large areas under government protection and management; classing forests in environmentally sensitive areas as government-managed and protected forests; and classifying the remaining small patches and narrow fringes as potential community and leasehold forests. Third, it sought the commercialization of resources through a profit-oriented joint venture of Nepalese and Finnish private companies. Lastly, it did not address the issue of encroached forests in the *tarai* (Pokharel and Amatya, 2001). Strong opposition from civil society groups, including the Federation of Community Forest Users, Nepal (FECOFUN), ensued and the plans were shelved.

The Netherlands Development Organization (SNV)-Nepal is supporting the Biodiversity Sector Programme for Siwalik and Tarai (BISEP-ST) from 2002 to 2007, with a budget of about €12 million. The project covers eight *tarai* and inner *tarai* districts and mainly supports the piloting of collaborative forest management (CollFM) in this phase.

Table 6.2 *Donor involvement in the* tarai *forest sector*

Period	Project	Donor	Comment
1985–1994	*Tarai* Community Forest Development Project (TCFDP)	World Bank	Afforestation in social forestry paradigm – failed
1991–1998	Forest Management and Utilization Development Project (FMUDP)	Finnish International Development Agency (FINNIDA)	Technocratic and profit-oriented, production-oriented Operational Forest Management Plans – shelved after civil society agitation
1992–2005	Churia Forest Development Project (ChFDP)	German Agency for Technical Cooperation (GTZ)	Integrated Social Forestry Project: support to livelihood concerns of local people
from 2002	Biodiversity Sector Programme for *Siwalik* and *Tarai* (BISEP-ST)	Netherlands Development Organization (SNV)	Retreat from forest users-only approach towards 'consultation' and multi-stakeholder approach
2001–2011	Livelihoods and Forestry Programme (LFP)	UK Department for International Development (DFID)	Supporting all kinds of participatory forest management (PFM), including community forestry, leasehold forestry (LHF) and collaborative forest management (CollFM) under a sectoral approach

The enactment of the 1993 Forest Act and 1995 Forest Regulations – formalizing the handover process and recognizing CFUGs as self-governed and autonomous corporate bodies for the management and use of CFs according to an operational plan approved by the local district forest officer (DFO) – provided a legal basis for the implementation of CF. The 1993 Forest Act can therefore be taken as a turning point in terms of PFM in the *tarai*. Many of the forest protection committees (FPCs) formed during the political unrest were handed over as CFs after this act. The handover has, however, changed the basis for local forest product extraction in a complex way, and has often reduced the rights of the *tarai* people (particularly non-members of new CFUGs) to receive timber and firewood from forests for household consumption to which they were earlier entitled during certain seasons by paying nominal fees to the DoF.

Another effort being made is to look at smaller patches of CFs or leasehold forests not merely as units of forest management, but rather as landscape planning for forest management where different PFM processes/models will complement each other at different levels and units. Such efforts are being envisioned and promoted by projects such as the *Tarai* Arc Landscape project (TAL) and the Western *Tarai* Landscape Building Project (WTLBP).

Table 6.3 provides a brief history of forestry and the emergence of participatory forestry in the *tarai*.

Current policies and major policy issues

The situation in the *tarai* is significantly different from that in the hills because of the different forest resources and their very high value and easy access, as well as the tremendous market for forest resources. The socio-economic and institutional settings are also quite different. For the reasons already detailed, effective forest management according to the provisions of the 1976 National Forestry Plan and the 1989 Master Plan for Forestry Sector has not been implemented. The livelihood concerns of indigenous communities have hardly been considered, and consideration, where given, has mainly addressed the concerns of the more politically influential migrant populations.

The 1993 Forest Act and 1995 Forest Regulations do not differentiate between the hills and *tarai* with regard to CF. This has caused a lack of policy clarity over the implementation of PFM in the *tarai*. Some government officials have promoted CF in the *tarai*, whereas others have tried to restrict it. As a result, the pace of implementation of the CF programme in the *tarai* has been slow.

The government imposed a tax on timber sales from *tarai* CFs that was challenged by civil society in the Supreme Court, on the street and in policy debates. The 2000 Revised Forestry Sector Policy supports a less participatory alternative model of PFM in the *tarai* – that is, CollFM, in which distant users and local forest user groups (FUGs) can acquire forest products, with royalties shared between local government and the DoF. The district forest office retains power over forest management, but involves multiple stakeholders in a consultative forum process. CollFM was conceptualized to address the issue of distant users in the *tarai*, while also maintaining the DoF's stronghold on resources.

PFM implementation processes and practices
Most of the remaining *tarai* forests are being managed as government forest. The continuous loss and degradation of these forests has led to the realization that people's participation in their protection, as well as coordination and communication between different stakeholders, is essential.

The implementation of the PFM in the *tarai* began in earnest after the 1993 Forest Act, which devolved authority to DFOs to hand over government forest as CFs to local communities. Several innovative DFOs working in *tarai* districts, encouraged by the success in the

Table 6.3 *History of forestry and the emergence of participatory forestry in the* tarai

Period	Major events and policies
Until 1951 – consolidation of the state	Extensive forests inhabited by indigenous/tribal groups at low density. *Tarai* forest lands granted by rulers to clients for services rendered. 1925 establishment of *Kath Mahal* (government Timber House) and extraction of *tarai* timber for supply to British India. 1942 formation of Department of Forests (DoF).
1951–1975	1957 formulation of Forest Nationalization Act. 1961 Forest Act and 1967 Forest Protection Act (the so-called 'Shoot the Bullet Act') strengthened the powers of the DoF. Government resettlement programme under the Nepal Punarvas Company encouraged clearance of *tarai* forests for distribution to hill people.
1976–1992	1976 National Forestry Plan. 1978 *Panchayat* forests and *panchayat* protected forests regulations enacted. 1985–1994 World Bank-funded *Tarai* Community Forest Development Project (TCFD) 1989 Master Plan for Forestry Sector (MPFS) 1991–1998 Finnish International Development Agency (FINNIDA)-supported Forest Management and Utilization Development Project (FMUDP) 1992–2005 German Agency for Technical Cooperation (GTZ)-supported Churia Forest Development Project's (ChFDP's) Participatory Forestry Project in eastern *tarai*
1993–1999	1993 Forest Act and 1995 Forest Regulation provided legal basis for the implementation of CFs with CFUGs as institutions 1998 Operational Forest Management Plans for 18 *tarai* and inner *tarai* districts prepared and approved, but not implemented 1998 Local Self-Governance Act (LSGA)
2000–2005	2000 Government Order: provision to include growing stock and prediction of sustainable harvest of forest products in the making of operational plans 2000 Government imposed 40% share of revenue in sale of sal and khair species from *tarai* and inner *tarai* CFs (if sold in market or to non-members of their CFUG) 2000 His Majesty's Government of Nepal (HMGN) Revised Forestry Sector Policy introduced collaborative forest management (CollFM) for the management of *tarai*, inner *tarai* and *Siwalik* forests 2002 Netherlands Development Organization (SNV)-Nepal-supported Biodiversity Sector Programme for *Siwalik* and *Tarai* (BISEP-ST) and UK Department for International Development (DFID)-funded Livelihoods and Forestry Programme (LFP) assist in initiation of CollFM 2002 CollFM directives and guidelines developed and approved by the Ministry of Forests and Soil Conservation (MoFSC) to conduct the CollFM pilot programme in a few *tarai* districts 2004 Government reduced the share of royalties in CF from 40% to 15% of revenue from the sale of sal and khair timber

Source: Hobley and Malla (1996); Bhatia (2000); Malla (2001); Bhatta and Tiwari (2001); Parajuli (2003)

hills, organized CFUGs in their districts and handed CFs over to them. Since there were no clear guidelines or experience related to user identification in the *tarai*, they tried following the guidelines for the hills. It was easier to organize the hill migrants since many of them were already aware of CF policies and practices.

Handing over forests to nearby communities rewarded recent settlers and encroachers, many of whom claimed land by faking landlessness (JTRC, 2000). Most of the people living in the northern *tarai* are migrants who either cleared the forests themselves or bought land from others who had cleared them, and the government has legalized their occupancy in most parts of the country (Bhatta, 2002).

Land entitlement in the *tarai* has always been a serious political issue in Nepal, especially in illegally converted forest land. The government provided land titles to hill migrants who had illegally encroached on forest land and occupied and used it for several years through several resettlement commissions. This set a bad precedent and indirectly motivates encroachers to encroach further into *tarai* forest land. The DoF's efforts at controlling encroachments have not been very successful.

In spite of a clear policy stipulation that forest land must not be used for other purposes, it has been given to universities, schools, hospitals, *sukumbasi* (landless people) and victims of natural disasters. Illegal settlers near and around highways and in forests are a major problem in the *tarai* as they often degrade and reduce forests by expanding the area that they occupy. In the whole process of encroachment and land entitlement, native indigenous peoples such as the Tharu, Danuwar and Darai have lost their access to forests and also the title to the land they were tilling. This has resulted in the significant exclusion of traditional, now distant, users who (as forest frontiers receded) found themselves further from the existing forest, with the result that conflicts arose between 'near' and distant users. The poorer and marginalized people have been excluded, their traditional rights completely ignored and their livelihoods threatened. The deficit of forest resources in the southern area and the imbalance of access to resources and unequal distribution of use rights to distant users are now pronounced (Kanel, 2000; Pokharel, 2000; Statz, 2003).

Not all nearby users were included as members in CFUGs. Many were excluded, mainly due to the arbitrary delineation of communities during the identification of users. The stratification and exclusion of forest users have negatively affected the livelihoods of many households, mainly poor families who were greatly dependent on forests for their livelihoods. It is, therefore, both absolutely essential and extremely challenging to address the issue of distant users (especially the ultra-poor communities) and their need for and dependence on forest products and services.

This issue of inclusion/exclusion is the main reason CF in the *tarai* has received serious criticism (Baral and Subedi, 1999; Takimoto, 2000; IDEA, 2003). Any stereotypical singular policy approach to the governance and management of *tarai* forests is unlikely to address the legitimate concerns of the people in the southern part of the *tarai* (Adhikari et al, 2005). In the hills, the issue has emerged, albeit on a smaller scale, and has sometimes been resolved by giving distant users secondary status within CFUGs. A similar stratified membership approach may well be appropriate for the *tarai*.

The 1998 Local Self-Governance Act provides village development committees (VDCs) and district development committees (DDCs) with the rights and responsibilities to manage local natural resources, including forests (SchEMS, 2003). It also stipulates that DDCs are authorized to develop and implement plans for the conservation of forests, biological diversity and promotion of the environment, and to levy taxes on resin, herbs, slate, sand and animal products, such as bones, horn and others (HMGN, 1999). The provisions of the 1998 LSGA and CF (which cede ownership of forest resources from their forest to CFUGs as long as they are managed following the approved operational plan) contradict each other in terms of their authority over the same resources. These conflicts could be resolved through various stakeholder consultations and by amending the conflicting provisions of the relevant acts. There is also a need to take precautions in future to ensure that such policies/acts can only be developed in wider consultation with all of the stakeholders.

In this context, the 2000 Revised Forestry Sector Policy proposes specific policies for the management of the *tarai*, inner *tarai* and *Siwalik* forests (HMGN, 2000), demarcating large blocks of forests for management by the government, in collaboration with local bodies and people. The government is seeking the 'collaboration' only of local people and local institutions for the CollFM of forest blocks, with income shared among collaborating partners (25 per cent to DDCs, VDCs and local FUGs, and 75 per cent to the government treasury). CollFM is being piloted in three districts: Bara, Parsa and Rautahat. At district level, this programme is guided by the District Forests Coordination Committee (DFCC), a coordinating body comprising DDCs, line agencies, VDCs, FUGs, major political parties, non-governmental organizations (NGOs) and forest-related traders.

In 2004, the MoFSC developed and approved non-government service provider (NGSP) guidelines, which promote the participation of the private sector and NGOs/community-based organizations (CBOs). However, the actual involvement of these and their contribution to PFM practices have yet to be seen.

CFUGs/self-initiated forest protection groups

At present, there are 2157 CFUGs in the *tarai* and inner *tarai*, with an average area of 142ha per CFUG (DoF–CFD, 2006). During the early and mid 1990s, the initiative for CFUG formation was taken mainly by DFOs. In some districts, (e.g. Jhapa), local people took the initiative, whereas in others (Saptari, Siraha, Dang, Banke, Bardia and Kailali), CFUGs originated from donor-supported projects. By the late 1990s, local people took on the protection of forests as their CFs and demanded that they be handed over to them. All *tarai* districts have CFs (see Table 6.1). This shows that donor-funded projects have not been a critical factor in handing over CFs in the *tarai* as there have been significant handovers in districts with no donor-supported projects.

There are a few examples of community-initiated groups starting to protect and manage forests in the *tarai*. For example, in 1980, in Siraha district in the eastern *tarai*, a group of 80 households started protecting the forest in the nearby Churia hills, which had been handed over to them as CF in 1991. The CF, known as Jhauwai Kholsi Chure Danda CF, covers an area of 274.5ha. It has natural bamboo and sal forests. The people there, having experienced the decline of nature due to deforestation and forest degradation, realized the need and benefit of forests for their survival and started protecting them. This has now become one of the best CFs in the region. Another example is in Dang district in the western inner *tarai*, where a local leader, Chyang B. Thapa, organized and persuaded the people of his community to protect a nearby forest in the mid 1970s, even before CF regulations were introduced in Nepal. In 1993, the 212ha sal forest, Bhawani CF, was legally handed over to the group (Ministry of Agriculture, 1995). These self-initiated CFUGs are much better in terms of forest management and sharing of responsibilities and benefits than the outsider-promoted CFs, showing that if local communities become aware, take the initiative and take charge of forest management, PFM is much more successful.

Community forest management planning: Operational plan and priorities

There are no specific CF guidelines for the *tarai*, and in most cases guidelines from the hills and social mobilization processes have not been properly followed. The formation of CFUGs and the development of operational plans have generally been led by district forest office staff and the elites of the community. Although they were prepared by a few executive committee members, the district forest office staff played the lead role. Only a few CFUG executive committee members were aware of, and understood, the documents and processes. Very few female executive committee members and other general members of the CFUG had access to these documents. Many general members, especially from poor occupational castes and marginalized groups, expressed dissatisfaction with the stipulations of these documents, mainly due to the fact that there were no special provisions for poor users and prices for allocations of forest products were high. In some *tarai* CFs, because of the high commercial value of timber and other forest products, timber mafias influence the processes, particularly timber harvesting and sales such as in Jhapa.

Spatial and temporal variations and policy directions

Having reviewed the emergence of PFM in the *tarai*, we might conclude that PFM is not suited to a difficult political and economic environment fraught with ruthless aggrandizement and expropriation by the elite. Maintaining forest land use and protecting the forest appear to be a losing battle. The CF model transferred from the hills has played into the hands of newly arrived local elites, excluding indigenous forest users and consigning them to poverty.

Despite these difficulties, forest policy may appear to be moving in the direction of a more stabilized forest frontier and more participatory approach to forest management. Community forestry, leasehold forestry and CollFM are examples. Even in the case of *tarai* national parks and protected areas, buffer zone community forests (BZCFs) are being promoted. Controversy and conflict persist, however, between different stakeholders over the different models of PFM. Some stakeholders (such as FECOFUN) only favour CF, whereas others believe that having more PFM models is helpful.

Sample villages and community forests

This study seeks to understand the ways in which PFM policies have been implemented, as well as their field impact. The methodology involved purposively selects two contrasting districts to cover the *tarai* proper, *Siwalik* (Saptari district) and the inner *tarai* (Dang district). Data was collected at district level. For more detailed study, three CFUG villages in each district were selected using a stratified random sample from clusters of active and medium-active groups, as categorized by the district forest office.

Within the CFUGs, a household survey was conducted after selecting households on a stratified random basis. The households were categorized into four wealth ranks. A sample of 20 households was randomly chosen where the total number of households was below 200, and one of 10 per cent where the CFUG exceeded 200 households, so that all wealth categories were represented. The distribution of households and population in the district, the number of CFUGs, the CF area, the number of households in CFUGs and the average income of CFUGs are presented in Table 6.4.

Table 6.4 *Distribution of households, population and forest areas according to districts*

District	Forest area (ha)*	Number of CFUGs**	Total CF area (ha)**	Total number of households in CFUGs**	Total number of households in the district***	Total population in the district***
Dang	161,900	373	66,250	62,468	82,495	462,380
Saptari	27,200	105	10,087	17,081	101,141	570,282

Source: * DoF (2002); ** DoF–CFD (2006); *** CBS (2002)

There were fewer CFUGs in both number and area in Saptari (105; 10,087ha) than in Dang (373; 66,250ha) because of the great difference in the total areas of forests in the two districts: 27,200ha in Saptari compared to 161,900ha in Dang. The average area of CFs in Dang (177ha) is much higher than those in Saptari (96ha).

Table 6.5 *Details of study community forest user groups (CFUGs) in Dang and Saptari districts*

District	Study CFUG name	Forest area (ha)	Major forest type	CFUG formation year	Total number of households	Number of ethnic/caste groups	Men/women in executive committee
Dang	1 Namuna Mahila	207.0	Sal	1995	240	5	7/5
	2 Jansewa	122.62	Sal	2000	115	12	6/3
	3 Gadibarah	216.0	Sal	1994	316	9	9/4
Saptari	4 Malati Mahila	84.1	Sal mixed	1993	145	9	0/16
	5 Gangajali	221	Sal	1995	206	11	13/2
	6 Churiya Mai	171.2	Khair	1997	222	17	12/3

In all, six of the studied CFUGs in Saptari and Dang (57 per cent of the households) were poor or very poor. Poor and very poor households were in higher proportion in Saptari (64 per cent) than in Dang (47 per cent). This categorization was based on the wealth criteria set by the communities concerned. Generally, rich households had food surpluses, sufficient land for farming and secure earnings through their jobs or business, etc. The medium-rich households mainly comprised well-to-do subsistence farmers, while the poor and very poor households earned their livelihoods through wage labour and occupational activities. The overall number of unemployed was very high in both districts.

Table 6.6 *CFUG member households by wealth rank*

Name of CFUG	Total households in study CFUGs	Households by wealth rank			
		Rich	*Medium rich*	*Poor*	*Very poor*
1 Namuna Mahila	240	30	70	70	70
2 Jansewa	300	50	90	60	100
3 Gadibarah	120	6	73	22	19
4 Malati Mahila	136	25	27	26	58
5 Gangajali	206	20	81	62	43
6 Churiya Mai	219	20	33	50	119

In all six CFUGs, households followed diverse strategies for their livelihoods. Farming was the most common strategy, the main farm product being paddy, followed by maize, millet and wheat. Oil crops, vegetables, potatoes and pulse crops were also grown in the area for domestic use and as cash crops. Other main on-farm livelihood activities involved live-stock-keeping, dairy activities and agricultural wage labour. Off-farm work involved trade, services, foreign migration for labour work, and occupational skills such as blacksmith, tailor and cobbler work. Production and sale of bamboo furniture, baskets and brooms was another income source for a small number of households in Saptari.

In Dang, the main tree species were sal, asna, khair and simal, whereas in Saptari, it was sal alone. Saptari had fewer forest products than Dang, where the studied CFUGs linked dairy production with forestry, using the CFs to produce fodder. In Saptari, due to the smaller area of forest, timber, firewood and fodder were collected from government forests in the Churia region. Very few NTFPs were collected and traded from Saptari.

In Dang, forests covered more than 70 per cent of the land. *Sungandhawal, Timur, Chiuri, Jhyau, Kurilo,Ritha* and *Babiyo* were the major species of NTFPs collected and marketed from Dang (DFO, 2003). These were sold to brokers who brought them to Nepalgunj and sold them on to big traders, who dictated the price and received a major portion of the profit, whereas the collectors were paid very low prices.

Outcomes and impacts

In this section, the outcomes and impacts of CF are presented in three parts: CFUG forest management institutional processes, the household livelihoods and forests.

Processes: Local forest management institutions

Of the 2157 CFUGs formed in the *tarai*, 478 were in the study districts. In both study districts, district forest offices initiated CFUG formation with the assistance of donors. The GTZ assisted the Saptari District Forest Office in supporting 105 CFUGs, and CARE-Nepal, with support from the US Agency for International Development (USAID), assisted the Dang District Forest Office.

In Saptari, CFs were accessible to only 17 per cent of the total population, so only a small fraction of people living near the forest (mostly hill migrants) in Saptari enjoyed their bene-fits and distant users (indigenous *tarai* inhabitants) were marginalized. Only 37 per cent of the total forests of Saptari were allocated as CFs because the remaining forests were located too far from communities to be eligible. Virtually all of the CFs had been handed over to the people.

In Dang, CFs were accessible to 76 per cent of the total population. The remaining 24 per cent of the people were not members of CFUGs, but had some access to forests or potential CFs. Most of the accessible forests close to the village had been handed over to nearby communities; therefore, the formation of new CFUGs was low (see Figure 6.1).

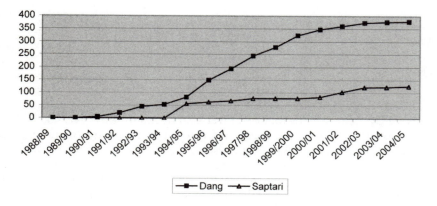

Figure 6.1 Trend of formation of community forest user groups (CFUGs) in Dang (inner tarai) and Saptari

Source: DoF (2004)

Saptari reflects the pattern typical across most *tarai* districts as opposed to inner *tarai*. A large population of hill migrants has recently settled in the southern fringes of forests, and many of these migrants have encroached on forests, legally or illegally. Many have also demanded that the remaining forests be handed over to them as CFs. In Saptari, CF forma-tion started in 1992, accelerating after the enactment of the 1993 Forest Act.

When the CFUGs were being formed, traditional users were either not informed or were excluded on the grounds that they lived far from the forest and could not contribute to managing and protecting it. Thus, almost all distant users were excluded and prevented from legitimately collecting livelihood-related forest products. Many nearby users (within the vicinity of 3km) were also excluded. This has badly affected the livelihood base of the poor and marginalized people in the *tarai*.

'Distant' forest users used many forests whose management had subsequently been handed over to forest-adjacent villages. Those coming from 3km to 7km away, and as far as 15km to the south, used forests occasionally, but were heavily dependent on forest resources for crucial livelihood inputs. According to the GTZ-funded Churia Forest Development Project staff, the project and district forest office originally perceived distant users as tradi-tional users who were unable to be involved in active forest protection due to their remoteness, but after five years the project realized that distant users could participate in the CF programme by joining existing CFUGs as secondary users, or by forming new CFUGs consisting entirely of such distant users.

By 2004, there were eight such CFUGs consisting entirely of distant users who had been given forests far from their villages. These new CFUGs of distant users had to come all the way from the south and pass through other CFs to reach their own CFs in the Churia hills. An inappropriate user identification process had led to such a problem.

Apart from Saptari, the Bara and Dhanusha districts were also considered in light of households excluded from CFs. In Bara, only nine CFUGs had households along the East–West highway as members. Households adjacent to or far from the CF had not been consulted about becoming members and were deprived of their use rights. In most cases, the native inhabitants of the *tarai* had lost their access to forests. The case of Haraiya village (1352 households) in Bara district illustrates households that are excluded from timber and fuelwood collection (see Table 6.7 and 6.8) in *tarai* CFs.

Table 6.7 *Sources for annual consumption of timber (Haraiya village in Bara district – 'close' to forest)*

Sources	30 years ago	20 years ago	10 years ago	Today
Private forest	0.00	3.08	6.92	15.77
Collaborative forest management	0.00	0.00	0.00	1.15
Government forest	100.00	89.62	82.31	63.85
Community forest	–	–	0.00	0.00
Purchased	0.00	7.30	10.77	19.23

Until about 30 years ago, the main source of timber for the people of Haraiya village was the government forest. Today, they have to either purchase timber or grow it on their own land. They do not have access to nearby CFs, which affects the poor most as they are unable to purchase or grow timber. Table 6.8 shows the proportions of their consumption from vari-ous sources. Evidently, poor people are most dependent on firewood from the forest and suffer most when forest access is denied them.

In Saptari, Lohajara, Bakdhuwa and Mainakaderi villages were denied access to forests, although CFs were within 3km of these villages. A total of 450 households had been excluded from CFs despite their traditional use of forests. The issue of distant users is promi-nent in *tarai* districts from Jhapa to Banke.

Table 6.8 *Sources for annual consumption of fuelwood (Haraiya village in Bara district – 'close' to forest)*

Sources	30 years ago	20 years ago	10 years ago	Today
Private forest	1.15	2.31	2.69	12.31
Collaborative forest management	0.00	0.00	0.00	0.77
Government forest	95.77	92.69	84.23	64.62
Communal plantation	0.00	0.00	0.00	0.00
Community forest	0.00	0.00	0.00	0.00
Purchase	3.08	5.00	13.08	22.31

Formation and mode of operation of CFUGs

All the six CFUGs studied were managing forests and providing forest-related income and social benefits to their members. Forest users in these CFs were mobilized through three different approaches. A unique approach in Malati Mahila CF divided the forest into plots, allotting one to each member to manage and harness the benefits; the most wide-spread form of operation was through users' participation in forest management activities on a voluntary basis, and the third form employed paid forest guards to protect the forest. Forest management and all other community development activities were carried out through paid labour, and even executive committee members were paid for their contribution. All of the CFUGs studied in Dang employed forest guards to protect forests, mainly from the activities of illegal collectors of forest products from nearby villages. In Saptari, only one of the three CFs studied employed a forest guard. Executive committee members of the three CFs were paid for their services, which included managing the auction of forest products, distribution of forest products within CFUGs, implementation of the rural infrastructure development plan and so on.

Preparation of operational plan and forest management

Forest management practices, as enshrined in operational plans, varied according to local conditions. Many local communities, particularly the poor and marginalized, complained that operational plans did not reflect their priorities. Although operational plans should ideally be prepared by the whole CFUG and endorsed at a general assembly, in practice, DoF staff and executive committee members played the leading role in preparing them.

All studied CFUGs had constitutions and operational plans prepared by their executive committees with the guidance and assistance of district forest office staff. However, executive committee members and district forest office staff claimed that their operational plans had been prepared and finalized by consensus in the general assembly, and the district forest office had just facilitated the process. In Dang, only 3 male executive committee members out of 20 and none of the female members had read their CFUG's constitution or operational plan. In Saptari, 24 of the 42 male executive committee members and one of the five female members had read their constitution and operational plan. Only in one of the six CFUGs was the operational plan strictly followed. An overwhelming majority of general users were not even aware of the constitution or the operational plan, and the executive committee members who had read them were not satisfied with them. The reasons for their dissatisfaction were lack of provision for poverty alleviation, the high prices of forest products and the impracticality of the constitution and the operational plan.

Table 6.9 *Income, expenditure and fund transparency in CFUGs*

CFUG name	Fund generated (1990–2004) (Nepali rupees)	Income (2003) (Nepali rupees)	Fund in CFUG at present (Nepali rupees)	Assessment of transparency
Namuna Mahila	177,000	12,000	101,000	Poor
Jansewa	421,000	121,000	201,000	Moderate
Gadibarah	195,000	15,000	150,086	Poor
Malati Mahila	349,000	30,000	80,000	Good
Gangajali	397,000	120,000	13,000	Poor
Churiya Mai	308,000	35,000	19,910	Poor

In the six CFUGs studied, funds generated in the year 2003 varied between 12,000 Nepali rupees and 35,000 rupees. They exceeded 100,000 rupees in only two CFUGs. Although the transactions and savings in their funds did not involve large amounts, transparency in fund utilization and mobilization was poor in four CFUGs, moderate in one and good in the last. Poor transparency indicates poor participation in decision-making processes and low accountability of executive committee members to the CFUG.

CFUG governance process: Transparency, participation and accountability

The size of the forests and the number of users were much larger in the *tarai* than in the hills (some CFUGs include more than 2000 households), and often in these large CFUGs the executive committee members took decisions representing all members and sought their formal endorsement. This practice is likely to lead to abuse of authority and lack of transparency. Executive committee members were managing the forest in close interaction with forestry staff, contractors, district-level federations and other elites. They kept important information to themselves, rather than announcing it to all CFUG members by public notice. Often the general bodies of CFUGs were not aware of changes in the rules and therefore found themselves liable to pay fines. Out of the 83 poor households in both districts in 2004, 11 (over 13 per cent) had paid fines ranging from 65 Nepali rupees to 101 rupees on charges of stealing forest products and cattle grazing. In Saptari, both CFUG members and non-members had paid fines for these offences. More users had paid fines in Saptari than in Dang.

Community forests had increased local people's participation in forest management in the *tarai* and inner *tarai*. However, participation in decision-making was limited and mostly influenced by elites, especially males.

Transparency regarding CFUG funds and decision-making processes were very low. Information, particularly on fund generation and mobilization, was not shared with all CFUG members. Funds were misused and applied in favour of elite members. Group funds were used for either infrastructure development in the areas where elites lived or for their benefit. For example, in Gangajali, Gadibarah and Namuna Mahila groups, funds raised from the sale of timber were spent on rural infrastructure development, such as the construction of roads, trails, temples, and CFUG office buildings by decision of the local elite.

Similarly, executive committee members had become less accountable towards forest users. There were nexus, particularly between committee members, forestry staff and timber contractors, which contrived operational plans that showed allowable cuts of timber and firewood greater than the amount approved by the DFO. With this approval, wood was harvested and timber and firewood sold at nominal prices. Many users could not even get their basic subsistence needs, while the forest products easily reached big markets, such as Kathmandu.

Representation, decision-making, management, relations between the village and the DoF, and other institutions and social cohesion

The forest management systems followed by most CFUGs emphasized the production, harvesting and distribution of high-value timber. The outcomes were often pro-rich for a number of reasons (Bhatta and Dhakal, 2004; Iverson et al, 2006). Poor people could not afford the prohibitive charges or sawing costs, and there was virtually no special provision for pro-poor distribution systems that might waive charges and allow the poor to pay the cost of sawing by selling a portion of the sawn timber. Furthermore, the growth period of a tree (often sal species) for timber production is very high and does not support the short-term recurrent livelihood needs of the poor. There were no rules for directly sharing cash generated from the sale of timber among households. Rather, it was kept in the group fund, which was often used for infrastructure development to reflect rich households' priorities. Low-caste poor households were generally allowed to collect only fodder, dry twigs and leaf litter for free, and had to pay for other forest products, which were unaffordable. The respondents of all six CFUGs expressed that a more equitable approach must be found so that PFM benefits could reach the neediest and most vulnerable groups of the community.

Box 6.1 Malati Mahila Community Forest: Silvi-pasture combined with dairy marketing

Om Prakash Dev

An example of good participatory forest management (PFM) practice comes from the Malati Mahila CF of Mohanpur, Bakdhuwa village development committee (VDC), in Saptari. This community forest user group (CFUG) was formed in 1993 by the district forest office and largely comprised forest users who had migrated from the hills. A total of 84ha of forest land were allocated to 145 households. Community forest users had limited landholdings (average 0.2ha per household), and their main mode of livelihood was livestock-rearing for milk production. They linked fodder production from the CF with their cattle. As the production of milk increased, the CFUG also contributed to the formation of a milk marketing co-operative, which established a milk-chilling centre. It initiated a process of replacing the unproductive native livestock with hybrids, which produce more milk with less fodder consumption. This benefited the users of this community forest economically, and they were now better off compared to many other *tarai* CFs. The CFUG also provided loans to its members at 12 to 18 per cent interest. Savings and credit organizations were also founded by the local people and various non-governmental organizations (NGOs), which provided credit at interest lower than that of moneylenders and co-operatives. This significantly improved members' access to credit. However, rich and elite groups were on the decision-making bodies and thus expropriated higher benefits than poorer members. Conflict arose due to the fact that the forest land had already been divided and allocated to individual households: some households had split and were claiming individual plots for themselves.

Most of the decisions related to CF were made in executive committee meetings. CFUGs seemed to have adopted certain patterns of decision-making consistently across institutions, but the frequency of meetings varied. In some cases, they met once a month (Dang), in others, whenever executive committee members felt it necessary (Saptari), and in one case no meetings had been held for a long time. All CFUGs studied kept the minutes of meetings. The poorer groups' participation in executive committee decision-making was low; but it was good in general body meetings. Table 6.10 illustrates user participation in different CFUG activities.

Table 6.10 *Participation of forest user households in different CFUG activities by wealth rank*

Activities	Percentage of households participating			
	Rich (n = 19)	Medium rich (n = 46)	Poor (n = 37)	Very poor (n = 43)
Forest user assembly	90	89	81	83
Executive committee meeting	47	48	22	11
Forest protection	63	48	47	70
Silvicultural operations (*Godmel*)	42	39	36	36
Training/workshop	16	16	11	13
Study tour	2	4	0	2

External facilitators from donor-funded projects emphasized the concept of regular meetings and other institutional activities. However, in this study the CFUGs had adapted processes to their own local situations, as forest management is a long-term enterprise and members had other pressing workload issues in agriculture and other livelihood activities.

The views of CFUG members on decision-making power reflected the belief that the CFUG committees held majority power in most CFUGs.

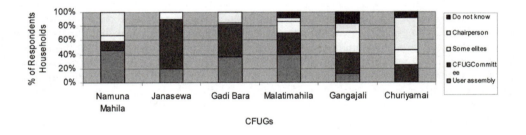

Figure 6.2 Who makes decisions in community forest user groups?

Source: Bhatta et al (from survey data)

Most decisions made in CFUGs were related to planning and forest management, covering forest conservation, use and protection. Some decisions were related to community infrastructure and rural development.

Six major decisions were made last year in each of the three Saptari CFUGs studied, and 15 in each of the Dang CFUGS; but in each case only one of these was a specific provision to benefit poor households. Thus, poor groups' needs and priorities received little consideration in CFUG decisions. Furthermore, even if decisions were made in favour of poor people, they were not always implemented. Instead, it was the decisions of interest to the elites (who dominated the executive committee), and which were related to rural infrastructure development, that were largely implemented.

In most of the groups, executive committee members and some elite groups dominated decision-making (see Figure 6.2), even though members were elected by consensus at the general assembly. Transparency was often lacking in CFUGs as decisions were normally made by a few executive committee members (mainly the chair, vice-chair and secretary). Similarly, in all cases except in the two female-headed CFUGs in this study (Malati Mahila and Namuna Mahila), women were excluded from policy- and implementation-level deci-sion-making processes. Unless there is a fairly broad base for decision-making, it is unlikely that the interests of all groups will be represented, and in such cases there is a high proba-bility of some groups not complying with management plans.

Poorer communities, including Mushahar, Sardar, Chamar and the blacksmith commu-nity, could not afford the time to actively participate in CF activities, different fora and occasions, although they participated widely in forest protection, forest operation and tree planting. In both districts, some women-only CFUGs were formed and empowered to take on strategic community development responsibilities.

Relations between DoF staff and CFUGs

The relationship between DoF staff and CFUGs was very poor in the *tarai* compared to the hills. DoF staff were still engaged in policing and patrolling, and they tried to keep people away from government forests. There were several cases where *tarai* inhabitants, particularly those living far from forests, had no access, and if they went to forests to collect timber, fire-wood or other forest products they were harassed in various ways (e.g. they were asked to pay bribes, were physically tortured or molested, or put in custody).

The effect of insurgency on decision-making and forest management

In many *tarai* and inner *tarai* districts, CFUGs had been negatively affected by the Maoist insurgency. The government's armed forces, for instance, had established security check-points across the *tarai* road network and had banned the harvest and sale of timber and firewood for commercial purposes in some districts (Dang, Bardiya and Jhapa), suspecting that part of the proceeds might be spent on supporting the insurgency. In Banke and Bardiya, the DFO with the instruction of the army had frozen the bank accounts of all CFUGs, suspecting them of financially supporting the insurgents or suspecting the insur-gents of extorting money from them. CFUGs in both hills and *tarai* were being forced to pay funds to Maoists. In Dang, the army occupied Gadibarah CF despite protests from the CFUG, and regular searches and patrols by the army inside the forest had made it difficult to harvest forest products.

Impact on forests

This section considers the overall impact of the changed management arrangements on the forest itself. Although we were not able to compare changes in forest conditions due to lack of baseline data, we could assess the current regeneration status as an indicator of improve-ment.

A forest inventory was carried out in all CFs studied, which included measurement of diameter at breast height (dbh), estimation of height, and a count of tree, shrub and herb species following a standard sampling procedure. To calculate the volume, a form factor of 0.60 was used, and to calculate the mean annual increment (MAI), the age of the tree species was estimated by local people. Therefore, the volume and MAI provide a general idea for comparison, although they must be treated as approximations rather than precise figures.

Due to the absence of baseline forest inventory data before the formation of CFs, the current data on growing stock cannot be compared to ascertain any increase or decrease. Instead, local people were interviewed to give a qualitative assessment of the pre-CF forest status. It was inferred that the growing stock of forests had significantly increased after the formation of CF due to the protection of forests, as well as reduced grazing, encroachment and fire. This restriction on forest use appeared to have stemmed from the feeling of the local people that they now had ownership of the forest. CFUG members perceived that forest condition had improved or was improving in almost all CFs.

The method used to determine current forest condition was based on that employed by Yadav et al (2001) and the condition of growing stock on the *Forest Inventory Guidelines for Community Forests* (CFD, 2004). The results, shown in Table 6.11, allowed us to roughly classify forest condition into good, medium or poor.

Table 6.11 *Details of forest condition based on growing stocks in all community forests*

Name of CF	Regeneration per hectare (new shoots up to 30cm high)			Saplings per hectare (31cm to 2m)			Growing stock (pole and tree) (>2 m)			Forest condition assessment
	Good	*Medium*	*Poor*	*Good*	*Medium*	*Poor*	*Good*	*Medium*	*Poor*	
	>5000	*2000–5000*	*<2000*	*>2000*	*800–2000*	*<800*	*>198.36m³/ha*	*56.67–198.36m³/ha*	*<56.67 m³/ha*	
Namuna Mahila	13,400			2724				99.283		Good
Jansewa	6511			6076				119.49		Good
Gadibarah		3566			1386		314.1			Medium
Malati Mahila	5977					508		154.82		Medium
Gangajali	19,766				4297		207.9			Good
Churiya Mai			1856			566			99.28	Poor

The results obtained show that forest had regenerated significantly in all CFs studied except Churiamai CF (Table 6.11). Most CFUGs wanted to harvest the most valuable timber for domestic use and for sale even when it was not harvestable from a management perspective. Selective felling of good-quality trees for firewood was also carried out, which resulted in over-harvesting in Churaimai CF.

Forest condition had improved significantly in three CFs and moderately improved in one. In Churiamai CF, forest condition was medium to poor (due to a lack of capacity to regulate forest product supply), and in Gangajali CF it was deteriorating (due to over-harvesting of timber). The Malati Mahila CFUG had even received the Sarbamanya Ganeshman Singh Award, a national award given to the best CFUGs for their contribution to forest management. Here, individual households to whom plots of forests had been allocated had significantly improved forest condition through their protection and care of plants and grasses.

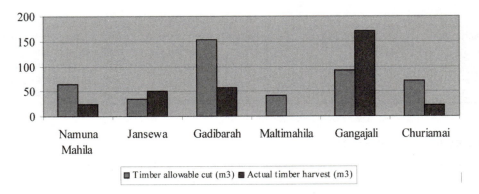

Figure 6.3 Details of allowable and actual timber harvest in community forests studied

Source: Bhatta et al (from data collected during study)

A major change reported by people was that supplies of fodder and firewood had increased significantly in most CFUGs and the condition of the CFs had generally improved, but that the condition of adjoining and nearby national forests had deteriorated. These changes, their reasons and the issues involved are shown in Table 6.12.

The actual timber harvest was higher than estimated allowable cut (see Figure 6.3) in two of the six CFs studied, and considerably higher in Gangajali CF, which may lead to poor stocking and degradation of the forest. In other CFs, the actual timber harvest was much lower than the estimated allowable cut, which shows that these forests will become over-stocked and the growth of growing stock may be retarded. Firewood harvest was almost equal to the allowable cut, exceeding it in Churiyamai and Jansewa CFs. This demonstrates a higher demand for firewood than for timber.

Table 6.12 *Reasons for changes in forest condition*

Name of CFUG	Changes	Reason	Further issues
Namuna Mahila	Improvement in community forest conditions	Due to their dependence on the forest, people protect it very well	Scarcity of forest products leads to better forest protection
Jansewa	Significant improvement in forest conditions	Extraction of forest products from adjoining national forests and conservation of their CFs	Protection of adjoining national forest
Gadibarah	Significant improvement in forest conditions and in *panchayat* forest	Increase in private tree plantation on *bari* (upland) edge due to forestry awareness raised by CF	Scarcity of forest products leads to better forest protection
Malati Mahila	Increase in tree fodder availability	Allocation of forest land to individual household members by CFUG in order to cultivate grass and tree fodder to stall-feed livestock for dairy business	Only fodder subsistence needs are met; but there is a shortage of other forest products
Gangajali	Deterioration in both CFs and national forests	Increasingly high pressure on forests due to over-harvesting by CFUG to accumulate money	Fodder and firewood not sufficient
Churiya Mai	Deterioration in forest conditions	Inability of CFUG to protect the forest from outsiders as the pressure from distant and excluded users is very high	Distant users from the south collect forest products from the CF without restriction, and CFUG unable to stop them

Impacts on livelihoods

Understanding the very complex impacts of implementing PFM in the *tarai* is an extremely challenging task, not least because it affects CFUG members *as well as* many others whose use of forests has changed over time. Here, we seek only the most elementary analysis of CFUG members (not distant forest users) according to a modified sustainable livelihoods framework. Four main indicators were used to assess this:

1   changes in direct household assets and entitlements to collective assets and opportuni-
    ties (loss or gain of forest land from cultivation, grazing, firewood, etc);
2   income and expenditure changes;
3   vulnerability and well-being; and
4   sustainability of the impact and causal links between PFM and the impact.

CFUG households followed a diverse range of livelihood activities. The most frequent livelihood pattern was, primarily, agriculture-based, augmented by livestock, forest, and other
non-forestry-related activities and income from non-agricultural sources. The occupations of
more wealthy households were often in trade or service, some members of the family migrat-
ing for foreign labour. Poorer households, lacking sufficient agricultural land or livestock,
typically engaged in agricultural wage labour and off-farm work of various types, including
artesan production and bamboo crafts, or as blacksmiths, tailors and cobblers.

Direct asset and entitlement changes

Change in household physical assets
Although attributing livelihood changes to particular causes is very difficult in this complex
and dynamic environment, the implementation of CF certainly seems to have been a turning
point in the livelihoods of many CFUG members. The difference between before-and-after
CF scenarios in terms of the physical assets (house, cattle sheds and others) belonging to
CFUG members indicates tangible improvement. Nevertheless, although household physical assets had improved for all wealth categories, a much larger proportion of wealthier
households (84 per cent) had improved their physical assets compared to the poor households (34 per cent). Before CF, 11.5 per cent of households did not own permanent houses,
whereas since its establishment, as one of the most expensive components in house construction – timber – became easily available at cheaper rates, 99.3 per cent did. The rich benefited
more, however, as many poor and very poor households could not afford to buy timber. The
housing conditions of poor and rich differ: the rich use good-quality timber whereas the
poor use poor-quality poles and posts. There was a high subsidy on timber, or even free
timber, for house construction in Dang, but not in Saptari. Thus, in Dang the housing conditions of the poor had also improved after CF was introduced. However, in Gangajali and
Churiamai, poorer groups such as the Mushahars, a very poor and mostly landless indigenous group, still lived in huts as they had not been given a subsidy on timber. Similarly, only
15.5 per cent of the households had access to water taps before CF, and this increased to 44
per cent after its introduction. Income from CFs was invested in improving the drinking
water supply and increasing the number of taps. Here, also, the rich benefited more than the
poor.

Changes in livestock number and livestock-related income since the introduction of community forestry
Table 6.13 shows significant changes in livestock-related earnings in Malati Mahila CF,
whereas there is no clear pattern in other CFUGs, probably due to many other factors acting
on this variable.
    Malati Mahila CF users perceived CF, livestock-rearing and milk production as their
main livelihoods, and all wealth categories had increased their incomes exceptionally well.
They had developed their CFs into a kind of silvi-pasture, where they grew fodder and
grasses to feed their livestock. In other CFs, users perceived agriculture rather than livestock-rearing as their main source of livelihood. Livestock played a supportive role; but its
contribution was not directly visible.

Table 6.13 *Changes in annual income from livestock of CFUG members in Nepali rupees (2004)*

Name of CFUG	Before community forestry				After community forestry			
	Very poor	Poor	Medium rich	Rich	Very poor	Poor	Medium rich	Rich
Namuna Mahila	617	3914	2129	6067	783	4386	2671	6133
Jansewa	1750	2875	4267	14,000	1500	4425	2400	6000
Gadibarah	2600	1500	6846	14,625	5760	3660	8769	14,438
Malati Mahila	7880	7200	4600	33,333	24,000	36,000	87,920	67,667
Gangajali	9083	911	2650	2500	2500	3433	6875	2000
Churiya Mai	9909	18,160	2000	0	12,909	10,120	5625	0

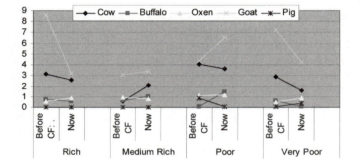

Figure 6.4 Changes in livestock numbers in Saptari district CFUGs (per household in 2004)

Source: Bhatta et al (from data collected during study)

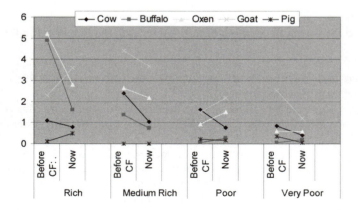

Figure 6.5 Changes in livestock numbers in Dang district CFUGs (per household in 2004)

Source: Bhatta et al (from data collected during study)

Figures 6.4 and 6.5 show the changes in the number of chief livestock per household since CF in different wealth categories in Saptari and Dang CFUGs, respectively. In Saptari, livestock such as cows, buffaloes and goats were slowly being converted to improved breeds through hybridization, and this was the major reason for the reduction in the number of these livestock for the rich category. On the other hand, the poor and very poor had difficulty in keeping large herds since free grazing space was limited in CFs and they could not afford to stall feed them with their limited workforce. However, rich people in Dang were reducing their livestock numbers as they had difficulty managing large herds due to restrictions on free grazing in forest areas. The oxen population was decreasing in the rich and increasing in the poor categories mainly because many rich people had started using tractors and other machines for draught power, whereas poor people were still using oxen (as their status improved they acquired more oxen). This also indicates that the reduction in the number of livestock is not an indication of lost assets, especially for the rich categories. In the case of rich categories, although the number of livestock had decreased, quality and productivity had not suffered with CF.

Sources of forest products for households

The question of whether forest product flows have changed due to CF is complex – immediately after the establishment of CF, regulations tend to be restrictive and only with time do benefits accrue. Furthermore, open-access extraction can allow higher off-take levels for a limited time before resources are exhausted, and so a more sustainable system may require lower off-take levels in the long term.

Although CF was a major source of forest products for members, it had not been able to supply all of the requirements of users, who had to use various other sources. Community forests were the main source of timber, but not of firewood or fodder. Different wealth categories fulfilled 40 to 50 per cent of their firewood demands from CFs. National forests were still the main source of several forest products, such as firewood, timber and fodder. Since CF, poor and very poor categories depended more on national forests for various products, particularly firewood and fodder, and to support their livelihoods since they collected these for their own use as well as to sell. Unlike the rich categories, they could not supplement their requirements for firewood and fodder from their own farms as they had very small landholdings. The medium rich and rich supplemented their requirements by purchasing firewood, mostly from national forests, and collected by poor people in the community whose livelihood mainly depended on this. Alternatives to firewood (biogas and liquid petroleum gas, or LPG) were in limited use and only occurred in a few communities.

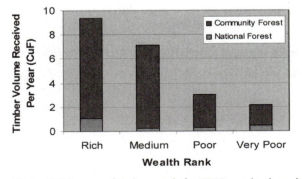

Figure 6.6 Sources of timber supply for CFUG member households

Source: Bhatta et al (from data collected during study)

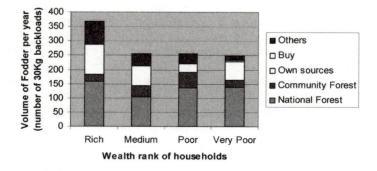

Figure 6.7 Sources of fodder supply for CFUG member households

Source: Bhatta et al (from data collected during study)

The amount of tree fodder collected from different sources varied in different CFs. Only in Malati Mahila CF did all categories of users fulfil their demand for fodder (mostly from the CF). The variations in the supply of fodder from different sources depended on the wealth category of users; the size of their private land; the size of the CF and national forest; the distance of the forest from the settlement; the condition of the forest; the availability of fodder; and the CFUG's operational plan. Poor and very poor categories were more dependent on CFs for fodder, but very little attention was given to fodder management in operational plans (except in Malati Mahila), which negatively affected the livelihood improvement opportunities of these wealth categories.

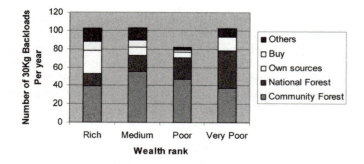

Figure 6.8 Sources of firewood supply for CFUG member households

Source: Bhatta et al (from data collected during study)

Household time spent on firewood collection varied in different CFs for all wealth categories. It decreased slightly after CF in Dang, whereas in Saptari it increased (in Gangajali and Churiyamai CFs) because households could not go to their forests. Instead, they had to go to adjoining national forests that were open access for them, where they competed with other distant users. Churiyamai CF had mostly khair (*Acacia catechu* spp) forest, and there was no formal rotational management system for firewood collection. Similarly, in Gangajali CF, executive committee members emphasized harvesting and selling firewood, rather than distributing it among users. Those in the poor and very poor categories suffered most from this.

Household income and employment through community forestry

Poor households lost some income opportunities, but gained others. CFUGs generated employment as forest guards or labour in forest management activities, but more labour-intensive forest management in CF was still far off. Some CFs paid in kind, such as with timber or firewood that could be sold for cash. Community forestry provided income opportunities through the collection and sale of NTFPs. Some CFUGs provided loans at nominal interest from their group funds for income-generating enterprises to support poor households. Because such opportunities were limited, users competed for them. There were direct costs involved in becoming members of CF, such as the voluntary contribution of time to protect and manage the forests, and some CFUGs collected periodic fees from their members for the group fund, as well as occasional contributions in cash, kind or labour towards community development activities.

In the 6 CFUGs studied, a total of 11 people (mostly medium rich) were employed in CFs on a seasonal basis: 9 as forest guards and 2 as office secretaries. For example, in Gangajali CFUG, three individuals were involved as forest guards and one as office secretary. A few people from poorer groups were employed in the collection and sale of forest products and other forest management activities as seasonal labour, earning between 1000 Nepali rupees and 5000 rupees per month for six months of the year. However, 10 (mostly poorer) respondents out of 148 had lost the DoF employment in forestry development projects and/or seasonal wages through timber contractors that they had before CF was established. Very poor groups also lost self-employment opportunities related to livestock (since grazing was now controlled) and the collection and sale of forest products in local markets, and occupational groups such as blacksmiths lost their jobs due to a lack of charcoal supply from CFs. Very poor members of Namuna Mahila CFUG complained that the executive committee was not providing them with free firewood, which they had obtained before CF.

In Saptari, traditional livelihood patterns of some of the poorest castes had changed with CF. For example, before the advent of CF, Mushahars cut and sold timber and firewood for their livelihood. Now such opportunities were limited to annual timber harvesting; but these were constrained by the prescriptions of the operational plan, and by executive committee decisions about who to employ and the new auction system. Many CFUGs (e.g. in Gangajali) were beginning to sell timber by auction to timber contractors, who used their own trucks and brought labourers from outside the village. Thus, the Mushahar groups in the Gangajali and Churiamai CFUGs had lost most of their forest-based seasonal work. Since there were few other livelihood opportunities for these groups, they were now engaged in illegal timber felling in national forests and whatever other wage labour they could find.

The greatest benefits from CFs went to the medium and rich households and to those settled closest to forests. Distant users were excluded and alienated from the forest resources that they had been using before the advent of CF. The poorest of the distant users became more vulnerable and had to migrate seasonally to cities in Nepal and India in search of employment. Even those poorer groups who were CFUG members, such as the Mushahars of Saptari, experienced only negative impacts from CF on their livelihoods.

Donor projects, in particular, have promoted CFUG-level training as an intervention to improve the skills of the local people, thereby leading to improved employment and income prospects. However, these are not generally successful as training is only one of the necessary inputs for transforming individuals' economic activities, and without capital and market linkages many of the trainees soon forget what they have learned. Indeed, travel allowance and daily subsistence allowance, the daily expense payments for attendance, are often treated as the main benefit of attending. Elites and their networks also tend to dominate participation. Donors provided training related to income generation activities, such as NTFP collection, beekeeping, mushroom cultivation, bamboo craft and vegetable farming.

Several of these skill-based trainings have, indeed, helped people to develop their skills to manage what they have learned. However, even people who used the skills effectively were not able to improve their livelihoods significantly – often such training is designed to help develop participants' skills, whereas to benefit from these they also need to understand and utilize market forces (participants lacked a clear understanding of this). Successful enterprises need capital, raw materials and other requirements in time, which are often not available. Furthermore, the market for certain products is flooded as many trainees start producing the same products at once.

Community development of physical assets through community forestry

Community forestry has been a source of development and repair of communities' physical assets. Such assets include trails, small irrigation channels, drinking water systems, school buildings and furniture, health posts, community buildings, *chautara* (village meeting and resting platforms) and temples. Improvement of such infrastructure was financed fully or partly from CFUG funds, supplemented from other sources, such as voluntary work and timber.

Infrastructure development in a community implies improvement in the potential environment for generating livelihoods. However, this has also imposed disproportionate burdens on poorer households as they have to contribute labour and share other burdens (as a member of a CFUG, every household is asked to contribute equally), sometimes foregoing their urgent livelihood needs. Poor members felt exploited in this regard – they frequently had to share the burden of building something that they might not use in practice, such as irrigation canals (since they did not own irrigable land) or a school to which they could not afford to send their children (see Table 6.14).

Table 6.14 *Development of community infrastructure after community forestry*

	Tree plantation (ha)	Trail/road (km)	Temple (numbers)	School building (numbers)	Water supply schemes (numbers)	Canal improvement (km)	Contribution (rupees)
Namuna Mahila	1	1	–	1	1	–	76,000
Jansewa	1	2	–	1	1	35	220,000
Gadibarah	25	–	–	–	1	–	45,000
Malati Mahila	60	1	–	1	–	–	269,000
Gangajali	5	2	1	1	–	–	384,000
Churiya Mai	8	–	–	–	–	–	17,000

The three Dang CFUGs spent about 500,000 Nepali rupees on community infrastructure development. In Saptari, Gangajali CFUG spent its funds on temple construction since most of its infrastructure had already been developed by government agencies. In both districts, most of the CFUG funds accumulated from the sale of forest products were used for developing community infrastructure, rather than for income-generating programmes.

Changes in vulnerability and the well-being of households

When these six CFUGs were formed, the livelihood concerns of poorer and marginalized families were not considered. Although the poorest groups had particular livelihood needs and were most vulnerable to changes in the management of common property resources, their concerns were seldom addressed. They had neither the opportunity nor the capacity to strongly influence decision-making as executive positions were mainly held by the elite. The need for their participation in decision-making was slowly being recognized by service providers (the district forest office and NGOs), donors and policy-makers, and their vulnerability and concern for their well-being had started to attract attention during planning and strategy formulation. As a means of reducing their vulnerability and enhancing their well-being, credit facilities and income-generating activities were identified as pro-poor measures that CFUGs could introduce. Since 1990, different NGOs and CBOs had started savings and credit schemes and co-operatives with the support of NGOs funded by international funding institutions (IFIs). These activities had inspired PFM projects to pursue savings and credit and community development activities in both districts. Poor people lacked access to easy credit; but local moneylenders charge very high compound interest rates, and once trapped in debt it is very difficult to escape. The CFUGs spent most of their funds on rural infrastructure and salaries for forest guards and other wage workers. Only two of the six CFUGs had set aside funds –10,000 rupees (0.01 per cent of total income since the formation of the CFUG) in Namuna Mahila and 50,000 rupees (14 per cent of total income) in Malati Mahila CFUG – as separate revolving funds to provide credit to poor group members at a low interest rate (18 per cent) and without collateral. Such credit was provided for income-generating activities, as well as for family emergencies such as medical treatment or the construction or repair of a house.

Community forestry policy did not specifically address the issues of poverty and vulnerability. To fulfil the stated objectives of government policy (e.g. the Tenth Five-Year Plan), specific provisions for vulnerable, poor and marginalized people could make a great difference, at least in ensuring that these practices were more widely followed in CFUGs.

NTFP marketing

Activities to promote NTFP were not a priority of the six CFUGs. All, except Malati Mahila, focused mostly on the marketing of timber, rather than other forest products. Malati Mahila CFUG focused more on fodder and forage production. However, there were ample opportunities and potential to promote NTFP cultivation, management and marketing. Sal leaf plate-making as a small-scale cottage industry is a good example. Many similar opportunities could not be utilized, mainly due to the lack of support from service providers or very low, or a total absence of, processing and marketing technical knowledge. NTFPs such as fodder, sal leaves and other medicinal plants were mostly collected for domestic use. Commercial use of NTFPs was insignificant.

People's commitment to PFM could be clearly seen, especially where there were direct material benefits. But they were not very sure about the future of PFM and whether they would remain the owners of the forests that they had developed and maintained.

Explaining policy, practice and impact:
Governance and policy process

Policy and governance processes

In contrast to hill CFUGs, which have greater autonomy, in the *tarai* there has been a history of far less cooperation from community/forest department staff. Prior to the introduction of

CF, most forestry staff were policing, patrolling and licensing, rather than working with local stakeholders. There was, thus, very limited participation of DDCs, VDCs, local people and communities. Inevitably, there has been a similar lack of involvement of wider stakeholders in the forest-sector policy and strategy formulation processes.

Often, bureaucrats prepare policy drafts without consulting other actors. Policy dialogue is lacking since only a few active parliamentarians have been sensitized to PFM issues. The proposed policies are considered and generally approved by both the lower and upper houses of parliament and then sent to the King for final approval. There is no climate for multi-stakeholder deliberation and donor coordination, and there are limited interactive and reflective processes of policy formulation and coordination among stakeholders. Large donors such as the Danish International Development Agency (DANIDA) have withdrawn their support from PFM due to political crises in the country.

Since the restoration of democracy in 1990, multi-stakeholder deliberations and donor coordination have gradually been emerging. For instance, during the formulation of the Tenth Five-Year Plan in 2002, a wider stakeholder consultation was conducted. Similarly, annual planning and project policy-making were often carried out through limited or wider consultation of bureaucrats and project personnel at various forums such as forest-sector coordination committees (FSCCs), bilateral forestry projects' own fora, project coordination committees and regional planning workshops.

At district level, limited discussion and sharing of information have been taking place among stakeholders through district forests coordination committees (DFCCs) in 11 *tarai* and inner *tarai* districts where the Biodiversity Sector Programme for *Siwalik* and *Tarai* (BISEP-ST) and the Livelihoods and Forestry Programme (LFP) support the PFM process. This forum has the representation of local governments (DDCs; VDCs; CFUGs; collaborative forest management groups (CFMGs); district-level major political parties; private-sector forestry-related enterprises; government line agencies; district-based NGOs; and others). Although this sort of forum may seem effective, in most cases bureaucrats and project personnel, the elite or people who have been able to capture the benefits from forests have influenced the decisions made in DFCCs. DFCCs face many challenges, such as *ad hocism* in decision-making and proper service delivery to CFUGs. Issues relating to poverty alleviation among the poor and very poor groups, especially those who have very limited access to forest products, are rarely raised.

Donors have shown little transparency about their project support, despite frequent claims of exemplary transparency. Bilateral projects in the *tarai* forestry sector, such as German Agency for Technical Cooperation (GTZ)-supported Churia Forest Development Project or the UK Department for International Development (DFID)-supported Livelihoods and Forestry Programme have their own institutional implementation setup parallel to that of the DoF. The MoFSC, DoF and their line agencies are partners in these projects, one of the main objectives of which is to strengthen DoF capacity. There are two ways to fund projects: through the government fiscal system and by direct donor funding to service providers and CFUGs. The direct funding system tends to limit the transparency of allocations because project records and the progress of direct expenditures are not recorded in the DoF's reports to the National Planning Commission (NPC) and the Auditor General's office.

NGO involvement with regard to both advocacy and service provision roles has been very limited in the forestry sector compared to other development-related sectors. The Federation of Community Forest User Groups Nepal began intervening in the *tarai* only from 1997, mainly due to its active opposition to the Bara forest and associated operational forest management plans, which it felt to be against the interests of the local people. In more recent years, NGO activities have increased in the *tarai* because several projects have started subcontracting various implementation tasks to them since it has become increasingly

---

Box 6.2 Sawmills, timber and law in the *tarai*

Binod Bhatta and Akhileswar Karna

There are numerous (mainly commercial) timber industries in the *tarai*. Although the state sawmills are in a sick state due to losses, private sawmills are operating profitably. It is difficult for CFUGs to establish their own sawmills, largely due to the rule that any sawmill or forest-based industry must not be within 5km of forests. The aim was to minimize illegal harvesting of timber by timber mafias. Although this rule has been a hitch in establishing CFUG sawmills, a study of community sawmills in Kabhrepalanchok and Sindhupalchowk districts concludes that economic and financial returns from log sales are likely to be greater than those from community sawmills (NACRMLP, 2003).

Private sawmills obtain wood from government-managed forests and also from CFs. The tasks of tree counting and marking, logging, log collection and transport from government forests and CFs are supervised by district forest office staff.

Furniture and small-scale wood-based cottage industries must have a licence from the District Cottage Industries office to operate, and people are also asked to submit a 'no objection' certificate from the district forest office. It often takes two or three months to fulfil these requirements.

If the forest is within 5km of the sawmill, applicants cannot get a licence. However, many private sawmills within this threshold openly flout this rule. When the Department of Forests (DoF) started taking action against them by seizing their machinery or cancelling their licences for processing timber, many sawmill owners obtained a stay order against this policy from the Supreme Court through their network, Timber Entrepreneur Association-Nepal, and continued to run their sawmills.

The policy still exists, but small-scale forest-based industries using forest products from their own or other sources have suffered the most since they find it very difficult to get a licence. Many main highways pass within 5km of the forest, and market centres are established along them. It is difficult for such industries to operate from places that are inconvenient as well as uneconomical. Proper monitoring and auditing of forest products, rather than enforcing the impractical 5km distance policy, may resolve this issue.

---

difficult to operate amidst the insurgency. At the central level, the involvement of NGOs in policy debate has been limited to *tarai* forest management issues, although FECOFUN has strongly lobbied for its CF member constituency.

During the early years, FECOFUN took time to develop robust institutional arrangements. Recently, it has been playing a strong and independent advocacy role. Despite the conflict, it has effectively mobilized CFUG members behind wider demands for the resumption of democracy. However, the interests of CFUG members in the *tarai*, where distant users are excluded from membership and stakeholders have varying interests, are not always consistent with those of the wider society. FECOFUN has repeatedly advocated CF in the *tarai*. However, it has been rather silent about the implications for distant users.

The national forest policy related to CF in the *tarai* is ambiguous and has been left to the discretion of DFOs. In some districts, DFOs hand over CFs according to potential CF areas identified in the district plan, whereas in others they hesitate even to hand over identified potential areas. The 2000 Forest Policy brought in the concept of CollFM implemented through multi-stakeholder participation, which is being piloted in three districts.

Governance at village level has been as problematic as at regional level, probably due to the very high value of the standing forest when converted to timber. Despite great variations and many exemplary cases, there is little transparency in decision-making processes and CFUG funds within *tarai* CFUGs. As well as the monopoly of elites in decision-making and the exclusion of the poor from benefits, cases of embezzlement are frequently reported. Iverson et al (2006) report 'hidden subsidies' to the elite in timber distribution arrangements.

As already mentioned, in some areas, particularly where extensive forests remain (as in Dang), a nexus often exists between executive committee members, timber contractors, private enterprises, CFUG networks and even some DoF staff members, particularly those involved in the preparation and approval of operational plans. The executive committees frequently prepare operational plans that show a higher allowable cut for timber and firewood than the actual amount, and get them approved by the DFO, after which they sell the products at reduced prices to favoured contractors.

However, not all *tarai* CFUGs should be tainted by the poor practices of some. There are, equally, many exemplary management and governance practices of CFUGs. For example, many groups have developed their constitutions and operational plans in a democratic manner, conduct regular inclusive and transparent meetings, and have formed separate female groups, which has empowered women and facilitated their participation in decision-making processes. Many CFUGs promote the livelihoods of some of their members through a range of support activities. A small number of CFUGs have successfully experimented with forest land allocation on a household basis for fodder, grass, inter-cropping and agro-forestry practices. A few CFUGs have promoted group-based livelihood practices, such as cattle farming for milk and meat production, NTFP collection and sale, and loan facilities for poor households. However, these livelihood development programmes have been limited to a few CFUGs.

About 50 per cent of the CFUGs have generated substantial funds, mainly from the sale of timber, which have largely been spent on rural development activities that favour wealthy members. These funds have primarily been mobilized under the control of the elite. In most of the groups, excessive expenditure is occurring in office management, administration costs and daily allowances for executive committee members.

Tarai CFUGs are poor in institutional governance processes; but the protection and management of forests is better than in the past. Many poor distant users have to pay a levy to the Maoists to collect forest products. Maoists have been claiming a certain percentage of revenue from the sale of forest products.

The MoFSC has occasionally been in the limelight in relation to corruption cases. A few years ago, the Commission for Investigation into Abuse of Authority (CIAA) fined the secretary of the MoFSC approximately 150,000 Nepali rupees for allowing timber contractors to harvest and transport timber from Dadeldhura district even after the expiry of their contract. It was alleged that he had done this in return for a heavy bribe paid by the contractor, who, in turn, harvested much more than was officially agreed.

Availability of forest products and coping strategies of distant users in the *tarai*

The analysis of the availability of forest products from government-managed forests before and after CF revealed that about 70 per cent of the local population of the *tarai* had been denied access to CFs by being excluded from CFUG membership. They were also denied access to government forests, which had made it difficult for them to meet their forest-based livelihood requirements. Tables 6.15 and 6.16 illustrate the availability of forest products before and after CF and the coping strategies used.

Before CF, although the villagers living close to forests collected more forest products than distant users, the distant users also collected forest products from government forests in order to fulfil their annual requirement by paying royalties to the government. However, in the present institutional setting of CFs in the *tarai*, distant users were unable to collect these products because forests were handed over to local communities. Table 6.15 illustrates the availability to distant users of forest products from government forests before and after CF.

Table 6.15 *Annual mean availability of forest product to households of excluded forest users before and after community forestry*

District	Geograph-ical location in relation to forest*	Village development committees	Timber (VDCs) (cubic feet)		Firewood (*bhari*)		Fodder (*bhari*)	
			Before CF	*After CF*	*Before CF*	*After CF*	*Before CF*	*After CF*
Saptari	Close	Lohajara	10	2	150	0	0	0
	Mid distant	Mainakaderi	6	0	50	0	0	0
	Far	Koiladi	10	0	0	0	0	0
Dhanusha	Close	Digambarpur	5	0	3	0	0	0
		Bengadabar	8	0	2	0	0	0
	Far	Duhabi	6	0	1	0	0	0
Bara	Close	Dumarbana	20	4	4	2	30	10
		Jitpur	20	0	5	2	15	2
		Nijgadh	25	0	1	1	30	30
	Mid distant	Ganj Bhabanipur	8	0	6	3	0	0
		Parsauni	12	0	4	0	0	0
		Rampur Tokani	10	0	7	0	0	0
	Far	Telkuwa Benauli	9	0	4	0	0	0
		Piparpati Jabdi	6	0	4	0	0	0
		Pattarhati	11	0	7	0	0	0

Notes: * Close (adjacent to forests) = below 5km; mid distant = between 5km and 10km and those who use forest products either by going themselves to forests or by purchasing forest products; far distant = above 10km (users rarely go to forests and use few forest products from government-managed forests).

Table 6.15 clearly shows that CFUG committee members had control over forests after their handover to adjacent users as CFs; mid- and far-distant users had been deprived of their use rights, and quite often conflicts arose between them. Therefore, many users of forest products had lost their use rights and been forced to find alternative sources to fulfil their requirements.

Distant users' alternative energy sources and coping strategies

The availability of firewood had clearly declined due to the introduction of CF, and a large population of forest users, particularly those living far from the forest, had been forced to find alternative sources. Most households in all geographical locations still used firewood as a major cooking energy source (see Table 6.16); however, villagers close to forests were able to collect firewood from government-managed forests, whereas distant users mostly relied on their own scarce sources, which provided very low-quality firewood. Poor households often faced difficulties in meeting their cooking energy needs. They collected twigs, dry branches and dry leaves, sweeping the floor of government forests, other private woodlots and horticulture gardens to collect sources of cooking energy. The majority of distant villagers used cow-dung cake as their chief source of cooking energy, as well as dried bamboo, kerosene, leaf litter and jute residue. Only a few households settled near the forests were able to establish biogas plants as a cooking energy source.

Table 6.16 *Households' alternative sources of cooking fuel (number of months per year)*

District	Geograph-ical location in relation to forest	Village development committees (VDCs)	Fire-wood	Cow dung	Others*	Leaf litter	Biogas	Kero-sene	Wheat straw	Paddy straw	Total
Saptari	Close	Lohajara	3	6	0	0	0	0	1	2	12
	Mid distant	Mainakaderi	2	6	1	0	0	0	1	2	12
	Far	Koiladi	1	7	1	0	0	0	1	2	12
Dhanusha	Close	Digambarpur	3	3	5	1	0	0	0	0	12
		Bengadabar	8	2	2	0	0	0	0	0	12
	Far	Duhabi	2	5	3	1	0	0	0	1	
Bara	Close	Dumarbana	12	0	0	0	0	0	0	0	12
		Jitpur	8	1	0	0	1	1	1	0	12
		Nijgadh	7	1	0	2	2	0	0	0	12
	Mid distant	Ganj	3	7	1	1	0	0	0	0	12
		Bhabanipur Parsauni	6	2	2	1	0	1	0	0	12
		Rampur Tokani	6	3	3	0	0	0	0	0	12
	Far	Telkuwa Benauli	7	3	1	1	0	0	0	0	12
		Piparpati Jabdi	6	4	2	0	0	0	0	0	12
		Pattarhati	6	3	2	1	0	0	0	0	12

Notes: * Other agricultural residues include maize stalks, rahar stalks, masura stalks, dried sugarcane stalks, bamboo, dried branches and other parts of jute.

Rich and medium households from distant villages bought timber from illicit cutters or obtained it in limited quantities from private source trees and woodlots. In Dhanusha, timber requirements were met mostly from private sources, although villages were close to government forests. However, despite being close to government forests, the villagers of Digambarpur and Bengadabar, and all distant villages (including Duhabi), substitute their timber requirements with low-cost, low-quality cement pillars and inferior species of timber from their own farms. Poor, very poor and medium-class households often use bamboo for house construction. Distant users depend on bamboo leaf, ground grasses, *janera*, *sawa* and agricultural residues such as straw from their own or communal land for forage.

Earlier, most of the *tarai* population had been meeting their timber and firewood requirement from forests. There was a system of *Gharsangha purji* for timber and *Daura purji* for firewood, which required people to pay royalties to a government ranger on the spot to collect timber and firewood from forests. For this purpose, they normally used bullock carts, bicycles or head-loading. Even after the *purji* system was discontinued, people continued collecting dried wood, bushes, leaf litter and other small branches. However, this was not sustainable, and resources gradually became scarce. With the new act in place, the DoF also started to be stricter about limiting such practices.

Following the formation of CFs, CFUGs started claiming that they were managing their forests sustainably. Furthermore, the conservation of forests had improved, and illegal felling of trees and encroachment had, to a large extent, been controlled. However, several other stakeholders (mainly distant users and poor households) claimed that the benefits were now limited to the elite households of CFUGs; many stakeholders were either denied or given only limited access to forest products and benefits from the CF. This raised serious concern about whether conservation and the sustainable use and management of forest resources are possible. If equity is ensured in the decision-making of CF management, conservation and sustainable use can be achieved.

Tenure issues

The indigenous inhabitants of the *tarai*, including the Tharu, Satar, Rajbansi, Danuwar, and Musahar, have largely lost their use of forests through their *de facto* exclusion from CFs by new settlers. Similarly, in the Maithili-, Bhojpuri- and Abadhi-speaking belts, the major populated regions of the *tarai*, as the forests have shrunk northwards due to their conversion to settlements, roads and development, villages once close to forests have become distant settlements. All people enjoy landownership claims on ancestral land; but as the population grows, agricultural land is distributed between brothers and therefore becomes fragmented – thus, the poor become poorer. Poverty often forces poor families to sell their land and move elsewhere, losing ownership of their ancestral land.

The settlement of migrants is a common feature in all *tarai* and inner *tarai* districts (Ghimire, 1992). DoF efforts to control the settlement of such illegal migrants have met with very little success so far because of poor commitment and support from politicians and policy-makers, inconsistency in policy due to frequent changes, and because most past governments (until the *panchayat* era) liked to create settlements near *tarai* forests.

In contrast with the current forestry policy of not converting forests to other uses, forest land has been granted to other social development organizations, such as universities, hospitals and *sukumbasi* (landless people). Illegal settlers near highways and in forests have degraded nearby forests, and the government's past efforts at providing land titles to such encroached land has set a precedent that is hard to reverse.

Major policy actors, their changing characteristics and influences at different levels

The major policy actors in *tarai* forestry at the national level are the MoFSC, the Ministry of Local Development (MoLD), bilateral donors and other IFIs, forest-user federations such as FECOFUN, forest contractors and their associations, private forest product-based business companies, educational and research institutions, and the judiciary.

Under the MoFSC, there are five departments at the centre and five regional directorates in each of the five regions. The DoF is the major actor that facilitates, directs, supports and monitors district forest offices, which exist in 24 *tarai* and inner *tarai* districts.

The story of PFM in the *tarai* is quite different from that in the hills. Forest resources have generally been controlled by the DoF and elites. Since the value of forest products is very high, many of the more powerful actors pursue their own interests, rather than the collective interests of sustainable forest management and livelihood development of the wider community.

The regional directorates monitor and evaluate forestry activities in the district. They organize annual regional planning and review meetings in each region, where donors' representatives and government officials from district, regional and central levels sit together for two or three days and share past programmes and plans for the coming year. At this forum, stakeholders also put forward their ideas of further policy support needed to implement PFM activities, and targeted programmes and plans.

Many actors with significant involvement in forest management are not part of this regional directorate-facilitated forum. These include NGOs, business enterprises, FECO-FUN and PFM institutions (such as CollFM, CF, leasehold forestry, DDCs and VDCs) that play important roles. On the other hand, the DFCC has emerged in some donor-supported *tarai* districts (such as eight BISEP-ST-supported districts and three LFP-supported districts) as a more inclusive forum for these groups. It is chaired by the DDC chairperson, and many actors come together to share their views. However, marginalized groups have not yet found their voices in these discussions.

Box 6.3 Marketing private tree products

Akhileswar Karma and Binod Bhatta

Many rich *tarai* farmers had private woodlots (of less than 1ha in extent in proper *tarai* and 1ha to 5ha in dry land in the southern Churia hills), consisting mainly of sissoo plantations, mango orchards and trees around the homestead and fishponds. Sissoo planting started in the 1980s with the introduction of the tree plantation concept by the Department of Forests (DoF), supported by the World Bank. Other trees and woodlots have been created traditionally without external assistance. Poor people have little or no land, and are therefore deprived of trees and woodlots. *Tarai* people, especially the rich, have been planting trees on their private land for more than 60 years.

Private forests offer an alternative source of forest product supply. However, the existing process is not farmer friendly and discourages people from planting trees on their private land. There is, currently, a very slow and complicated process for marketing forest products from private forests or woodlots, based on the 1995 Forestry Regulations where owners of trees must get the permission of the district forest office to harvest for domestic use or sale. In two *tarai* districts we found an incredible eight stages in the marketing of private timber:

1   The owner submits applications to the district forest office with a land registration certificate, land tax receipt, citizenship certificate, and letter of recommendation from the village development committee (VDC)/municipality.
2   The application is sent to *Ilaka* or range post office for identification of the tree(s) and a detailed investigation.
3   After the investigation, the range post office or *Ilaka* sends its report to the district forest office.
4   Based on the report, the district forest officer (DFO) asks a ranger to make *Chhapan* (mark the trees for felling), prepare *Chhapan Muchulaka,* and submit the quantity of timber that could be extracted from the particular trees that farmers have requested for harvesting and sale to market.
5   Once the ranger completes the process, he again sends detailed reports to the district forest office and, based on those reports, the DFO may provide a letter giving the farmer permission to harvest the trees.
6   After the trees are harvested and cut into different sizes, the farmer asks the ranger to check the size and number of logs, for which he has to go again to the district forest office.
7   The ranger submits detailed reports to the DFO, and after that tree-owners have to pay value-added tax on the basis of species and volume of timber/wood extracted.
8   The DFO provides a release letter to the farmer allowing him or her to transport the timber to any sawmills or wherever he/she wants.

Within these processes, farmers face many challenges and difficulties, and middlemen frequently play a vital role in ensuring the receipt of the release letter from the DFO. A similar process has to be followed for domestic use; but, in reality, farmers do not follow it. Some farmers and middlemen paid 'extra money' (bribes) to those in authority on the study sites, otherwise the process can take more than three or four months. Therefore, the existing process clearly discourages people from planting trees on their private land.

A noble provision to make forest products available to market users and rural distant farmers of the *tarai* is the District Forest Product Supply Board (DFPSB), which has been formed in all *tarai* districts to ensure the supply of forest products to distant villages. Similarly, the Timber Corporation of Nepal (TCN) has been maintaining the timber supply to towns and cities. Both have been focusing on the *tarai*. However, most distant users do not know of the existence of the DFPSB. Households are rarely able to receive firewood and timber through it. The TCN, on the other hand, has been providing timber to towns, but during recent years has been unable to perform its stipulated functions.

Coordination among development support agencies – such as government line agencies and departments, civil society and donor-supported projects and programmes, along with forestry development activities – is important to support the diverse needs of CFUGs/CFM groups. The Local Self-Governance Act gives the DDC the role of coordinating ministerial line agencies and district-level NGOs in order to integrate programmes; however, the implementation of this is very difficult as coordination between different organizations and actors is often lacking.

There has been less donor support for the forestry sector in the *tarai* region than in the hills. Donor-supported plans and programmes have been unable to create a supportive environment for the successful and equitable involvement of the local people. For example, despite raising attention to *tarai* forest management challenges, three major projects, including the *Tarai* Community Forestry Development Project (TCFDP), supported by the World Bank, and the Forest Management and Utilization Development Project (FMUDP), supported by the Finnish International Development Agency (FINNIDA), could not develop politically acceptable and, therefore, successful models for *tarai* forestry issues.

The Churia Forest Development Project (ChFDP), supported by GTZ, has recently been phased out from some of the VDCs of three districts of the eastern *tarai* after more than a decade. The majority of the livelihood-oriented activities were implemented by this project without DoF coordination. Many are now proving unsustainable due to lack of continued support. For example *bel*-squash making was started through local NGOs in Saptari; but this gradually stopped after the termination of the project when project-supported NGOs withdrew their services from villages.

More recently, BISEP-ST (supported by the SNV) and LFP (DFID supported) have emerged as large projects in the region. The former is involved through the DoF in piloting the CFM concept, and the latter works through NGOs and the DoF in all types of forestry such as CF, CFM and leasehold and private forestry.

Politics of information and knowledge of forest management in the *tarai*

There is also an attitudinal issue. Most forestry staff trained in forestry colleges on a 'scientific' basis think that local people have little or no knowledge of forestry and must be taught how to raise seedlings and where and how to plant them, which species are suitable for particular sites, how to carry out silvicultural operations, and how to harvest trees. The government has maintained its control over *tarai* forests and frequently changes the rules, often through circulars and directives (e.g. the ban on green felling and on the harvesting of certain species). This has resulted in inconsistencies in policies, and people have become confused about state policy in the *tarai*.

Although the MoFSC has realized that *tarai* forests cannot be managed without local people's involvement, which contributed to the evolution and implementation of PFM, its policy has been criticized for being developed through 'command and control', rather than through the deliberative involvement of different actors. Although some policy documents, such as CFM, were prepared in consultation with stakeholders, particularly at district level, the majority of stakeholders were not consulted, or even if they were their ideas and experience were not incorporated within policy documents. Moreover, in meetings and workshops, central-level bureaucrats tend to pay more attention to the views of DFOs rather than other stakeholders. The perception tends to be that DFOs know the practical field-level difficulties of implementation and that they also represent the voices of field staff and local people. Unfortunately, this is often not true, as many DFOs pay little attention to their field staff. Therefore, there is often little knowledge and information-sharing between the DoF and other actors/stakeholders.

There is a conflict of ideas and knowledge between different donor-supported projects at the national level. Bilateral projects bring with them their own field experience; but there is little sharing with other project and government staff. There is a tendency for donor-supported projects to seek status by generating new ideas and concepts, which can, in turn, create confusion for field practitioners. Each nursing his or her own brand concepts can lead to less chance of learning from the other.

Regular sharing of field experience can help to transfer innovations from one place to another. Some of the more innovative DFOs have initiated regular interactive field staff meetings aimed at building positive attitudes towards local people and creating a more participatory environment in the district. However, many DFOs have not been conducting regular staff meetings in the *tarai*, and even when they do the general trend is that DFOs direct their subordinates rather than listen to them. Similarly, in CFUGs, decisions are mostly made by executive committee members. The users' assembly is often organized as a formality and to show, on paper, that members have been consulted in a top-down manner. In many CFUGs, the majority of user group members participate in assembly meetings; but the elites expropriate the benefits and many households are excluded from them. Illiterate people, especially women and other marginalized groups, are dominated by executive committee members because they cannot read the operational plan, the constitution or decisions recorded in the minutes. All of these locally available documents are accessible only to literate people.

In Dang, fund management in Namuna Mahila and Gadibarah CFUGs is transparent, thanks to the public audit process initiated by CFUGs with FECOFUN's support. This emerged after conflict between group members over the low level of transparency. A public audit system was subsequently adopted so that all CFUG households could know about their fund investment. FECOFUN and the Strengthening the Role of Civil Society and Women in Democracy and Governance project (SAMARPAN) organized training for female executive committee members. The executive committee has proved its capability to manage CF for years, and is making further improvements to its account- and record-keeping systems. Namuna Mahila CFUG has also constructed a road to link the village with the market, built more water supplies and taken landslide reduction initiatives.

Actions for policy change

The findings of this study were shared and discussed with CFUGs, village-, district-, regional- and central-level stakeholders and policy-makers. The study and subsequent discussions revealed the following important suggestions related to policy change.

One of the key issues is the concerns of distant users, whose forest-based livelihoods have to be addressed in policy or in practice. The integrated approach to sustainable forest resource management with livelihood enhancement for the poor should be the primary objective of *tarai* forest management, coupled with the devolution of power to manage resources to local people.

The *tarai* forestry issues can be resolved only through a comprehensive and integrated participatory approach to forest management, such as by promoting CF, CF management, leasehold forestry and government-managed forests, as well as the promotion of private and communal forests, proper management of canal-side and roadside plantations, management of protected areas, control of illegal encroachment, preventing forest land distribution, breaking the timber mafia and corruption nexus, and Churia conservation (both upstream and downstream). There is a need for the full commitment of politicians, policy-makers and high-level bureaucrats in this process. Furthermore, a landscape approach to planning and managing resources is needed.

Conclusions

The Nepal *tarai* is a unique region where those responsible for managing forests have been in direct competition with those who seek to convert them to agricultural or other uses. The state has often supported conversion of forests. Moreover, the incentives of illicit felling are so great that many within the forest department engage in it.

There is no single way to tackle *tarai* forestry issues. Although CFUG representative groups such as FECOFUN often advocate over-simplistic solutions by applying CF, the issue will have to be solved by applying comprehensive approaches (such as different forest management systems, including CF, collaborative forest management, leasehold forestry and the promotion of private and communal forests). The basis of such a comprehensive approach could be a 'district forest sector plan', in which the management of each site and location is clearly mentioned, including the management regime of particular locations. This must be prepared using the participatory approach with all stakeholders involved. A land-scape approach to planning and managing resources, rather than looking at patchworks in smaller areas, is needed.

A number of management challenges in the *tarai* remain, such as:

- managing protected areas;
- controlling illegal encroachment and stopping the distribution of forest land for non-forest purposes;
- breaking the nexus of timber mafias, corrupt politicians and forestry officials;
- watershed management and Churia conservation, both upstream and downstream; and
- appropriately dealing with the forest-based livelihood issues of close-, mid- and far-distant people throughout the tarai region.

The challenges persist because the forest management authorities have not been sufficiently strong to overcome the incentives to illicit forest product harvesting and forest land settlement and conversion.

The issue of distant users needs to be fully addressed through a holistic approach. The most affected poor and marginalized groups whose forest-based livelihoods have not been addressed need to be properly considered in policy and practice. An integrated approach, with sustainable forest resource management and livelihood enhancement for poor people, should be the primary objective of *tarai* forestry. Priority should be given to the 'people first' concept, with devolution of power to local people (including distant users) to manage resources by involving actors, each with clearly demarcated roles and responsibilities.

All the above-mentioned actions should be taken through participatory processes, and the lead role must be adopted by local people who have suffered the most from the fragmentation of the forest. Replicating the hill model of CF is not the solution for the *tarai*; therefore, an inclusive modification has to be made to suit local reality and needs.

Again, the large population, which has often been marginalized and never listened to, must be brought under the umbrella of overall *tarai* forest management. The role of DoF staff needs to be considered as one of facilitation rather than of control. Field staff should be oriented to their role of extension agents so that they develop good relations with local people. The focus should be on enhancing the livelihoods of the poor. Forest product distribution and benefit-sharing among stakeholders must be emphasized.

As a result, all stakeholders must realize that *tarai* forest management cannot be achieved simply through one type of management regime alone, but that what is needed is a comprehensive approach where one effort complements the other. In the absence of such a realization, tremendous potential benefits from *tarai* forest are being wasted.

The contradictory provisions of the 1998 Local Self-Governance Act and CF could be resolved through consultation with all stakeholders and amendment of the conflicting provisions. There is also a need to ensure that, in future, such policies and acts are developed only after wider consultation with all stakeholders.

Self-initiated CFUGs have proved much better than outsider-promoted CFs at forest management and the sharing of responsibilities and benefits, and also in terms of sustainability. Therefore, in view of these advantages of a self-initiated approach over a government-initiated one, self-initiated groups should be encouraged, supported and facilitated in PFM as far as possible.

A major change reported by the people was that the condition of CFs had generally improved, although the condition of adjoining and nearby national forests had deteriorated. However, further empirical studies should be carried out on the relationship between the improvement of CFs and the deterioration of adjoining and nearby national forests before reaching any definitive conclusions.

The CFUGs believe that the formation of CFs has helped them to manage their forests in a systematic manner and that the conservation of forests has improved. However, several stakeholders from resource-poor households have claimed that the benefits are now limited to the elite households of CFUGs; many resource-poor households are either totally denied or given only limited access to forest products and benefits from the CF. Therefore, by ensuring equity in decision-making in CF management, better conservation and sustainable use of forests can be achieved. In other words, equitable participation and transparency in PFM management decision-making can ensure both conservation and the sustainable use of forest.

References

Adhikari, B. R. (2002) 'Ban Atikarman:Samsya ra Samadhan ka Paryas', in Pokharel, B. K., Nepal, S. M. and Kafle, R. (eds) *Hamro Ban*, Kathmandu Department of Forests, His Majesty's Government of Nepal

Adhikari, J., Dev, O. P. and Dhungana, H. P. (2005) 'State and forest: A historical analysis of policies affecting forest management in Nepal *tarai*', Unpublished paper, Kathmandu, Resources Development and Research Centre (RDRC)

Bajracharya, K. M. (2000) 'Intensive management of the *tarai* and inner *tarai* forests in Nepal', in *Management of Forests in Tarai and Inner Tarai of Nepal*, Proceedings of the National Workshop Organized by the Nepal Foresters Association, 11–12 February, Kathmandu

Baral, J. C. and Subedi, B. R. (1999) 'Is community forestry of Nepal's *tarai* in right direction?', *Banko Jankari*, vol 9, no 2, Presented at the National Workshop on the Management of the *Tarai* and Inner *Tarai* Forests, NFA, 11–12 February 2000, Kathmandu

Bartlett, A. G. and Malla, Y. B. (1992) 'Local forest management and forest policy in Nepal', *Journal of World Forest Resource Management*, vol 6, pp99–116

Bhatia, A. (ed) (2000) *Participatory Forest Management: Implications for Policy and Human Resources' Development in the Hindu Kush-Himalayas, vol V, Nepal*, Kathmandu, International Centre for Integrated Mountain Development

Bhatta, B. (2002) 'Access and equity in the *Terai* Community Forestry Programme', Paper presented at Human–Institutional–Natural Resources Interactions Workshop, 27–28 March, Organized jointly by IOF, IFRI and NFRI, Pokhara, Nepal

Bhatta, B. (2002) *Access and Equity Issues in the Terai Community Forestry Programme*, Nepal, Winrock International

Bhatta, B. and Dhakal, B. (2004) 'Forestry sector's role in Nepal's socio-political stability: A critical analysis of problems, prospects and potentials', in *Proceedings of the Fourth National Workshop on Community Forestry*, 4–6 August, , Kathmandu, Department of Forests

Bhatta, B. and Tiwari, S. (2001) *Forest Restoration Policy and Practices: A National Assessment from Nepal*, Report submitted to International Union for the Conservation of Nature, Nepal

Bhattarai, K., Conway, D. and Shrestha, N. R. (2002) 'The vacillating evolution of forestry policy in Nepal', *International Development Planning Review*, vol 24, no 3, pp315–338

BISEP-ST (Biodiversity Sector Programme for *Siwalik* and *Tarai*) (2002) *Stakeholders' Consultation Workshop Report*, Kathmandu, BISEP-ST

Bista, D. B. (1991) *Fatalism and Development: Nepal's struggle for Modernisation*, New Delhi, Orient Longman

CBS (Central Bureau of Statistics) (2000) *Nepal Population Report*, Khatmandu, His Majesty's Government of Nepal (HMGN)

CBS (2002) *Population Census 2001*, Kathmandu, HMGN

CBS (2004) *National Sample Census of Agriculture Nepal 2001/2002 Tarai*, Kathmandu, HMGN

CFD (Community Forest Division) (2004) *Forest Inventory Guidelines for Community Forests*, Kathmandu, CFD

DFRS (Department of Forest Research and Survey) (1999) *Forest Resources of Nepal (1987–1998)*, Forest Resource Information System Project Publication No 74, Kathmandu, DFRS

District Forest Office (2003) *Forest Working Scheme, DFO, Dang 2003–2007*, Dang, Nepal, District Forest Office

DoF (Department of Forests) (2004) *District Five Year Forest Work Plans – Dang and Saptari District*, Kathmandu, Department of Forests

DoF (2005) *Forest Cover Changes Analysis of the Tarai Districts (1990/91–2000/01)*, Kathmandu, DoF

DoF–CFD (2006) *Forest User Group Database (1990–2006)*, Kathmandu, Community Forestry Division (CFD) and Department of Forests (DoF)

Fisher, R. J. (1989) *Indigenous Systems of Common Property Forest Management*, Hawaii, East–West Centre

Forest Research and Survey Centre (1994) *Deforestation in the Terai Districts 1978/79–1990/91*, Project Publication No 60, Kathmandu, HMGN, Forest Research and Survey Centre, Forest Resource Information System

Ghimire, K. (1992) *Forest or Farm? The Politics of Poverty and Land Hunger in Nepal*, Delhi, Oxford University Press

Gilmour, D. A. and Fisher, R. J. (1991) *Villagers, Forests and Foresters: The Philosophy, Process and Practice of Community Forestry in Nepal*, Kathmandu, Sahayogi Press

GTZ (Deutsche Gesellschaft für Technische Zusammenarbeit) (2004) *Restoring Balances: Milestones of the Chura Forest Development Project in Eastern Nepal*, Kathmandu, GTZ

Gyawali, D. and Koponen, K. (2004) 'Missionary zeal on retreat, or the strange ephemerality of the Bara Forest Management Plan', in Sharma, S. (ed) *Aid Under Stress: Water, Forests and Finnish Support in Nepal*, Kathmandu, Himal Books

Hill, I. (1999) *Forest Management in Nepal*, World Bank Technical Paper No 445, Washington, DC, World Bank

HMGN (His Majesty's Government of Nepal) (1976) *The National Forestry Plan*, Kathmandu, Ministry of Forests and Soil Conservation

HMGN (1989) *Master Plan for the Forestry Sector Nepal*, Kathmandu, HMGN/Asian Development Bank/FINNIDA

HMGN (1995) *Forest Act 1993 and Forest Regulation 1995 (Official Translation)*, Kathmandu, Ministry of Forests and Soil Conservation

HMGN (1999) *Local Self-Governance Act 1999*, Kathmandu, Law Books Management Board, Ministry of Law and Justice

HMGN (2000) *Revised Forestry Sector Policy 2000*, Kathmandu, Ministry of Forests and Soil Conservation

HMGN (2001) *The State of Population Nepal, 2000*, Kathmandu, Ministry of Population and Environment

HMGN (2002) *Nepal Population Report*, Kathmandu, Ministry of Population and Environment

HMGN (2003) *Tenth Five-Year Plan (2002–2007)*, Kathmandu, National Planning Commission, www.npc.gov.np

Hobley, M. and Malla, Y. B. (1996) 'From the forests to forestry: The three ages of forestry in Nepal – privatisation, nationalisation and populism', in Hobley, M. (ed) *Participatory Forestry: The Process of Change in India and Nepal*, London, Overseas Development Institute, pp65–82

IDEA (Innovative Development Associates) (2003) *Study Report on Socio-Economic Opportunities of Tarai Forest Management*, Unpublished consultancy report submitted to BISEP-ST, Kathmandu

Iversen, V., Chettry, B., Francis, P., Gurung, M., Kafle, G., Pain A. and Seeley, J. (2006) 'High value forests, hidden economics and elite capture: Evidence from forest user groups in Nepal's *tarai*', *Ecological Economics*, vol 58, no 1, pp93–107

JTRC (Joint Technical Review Committee) (2000) *Community Forestry Issues in Tarai,* Proceedings of Workshop on Community Based Forest Resource Management, Kathmandu, JRTC

Kanel, K. (2000) 'Management of the *tarai*, inner *tarai* and Churia Forest resources: A reflection and perspective', Paper presented at the National Workshop on the Management of the *Tarai* and Inner *Tarai* Forests, NFA, 11–12 February, Kathmandu.

Mahat, T. B. S., Griffin, D. M. and Shepherd, K. R. (1986) 'Human impact on some forests of the middle hills of Nepal: 1. Forestry in the context of the traditional resources of the state', in *Mountain Research and Development,* vol 6, no 3, pp223–232

Malla, Y. B. (2001) 'Changing policies and the persistence of patron–client relations in Nepal: Stakeholders' responses to changes in forest policies', *Environmental History,* vol 8, no 2, pp287–307

Ministry of Agriculture (1995) *Agriculture Perspective Plan (APP) of Nepal,* Kathmandu, Ministry of Agriculture

NACRMLP (Nepal Australia Community Resource Management and Livelihoods Project) (2003) *Marketing of Timber Products from Pine Plantations,* Report prepared by URS Sustainable Development for AusAID, Canberra

ODG (Overseas Development Group) (2003) *Social Structure, Livelihoods and the Management of Community Pool Resource in Nepal,* Norwich, ODG, University of East Anglia

Parajuli, D. P. (2003) *Evolution of Forest Policy in Nepal,* Unpublished report, Kathmandu, Ministry of Forests and Soil Conservation (MoFSC)

Pokharel, B. and Amatya, D. (2001) 'Community forestry management issues in the *tarai*', in *Community Forestry in Nepal: Proceedings of the Workshop on Community Based Forest Resource Management,* 20–22 November, Godawari, Lalitpur, Kathmandu, Joint Technical Review Committee, pp167–188

Pokharel, R. K. (2000) 'Participatory community forestry: An option for managing *terai* forest in Nepal', Paper presented to the National Workshop on the Management of the *Terai* and Inner *Terai* Forest, 11–12 February, Nepal

SchEMS (2003) *The Compilation of Baseline Survey Data Related to Private Forestry, Agro-Forestry and Tree Planting in General,* Consultancy Report to the Livelihood Forestry Programme, Kathmandu

Statz, J. (2003) *Community Forest Management Demonstration Programme: Integrated Planning Processes for Natural Resource Management and the Distant User Approach,* Draft report of Churia Forests Development Project, Lahan, Nepal

Takimoto, A. (2000) *Impact of Community Forestry in Banke and Bardia Districts of Forestry and Partnership Project (FPP),* Kathmandu, CARE-Nepal

Winrock International (2002) *Emerging Issues in Community Forestry in Nepal,* Nepal, Winrock International

Yadav, N. P., Poudel, K., Acharya, M. and Subedi, R. (2001) *Forest Inventory and Yield Estimation for Community Forestry: A Guide Book for Field Practitioners,* Nepal, Nepal-UK Community Forestry Project (NUKCFP)

7

Joint Forest Management in West Bengal

Ajit Banerjee

Synopsis

We now turn our attention to Indian states, and first of all to West Bengal, the origin of joint forest management (JFM) policy, which has become the mainstream model for state participatory forest management (PFM) implementation in India.

West Bengal has been selected for consideration in this study for several reasons, not least because it was the first state to pilot what has now become JFM, but also because it has several distinguishing features that are complementary to PFM promotion at the grassroots level. West Bengal was the first state to carry out administrative decentralization by establishing elected *panchayats* down to the village cluster level in 1978, long before the 73rd Amendment to the Constitution of India established the elected three-tier *panchayats* framework (Chattopadhayay and Dunflo, 2004). It is also the pioneer of large-scale redistribution of land from large landholders to marginal cultivators and landless rural people. Bardhan and Mookherjee (2004) estimate that one in three landless families have received non-tradable land in this manner. The state is also the only one in the country that has had a leftist coalition government for two decades that has provided permanent tenancy rights to sharecroppers under the Operation *Barga* programme. These initiatives allowed the quick spread of PFM during the 1990s.

The deforestation and degradation of state forests, particularly those acquired from landlords in the early 1950s, continued up to the middle of the 1980s. The causes were many, but the major one was the failure of the forest department to protect the forests in the absence of the participation of the forest-fringe people in forest conservation and management. Joint forest management was thus introduced in order to involve rural people in the protection and management of their local forests (villages generally within 3km of the forest), jointly with the forest department, in return for entitling them to access to forests in order to collect subsistence-related forest products and to receive a share of the net income from timber sales. Starting in 1990, by the end of 2001 more than 44 per cent of all forests in the state had come under JFM, with its maximum development in the south-west of the state.

The author undertook an action-oriented research programme in South-West Bengal to assess the outcome of JFM, particularly the impact that it had had on the livelihoods of the associated forest-fringe people. The broad findings are discussed here in detail. These include some positive outcomes, such as improvement in the relationship between people living on forest fringes and the forest department, betterment of forest quality and quantity, and entitlement of the people to a share of the income from forest timber and collection of non-timber forest products (NTFPs). On the negative side, villagers have not gained decision-making powers through JFM; earn far less income than they potentially could under

JFM management due to the non-application of appropriate technology by the forest department; suffer a lack of transparency in investment and a lack of democracy within the village community; and there is little participation by the poor and landless sections and female members of the community (Banerjee and Springate-Baginski, 2005).

The context of joint forest management (JFM) policy in West Bengal

The implementation of JFM in West Bengal was the antecedent to the wider application of PFM across India. The increased commercialization of the timber market, the development of the forest department and the associated reservation of land, and the concomitant extinguishing of the people's rights to forest products led to dispossession and intense conflict with local people. By the early 1970s, relations between the forest department and local people had deteriorated badly, and extreme left-wing groups were gaining in influence against the state structures. At that time, small-scale initiatives by foresters and communities demonstrated that forest protection could be dramatically improved if forest-adjacent people were involved in return for a share of forest benefits. An experiment in Arabari in Medinipore district (the Arabari Socio-Economic Experiment) from 1972 onwards involved 618 families in the protection of 1186ha of degraded forests (GoWB, 2001). It was started by the author with a view to discovering whether the degraded forests could be naturally rehabilitated, as surviving tree stumps coppiced on their own under protection, through giving local people the authority and incentive to protect them. The incentive was provided by entitling them to collect NTFPs and promising a 25 per cent share of the net income when the standing timber was felled and sold. In only a few years the degraded forests visibly improved and the people participating fulfilled their household needs for fuelwood and fodder, and additionally earned some cash through the sale of forest products in the local market. This model and word of its impact on the forest and on household income spread rapidly among local people, who also started to protect forests close to them. Some forest officials assisted them in this and a number of forest protection committees (FPCs) were thus informally created. This happened contemporaneously with the ongoing government social forestry programme, which was promoting tree planting outside the forest area. After over a decade of informal developments and subsequent to the Government of India's (GoI's) 1988 Forest Policy recommending participatory forestry, the West Bengal State Government (WBSG) issued a Government Order in 1989 endorsing and encouraging the formal creation of FPCs. It included provisions for people to take over the task of protection and joint management of the forests adjoining their village and to receive forest products and a share of the revenue from final felling. Since this order, and with the support of the second World Bank project (of 1992 to 1997, extended by two more years), 3614 FPCs in JFM areas were formed by 2001 and numbers were increasing in West Bengal. FPCs are now involved in the joint management of over 81 per cent of forest land in South-West Bengal.

What has the impact of the implementation of JFM been on the livelihoods of local people and on the forests of West Bengal? Why has the process not been more beneficial to the poorest? Several earlier studies have looked at some aspects of these questions (Chandra Satish and Poffenberger, 1989; Poffenberger and McGean, 1996; Singh et al, 1997; Dutta et al, 2004; Mishra et al, 2004); but the research project on which this chapter is based has sought comprehensive analysis and answers, looking, in particular, at the field situation in the south-west of the state. For further details, readers are referred to two other project reports (Banerjee and Springate-Baginski, 2005; Banerjee et al, 2007).

Socio-economic profile of the state and its politics

West Bengal is a relatively small state (88,752km²) in the eastern part of India; but due to its population of 80.2 million (GoI, 2001) it has one of the highest population densities of any state of India (903 persons per km²). Cultivation extends over 62 per cent of the total land area (54,710km²), and although 13.38 per cent (11,879km²) of land was designated 'forest' in 1999 (GoWB, 2001), the actual forest cover (ie where tree cover is at least 10 per cent) is only 9.42 per cent (8362km²) of the state's area (GoWB, 2001). The extent of state forests has decreased since 1901 when they covered as much as 13,491km², or 18.84 per cent of the land (GoWB, 2001), although the decline stabilized after 1981 and has consistently increased since the late 1990s. The Forest Survey of India (FSI) reports a continuing increase, reaching 12,343km² at the 2003 assessment (FSI, 2005, Table 3.02). However, the increase from 2001 to 2003 represents a loss of dense forest cover (canopy cover 40 per cent and above) of 301km² and a gain of 1951km² of open forest (10 to 40 per cent). Seventy-two per cent of the total population of West Bengal is rural, and the majority are agriculture dependent.

The political climate of West Bengal is almost unique in India. The state holds regular elections; but unlike other states, for the last 25 years the people have consistently voted into power a left-front coalition constituted of the (Marxist) Communist Party of India (CPM) and a number of other parties.

Table 7.1 summarizes some parameters of the forest and livelihood data for West Bengal and India.

Table 7.1 *West Bengal and India: Forests and livelihoods summary (data for 1999 to 2005)*

Indicators		West Bengal	India
Land and forest	Geographic area	88,752km²	3, 287, 263km²
	Area of forest land (FSI, 2005)	12,343km² (2003)	774,700km²
	Actual standing forest area (includes	8362km²	637,293km²
	dense, open and mangrove forests) (GoWB, 2001)	(9.42%)	(19.39%)
	Rainfall (state average)	1700mm	–
Population	Total population (2001) (GoWB, 2003)	80.1m	1027.0m
	Population density	903 per km²	32.4 per km²
	Percentage working in agriculture	52.96%	60.60%
	Forest-dependent population	8.3m	Not available
	(GoWB, 2001, Table 4.2)		
	Scheduled tribe population in 2002	4.6m	Not available
	(GoWB, 2003, p9)		
Social development	Death rate (2002)	6.7 per 1000	8.1+ per 1000
	Literacy (seven years and older) (2001)	68.64%	65.38%
	(GoWB, 2003 Table 1.0)		
	Population growth rate (percentage)	1.77% per annum	2.13% per annum
	(1991–2001)		
	Infant mortality rate per thousand (2002)	49	63
Joint forest management (JFM) institutions	Number of forest protection committees (FPCs) (2003)	3892	84,632
	Percentage of forests under FPCs	44.61%	27.19%
	Percentage of forests under FPCs in	81%	
	South-West Bengal		
	Proportion of FPCs in state 'functioning well'	50%	20%
	(author's estimate)		

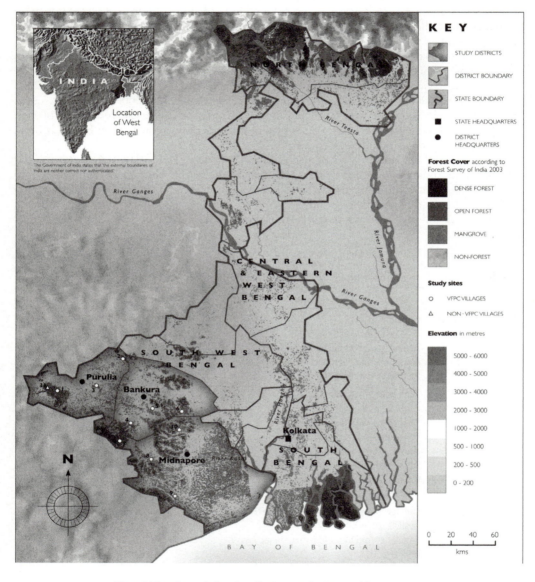

Map 4 West Bengal showing districts, study sites and forest cover

Note: Refer to Table 7.6 on p231 for study site names.

Source: Jonathan Cate

Demography and economy

The economically active population in West Bengal India forms 36.77 per cent of the total population, compared to 34 per cent of India as a whole. Of these, 53 per cent are involved primarily in agriculture, compared to only 3 per cent in livestock, forestry, fishery, fishing, hunting and plantation and allied orchards. According to available statistics (GoWB, 2003), forestry contributes only 2.6 per cent of the amount contributed by agriculture to the state domestic product. However, the forest's contribution is highly understated as it excludes substantial non-cash livelihood benefits from the forest and the forest's indirectly beneficial contribution to the ecosystem.

Overview of forests in the state

The state's forests are commonly categorized according to six major forest types, as shown in Table 7.2. The three main types account for about 97 per cent of all forests in the state.

Table 7.2 *Forest types of West Bengal in millions of hectares*

Forest Type	Extent		Region
	(millions of hectares)	% of total	
Subtropical broadleaved hill forests	0.005	0.4	North Bengal Himalayan lower hills
Tropical moist deciduous forests	0.459	38.8	North Bengal plains
Montane wet temperate forests	0.005	0.4	North Bengal Himalayan middle hills
Alpine forests	0.005	0.4	North Bengal Himalayan high hills
Tropical dry deciduous forests	0.430	36.4	South-West Bengal plains and plateaux of (mainly) Bankura, Medinipur and Purulia districts
Littoral and swamp forests	0.279	23.6	Sundarban coastal
Total	**1183**	**100%**	

Source: GoWB (2001)

The three major forest types (tropical moist deciduous forests, littoral and swamp forests, and tropical dry deciduous forests) have been subjected to different management practices, resulting in their currently having diverse status with regard to quality, quantity and vulnerability to biotic interference. In this chapter, however, we are dealing with JFM in South-West Bengal and, hence, only details of the management of the dominant forest type of the region are discussed – namely, tropical dry deciduous forests.

Before independence, the extensive forests of South-West Bengal were privately owned. The owners, usually *zamindars* (landlords), sold forest timber and firewood from time to time to bulk buyers and to local people at nominal prices. The local people hunted, and gathered dry firewood and NTFPs for their own use without restriction. The government took over all private forests under the 1953 Estate Acquisition Act and, finally, from 1956 onwards, declared most of these forests protected under the 1927 Indian Forest Act. During the legal processes, all customary rights of local people to the forests were extinguished. It is not clear how appropriately the Indian Forest Act regulations were followed to do so (see Box 7.1 on tenure).

When the government was appropriating private forests, private owners sought to take whatever they could, which led to a great deal of timber cutting, resulting in much forest loss.

Thereafter, the mixed sal forests, generally of low height and with open to dense canopy cover, were subjected to a single standardized state silvicultural management system: 'coppice with standards', where the coppice rotation was about ten years and the rotation of the 'standards' (the larger trees left in each area to provide shade for the regenerating shoots), 30 years. The coppiced poles were sold at auction to contractors, who sold them on to coal mines as pit props, for urban construction or to private buyers. The government also raised plantations of eucalyptus and akashmani (*Acacia auriculiformis* spp) on extensive areas of open forest and bare land. At maturity, usually after ten years, the forest department sold the wood to paper mills at concessionary rates. In this way the local people who had customarily used the South-West Bengal forests suddenly found themselves deprived of the use of forest and grazing land.

On acquiring the forests and until the 1980s, it was difficult for the government to protect the forest from over-cutting by contractors, over-collection of firewood and NTFPs, and unrestrained cattle grazing by local people. Improvement of the forests through 'scientific management' had failed, and forest quality and quantity had deteriorated to such an extent that annual forest felling had to be reduced by 1987 to 1988 to just 181ha, from an area that for decades previously had been around 2500ha (Palit, 2004).

After the initial village forest protection experiments, JFM was gradually introduced over the 1990s and has since flourished in South-West Bengal compared to North Bengal and Sundarbans. The relationship between the forest department and local people, which had become so difficult, is relatively easier with JFM in the sense that forest officials have reasonable rapport with local people.

Role of forests in agrarian livelihoods

From 1864 to the 1990s, the forest department consistently treated reserved and protected forest property as its exclusive estate, policing the boundaries between the forest and villages. However, the forest-fringe people, whose livelihoods were jeopardized, continued to make their customary claims on the forests, albeit surreptitiously and without taking responsibility for its sustainable use. In fulfilling their sustenance needs, these local people daily brought about potential confrontation with the government staff trying to obstruct them. The relationship between people and the forest department became extremely poor; consequently, the forest degraded to bare land at many sites.

Other than providing limited employment in plantation and harvesting activities and overlooking a little subsistence extraction of forest products, the forest department did not assist the fringe- and forest-dwelling populations with livelihood opportunities until 1990, when JFM and other livelihood schemes were introduced.

Many households (estimated by the forest department to be 22.6 per cent of the villages in the state, or 8571 villages with 8.3 million people) currently depend on forests for a substantial part of their livelihoods. Their uses include grazing cattle, collecting firewood, small poles and timber for house construction, cart-making, constructing plough pieces, and fencing kitchens and home gardens. Some people earn cash by collecting and selling forest products such as firewood, leaves, medicinal plants, fruits, nuts and berries, fibres, gums and resins, and silk and tassar cocoons locally. In addition, the villagers are involved in scarce local forest development work on a daily wage basis when they can.

Review of forest policy and implementation

Historically, local communities have managed and used their forests according to a diversity of practices. With the expansion of the market economy and trade, and the increasing power of elites and colonial rule, the central control of forest resources grew, leading to the creation of the forest departments, and the forest estate was formally put under their control.

After independence the GoI presented its new 1952 Forest Policy (revised in 1988). The Government of West Bengal (GoWB) more or less followed the national policy. However, between the two policy presentations, certain enactments, orders and policy decisions were made by the centre and the GoWB that sometimes adversely impinged on the policies (Saxena, undated). A particularly contentious point was that the 1952 policy categorically denied any special consideration of the needs of forest-fringe people. Largely because of this 'anti-people' policy, the relationship between the people living in or on the forest fringe and the forest department became difficult and sometimes violent. Twenty-two forest staff received physical injuries between 1966 and the mid 1980s, and a total of 22 villagers lost their lives in these clashes from 1969 to 1986 (Guhathakurta and Roy, 2000).

The social forestry (SF) policy was introduced in West Bengal in 1973 in an attempt to meet subsistence and marketable forest product needs from non-forest sources. It grew to a significant level of activity after a decade with the help of the World Bank Social Forestry Project from 1982 to 1991, which focused on promoting farm forestry (tree planting by farmers on their private land) and tree plantations on marginal public lands, such as roadsides and canal banks, with little attention to degrading forests. The Social Forestry Project has been criticized because it reinforced the anti-people policy, rather than shifting the emphasis to the more pro-people models emerging from Arabari at that time (Chatterjee, 1996). However, the project did lead to a large number of local people participating in a state tree-planting programme for the first time, and the achievements were remarkable: trees were planted over an area of 241,754ha. However, the species selected by foresters, rather than local people, were predominantly exotic: eucalyptus and *Acacia auriculiformis*. The tree survival rate was estimated at 53 per cent by the end of the project (Guhathakurta and Roy, 2000). Overall, the investment assisted a large number of landholders to make substantial profits from farm forestry. The marginal and landless people obviously could not take part.

The World Bank's forestry strategy in India has closely matched that of Indian states, as illustrated in Table 7.3.

Table 7.3 *Summary of forest policy evolution in West Bengal*

Policy activity	Indian state	World Bank	Major policy focus
Industrial forestry	1952–1976	Before 1978	Support wood-based industries (mainly private)
Social forestry	1976–1988	1978–1991	Wean local people from forest dependency and promote planting on private land
Environment and support for participatory forest management (PFM)	1988–present	1991–present	Support community in joint management of government-owned forests

Source: adapted from Kumar et al (2000)

The country and the World Bank persisted in supporting private forests outside government-owned forests until 1991, when they switched to supporting the community with JFM after the earlier strategies to halt deforestation and forest degradation failed.

Despite the large scale of the social forestry programme, the Arabari model had a more lasting impact. By the mid 1980s, its success and popularity was apparent, and the government issued a special order in 1987 fulfilling its promise to share the revenue from timber harvested in the once degraded forests with participating villagers. This order is recognized as a paradigm shift with regard to the entitlement of people from the forests in West Bengal (Poffenberger and McGean, 1996; Chatterjee, 1996; Singh et al, 1997; Guhathakurta and Roy, 2000).

In 1988, the central government issued a new forest policy, dramatically different to that of 1952. It recognized that the first charge of the forest was to tribal and poor people living in or near the forest, and that the forests should meet their needs. Based on the success of the Arabari experiment, in 1990 a national JFM order was issued by the Ministry of Environment and Forests (MoEF) and states gradually followed the guidelines and issued similar orders (Poffenberger and Singh, 1996, p65). This was the beginning of JFM in India, envisaging the participation of forest-fringe peoples in forest management and their entitlement to forest benefits. In West Bengal, the state government issued three separate JFM orders for South-West Bengal, North Bengal and Sunderbans in 1989, reissued in 1990 with amendments in 2004.

Table 7.4 *Timeline of major forest policy and participatory forest management (PFM) implementation developments*

Year	Issue	Comment
1952	Central Government Forest Policy	Focus on supplying national interest identified with defence of forests, communications and vital industries, with low priority given to local needs.
1953	West Bengal Estates Acquisition Act	State acquires all private forest lands, trees, forest products and intermediary interests in forests.
1973–1976	1976 Social Forestry Policy	1976 National Commission on Agriculture report promotes farm forestry by farmers on their private land to reduce pressure on government-owned forest land.
1972	Arabari socio-economic experiment begins	Action-oriented experiment promoting participation of forest-fringe people in protecting and developing the forest jointly with forest officials, in return entitling participants to collect forest products for home consumption and local sale, plus a 25% share of the net income from final felling.
1976	42nd Constitutional Amendment	Transfers forestry from state to concurrent list.
1980	Forest (Conservation) Act	Limits power of states to divert forest to non-forest uses.
1984–1991	World Bank Social Forestry Project	Promotes farm and community forestry (CF) in private and community land, respectively (total investment of 639 million rupees).
1987	West Bengal Government Special Order – Arabari benefit-sharing	No 118-For/D/6M-76/65, dated Calcutta, 7 March 1987, agreed as a special case to allow 25% of the net income of the first rotational timber felling to be distributed to local village participants in the experiment.
1988	Forest policy	Conserve environment and soil and supply subsistence needs of the local people as a priority (a dramatic change from 1952 policy).
1990	Ministry of Environment and Forests (MoEF) Joint Forest Management	Notification Agrees to the participation of the local people in forest management jointly with the forest department in return for forest benefits, including a share of timber revenue.
1992–1999	World Bank-West Bengal Forestry Project	US$39 million 'soft' loan; emphasis on promoting JFM and restructuring the forest department.
2004	Joint forest management (JFM) amendment	JFM Executive Committee enlarged with representation of more forest protection committee general members, particularly women.

The World Bank again responded to the state policy, rather than promoting policy change itself. It initiated a second project in 1992 entitled the West Bengal Forestry Project (funded through a US$39 million 'soft' loan). The project, initially for five years and later extended to eight, focused on supporting the promotion of JFM in West Bengal in order to improve the protection and productivity of the forests and to restructure the forest department. The result of technical improvements to the forests was a mixture of success and failure. The degraded forests showed some improvement; but the survival of the new plantations was low at only 46 per cent (Guhathakurta and Roy, 2000).

The basic features of JFM policy are found in the 1989 and subsequent JFM resolutions in West Bengal. These are as follows for South-West Bengal (SPWD, 1998).

Constitutional aspects
- JFM to be offered only in degraded forest areas.
- JFM open to all families in the village. In each family, both husband and wife are automatically made members of the FPC.
- An executive committee with a term of one year to be composed of:
  - representatives of the *panchayat*;
  - not more than 6 elected representatives of the FPC (recently increased to 14);
  - the local beat officer (the divisional forest officer's field staff) as member secretary.
- The forest department and FPC to jointly write a micro-plan of the tasks to be carried out in the JFM over a specified period, usually five years.

Responsibilities of the local people
- The FPC would be expected to ensure the smooth and timely execution of all forestry work taken up in the area and to protect forest/plantations (note that the forest department forest working plan would not be affected by the FPC micro-planning process).

Entitlements of local people
- FPC members are entitled to most NTFPs free, and 25 per cent of specific NTFP harvests, including cashew nuts, sal (Shorea robusta spp) seeds, tendu (Diospyros melanoxylon spp) leaves, honey and wax on approved tariff. Members must sell tendu leaves and sal seeds at approved rates to the Large-Scale Adivasi Multi-Purpose Society (LAMPS), a local unit of the West Bengal Tribal Development Co-operative Corporation Ltd, which has monopoly rights to these products.
- Members are entitled to 25 per cent of net income from timber sales.

JFM implementation processes and practices
In the West Bengal Forest Department's (WBFD's) implementation of JFM, it became apparent that the major change was in responsibility for forest protection, which was passed to the FPCs. Other management issues would continue to be handled by the forest department – namely, forest management planning, species selection, allocation of funds to different operations and 'investment', planting and afforestation, forest harvesting, and use of powers to punish.

It is obvious from the above that JFM favours the forest department as its regulator. However, it was also an instrument to involve the people in forest protection and, in return, to provide incentives in the form of entitlements that had been denied them since independence in private forests and for more than 100 years in reserved forests.

When the JFM guidelines came out in 1989, there were already 600 FPCs; this number rapidly grew to 3593 by 2001.

Table 7.5 *Forest protection committees (FPCs) by district (2001)*

Region	District	Forest area (ha)	Number of FPCs	Area under joint forest management (JFM) (ha)	Percentage of district forest under JFM
South-West	Midnapore	170,900	1131	143,539	83.99
Bengal	Bankura	148,200	1256	112,072	75.62
	Purulia	87,600	669	78,060	89.11
	Burdwan	27,700	77	18,579	67.07
	Birbhum	15,900	114	9068	57.03
Sunderbans	24 Paraganas	4,263,00	33	56,732	13.30
North Bengal	Darjeeling	120,400	201	57,832	48.03
	Jalpaiguri	179,000	90	44,767	25.01
	Coochbehar	5700	22	3405	59.73
Total		**1,187,900**	**3593**	**524054**	**44.11**

Sample villages and their forests

In this study, three districts were purposively selected in the south-west of the state to reflect the main agro-climatic and political regions where JFM has developed. Then ten FPCs, as well as one non-FPC village, were selected for detailed study. The target population for village-level study was the range of forest-dependent communities who had been both directly and indirectly affected by the recent implementation of the JFM policies. Bankura and Purulia districts were allocated three FPCs each and Midnapore District four, based on the area of FPCs under JFM management in each of the three districts.

The selection of FPCs from each district was performed by random sampling, stratified by block. The first stage was the selection of administrative blocks (discounting non-active blocks). The team went to different forest divisions in the three districts and collected FPC names and characteristics. Three blocks in Bankura, three in Purulia and four in west Midnapore were then randomly selected. The second stage was random selection of FPCs within the selected blocks. The list of selected blocks and FPCs is given in Table 7.6, along with the number of households and number of FPC members in each. One non-FPC village was selected at random in Purulia district in order to get an indication of the nature of the difference in development between FPC-managed and non-FPC forests. It is true that one village is a small sample; but the area of non-FPC forest in South-West Bengal is only 19 to 81 per cent forests covered by JFM.

The number of households differs from the number of FPC members in the sampled villages. In two cases there is only a small difference due to splits in extended families or newcomers from outside increasing the number post-FPC formation. In three cases (Dandahit, Taldangra and Keundi) there is a more significant difference. Some families in Dandahit were not interested in joining at the beginning, but have now applied for membership. Taldangra, being near a highway, has many families working outside the village who are not prepared to take responsibility for the forest. Keundi is a religiously heterogeneous village. Although the groups are harmonious, one large group does not participate in the FPC.

The mean forest area per household across the study sites is 0.9ha, varying between 0.43ha in Keundi FPC to 2.77ha in Dudhpania FPC. In the non-FPC village, it is 2.08ha. Indicators of the degradation status of the forest, as measured by the extent of open forest

Table 7.6 *Selected blocks, FPCs and characteristics*

District	Block	Name of selected forest protection committee (FPC)	Date FPC registered	Forest area (ha) (as surveyed by research team)	Number of house-holds	Number of FPC member house-holds	Percentage of village house-holds in FPC	Forest area per FPC house-holds (ha)	Degradation status (percentage of PFC forest)
Purulia	Arsa	1 Chakedabad	2001	49.15	29	29	100.0	1.69	53
	Raghunathpore	2 Dandahit	1990	127.91	184	86	46.7	1.49	38
	Hura	3 Dudhpania	1999	49.78	18	18	100.0	2.77	41
Bankura	Ranibundh	4 Raotara*	NA	NA	233	227	97.4	NA	NA
	Taldangra	5 Taldangra	1994	79.66	128	98	76.6	081	63
	Joypore	6 Katul	1990	159.9	93	93	100.0	1.72	11
West Midnapur	Belpahari	7 Gohalbera	1993	91.8	80	80	100.0	1.15	44
	Jhargram	8 Kesia	1996	29.04	56	47	96.4	0.62	13
	Nayagram	9 Sialia	1988	127.55	110	110	100.0	1.36	37
	Chandrakona	10 Keundi-Jamboni	1988	180.2	816	416	51.0	0.43	Plantation§
	Total	**10 FPCs**			**1747**	**1204**	69.3	0.90	35
Purulia	Bagmundi	11 Saharjuri (non-FPC)	Not formed	229.21	110	–	0	2.08	85

Notes: * Data for Raotara village forest not available due to insurgency interrupting the study.

§ Plantation only, therefore not categorized.

plus bare land as a percentage of the total FPC forest, is also variable, the highest being 85 per cent in the non-FPC forest and the lowest 13 per cent in Kesia.

Household selection
The total number of FPC households in the 10 villages was 1204, and it was decided to carry out a household survey on a sample of 14 per cent, or 167 households. Twelve households from the non-FPC village of Saharjuri were also selected. Households in the village were stratified in terms of four wealth ranks: rich; medium rich; poor; and landless and poor. Of the 167 households interviewed, 27 were rich, 59 medium rich, 60 poor, and 21 landless and poor. In Saharjuri, we interviewed 2 rich, 4 medium rich, 5 poor and 1 landless and poor households out of 110 families. The classification was based on two criteria: family landholding and income level. Data for these criteria is available from the *gram panchayat*, categorized as large peasant, middle peasant, poor peasant and landless (for our purposes rich, medium rich, poor, and poor and landless, modified where necessary by the perception of the villagers during the participatory rural appraisal (PRA) at the beginning of the survey; this correction was necessary because, in a few cases, farmers with little or no land may still belong to the rich category if one or more members work in a profession or at an office job).

The research findings are based on reconnaissance of the selected villages, PRA and interviews. Village reconnaissance took a day, PRA one or two days, and interviews, depending on the number conducted, two to five days.

Table 7.7 *Disaggregation of village households by wealth rank*

District	Forest protection committees (FPCs)	Total households	Rich (%)	Medium (%)	Poor (%)	Landless and poor (%)
Purulia	Chakedabad	29	13.8	65.5	20.7	0.0
	Dandahit	86	7.0	16.3	30.2	46.5
	Dudhpania	18	22.2	38.9	38.9	0.0
Bankura	Raotara	227	28.6	18.5	38.8	14.1
	Taldangra	98	13.3	26.5	53.1	7.1
	Katul	93	0.0	57.0	21.5	21.5
Midnapore	Gohalbera	80	33.8	25.0	32.5	8.8
	Kesia	47	0.0	42.6	42.6	14.9
	Sialia	110	0.0	78.2	14.5	7.3
	Keundi Jamboni	416	14.2	31.0	42.8	12.0
	Total (FPC)	*1204*	*14.8*	*34.6*	*36.5*	*14.2*
Purulia	Saharjuri (non-FPC)	110	16.4	33.6	41.8	8.2

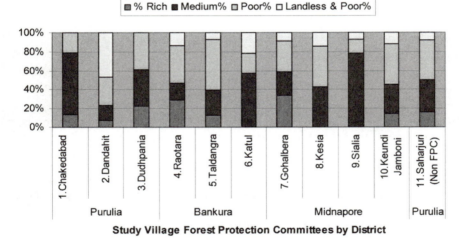

Figure 7.1 Composition of study villages by wealth rank of households

Source: Ajit Banerjee (from survey)

Outcomes and impacts

Here, we consider how the implementation of JFM in South-West Bengal has affected the local 'political ecology' in terms of institutional change, changes to the forest and changes to household livelihoods.

Local forest management institution in the sampled villages

The formation and implementation of JFM in South-West Bengal at the field level involved up to four actors. The possible actors at village level are a few individuals from the village, the village community, the forest department staff – in particular, the local ranger, forester, forest guards and the *bana mazdoor* (forest worker, a category in the staff hierarchy mainly involved in the protection and supervision of forest development activities) – sometimes a non-governmental organization (NGO) and, rarely, the *panchayat*. In some places, all four actors were equally active, while in others, one, two or three played the leading role. The World Bank also provided, at the state level, consultative assistance, much needed funding and motivation for the state to act urgently.

Among the sampled villages, Dudhpania is an example of a village group initiating action to protect the forest during the early 1990s. In the mid 1990s, the forest department helped it to form a forest protection committee after the group had approached it for support. In the majority of cases, however, the forest department was the main actor. For example, the forest department introduced the concept of FPC in the village of Kesia as early as 1981 and formed a group there in 1982, even before government orders were issued, on the basis of stories circulating about the Arabari participatory model. It only started functioning well in 1996, however, when the forest associated with the village was almost totally degraded. The local forest department forester was the initiator of Sialia FPC, formed in 1988 to 1989. There are very few cases in South-West Bengal where NGOs have been responsible for initiating FPCs, although some have contributed to its successful progress – for example, Chakedabad FPC was assisted by the Ramakrishna Mission Lok Shiksha Parishad.

FPC operation

Executive committee meetings, which are meant to be convened at least once a month by the forester, do not take place regularly in all the villages under survey. We found that they were only regularly held in Chakedabad (arranged by an NGO), Katul, Dandahit, Sialia and Keundi-Jamboni. They are rare and sporadic in Raotara, Taldangra, Gohalbera and Kesia, and used to be regular in Dudhpania; but now forest department officials are rarely seen.

Therefore, from our sample we might estimate that for JFM as a whole in South-West Bengal, there is the administrative structure of 50 per cent of FPCs' function, while in the other 50 per cent, the formal structure is not working.

Table 7.8 indicates members' perception (generally the male member of the household) of their participation in decision-making processes in the sample villages.

Table 7.8 *Interviewee attendance and perception of their role in decision-making processes (all figures in percentages and rounded)*

Question	Positive response (%)				
	Rich	Medium rich	Poor	Landless and poor	Average
Percentage of total number interviewed (n = 167)	*16*	*35*	*36*	*13*	*100*
Is forest management important?	100	100	100	100	100
Have you attended a forest protection committee (FPC) meeting in the last 12 months?	11	34	22	8	19
Do you feel that your say in FPC decisions is accepted?	6	24	13	7	11

Less than 20 per cent of households attend meetings, the poorest and the landless being the least participatory, followed by the rich. It was interesting to find that only 11 per cent of the richer section of the community and 8 per cent of the poor and landless had attended FPC meetings in the last 12 months. It is the general belief that the rich are in league with officials to influence decisions; but this does not seem to be the case in the samples. In West Bengal, with the left front in power for more than two decades, the influence of the rich has been somewhat reduced and passed on to the medium rich and a section of the poor, who took part in the meetings in the largest numbers. This is an interesting contrast with the other case studies, which all point to the political dominance of elites.

Product distribution/benefit-sharing

Based on the memorandum of understanding (MoU) that the forest department signs with the FPC, all participating households are expected to enjoy the following benefits equally:

* scope to air views on the formulation of forest management, development and benefit distribution policies (i.e. the democratic functioning of the FPC and the executive committee);
* entitlement to forest product collection, with a few restrictions concerning firewood and NTFPs;
* allocation (if available) of small timber for own use if the family's requirement is accepted as authentic;
* 25 per cent of share in kind of firewood produced by thinning;
* 25 per cent share in cash of net income when crops are finally harvested and sold; and
* investment of funds by the forest department according to micro-plan and employment in carrying out the programmes.

The following sections discuss how MoUs are implemented in practice. It will be apparent that while most obligations are fulfilled at a minimum level, the quality and quantity of the outcomes are far below their potential.

Democratic forest governance

The promise of democracy in FPCs has not been fulfilled, nor is there much indication of improvement towards that goal. An approximate power hierarchy in JFM from the perception of interviewees, the attitudes and expressions of various people in PRA exercises, and sometimes one-to-one discussions with liberal individuals in the community, place forest department field staff at the top, followed by the medium-rich and the poor, with the landless at the bottom. The rich are near the top, but only in respect to specific areas of JFM – for example, in business matters and trading, service and project implementation associated with JFM, but not so much in its day-to-day operations. Necessary labour, however, is drawn first from FPC members, whoever the contractor may be.

FPC decisions are increasingly influenced by political parties, sometimes through the *panchayat* system (e.g. Dandahit FPC, where influence has been exercised by the *panchayat* and favourably accepted by the FPC). Political intervention is also apparent in Gohalbera, Keundi and other FPCs. We may, therefore, add a position in the hierarchy with a question mark towards the top for political parties.

Entitlement to collection of forest products

PRA exercises and survey data revealed that members have undoubtedly acquired the entitlement to collect dry firewood, NTFPs such as sal leaves, eucalyptus, dry sweepings from the forest floor, medicinal plants, and edible plants and fruits without hindrance or without having to pay royalties to the forest department. The forest department does not object if members collect these products for sale at the local market, although privately they believe that this is not legal. The rules are unclear about this, however. Moving the products to distant markets requires a transit permit that is granted only to intermediaries and is not available to village collectors.

Most families periodically need timber for house repairs, construction of plough pieces, bullock carts and other agricultural implements, and to burn their dead. In a few villages, the executive committee itself approves this allocation; in others, it has to be further endorsed by the forest ranger. Except in the case of wood for burning the dead, which is approved immediately, applicants have to wait for the approval of the FPC or for the executive committee to meet, convened by the beat officer.

Share of thinnings

The working plan prescribes thinning of closed-coppice sal crops and plantations in their fifth year. In practice, this is not regularly performed. When thinning is carried out, members are entitled to 25 per cent of the thinned material for firewood. It is not feasible to separate 25 per cent in kind because this would be a very lengthy process; instead, all member householders are invited to collect a cartload or a few headloads from the thinned area. The remaining thinnings are then sold to anyone interested at the government royalty price, and 25 per cent of the net income then goes to the community. The more prosperous householders often buy thinnings to store at home.

Twenty-five per cent of net income of the final harvest

The forest department carries out final harvesting only when the FPC has protected the forest for at least five years and the crop is ten years old. Felling is subject to prescription in the divisional working plan. In addition, forest department staff are arbitrary in selecting *coupes* each year. In the researchers' interaction with forest department officials, we understood that market conditions are also an important factor in deciding on the *coupe* area for the year. Table 7.9 shows the share of profits received by FPCs in the sampled villages.

Despite the fact that the average age of the 10 FPCs is 12 years and the coppice rotation period is 10 years, only 3 of the 10 villages – Dudhpania, Sialia and Keundi-Jamboni – have felled final crops. No felling had been carried out in the other sampled villages. The distribution of the total income of 1,205,330 rupees, generated by just three of the villages, between the rich, medium rich, poor and landless in these three villages can be seen in Table 7.9. The average yearly amount per household across all of the villages is 78 rupees, a small sum that has virtually no impact on livelihood.

Local labour employment

The forest department employs local labour for all of its investments in the FPC. Table 7.10 shows operations carried out by the forest department.

Table 7.10 shows the forest management operations carried out in the sample FPC forests. It has been estimated (based on communication with the range officer in Jhargram range) that 1ha of coppicing work generates 76 person days of work, and 1ha of plantation generates 25 person days of employment. Based on these results, we can estimate how much employment was generated in the ten sampled FPCs since their formation in (1994 to 2001):

- coppice harvesting: 70ha;
- employment generated: $76 \times 70 = 5320$ person days;
- plantation harvesting done: 30 ha;
- employment generated: $30 \times 35 = 1050$ person days.

Table 7.9 *Per village net income in rupees from final harvest in the sampled FPCs (to 2004)*

District	Forest protection committee (FPC)	Year registered	Year of final felling and area (ha)	Amount received in rupees per FPC household	Total amount received by village FPC since date of registration (rupees)	Mean income per household per year (rupees)
Purulia	Chakedabad	1997	None	0	0	0
	Dandahit	1990	None	0	0	0
	Dudhpania	1999	2000 (20ha)	2200	52,200	580
			2002 (10ha)	700		
Bankura	Raotara*	–	–	–	–	–
	Taldangra	1994	None	0	0	0
	Katul	1990	None	0	0	0
Midnapore	Gohalbera	1993	None	0	0	0
	Kesia	1982	None	0	0	0
	Sialia	1988	1998	230	77,880	44
			2001	478		
			2004	*(Not yet disbursed)*		
	Keundi-Jamboni	1988	1999	996	1,075,250	158
			2000	638		
			2003	896		
	Total				**1,205,330**	**78**
Purulia	Saharjuri (non-FPC)	Not	No activity	None		

Note: * In Raotara FPC, social conflicts and naxalite (i.e. Maoist insurgency) activity meant that data collection had to be stopped.

It is clear that employment generated by the harvesting of forests and plantations since the formation of the 10 FPC forests totalled only 6370 days of work, a small figure indeed. The major employment generation is in the collection, use and marketing of NTFPs.

Forest department–village relationship

Although the relationship between the forest department and FPC members improved from the very low point it had reached before JFM, this does not mean that the two parties are satisfied with each other's performance. The FPCs' performance in forest protection is far below that expected by the forest department. Grazing remains generally uncontrolled and only a few members come forward to put out forest fires. Similarly, the FPC had expected assistance from the forest department in forest protection, greater investment in entry-level activities, a greater share of products from harvesting and thinning, and a share of at least 75 per cent of the net income of the final harvest, rather than 25 per cent.

In spite of this, a relatively good relationship is growing between the parties, except in some villages where political interference has complicated relations, as evidenced by complaints from the FPC members of Gohalbera and Keundi. In most places, however, there is a growing closeness between village people and local forest department officials. However, this often takes the form of a dependent parent–child relationship whereby the

Table 7.10 *Silvicultural operations carried out after FPC formation (to 2004)*

Forest protection committee (FPC)	Closed forest: harvesting area (ha)	Closed forest: post-harvest operation	Plantation (number of plants)	Post-planting operations (e.g. cleaning and replacing dead seedlings)	Other operations	Fuelwood and non-timber forest product (NTFP) collection	Remarks
Chakedabad	None	None	30,000 in 1995 and 3500 later (mainly eucalyptus and *Acacia auriculiformis*)	Yes	Gully protection	Lac cultivation and NTFP collection	Well vegetated
Dandahit	None	Multiple shoot cutting	30,000 *Arjun* and eucalyptus	Yes	1km earthen dam	NTFPs by neighbours	Well vegetated
Dudhpania	20ha in 2000; 10ha in 2002	Yes	None	None	Three ponds and 1km approach road	Firewood and NTFPs	Well vegetated
Raotara*	NA						
Taldangra	None	None	None	NA	NA	Sal leaves, mushrooms and dry firewood	Damaged in parts
Katul	None	None	None	None	None	Sal leaf, mushrooms and firewood	Well vegetated
Gohalbera	None	None	None	None	Four earthen dams	NTFPs, *patharkhadan babui* cultivation and tassar cocoons	Well vegetated
Kesia	None	None	10,000 eucalyptus	Yes	None	Firewood	Well vegetated
Sialia	1998, 2001, 2004; about 40ha (actual figures unclear) Yearly cashew harvesting				One well	Firewood, NTFPs and foods, including three types of edible mushroom	Well vegetated

Table 7.10 *continued*

Forest protection committee (FPC)	Closed forest: harvesting area (ha)	Closed forest: post-harvest operation	Plantation (number of plants)	Post-planting operations (e.g. cleaning and replacing dead seedlings)	Other operations	Fuelwood and non-timber forest product (NTFP) collection	Remarks
Keundi-Jamboni	Plantation felling in 1999, 2000, 2003 (about 30ha) Annual cashew harvesting	Yes	None	None	Approach road and well		Plantation only
Saharjuri (non-FPC)	No activity	–	–	–	–	–	Poorly vegetated

Notes: * In Raotara FPC, social conflict and naxalite (i.e. Maoist insurgency) activity meant that data collection had to be stopped.

NA = not applicable

FPC defers to whatever decision or recommendation forest department staff make. This leads to concurrence by the FPC in all that is proposed by the forest department, instead of local communities reaching the goal of independent action.

Forest protection
Although FPC members have taken on the major responsibility for forest protection, forest department staff are expected to assist when the FPC needs help. In the sampled villages, members protect the forests through patrol parties composed of representatives of households by rotation. Voluntary participation in protection worked successfully for a few years after registration. Although patrolling continues in a few places, FPC members are gradually becoming less vigilant.

The general complaint is that the forest department does not help when its assistance is required. Since members do not have the power to punish forest offenders whom they apprehend, they take them to forest department officials, who, the people said, are often lenient.

Overall, there is no doubt from the observations of the surveyors and the perception of villagers that the forests have improved since JFM, implying that the extraction of resources has decreased. However, the general impression gathered from FPC members (e.g. from Gohalbera) is that regular unauthorized small-scale collection occurs, particularly by other villages and occasionally by smuggler groups.

Micro-planning, decision-making and implementation
The micro-plan is the main document that sets out the activities that the FPC plans to carry out over a period of five to ten years. The plan has to be written to a preordained format. It does not impinge on standard forest department working plan provisions for the forest, and

therefore villagers do not have the power to control the main forest management decisions, such as species selection and rotation. The plan usually lasts five years.

The micro-plan is written jointly by forest department field staff and the FPC general body and is sent to the divisional forest officer (DFO) for approval. To simplify matters, the plan is written in tabular form with numbers only and without justifications or explanations. The sampled villages do have micro-plans; but most of them were filled out by the forest department. During the house-to-house survey, it was found that very few of the sampled householders knew whether a micro-plan existed.

Decision-making is seldom participatory and there is a lack of transparency in decision-making processes. The forest department is the major player in making decisions, being the controller of funds and owner of the forestland and its products. In addition, the forest department representative is the convenor of the FPC and executive committee meetings and, hence, has the upper hand. The *panchayat* and political leaders are also active; for example, in Sialia, Gohalbera and Keundi, they influence decisions with regard to the distribution of benefits. The 25 per cent cash income from cashew harvesting in Keundi-Jamboni due to the members was not distributed to individual householders on the justification that the amount is being spent on community infrastructure development. It remained unclear to the sampled householders who had been involved in taking this decision and what had been decided. During the interviews, it was understood that the only contributions by FPC members to decision-making processes during micro-plan preparation were those linked to entry-point activities, when members put forward suggestions that are often accepted in the micro-plan, and those linked to forest protection, as members like to organize forest patrols in a regular rotation.

Impact on households

The direct impact of JFM on individual households was measured in changes in assets from pre- to post-JFM; income in cash and kind; well-being/vulnerability of householders; and NTFP marketing.

Changes in assets

The major assets related to livelihood in FPC areas are livestock and land. Householders often do not remember how many head of cattle they had before JFM, which, in most cases, was eight to ten years ago. They are also vague about the quantity of land they own. Therefore, the information presented in Tables 7.11 and 7.12 and their accompanying conclusions are indicative and not conclusive.

The data in Table 7.11 show that there is little consistency between villages in the increase or decrease of family livestock assets. A simple average of the nine villages for which data is available makes it clear that greater numbers of livestock have not been acquired since JFM:

- number before JFM: 1.90 (cattle); 2.46 (goats); 4.32 (chickens);
- number after JFM: 1.80 (cattle); 2.49 (goats); 4.24 (chickens).

Table 7.12 shows a predictably skewed pattern of landholdings, although in interviews it was found that the rich have comparatively less land in South-West Bengal than before the legislation on land ceilings, with 5.08 and 4.64 acres being the highest. None of the interviewed families had bought land with additional earnings due to JFM. JFM has therefore made little difference to the landholdings of FPC members.

Table 7.11 *Number of livestock per household pre- and post-joint forest management (JFM)*

		Before (B)/after (A)	Cattle	Goats	Chicken
Purulia	Chakedabad	B	0.22	0.88	8.88
		A	*1.66*	*1.33*	*7.66*
	Dandahit	B	0.69	3.30	8.30
		A	*0.92*	*1.61*	*5.07*
	Dudhpania	B	1.20	0.80	7.40
		A	*3.20*	*3.40*	*7.80*
Bankura	Raotara	B	0.77	0.45	0.28
		A	*1.62*	*1.37*	*1.14*
	Taldangra	B	NA	NA	NA
		A	NA	NA	NA
	Katul	B	1.28	2.28	2.57
		A	*2.71*	*2.00*	*4.14*
West Midnapur	Gohalbera	B	3.16	3.66	0.83
		A	*1.91*	*3.58*	*2.00*
	Kesia	B	1.50	1.62	1.62
		A	*1.37*	*4.50*	*4.62*
	Sialia	B	7.42	5.85	6.07
		A	*1.85*	*2.50*	*3.00*
	Keundi Jamboni	B	1.47	3.38	2.95
		A	*1.11*	*2.19*	*2.78*
Purulia	Saharjuri		NA	NA	NA

Notes: B = two years before JFM and A = post-JFM (2003–2004).

NA = not applicable

Table 7.12 *Landholding per household for different wealth-ranking households by district (acres)*

		Homestead land (acres)	Own cultivation (acres)	Own cultivation fallow land (acres)	Encroached land (acres)	Leased in (acres)	Leased out (acres)	Wasteland (acres)	Share-cropping (acres)	Total (acres)
Purulia	Landless	0.02	0.01	0.00	0.00	0.00	0.00	0.00	0.00	0.03
	Poor	0.04	0.25	0.12	0.03	0.00	0.00	0.00	0.00	0.44
	Medium	0.08	0.86	0.13	0.04	0.00	0.00	0.004	0.07	1.98
	Rich	0.09	1.38	0.81	0.01	0.00	0.63	0.18	0.00	5.08
Bankura	Landless	0.02	0.00	0.01	0.00	0.00	0.00	0.00	0.00	0.03
	Poor	0.03	0.10	0.10	0.01	0.01	0.00	0.00		0.28
	Medium	0.07	0.42	0.22	0.04	0.03	0.00	0.00	0.01	1.07
	Rich	0.25	1.70	0.03	0.38	0.00	0.00	0.00	0.20	2.44
West Midnapur	Landless	0.03	0.01	0.00	0.00	0.00	0.00	0.00	0.00	0.04
	Poor	0.06	0.24	0.01	0.01	0.13	0.01	0.01	0.00	0.47
	Medium	0.05	0.56	0.05	0.01	0.01	0.01	0.02	0.00	0.82
	Rich	0.22	1.57	0.16	0.00	0.02	0.04	0.09	0.00	2.91
Purulia (non-FPC)	Landless	0.10	0.00	0.00	0.40	0.00	0.00	0.00	0.00	0.10
	Poor	0.14	0.38	0.04	0.28	0.00	0.00	0.04	0.00	0.88
	Medium	0.25	0.50	0.22	0.12	0.00	0.00	0.22	0.00	2.19
	Rich	0.35	1.45	0.20	0.05	0.00	0.00	0.40	0.00	4.64

Self-help groups

JFM, with support from NGOs, has been responsible for assisting women to introduce self-help group (SHG) credit systems in some of the sampled villages, including, for example, in Chakedabad and Dandahit. The women deposit their own money in a fund account on which they can draw at low interest in times of need, providing a safety net and a way of avoiding exploitative borrowing arrangements. The involvement of forest department officials and a few NGOs in JFM activities that could not have taken place pre-JFM has been instrumental in developing these self-help groups.

Income in cash and kind

These data are presented in the following sequence:

- income from all sources (forest + agriculture employment + service + monetary value of products collected from the forest and used at home) for different wealth-ranked households in each village;
- the above in percentages, separating forest income from other income;
- breakdown of forest income into different components, such as fuelwood, leaf collection, NTFPs, etc; and
- income–expenditure balance.

Table 7.13 *Average annual income per household from forest and non-forest sources on wealth-ranked basis for FPC and non-FPC villages (rupees)*

Forest protection committee (FPC)	Rich		Medium rich		Poor		Poor landless	
	Forest	Other sources	Forest	Other sources	Forest	Other sources	Forest	Other sources
Chakedabad	0	59,540	4342	29,386	4353	14,890	0	0
Dandahit	112	139,040	0	14,400	1968	16,088	2492	12,475
Dudhpania	4290	148,541	6305	25,930	6547	5575	0	0
Raotara	0	186,457	5848	60,130	7570	24,949	9975	8100
Taldangra	133	60,000	133	26,609	133	13,179	169	108,833
Katul	0	0	7788	22,782	6918	7997	8046	9500
Gohalbera	50	48,265	4503	17,153	5915	10,665	6840	11,450
Kesia	1250	56,405	0	21,593	3790	16,915	5040	13,200
Sialia	12,360	65,800	16,600	27,028	12,327	13,493	4800	15,090
Keundi Jamboni	3947	148,069	6041	40,548	5205	15,971	6745	14,983
Saharjuri	2400	40,400	2375	8383	1200	6860	1600	6000
Total	**22,142**	**912,117**	**51,560**	**285,559**	**54,726**	**139,723**	**44,107**	**95,681**

Table 7.14 *Percentage of average annual income from forest and non-forest sources of all inter-viewed households of all wealth ranks in FPC and non-FPC villages*

Forest protection committee (FPC)	Rich Forest	Rich Other sources	Medium rich Forest	Medium rich Other sources	Poor Forest	Poor Other sources	Poor landless Forest	Poor landless Other sources
Chakedabad	0.00	100.00	12.87	87.13	22.62	77.38	0.00	0.00
Dandahit	0.08	99.92	0.00	100.00	10.89	89.11	16.64	83.36
Dudhpania	2.80	97.20	19.55	80.45	54.00	46.00	0.00	0.00
Raotara	0.00	100.00	8.86	91.14	23.27	76.73	55.18	44.82
Taldangra	0.22	99.78	0.49	99.51	0.99	99.01	15.26	84.74
Katul	0.00	0.00	25.47	74.53	46.38	53.62	45.85	54.15
Gohalbera	0.10	99.89	20.79	79.21	35.67	64.33	37.39	62.61
Kesia	2.16	97.84	0.00	100.00	18.30	81.70	27.63	72.37
Sialia	15.81	84.19	38.04	61.96	47.74	52.26	24.14	75.86
Keundi Jamboni	2.59	97.41	12.97	87.03	24.58	75.42	31.04	68.96
Weighted mean	**0.88**		**14.90**		**26.82**		**32.10**	
Saharjuri (non-FPC)	5.61	94.39	22.08	77.92	14.89	85.11	21.05	78.95

Table 7.15 *Average annual income per household from forest and non-forest products of all ten sampled FPC villages (167 households)*

Sources of income (rupees)	Rich	Medium rich	Poor	Poor landless
Forest	1169	6116	5727	5441
Other sources	131,258	34,912	15,622	11,507
Total	132,427	41,028	21,349	16,948

Table 7.16 *Average annual income per household from forest and non-forest products in Saharjuri (non-FPC)*

Sources of income (rupees)	Rich	Medium rich	Poor	Poor landless
Forest	2400	2375	1200	1600
Other sources	40,400	8382	6860	6000
Total	64,400	10,757	8060	7600

Figures 7.2 and 7.3 clearly illustrate just how significant forest use is to poorer households; therefore, improvements to this component of their livelihoods can (potentially) significantly improve their overall income level.

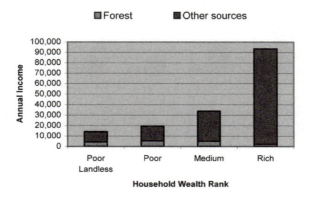

Figure 7.2 Total income by wealth rank from forest-related products and other sources

Source: Ajit Banerjee (from survey data)

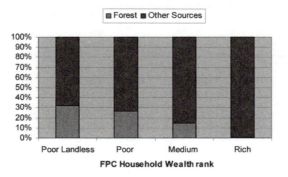

Figure 7.3 Percentage of forest-related income in total income by wealth rank

Source: Ajit Banerjee (from survey data)

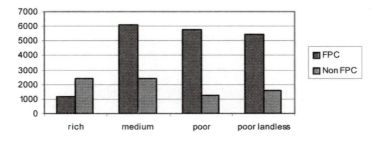

Figure 7.4 Average annual income (rupees) per FPC and non-FPC household (from forest sources only)

Source: Ajit Banerjee (from survey data)

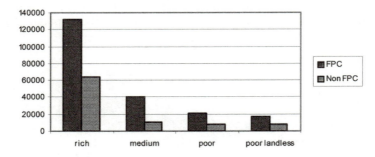

Figure 7.5 Total annual income (rupees) of sampled FPC and non-FPC households

Source: Ajit Banerjee (from survey data)

The income of the families of different wealth ranks from various elements of forest resources is presented in Table 7.17.

Table 7.17 *Income from various categories of forest resources in rupees and percentages per household for different wealth ranks for ten FPC villages*

	Share*		Employment		Sal leaves		Sabai grass		Fuelwood		
	Rupees	%	Rupees	%	Rupees	%	Rupees	%	Rupees	%	Total rupees
Rich	67.0	5.7	24.5	2.1	292.2	25.5	151.9	13.2	625.4	53.5	1169
Medium	88.1	1.4	538.2	8.7	3094.7	50.6	1603.6	26.2	788.9	12.9	6116
Poor	65.9	1.2	424.3	7.4	2346.1	44.0	1350.8	25.3	1265.7	22.2	5727
Poor/landless	58.2	1.1	870.5	16.3	2519.9	45.5	421.2	13.9	1262.3	23.2	5441

Note: percentage figures refer to percentage of income earned from specific NTFPs.

* = share of final felling.

Table 7.18 *Income from various categories of forest resources in rupees and percentages per household for different wealth ranks in Saharjuri (non-FPC)*

	Share	%*	Employ-ment	%*	Sal leaves	%*	Sabai grass	%*	Fuel-wood (rupees)	%*
Rich	0	0	0	0	0	0	0	0	2400	100
Medium	0	0	0	0	0	0	0	0	2375	100
Poor	0	0	0	0	0	0	0	0	1200	100
Poor/landless	0	0	0	0	0	0	0	0	1600	100

Note: * = percentage of income from forest sources only.

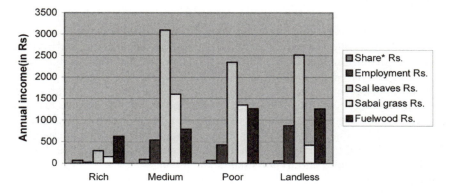

Figure 7.6 Annual income of households of different wealth ranks from different categories of forest resources

Source: Ajit Banerjee (from survey data)

Sal leaves, sabai and fuelwood are the main contributors of FPC households' livelihoods, while waged employment is moderate and the share of the net income from final forest harvesting is insignificant. Employment generated by NTFP collection is seasonal because of the seasonality of leaf growth, and is taken on only when more remunerative employment opportunities are not available.

Seasonal factors

Agriculture is the major occupation of households in most seasons. Table 7.19 provides a description of how NTFP collection, agriculture and forest department-supported employment are related by season.

Table 7.19 *Employment seasons*

Season	Agriculture	Specific non-timber forest product (NTFP)	Forest department-supported employment	Remarks
Monsoon (June–September)	Rain-fed planting	Medicinal plants, fuelwood (sal) and food collection	Forest plantation	Engagement and employment mainly in agriculture. Forest work clashes with agricultural work
Post-monsoon (October–November)	Cultivation of irrigated land	All species of fuelwood, sal leaves and sabai grass	Entry-point activities	Main engagement in all types of NTFP collection
Winter (December–February)	Harvesting and processing	All species of fuelwood and sal leaves	Entry-point activities	Engagement in all types of NTFP collection
Summer (March–May)	On irrigated land fuelwood	All species of fuelwood		Engagement in fuelwood collection only

Well-being and vulnerability

The introduction of JFM has come as a great relief to local people after the difficulties they faced between 1950s to the mid 1980s, particularly for landless poor and poor families, who, before JFM, had to collect firewood for home use and for cash income, and in doing so faced endless harassment from forest department officials who treated them as thieves. Since JFM, all FPC members have been entitled to collect NTFPs from the forests.

The other direct impact on the community is the allocation, by individual members, of time to different occupations. According to households interviewed, fuelwood collection used to average three to five hours a day, while it now takes only one or two hours three times a week. It is also estimated that one woman from each family allocates three to four days a week to collecting sal leaves. In addition, the family spends time drying and processing the leaves. These time savings add to potential family income.

However, members also have had to allocate more time to protecting the forest, attending meetings, training and sometimes visits from senior personnel (as well as researchers such as us), although time spent on forest protection is decreasing as rule-breaking declines. Patrolling, when done, requires one to four individuals on a monthly basis, totalling 12 to 48 days a year per household. Based on the present-day wage rate of 60 rupees, the rough opportunity cost to the family is 720 to 2880 rupees (based on the weak assumption that employment is available year round, which is not the case). Compare this with 5090 rupees: the average income from forest-related activities.

One of the more important impacts of JFM has been the improved social capital of the villages concerned. In the pre-JFM days, women would go out to collect firewood together to ward off harassment by forest department law enforcers; but now they go together by choice and collaborate with each other. Visits to the forests when groups are collecting dry leaves, sal leaves, firewood, fodder and medicinal plants bear this out. It was unheard of in pre-PFM days (although it is a tribal custom) for women to sing loudly while collecting forest products because it would have alerted forest guards to their presence. This practice has been revived in some tribal villages, where groups of women sing in unison while collecting leaves (the *santhal* tribals have special songs), as we verified in Chakedabad.

In order to influence the community in favour of participating in JFM, the forest department, in the context of World Bank project funding, introduced what are referred to as 'entry-point activities'. Some community development activities not necessarily connected with the forests would be carried out at the outset (and also, later, annually as far as funds permit), including agricultural development, installing drinking water facilities, promoting roads, establishing forest product-based cottage industries, poultry and piggeries, providing vocational training, etc. Problems, however, remain. The total investment in the ten villages cannot be established; but the details of the entry-point activities show that the scheme is small and has had a minimal impact on the communities. The different types of entry-point assistance that villages have obtained from the forest department include repair of the approach roads to Dandahit, Dudhpania, Sialia and Keundi-Jamboni; construction of earthen dams/ponds in Chakedabad, Dudhpania and Gohalbera (now badly needing repair); and construction of ring wells in Sialia and Keundi-Jamboni. Apart from Katul and Kesia, all other villages have had some training on various topics, such as beekeeping, mushroom farming, tassar and lac cultivation, tailoring, veterinary practice and carpentry, sal leaf craft, surveying and so on. Interviewees in Taldangra and Kesia indicated that they had not had any such contribution other than some training that they did not value.

With JFM, inter- and intra-village conflicts have arisen. The inter-village conflict is now very low because the affected villages tend to take a lenient view of people from neighbouring villages collecting products from their FPC forest. But as the number of such offences increase, the affected FPC villages have been protesting and asking for greater powers to deal with them.

In general, the relationship between FPC members is good. In a few villages, this is manifest in villagers voluntarily working together – for example, to repair an approach road in Dudhpania or a temple roof in Taldangra, or in CF patrolling in Chakedabad, Katul, Dandahit, Dudhpania and Kesia (Banerjee et al, 2007).

NTFP marketing

In general, NTFP marketing is not problematic in South-West Bengal in terms of finding buyers; but the process at the village level is depressed. Forest foods, fruits and fibres all have local markets. The buyers for sal leaves are local intermediaries who resell them at urban markets and also buy sabai ropes in bulk. The major problem is the low price that the villagers obtain due to lack of roads and other infrastructure in the villages, and the absence of facilities to add value to the forest products.

Impact on forests

This section presents the extent, category, quality and quantity of different FPC forest products. However, in view of the absence of baseline data from before JFM days, it is difficult to ascertain the status of the forests and their impact on livelihoods during that time.

After the formation of a forest protection committee and identification of the forest area associated with it, it is the task of the forest department to demarcate its boundary by field survey. However, it was discovered that this survey had not been carried out in any of the ten villages. The area found by our study, according to boundaries indicated by the FPC members, differed substantially from the forest department's own estimate for the forest area. All of the study FPC forests are dry deciduous forest types composed of closed forest, open forest, plantation and bare land, as seen in Table 7.20.

Table 7.20 *Forest type in study FPCs*

Forest protection committee (FPC)	Number of FPC households	Total forest area (ha)	Forest (ha) per FPC household	Closed forest (ha)	Open forest (ha)	Plantation (ha)	Bare land (ha)
Chakedabad	29	49.15	1.69	0.00	23.14	23.01	3.00
Dandahit	86	127.91	1.49	78.27	43.35	6.29	0.00
Dudhpania	18	49.78	2.77	28.99	20.79	0.00	0.00
Raotara	227	*NA*	*NA*	*NA*	*NA*	*NA*	*NA*
Taldangra	98	79.66	0.81	29.38	8.85	36.79	4.64
Katul	93	159.9	1.72	142.76	11.76	0.00	5.38
Gohalbera	80	91.8	1.15	51.27	27.86	12.11	0.56
Kesia	47	29.04	0.62	25.14	1.53	1.87	0.49
Sialia	110	127.55	1.16	79.89	23.15	24.51	0.00
Keundi-Jamboni	416	180.2	0.43	0.00	0.00	166.68	13.52
Total	*1204*	*845.83*	*0.70*	*435.70*	*137.29*	*248.25*	*24.59*
*Average**	*120.4*	*105.8*	*0.88*	*54.50*	*17.20*	*31.00*	*3.10*
Saharjuri (non-FPC)	110	229.21	2.08	33.49	193.50	0.00	2.22

Notes: 'Forest area' describes forest as found by the survey team.
Closed forest = canopy closure of 40 per cent and more.
Open forest = 10 to 40 per cent canopy closure.
Bare land = less than 10 per cent canopy closure.
Plantation = man-made forest.
NA = not available.

* The average of forests associated with nine FPCs, excluding Raotara FPC, for which detailed information was not available. All forests are dry deciduous type with varying proportions of sal (*Shorea robusta*) and sal-associated miscellaneous species.

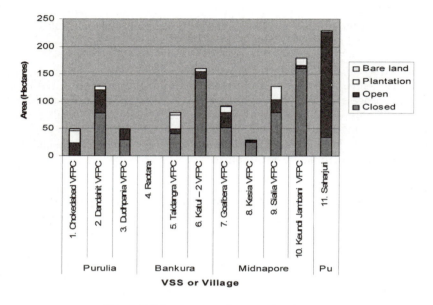

Figure 7.7 Forest area and status of study areas

Source: Ajit Banerjee (data collected for this study)

Table 7.21 *Tree density and timber volume of study areas*

District	Name of village forest protection committee (FPC)	Diameter class range (cm)	Stem per ha	Volume per ha (cubic metres)
Purulia	Chakedabad	05–19	894	19.064
	Dandahit	05–20	1446	25.340
	Dudhpania	05–50 and above	1285	33.598
Bankura	Raotara	NA	NA	NA
	Taldangra	05–40	1445	44.854
	Katul – 2	05–20	1351	58.036
Midnapore	Goalbera	05–40	1282	35.292
	Kesia	05–19	1887	51.414
	Sialia	05–20	1461	30.270
	Keundi-Jambani	05–30	475	24.285
Purulia	Saharjuri	05–50 and above	317	16.538

NA = not available.

Our survey of the village FPC forests shows that timber stock varies between a moderate 16.54 cubic metres per hectare in Saharjuri to as much as 58.04 cubic metres per hectare in Katul. The comparative advantage of some FPCs has not been exploited, however, as the forest department does not harvest the forests regularly at the end of its rotation. In Saharjuri village (non-FPC) there is a large open area (193.5ha) with only 16.54 cubic metres of timber volume per hectare, which reflects degraded forest.

The surveyed estimate of growing stock of shrubs and herbs, which includes most of the medicinal species, is shown in Figure 7.8.

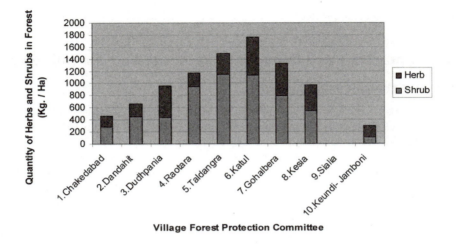

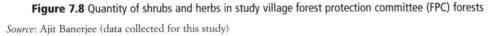

Figure 7.8 Quantity of shrubs and herbs in study village forest protection committee (FPC) forests

Source: Ajit Banerjee (data collected for this study)

The quantities of shrubs and herbs vary. Shrubs are lowest at 107kg per hectare in Keundi, where forest is constituted predominantly of eucalyptus plantation, and highest in Taldangra at 1150kg per hectare. Equivalent figures for herbs are 186kg per hectare in Keundi and 635kg per hectare in Katul. The small quantity of shrubs and herbs in Keundi was found to be due to the adverse effect of the eucalyptus trees.

During PRA, FPC members made some observations about changes to forest resources:

- There was an increased availability of firewood and dry leaves for cooking and sale in the market (Keundi), plenty of firewood (Kesia and Gohalbera), and an increased availability of dry leaves and twigs in Taldangra. The availability of small timber for personal use has become plentiful (Katul-2).
- NTFPs such as khamalu (potato tubers) and mushrooms have increased, as have medicinal plants, thus increasing income (Dudhpania). There was increased income from lac and NTFPs (Chakedabad).
- Illicit felling was reduced in Katul-2 and in a number of other FPCs.
- Due to better forest stocking, elephants and deer have been able to take shelter in the forests and invade cultivated areas, damaging crops (Chakedabad, Katul-2 and Gohalbera).
- Out-migration has been checked; but members ascribe this to the panchayat (Kesia) and not to JFM. Migration has also stopped in Taldangra.
- The water level in wells and shallow tube wells has gone down since JFM was introduced (Keundi-Jamboni, Sialia, Kesia and Taldangra). In Keundi, this may be explained by overuse of shallow tube well water for irrigation of boro paddy, potato and other crops, not only in the village but over the region as well. Forest management may have little to do with the diminishing water table.

- Summer temperatures have increased and rainfall has been infrequent (Sialia and Kesia). This also may not be related to forest management.

Thus, the impact of forest management had been mainly noticed in the improved supply of various forest products, particularly firewood and NTFPs, and better forest stocking.

Policies, practices and impacts: Governance and the policy process

Policy and the governing process

The forest department is the major partner in JFM and has devolved virtually no significant decision-making powers to FPC members. Within the FPC, the poor and women remain in the background, marginalized and unable to exert any influence. The forest department has not attempted to enhance the productivity of the coppiced forests, nor has it allowed the people to do so. In many FPCs, members would like to plant spices and fruits that grow in tree shade – for example, turmeric, pineapple and cardamom. However, the West Bengal Forest Department continues to be engrossed in its traditional practice of growing pole-size timber regardless of the preferences or demands of the people. Additionally, the forest department's interest in forest protection work has waned significantly, leaving it to the people but without providing them with the power to punish offenders. The beat officers generally allocate little time to convening executive committee or FPC meetings, with the result that discussions by the FPC members are commonly unofficial. Lack of forest department transparency with regard to financial estimates for forestry and rural work in JFM areas and some associated corruption are common. Added to this are the arrogance and high-handed attitudes of many forest officials.

In sum, the governance of JFM in West Bengal requires substantial improvement. The main victims of the currently bad state of governance are local households and the forest itself. Participating people are becoming indifferent and, in some cases, are in outright opposition to JFM.

Tenure and legal issues

Tenure issues were found to be problematic for JFM beneficiaries on two counts:

1 The customary rights in private forests pre-independence are no longer allowed, although JFM permits free collection of products for home consumption. In meetings with FPC members and in PRA exercises, older individuals would speak about the British system being more people oriented.
2 JFM orders by state governments are not backed by legislation. It is believed by all that these orders, which have given rise to so many JFM committees involving thousands of people, cannot simply be rescinded; but there is apprehension in the minds of the people as there is no legal security, and this issue is often justifiably raised by the more educated and the NGOs.

It is necessary to understand the reasons for these assumptions in order to get to the actual status of people's tenure rights in the state forests. The recent study sponsored by this research programme brings to light the following information (Ghosh et al, forthcoming) (see Box 7.1).

Box 7.1 Extinguishing forest rights in South-West Bengal: An historic betrayal of communities

Soumitra Ghosh

The way in which the state took over private forests in South West Bengal and extinguished all rights there is a typical example of the circumvention and deliberate misinterpretation of existing laws in order to deprive people of their historic rights. This has happened in many parts of India, but nowhere else as blatantly.

Taking the Jungal Mahal area (a former nomenclature that includes most of South-West Bengal's forests) as an example, after 1953 no survey was ever conducted. The state appointed no forest settlement officers, and no rights have ever been settled. All forestland in the area simply came under forest department control by executive fiat as it was taken for granted that all rights had been vested in the state.

Until today, working plans for forest divisions within the erstwhile Jungal Mahal area have maintained that there were no rights in the South-West Bengal forests at the time of takeover; the same working plans, however, say that forests were 'burdened' with rights before 1953 and quote the reports of various commissions and committees appointed by the colonial forest department from time to time. This pattern was similar for all forest divisions of South-West Bengal.

When the state took over the private forests of South-West Bengal in 1953 to 1954, it ignored a discourse on the rights and limits of government property in South-West Bengal forests that had been going on for about a century, as it also ignored the peculiar tenurial nature of the properties that it took over. Several million *raiyots* and under-*raiyots* (small and marginal peasants), *bargadaars* and under-*bargadaars* (share-croppers) were settled over land divided into thousands of estates. Each estate had its own arrangements with its *raiyots*, and many of the forest rights were, in fact, part of such tenurial arrangements. In almost all Jungal Mahal areas, rights included grazing and rights over various forest products including timber. Most of these rights were so secure that even the colonial forest department stated that its management objects were 'to provide for sustained supply of forest produce as required by the right-holders' and 'to provide regular and sufficient supply of firewood and poles ... for house-building and agricultural implements to the rights holders', amongst other things (this was recorded in the 1940 Working Plan for Matha and Quilapal Private Protected Forests, now in Purulia division of West Bengal).

Rights (free of charge) recorded in the plan included timber, firewood, various NTFPs and grazing. One of the rights was the use of forests (by pollarding) for tasar silk cultivation, something the villagers of the area continue to demand even today as part of the JFM arrangement.

One of the ostensible reasons for state acquisition of *zamindari* properties was to provide security to *raiyots* and under-*raiyots*, who were traditionally seen as the worst victims of permanent settlement, as the 1953 Estate Acquisition Act makes clear. It says that 'every one of the following rights which may be owned by an intermediary ... shall vest in the state ... all lands in an estate comprised in a forest together with all rights to the trees therein or to the produce thereof and held by an intermediary ... vest in the State' (Section 5(aa)).

But who were the 'intermediaries'? And whose rights shall be 'vested in the state'? The 1953 act answers, without ambiguity:

> *Encumbrance in relation to estates and rights of intermediaries therein does not include the rights of a raiyat or of an under-raiyat or of a non-agricultural tenant, but ... includes all rights or interests of whatever nature, belonging to intermediaries and other persons.*

Contrary to the forest department's claim, then, the rights of *raiyots* and under-*raiyots* in South-West Bengal's private forests were not vested in the state, and the state or its agencies responsible for the acquisition could not have *legally* notified these forests as government forests without first carrying out the due forest settlement process. And even with settlements, how legal or ethical would it have been? What makes a land 'government land'?

Source: Ghosh et al (forthcoming)

Major policy actors, their changing characteristics and influences at different levels

In addition to the people of the villages, there are a number of other actors in JFM. We discuss each separately, with the most important – namely, 'the people' – at the end.

The first is the *West Bengal Forest Department (WBFD)*. The forest department derives its substantial powers from controlling extensive forest lands. Despite JFM devolving one of the major responsibilities of field staff to village FPCs, the number of forest department field staff has not declined. Indeed, over the 1990s, the forest department's staff numbers and salary bill have increased:

> *The* State Report of West Bengal Forests *for the year 1990–1991 mentions the number of senior posts as 120 and 6345 posts of all other categories. In 2001, it reports 239 and 11,778. In other words, the department has increased its staff by about 100 per cent in ten years without any significant gain either in the forest area, the quality of forest and its ecosystem.* (Banerjee and Springate-Baginski, 2005)

In addition, during this period the West Bengal Forest Development Corporation Limited (WBFDC) has also been substantially involved in the JFM to a varying extent.

In this state, as in others in India, the abolition of government department posts is almost unheard of. One action that the GoWB has taken is to stop recruiting to fill forest guard and lower-level vacancies created by retiring personnel. But senior forest officials, instead of welcoming this strategy, complain that too many vacancies hamper appropriate administration. Because the forest department is such a major employer, there is a large and politically influential interest group of public employees who seek to maintain their relatively secure and well-paid positions in the face of declining revenues from forests.

The next most important officials in JFM are the beat officers, as convenors of the executive committee and FPC meetings. The DFO controls the beat officers through the range officer, who also controls FPC funds for disbursement. The DFO is the authority who approves micro-plans and applications for FPC membership and sanctions the use of funds. The DFO operates through his assistants and is the most important forest department player in JFM.

It seems that corruption is prevalent in the state; but its dimensions are not known. There are, however, press reports and court cases in which even some DFOs were found to have been involved. An Organisation for Economic Co-operation and Development (OECD)/World Bank report commenting on corruption in JFM in India states (and this can also be applied to West Bengal):

> *The evidence of informal conversations with staff, NGOs and beneficiaries suggests that the problem is widespread; but there is no evidence to indicate that corruption in the Bank projects is more or less than other [forest department] activities.* (Kumar et al, 2000, p67)

The pattern of corruption has changed over time and has now shifted more towards a 'cut' of the expenditure budget, rather than arrangements with timber smugglers and contractors. The amount of 'cut' is difficult to prove; it is said to vary from 20 to 35 per cent. There is, however, evidence of a *nexus* between some local forest officials and local elites, generally members of the executive committee of the FPC who monopolize most of the FPC supervision jobs. The main opportunity in West Bengal would be the recently established forest development authorities (FDAs), which have been created to channel central government funds for JFM. These funds pass through the conservators (i.e. the head of a 'forest circle' within which several divisions are managed by DFOs) that head different FDAs, to the range officers who are the disbursers (unlike in some states where FPC presidents and beat officers are joint signatories for fund management). Corruption retards the pace of development, apart from its impact on the morale of honest staff.

Donors, primarily the World Bank, have been active in West Bengal. The World Bank loaned funds for two projects up to 2003, although rather than promoting policy change it appears that it gave funds where policy had already changed to support its implementation. Donors and the government generally decide how to implement JFM. The opinions of local people involved were given very little attention. For example, micro-plans continued to be written by the foresters with hardly any input from members of any of the sampled FPCs.

The World Bank is an agency whose *raison d'être* is lending. Its forest-related loans to West Bengal total 2779 million rupees, on standard soft terms with a long maturity. The borrower is the central government, which lends to the state as part of central development assistance on standard terms and conditions fixed between the centre and the state. The state pays back part of the loan in local currency at a higher interest rate than the lender charges the central government.

Normally, at the *panchayat* level (usually a cluster of villages), a number of government departments, such as agriculture, irrigation, veterinary, public health, education, land revenue and a few others, are represented. However, the forest department has kept JFM implementation strictly in its own domain, with only minor consideration of the views of *panchayat* members. The latter are elected on party lines and are quite influential in building up pressure on various issues. In some villages, as discussed earlier, they play a significant role in supporting their own people, thus upsetting the democratic functioning of FPCs.

Mafias have appeared in South-West Bengal, although this is sporadic and only in places where FPC members are indifferent or unable to confront them. A specific example was related by Gohalbera FPC members, who said that they could not handle such large groups (who are also sometimes armed). There are occasional reports from some FPCs of smugglers appearing in trucks and cutting and carrying away the larger sal and planted teak or eucalyptus trees. We heard about this from Chandra FPC in west Midnapore district, which is, however, not one of the ten selected FPCs. Thus, the timber mafias are a nuisance rather than a significant menace in South-West Bengal so far.

Sometimes the forest department acts arbitrarily to extend forest boundaries, enclosing forest areas belonging to private individuals or to the community and creating distress and ill will (see Box 7.2).

Local people are meant to be the most important actors in JFM; but although there is no doubt that they play a significant role, they are generally under the direction of forest department officials. Furthermore, members are not equal. The poorer sections, tribal people and women are present as beneficiaries, but play very little part in decision-making. Only in a few women's FPCs have the women taken over as the primary activists, leaving the men in the background. We did not have any exclusively women's FPCs in our sample; but there were 17 such committees by 2001 (GoWB, 2001). For example, the *State Forestry Report* states that 'some FPCs are exclusively controlled by women groups and their positive participation has changed the complexion of working of FPCs in localized area' (GoWB, 2001, p64). Tribanka women's FPC in Vishnupur is one such case (author's personal experience).

We can thus see that in spite of the participation of many actors in FPCs, it is the forest department, to a certain extent the *panchayat* members and certain sections of the FPC membership that constitute the active core, with others playing minor roles.

One other active group in the forests of South-West Bengal is the left-wing insurgents who base their activities in forest areas. Their concentration is in the forests of the tri-junction of Purulia, west Midnapore and Bankura districts. The militants have not in any way

Box 7.2 Personal experience of researching sacred groves adjacent to joint forest management forests

Dr Debal Deb, Centre for Interdisciplinary Studies, Kolkata

Amid the exotic monocultures protected by the State forest department (FD) at the expense of natural mixed forests, I discovered some years ago a private plot of land in Bankura where a dispersed stand of mature sal (*Shorea robusta*) trees existed, surrounded by a tract of protected forest maintained by local Forest Protection Committees (FPCs). Apart from the fact that I myself was involved in studying the grand outcome of people's participation in restoring biodiversity and household economies in rural Bengal, the existence of a stand of mixed forest lying outside the state-managed forest tract fascinated me. From the ecological point of view, one could easily observe the ecological processes in the unmanaged forest stand compared to the state-managed (that is felled every 8 to 10 years) forest, and this is a rare opportunity to study deciduous forest dynamics in the State. So in March 2004 I obtained requisite permission from the land owner who lives some 7km from this land, built a temporary shed for my workers to study the forest, and started my ecological research – chiefly involving estimation of biodiversity, species turnover and food web structure. The research was initiated in collaboration with ENDEV, headed by Dr A. K. Ghosh, one of the country's most renowned biodiversity experts. The landowner, Sri Radha Raman Thakur, was in fact happy that his land, otherwise left unutilized so far, was being used for some social benefit.

In early March 2005, upon my return to the village after a month of work elsewhere, I received the jolting news that most of the trees on this private land had been felled by the FD, without even informing the landowner. During February–March 2005, the FD was felling trees in the adjoining patch of Protected Forest according to its District Working Plan for the year 2005, and in a felling spree, removed 130 trees standing on the adjacent private land as well. The FD officials (including the Beat Officer, the ex-officio Secretary of the FPC), in collusion with a handful of FPC executive members, did not pay heed to objections raised by some villagers who worked as forest labourers. The unauthorized felling was perpetrated in the presence of a handful of executive committee members of the FPC.

This incident is an example of how the power of the FPC in forest management is usurped by the rural elite. It also shows how the forest bureaucratic high-handedness has reduced the 'joint' and 'participatory' spirit of the much-lauded JFM to mere rhetoric, in which grassroots members of FPCs have no role in key decisions regarding forestry operations. In any case, I wrote letters to the PCCF, the CF (Central Circle) and the DFO, apprising them of this incident – on March 21, June 9, and August 13, 2005. The landowner himself wrote letters of complaint on March 12 to the DFO and the Range Officer, who haughtily brushed aside his complaint.

After my letters – and also a reminder – to the PCCF and CF (CC) failed to elicit any official response, I decided to file a Public Interest Litigation (PIL) before the Calcutta High Court. The Court issued order to the Beat Officer to appear before the court on Friday the September 2, 2005. On Saturday September 3, the news appeared in *The Hindustan Times*. On Monday the 5th the Range Officer and the local Beat Officer raided the house of Debdulal Bhattacharya, who happens to be a worker of *Vasudha*, my research station. The forest officials, accompanied by some policemen, seized 30 sal poles from his house in his absence, and issued a Seizure List stating that there were 80 poles of unusually large dimensions. Debdulal's father, an illiterate man, was coerced to put his thumb impression on the copy of the notice, not knowing the content of the seizure list.

Simultaneously, Debdulal was given a charge that he had been involved in stealing 15 sal poles that had been "discovered" from our temporary shed (built in March 2004) on the private land where the trees were felled (in March 2005). The apparently intended implication was that it was our Centre that felled all the trees and then tried to blame the innocent FD staff.

This clumsy attempt at criminalizing the accusers angered a section of conscientious villagers. About 30 people from 5 village mouzas signed a letter to the PCCF and DFO stating that almost every household in the same Beat had similar stock of wood bought, without any receipt, from the FPC in the presence of the Beat Officer; this practice has been the norm over the past 10 years. If the Beat Officer, himself being the Secretary of the FPC, sold the wood to villagers, how would the latter suspect that the sale was illegal? The Beat Officer took the money from the sale (in collusion with some top brass of the FPC), but the villagers unwittingly became law-breakers, whom the FD could sue at will and convenience. Today Debdulal is pilloried, tomorrow any other person will be, if he appears to be not so pleasant to the Hierarchy. The villagers demanded an official enquiry into the matter. When no answer came from the FD hierarchy, these villagers put up a Writ Petition before the Calcutta High Court on September 22, 2005.

Meanwhile, the FPC executive committee members felt threatened by the proceedings against the FD because they feared their collusion with the FD and their share of the booty from the illicit felling would be revealed. They put constant pressure on Debdulal and his family members to ask me to stop the litigation. In exchange, they offered, the FD would withdraw all allegations against Debdulal and his father, because they admitted, those were false accusations, only to pressure the main litigant. On September 24, 2005 they met with me to say that it was only for me that Debdulal and his family were suffering, and therefore I ought to withdraw the case and negotiate with the FD. When I asked them to tell the truth in a signed statement that Debdulal had actually bought his woodlot without receipt in the presence of the FPC and the Beat Officer, they were anxious as tthey said that the sale was illegal, and therefore they would not sign any such statement as that would implicate them all as accomplices to the crime.

On the one hand, the FPC top brass were wary about the exposure of their involvement in the wood sale, which they already knew was illicit, on the other hand, the grassroots people, for whom the sal forest had an immense subsistence value, were happy that the wrong was at last going to be redressed. Many of them boldly stood in front of media cameras, stating that the sal trees on the private land were felled in the presence, and on consent, of the Beat Officer as well as a few Executive members of the FPC, knowing fully well that the land did not belong to the mouza demarcated for departmental felling.

I must explain the origin of the 15 sal poles used in constructing the temporary shed at the felled site. I bought ten of them from two villagers, one of them member of a neighbouring FPC, on the understanding that they were selling those poles to me from their own stock which they had earned from the FD as their legitimate share from a previous departmental harvest. They also gave me two signed receipts. The remaining five poles had been lying *in situ* prior to my occupying the land on the March 1, 2004. They might have been felled either by the FD or by some unknown wood thief, who was unable to remove them after I occupied the land. I obtained permission from the landowner to use them in the construction of a temporary shed, on condition that I would leave them there upon completion of my study. Had I been guilty of theft of these poles, I would not have kept them there: I had more than 5 months in hand to conceal them before lodging my suit against the FD.

The villagers' Writ Petition was put up before Mr Justice Sirpurkar's bench on the September 30, 2005, and was summarily identified as a false litigation, fabricated only in order to save Debdulal Bhattacharya. The verdict: *The villagers ought to pay a penalty of Rs. 2000.00 as cost to the State.*

The moral: *In a legal battle against the State, the accusers must be prosecuted.*

And now the FD officials are all coming down on the villagers who signed the petition, and threatening to punish them. Even the DFO has asked some tribal youth be prepared to sign a document the FD may use in the litigation.

The PIL against the FD on my accusation of felling of the 130 sal trees have not been heard in the intervening 17 months. Meanwhile, the FD filed a new suit of illegal transit of wood against Debdulal, his father, the landowner and me before the Bankura district SDJM court at Bishnupur. We all had to appear before the court several times since, as suspect criminals, while the FD has dallied with the suit, unwilling to put up any witnesses to settle it. Our harassment continues at Bishnupur SDJM court. The FD's insinuation that I, or at least my organization, was busy stealing wood from the state forest in the guise of ecological research over years has been ostensibly entertained. The history of my active involvement for over a decade in conserving indigenous biodiversity and cultures, or my career as a field ecologist with a long list of scientific publications, or my collaboration in this particular research in question with ENDEV, matters little.

Finally, the Calcutta High Court heard the PIL on March 13, 2007 and **dismissed it on grounds that the petitioners of the PIL are accused in a criminal suit**. **The petition of the landowner** seeking compensation for illegal removal of trees from his land **was also dismissed**. The argument that the criminal suit was framed months after I had started the PIL, had no place in the verdict. The Calcutta High Court's Chief Justice is obviously inclined to believe that a citizen with no institutional affiliation must have some ulterior, criminal motive behind accusing any government officials for corruption.

We await the verdict of the SDJM court of Bishnupur. We have now gained enough understanding of the State's juridical system to that conclude that before the blinded eyes of justice, *all enemies to the state are disposable.*

Note: For further information see Deb (2007)

disturbed the functioning of the JFM; but the forest officials in some places have withdrawn from interior posts and seldom visit some FPCs (e.g. one of our sample villages, Raotara FPC).

The politics of information and knowledge about forest management

Since the forest department is one of the major players in JFM, it uses its own technical repertoire and forestry knowledge base as learned at government forest schools and colleges. This knowledge is based on classical forestry centred around timber production, mensuration, yield calculation, wildlife park management and so on, and is less concerned with issues of NTFPs and their marketing, the sociology, livelihood and gender issues of local people, self-help credit arrangements, and so on. This does not mean that the forest department officials are not dealing with these issues; but they do so as amateurs based on knowledge gathered through their forest-related field interactions.

Remarkably, even foresters' classical forestry knowledge is hardly ever applied to JFM forests. Apart from the divisional working plan (written for the whole division by a forest official and approved by government) that the DFO and local officials use for FPCs, there are no arrangements for forest surveys, mapping, the categorization of FPC forest and its micro-management to satisfy local needs or priorities. Hence, a plan meant for a macro-area is applied to a micro-plan, and this is inevitably inappropriate for local needs.

There is another knowledge reserve in JFM areas referred to as indigenous knowledge. This is the knowledge of the people, especially the tribal people, of forests and their wildlife and their relation to livelihood, culture, religion (sacred groves) and other aspects, and their methods of managing forest protection, forest closure, tree planting, NTFP management and so on, empirically gathered over generations. Forest department staff consider this knowledge inferior and incomplete, and it is excluded without exception from working plan prescriptions that 'wag' the FPC micro-plan 'tails' uniformly. This is a typical case of sectarian information politics generating uniformity, and rejecting diverse traditional and innovative local knowledge bases. The forest department officials' procedure for areas to be harvested, calculation of production quantity, 'lot' preparation of the products, and calculation of net from gross income is not transparent to local people, who are not initiated into how, or according to what information, decisions are made. The complicated procedures obscure the truth and, as such, obstruct accountability.

Actions undertaken by the research team to identify measures to improve processes

Many sociologists, NGOs and forest researchers, as well as other civil society groups, are concerned about JFM and the way in which it is implemented. These include the Ramakrishna Mission in Kolkata; Narayan Banerjee, working on women's empowerment; the West Bengal Forest Researcher Sivaramakrishnan (Sivaramakrishnan, 1999); Madhumati Dutta and associates (Dutta et al, 2004); and many others. These groups have sought to identify the problems and deficiencies of JFM through PRAs, household interviews and public hearings. There are, however, two lacunae that make their investigations incomplete in our view. One is that detailed information on how the practices affect households' livelihoods in the context of their entire income is hardly ever gathered. Second, local people's solutions to the problems they face are rarely considered. At present, the social activists propose solutions, albeit with the best intentions. We believe that this method has to change in favour of action research based on grassroots proposals of solutions.

A further issue here concerns policy-making. In a democratic country such as India, it is accepted that legislators represent the people and therefore should be the policy-makers.

The primary activity of state assemblies and central parliament is enacting legislation. Unfortunately, forestry policy formulation has never assumed importance for legislators. Executives have generally done the thinking and formulation, which is then more or less rubberstamped by the legislators with little discussion. This is obvious from the fact that a change as important in the forest management paradigm as JFM executive orders continues to be effective with no legislative back-up. This must change – legislators have to be motivated to take more interest in forestry policy-making, and it is likely that this will require the mobilization of the electorate to raise the issue with their democratic representatives.

In the first phase of the study, we collected data on the impact of JFM on the livelihoods of the people and the problems that the FPCs face in implementing JFM more successfully. In the second phase, after analysing the research findings, as discussed above, we have taken on the following tasks to respond to the points above:

- It is important to interact with the FPC members in the sampled villages in order to hear their suggestions for solutions to the problems that they highlighted during the first phase of the study. This has been carried out in a series of meetings in the sampled villages (11 in all) after first circulating booklets written in the local language and incorporating the phase 1 findings on livelihood impacts on FPC members.
- The problems enumerated by the people and the solutions they proposed are then listed and disseminated to local legislators, forest and administrative executives, who are urged then to make necessary FPC-proposed policy changes to bring about their implementation. In addition, two documentary films have been made about the action research at the grassroots level and its findings, which will be circulated in the appropriate places.

The findings are summarized as follows.

Main constraints to better achievement of JFM and possible corrections
The main constraints to PFM as identified by the local people can be summarized as follows:

- There is a lack of power in FPCs to plan and manage 'their' forests. Currently, FPCs are mainly patrolling forests that are managed to a very poor standard by the forest department according to very general working plans.
- Decision-making is perfunctory in that the local FPC has only limited discretionary power. The forest department holds power over micro-planning, dealing with offenders and the disposal of timber and firewood. At the village level, decision-making is typically dominated by the beat officer with the concurrence of the elites and middle-class village males. The women, poor and landless are marginalized in decision-making.
- There is a total absence of legal frameworks to provide security to the entitlements agreed in JFM orders and to reduce the forest department's discretionary powers.
- The forest department has managed the forests according to its standardized sal coppice rotation without any modifications according to local needs or opportunities. There have been no attempts to increase the productivity of NTFPs, and little technical guidance by the forest department to add value to NTFPs or in their marketing, despite the fact that these make up a significant part of the livelihoods of most households in the region.
- The lack of forest/ecological and socio-economic baseline studies of each village associated with FPCs does not allow comparative monitoring with time series data. Currently, there is a complete lack of systematic monitoring of the functioning of the FPC or the sustainability of the forest, either by the FPC or by the forest department.

- The forest department is characterized by poor governance. There is a lack of transparency with regard to the allocation and handling of development funds, apparently favouring the nexus of local elites, contractors and unethical politicians.
- There is a lack of inter-sectoral coordination. The forest department has failed to harness the expertise of other departments in village areas.
- Traditional/community knowledge and forestry practices have been neglected.

Recommendations

Policy changes

First, joint forest management must be legislated for with a few administrative resolutions. This enactment can be a part of the 1927 Indian Forest Act that is now being considered for revision. Second, an FPC member (or members) needs to be empowered to the status of a forest officer of rank, thereby allowing FPCs to deal with forest offences, planning and implementation of forest development work.

Lack of technical guidance to add value to forest products and to increase forest productivity

The forest department has failed where it could have been most useful – namely, in technical guidance and investment of funds to process forest products for sale. The micro-plan is principally an instrument to plan for the fulfilment of local family needs and the sale of those products that will yield the greatest sustainable income. We found that the largest part of the income comes from NTFPs, and the greatest family need is for fuelwood, grazing and fodder. None of the micro-plans include any of these features. The micro-plan preparation procedure needs to be completely changed. First, a participatory rural appraisal should be carried out in the FPC area, and the people's forest needs and sources of income must be quantified. The plan will then focus on the production of these products. If there is a need for fodder, then fodder plantation and rotational grazing in existing coppice forest should be introduced.

Some small tokenistic initiatives have been taken by the forest department, such as the establishment of a few sal plate-making machines (although these often seem to be stolen from the community and 'privatized' by village elites with the acquiescence of beat officers). Overall, the forest department has had little impact on value addition. The private sector has installed many of these machines; but the profits largely escape FPC members as the machine owners buy the leaves to make the plates themselves.

Similarly, state research divisions have carried out little research and applied knowledge to improve the productivity of coppiced forests and NTFPs.

Lack of baseline studies and monitoring of sustainability

The state of the forests in South-West Bengal, although improved, requires monitoring of trees, shrubs and herbs; biodiversity; and the role that the forest plays in soil and water conservation. The state should make inventories of all FPC forests and measure trees, shrubs and herbs (e.g. in laid-out plots with a global positioning system, or GPS, location so that successive measurements can be done). We have initiated this sort of work in the sampled FPCs as part of this study; but if progress is to be tracked in future, it would require a sustainably institutionalized monitoring system.

An independent monitoring unit (not part of the forest department, as at present) should be established. But before doing so, an in-depth consultative exercise should be carried out to find out how to monitor effectively and affordably, and which of the indicators incorporated within baseline studies should be used. Besides a baseline study of forest resources, a socio-economic baseline study of sample FPCs in different parts of West Bengal is also needed. We also suggest that the present experiment should be continued in other regions.

Poor governance by the forest department and lack of transparency

- It is necessary that forest officials shed their bureaucratic patterns of behaviour in dealing with village people, as well as their belief that the so-called classical forestry system is the correct line of approach in forest management. This would entail extensive training of forest officials at all levels in innovative technology and dealing with people effectively, fairly and democratically.
- The legal empowerment of FPC members to deal with offenders from other villages or mafia gangs seeking to extract trees and other forest products is necessary.
- The forest department is expected to dispose of timber, poles and firewood and share 25 per cent of the net income as per government orders. First, the harvesting itself is conducted in a very erratic and unsystematic manner by the forest department so that many FPCs have never experienced any harvesting. Furthermore, FPC members believe that the sales are mismanaged, resulting in products sold at less than market price. Timber merchants seem to be working in a cartel to underbid. When the net share is generated, it often takes many months before the cash is actually received by FPC members. There is also no transparency in the calculation of the FPC's share – members are unclear about the method and the forest department does not share its calculations. FPC members should be empowered to conduct their own disposal of forest products if they wish.
- Members are often ignorant of FPC decisions and about fund allocation and expenditure. The forest department should circulate its decisions and financial estimates for work to be done to FPC members well in advance. The best way is to print them on paper strips and paste them in prominent places in the FPC villages.

References

Banerjee, A., with Sen, S., Das, S., Das, A. K., Haldar, S. P., Bandyopadhyay, A., Ghosh, S., Das, S., Saha, S., Talukdar, J., Mishra, T., Maity, S., Banerjee, P. S., Dutta, S. C. and Nandy, H. (forthcoming) *Participatory Forest Management in West Bengal: Outcomes and Livelihood Impacts*, Norwich, University of East Anglia

Banerjee, A. K. and Springate-Baginski, O. (2005) *Summary Report on Action Programme on 'Impact of Joint Forest Management (JFM) on the Livelihoods of the People in South-West Bengal'*, Kolkata, Nari Seva Sangha

Banerjee, A. K. and Springate-Baginski, O. (2007) *Impact of Joint Forest Management (JFM) on the Livelihoods of the People in South-West Bengal*, publication forthcoming

Bardhan, P. and Mookherjee, D. (2004) 'Poverty alleviation efforts of *panchayats* in West Bengal', *Economic and Political Weekly*, Mumbai, 28 February, pp965–974

Chandra, S. N. and Poffenberger, M. (1989) 'Community forest management in West Bengal: FPC case studies', in Malhotra, K. C. and Poffenberger, M. (eds) *Forest Regeneration through Community Protection*, Kolkata, West Bengal Forest Department

Chatterjee, A. P. (1996) *Community Forest Management in Arabari: Understanding Socio-Cultural and Subsistence Issues*, Delhi, SPWD

Chattopadhyay, R. and Duflo, E. (2004) 'Women as policy makers: Evidence from a randomized policy experiment in India', in *Econometrica*, vol 72, no 5, pp1409–1443

Deb, D. (2007) 'Sacred ecosystems of West Bengal', in A. K. Ghosh (ed) *Status of Environment in West Bengal: A Citizen's Report*, ACB Publications, Kolkata (in press)

Dutta, M., Souvanic, R., Subhayu, S., Dibyendu, S. and Maity, S. K. (2004) 'Forest protection policies and local benefits from NTFP: Lessons from West Bengal', in *Economic and Political Weekly*, 7 February, Mumbai, pp587–591

FSI (Forest Survey of India) (2005) *State of Forest Report 2003*, Dehra Dun, Ministry of Environment and Forests

Ghosh, S. with Nabo Dutta, Hadida Yasmin and Tarun Roy (forthcoming) *Commons Lost and 'Gained'? Forest Tenures in The Jungle Mahals of South West Bengal*, Norwich, University of East Anglia

GoI (Government of India) (2001) *Census of India*, Delhi, GoI

GoWB (Government of West Bengal) (2001) *State Forestry Report*, Kolkata, GoWB

GoWB (2003) *Economic Review: Statistical Appendix 2002–2003*, GoWB, Kolkata

Guhathakurta, P. and Subimal, R. (2000) *Joint Forest Management in West Bengal*, New Delhi, WWF-India

Kumar, N., Saxena, N., Alagh, Y. and Mitra, K. (2000) *India: Alleviating Poverty through Forest Development*, Washington, DC, World Bank

Mishra, T. K., Maiti, S. K. and Mandal, D. K. (2004) 'Joint forest management in West Bengal: Its spread, performance and impact', in Ravindranath, N. H. and Sudha, P. (eds) *Joint Forest Management in India*, Hyderabad, University Press, pp160–180

Palit, S. (2004) 'West Bengal' in Winrock International (ed) *Root to Canopy*, New Delhi, Winrock International, pp181–192,

Poffenberger, M. and McGean, B. (1996) (eds) *Village Voices, Forest Choices*, New Delhi, Oxford University Press

Poffenberger, M. and Singh, C. (1996) 'Communities and the state: Reestablishing the balance in Indian forest policy', in Poffenberger, M. and McGean, B. (1996) (eds) *Village Voices, Forest Choices*, New Delhi, Oxford University Press

Roychowdhury, K. C. (1964) 'The forests of the southern circle: Its history and management', in Government of West Bengal (ed) *West Bengal Forests*, Calcutta, GoWB

Saxena, N. C. (undated) *Forest Policy in India*, New Delhi, WWF-India

Singh, S., Datta, A., Bakshi, A., Khare., A., Saigal, S. and Kapoor, N. (1997) *Participatory Forest Management in West Bengal*, New Delhi, WWF-India

Sivaramakrishnan, K. (1999) *Modern Forests, Statemaking and Environmental Changes in Colonial Eastern India*, New Delhi, Oxford University Press

SPWD (Society for Promotion of Wastelands Development) (1998) *Joint Forest Management Update 1998*, New Delhi, SPWD

8

Forests and Livelihoods in Orissa

Kailas Sarap

Synopsis

Orissa contains among the highest concentrations of forest-dependent populations, particularly of tribal groups, and some of the most acute poverty in India. In terms of forest management, it is unique in many ways. Communities took the initiative well before the forest department, and self-initiated forest protection committees have spread widely across the state since the 1960s. The forest department took the initiative in the formation of *Vana Samarakshyan Samities* (VSS), or forest protection committees (FPCs), only during the 1990s. About 9677 VSS have now been formed in different forest divisions of the state, many of which are simply formalized self-initiated groups.

The state implementation of participatory forest management (PFM) has been a haphazard affair, reflecting the weakness of the Orissa Forest Department as an institution. There has been a lack of proper participatory process, either at the outset or post-formation: local people's participation in the preparation of the 'micro-plan' is generally marginal, as the forester exerts major control over this. In VSS executive committee and general body meetings, important decisions have been taken by elites, including the forester (who is secretary). Self-initiated groups were found to be much more participatory than the VSS, although few women are involved in either type of management (VSS and self-initiated forest protection groups, or SIFPGs), and those who are have little power in decision-making.

Conflicts of various natures, including in and between villages, were found in the study areas, with consequential deforestation, mismanagement of resources and judicial proceedings. There are also conflicts over sharing benefits, usufruct rights, illegal felling, forest boundaries and with forest mafias. Mining, mostly located in forest areas, has led to conflict between forest-fringe communities and mining leaseholders.

The forest development agency (FDA) scheme has further created a rift between those VSS villages which receive FDA assistance and those which do not, as forest officials concentrate their resources and time more on FDA-assisted villages than on non-FDA villages. This acts as an incentive for many people in self-initiated community forest (CF) villages to convert to VSS in order to have the chance of financial help from the scheme.

Dependence on the forest for their livelihoods has been crucial for poor households in the absence of adequate resource endowment, such as land and access to the service sector. Collection of forest products (for own use and market sale) is a major labour allocation for the majority of poor households. Overall, 25 per cent of household income comes from forest-related activities. This supplements the total income of poor households; but absolute income from the forest is very low for this group. About 80 per cent of total forest income is generated during the summer season and around 12 per cent from the rainy season. Since a major portion of forest income in the rainy season is consumed, forests are fundamental to the poorest groups as insurance against starvation and hunger. Given their low total income, these households have to depend on loans to subsist. More than 40 per cent of FPC households have outstanding loans from private creditors such as shopkeepers. Indebted collectors are forced to sell their forest products to their creditors at depressed prices.

PFM has generally led to improved forest conditions and increased access to a variety of non-timber forest products (NTFPs), although there has been no improvement in market relations regarding NTFP sales, which are critical in determining the level of livelihood benefits (despite the licensing of NTFP extraction being transferred to *panchayats*).

Different stakeholders have contradictory interests and are uncoordinated; as a result, forest-related policies have not been effective in raising the quality of forests and the livelihoods of rural people. To achieve these two objectives there is a need for assured tenurial rights to the forest as an incentive to local communities; recognition and endorsement of the diverse local participatory management practices, including self-initiated groups; legitimization of subsistence forest use; promotion of forest product value addition and enterprise development; improvement of marketing networks through collaboration with *panchayats*, self-help groups (SHGs) and non-governmental organizations (NGOs); and increased investment in the forestry sector.

Participatory forest management (PFM) policy and context

Orissa has had a distinctive experience with PFM due to a range of factors, which are discussed in this section. Orissa's eclectic origins as a state and its extensive and relatively inaccessible forested areas have meant that governance and service provision have been particularly challenging. Coupled with high levels of livelihood forest dependency, this has led to communities taking their own responsibility for forest protection, and the state forest department struggling to keep up and maintain legitimacy.

Socio-economic profile of Orissa

Orissa is among the poorest states in India, with very high levels of income poverty. About 47 per cent of the population was classified as below the poverty line in 2000 (GoI Planning Commission, 2003). The percentage of poor people in southern and western Orissa is particularly high. 'Scheduled tribes' and 'scheduled castes' are among the poorest groups in the state; the tribal population, which accounts for about 22.13 per cent of the total population, is the poorest of all. Southern Orissa is one of the poorest regions in India. Of the total number of poor, 90 per cent live in rural areas, and poverty is particularly intense among tribal populations living in forest-fringe villages. The majority of tribal people live in southern and western Orissa, where most of the state's forests are located.

Geographical and historical perspective

The state of Orissa is located in the east of India, with a total geographical area of 15.57 million hectares. The population of the state is 36.71 million (3.6 per cent of the total population of India), of which 85 per cent are rural and 15 per cent urban. The average population density is 236 persons per km². The livestock population is 22.7 million, 4.8 per cent of that of the whole country. The present-day state of Orissa was initially formed in 1936 from parts of three provinces – Central Province, Bengal Presidency and Madras Presidency, as well as parts of Bihar. During 1948 to 1949, another 26 princely states were added. Orissa therefore inherited often contrasting styles of governance involving a complex mixture of different rights, concessions and forest rules.

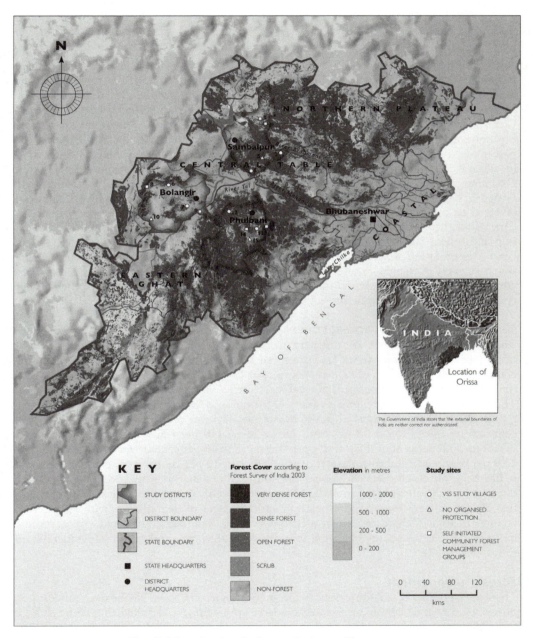

Map 5 Orissa showing districts, study sites and forest cover

Note: See Table 8.2 for village names.

Source: Jonathan Cate

The state of Orissa can be divided into four distinct physiographic regions: North Plateau, Eastern Ghat, Central Tableland and Coastal Plains (see Map 5). Extensive forests and tribal populations are mainly found in the first three regions. There are 52,000 villages in Orissa. Of these, 29,302 (56.35 per cent) record forest as a land use. Eighty-five per cent of villages, including more than 5000 forest-dwelling villages, have less than 100ha of forest.

Overview of forests

According to the 2003 Forest Survey of India (FSI) survey, Orissa ranks fourth among India's states and union territories in terms of area under forest cover. The total area of the state is 155,707km^2. The survey records the forested area as 58,136km^2, 37.34 per cent of the total geographical area of Orissa. Actual forest covers 30.2 per cent of geographical area. Clearly, about 7 per cent of forestland area in the state does not contain any tree cover. Furthermore, forest cover is distributed unevenly in different regions of the state. The percentage of forest area to geographical area is much below the state average (30.2 per cent) in coastal districts. Southern Orissa (Gajapati, Kandhamal, Koraput, Malkangiri, Nawarangpur and Rayagada) and western Orissa (Kalahandi, Bolangir, Sambalpur and Deogarh) have much greater forest cover, although this has decreased in recent years. Forest diverted to other uses between 1982 and 2002 totalled 27,055ha. The main reasons for conversion of forest have been mining (35 per cent), irrigation (22 per cent), human habitation and others (16 per cent), and infrastructure (railway, transmission lines, etc) (17 per cent). Deforestation of 18,106ha occurred during the period of 1971 to 1982. The three divisions – namely, Karangia, Sundergarh and Jeypore – saw very high deforestation; these are areas where mining and wood-related mafias are very active, which is likely to have contributed significantly to deforestation.

Nature and density of forest cover

The main types of forests are northern tropical semi-evergreen, northern tropical moist deciduous, tidal swampy forest and southern tropical dry deciduous. The state's forests contain a high proportion of valued species. For instance, sal (*Shorea robusta*, a high-value timber species) forms 30 per cent of total forest 'crop', and bamboo comprises 27 per cent. The forest composition provides a variety of products to local people.

Of the total forest, reserved forest constitutes 45.3 per cent, protected forest 26.7 per cent and unclassed forest 28 per cent. Dense forest (40 per cent or more crown density) makes up 55.6 per cent of the forest; open forest (10 to 40 per cent crown density) comprises 43.9 per cent; and mangrove (crown density less than 10 per cent) amounts to 0.5 per cent. It has been observed that many areas previously under high density forest are becoming open and scrub forest. Many reserved forests are similarly in serious decline. It is estimated that 50 per cent of reserved forests are in various stages of degradation, with 30 per cent severely degraded, characterized by a canopy cover of less than 20 per cent. Moreover, the moist deciduous forests are changing to a dry deciduous type and becoming vulnerable to fire (GoO, 2002).

Forest management

There are a number of unique aspects to forest governance in Orissa. Unlike most other states, due to the way in which Orissa was formed, large areas of forest, despite being under *de facto* forest department management, are not under its *de jure* authority. Forests are controlled by two departments – namely, the forest department and the revenue department. Although the forest department is responsible for protecting all types of forests, it only controls and manages about 45 per cent of forest area categorized as reserved forest. Undemarcated protected and unclassed forests are under the administrative control of the revenue department.

There has never been a proper survey of the state and its forest land, and rights, partic-ularly tribal rights, have never been settled across most of it. A land survey of much of what is now Orissa was carried out between 1922 and 1928 by the colonial government for revenue purposes, although this was incomplete and somewhat arbitrary. However, it contin-ues to be used as the basis of land records, meaning that many areas remained unsurveyed. Furthermore, although the tribal cultivation of forest areas has been recognized as an histor-ical practice, because their rights have never been settled they are treated as illegal 'encroachers'.

The role of forests in agrarian livelihoods

The state is characterized by high levels of inequality, evident in the skewed land ownership, levels of landlessness and ambiguous tenurial rights. A substantial proportion of the popu-lation, especially in southern and north Orissa, depend on forests for a large part of their livelihood.

The state has 8.145 million tribal people who constituted 22.13 per cent of the total population at the time of the 2001 census (GoI, 2001). The scheduled areas, found in 134 blocks out of 314, contain very high concentrations of tribal peoples, who have historically settled in forest fringe and forest areas to practice shifting cultivation. During the late 1980s, it was estimated that about 8 per cent of the state's forest areas were under active shifting or *podu* cultivation.

A significant proportion of the rural population depend on the forest for their liveli-hood. Local people collect a variety of leaves, mushrooms and bamboo shoots for food and medicinal plants, fruits and bark to treat sickness. The Bolangir district forests are famous for a variety of medicinal plants, which households use for home treatment as well as selling them locally and outside the district. Similarly, many households depend on sales of wood and bamboo articles, which are widely used locally. Given that many households have no surplus income, they find bamboo articles easy to procure at low prices. Thus, in the forest-fringe villages, forest products work as lifelines throughout the year in one way or another.

Historically, agrarian production practices have involved forest use as an adjunct to agri-culture in the lean season and as a safety net during hardship. This has changed little with the lack of emergence of more productive livelihood options in Orissa, one of the least economically developed states in India, which additionally has experienced a decline in per capita food availability, especially during the 1990s, due to the very low growth of the agri-culture sector.

The state is characterized by frequent natural disasters such as floods and droughts. The impacts of these disasters fall more heavily on the poor, due to their limited assets and lack of safety nets. In such a situation it is expected that they would forage in forests for food and resort to distress sales of a greater quantity of forest products in order to survive.

Dependence on the forest is also influenced by agricultural development, forest quality, existing policies and institutional arrangements. For instance, since the colonial period and especially from 1865 until the late 1980s, the forest was used by the forest department to generate state revenue from felled timber. Some of the local people's customary livelihood claims to the forest were restricted by the state during this period through the enforcement of various laws, as we see below.

In a situation unique in India, self-initiated forest protection groups emerged during the 1960s and 1970s across Orissa's forested areas, which had been degrading through over-extraction. The movement to regenerate, conserve and manage the forest on which rural people depended for their livelihoods spread as a voluntary community-based mechanism. Many intellectuals and other members of civil society helped the movement, and youth groups spread the concept throughout forested rural areas. It has been estimated that about 8000 self-initiated forest protection groups continue to protect the forests.

The state finally took note and sought to catch up through provisions in the 1988 National Forest Policy, which facilitated participatory management of the forest sector. In Orissa, the forest department sought to formalize these groups by passing a joint forest management (JFM) notification in 1993 and seeking to form a VSS. However, as the motivated villagers were already organized into self-initiated groups, the formation of a VSS, in practice, meant converting many self-initiated groups into what might be called 'company unions'. By the end of 2005, more than 9000 VSS had been converted in this way and also newly formed, although it is estimated that about 8000 non-formalized self-initiated forest protection groups continued to protect the forests independently of the government scheme.

Thus, both self-initiated forest protection groups and VSS manage the forests along with the forest department. The state's VSS have the apparent benefit of state legitimization to enforce their rules. However, the forests that they have regenerated come under the control of the forest department, and the share of revenue from felling only applies from the time that the groups are formalized. Moreover, the VSS have to accept a beat officer as the head of their committee. Self-initiated groups, on the other hand, while free from the often obstructive bureaucratic processes involved, lack the endorsement and financial input of the forest department.

Review of forest policy and major events

This section briefly discusses the state's forest policies. The evolution of various types of forest protection is shown in Table 8.1. Historically, forests were so extensive relative to population that there was little call for regulation, although it is believed that there were some customary provisions around villages.

Following the 1935 Government of India Act, forests were put on the provincial list. This empowered state governments to frame their own laws. However, the government of Orissa did not take any legislative action regarding forests until the 1972 Orissa Forest Act, applicable to all parts of Orissa except the districts of Ganjam, 'undivided' Koraput and Baliguda, and the G. Udayagiri *taluk* of Phulbani district, where the 1882 Madras Forest Act remained in force. Section 3 of the Orissa Forest Act empowers the state government to notify any land as reserved forest. It also sets out detailed procedures through which to declare the proposed area as reserved forest, including settling rights and privileges that exist. Furthermore, Section 33 of the act states that the state government may, by notification, declare any land protected forest which is not included in a reserved forest and may, by notification, declare any land 'village forest' for the benefit of any village community or group of village communities. Sections 27 and 37 of this act provide provisions under which any person can be declared a forest offender.

With regard to the customary rights and liabilities of forest dwellers, the state assumed wide powers to declare any land as forest. Any unauthorized dealings of the people concerning forest products were made an offence. The act provided ample power to forest officials to interpret what sort of rights and privileges the local people were entitled to from the forest.

In 1981, two NTFP trading-related acts were passed: the 1981 Orissa Forest Produce (Control of Trade) Act and the 1981 Orissa Kendu Leaf (Control of Trade) Act. Both asserted monopolies on the collection and sale of certain NTFPs by the state government through the Orissa Forest Development Corporation (OFDC). The forest department leased this right to the Tribal Development Co-operative Corporation (TDCC), which in turn delegates individuals and private agents to handle the collection and trade in leaves. However, there are a number of deficiencies in these organizations regarding the procurement of forest products, due to which private traders exploit collectors, who cannot therefore achieve proper prices for their wares.

The first phase of the Social Forestry Project (SFP) was started with the support of the Swedish International Development Agency (Sida) in 1983 and continued up to 1988 in all 13 districts. The main objectives of the SFP were to create sustainable forest resources so that the people could meet their requirements for fuelwood, fodder and minor forest products. The jurisdiction of this project was confined to village woodlots and plantations near village common land known as 'social forest'. There was no involvement of communities in the protection of natural forests under this project. The Orissa Village Forests Rules were framed in 1985 and the involvement of NGOs in the process brought people's participation into forest management. The second phase of the project was implemented between 1988 and 1996. The focus was the market orientation of activities with a view to generating cash income and the sustainable development of renewable resources. The state government has spent about 1650 million rupees on SFP through a loan from Sida. However, the achievements of the SFP were not very encouraging in terms of area coverage, choice of species and the participation of the poor and women in management (PCCF, 1991).

During the 1970s and 1980s, self-initiated CF protection groups emerged in many districts of Orissa. Even though these initiatives received little or no support from the forest department, central and state governments began to perceive their significance and acknowledged the need to recognize and legalize community efforts.

The 1988 Forest Policy represented a significant departure from previous policies because it emphasized that local people must be actively involved in forest protection, conservation and management programmes. Local people living in and around the forest were to be considered partners, not only in the protection and regeneration of forest, but also in sharing the usufructs and profit. The focus of forest management shifted from forest commercialization to conservation and the rights of the local populace to the forest products.

Following the 1988 Forest Policy, the GoO issued a resolution on 1 August 1988 according to which local communities were to be involved in the protection and regeneration of degraded reserved forests with the promise of certain benefits and concessions.

JFM Resolution of 1990: A participatory approach
On 1 June 1990 the Government of India (GoI) passed guidelines launching the JFM programme. Orissa's government modified certain provisions of the resolution in 1990, 1993 and 1996 in order to encourage participatory management. Formal VSS forest protection groups were formed and their numbers increased rapidly through the late 1990s and later.

Guidelines on JFM 2000
On 21 February 2000, the GoI issued guidelines on various JFM activities in response to several issues confronting FPCs, NGOs and forest departments. The major features of the guidelines were provision of legal status to JFM committees under the 1960 Societies Registration Act; increased participation of women in general bodies (at least 50 per cent to be women) and executive committees (at least 33 per cent); extension of JFM to less degraded forests; preparation of micro-plans for JFM areas by the forest department in consultation with user groups; recognition of self-initiated forest protection groups; and contributions by the VSS and the forest department towards the regeneration of resources. However, the VSS in the state were not registered under the Societies Act, being based only on an administrative order. As a result, more than 9000 VSS were formed in Orissa by the end of 2005. However, it has been found that many of the local VSS groups are extremely weak institutions beset with a number of problems. About half of all VSS are inactive due to various factors, including lack of tenurial security.

According to a status report brought out by the principal chief conservator of forests (PCCF) office in 2000, of the total 6768 VSS formed in Orissa by April 1999, memoranda of understanding had been signed by only 2617, or 39 per cent. Further micro-plans for forest management have been prepared for only 177 VSS (2.61 per cent). It is clear that there has been an increase in the formation of VSS in order to fulfil the target by the forest department. But this rapid increase without properly nurturing community participation and institution-building, and without preparing micro-plans, is open to all kinds of problems.

Formation of the forest development agencies (FDA), May 2002

The GoI passed a resolution in May 2002 for the consolidation of all types of village-level forestry-related programmes under *Samanvit Gram Vanikarm Samridhi Yojana* (SGVY – literally, integrated village afforestation and eco-development), to be implemented through forest development agencies (FDAs). Under this resolution, all forest-related programmes and village development programmes in forest-fringe villages would be implemented through JFM committees. The central government decided to transfer funds to the FDA directly. The help of NGOs and local people were to be sought in implementing the programme. The FDA would be registered under the Societies Registration Act. The new arrangement gave the forest department significant control over activities relating to the allocation of resources through the VSS.

Orissa Forest Sector Vision 2020

Orissa's government has prepared the Orissa Forest Sector Vision 2020, with financial assistance from the UK Department for International Development (DFID). A number of stakeholders have been involved in preparing the document. It is expected to guide policy-making and bring changes to forest management paradigms and practices in the state. It recognizes the usefulness of multiple participatory management models, including JFM and community forest management (CFM); forest-based subsistence and livelihoods; promotion of value addition and forest-based enterprises; assured and appropriate tenurial rights providing incentives for local participation and responsive and motivated forest departments; and bottom-up participatory approaches.

Orissa Forest Sector Development Society 2006

The Orissa Forest Sector Development Society was formed by the GoO. It will work under the direct control of the forest and environment departments and through JFM. The objective of the society is to oversee the implementation of programmes in order to restore degraded forest, conserve biodiversity and improve livelihood sources for the forest-dependent population. The Japan Bank for International Cooperation (JBIC) is supporting the state in the project with the financial assistance of 6000 million rupees.

Recent forest policy issues

Some provisions of the 1972 Orissa Forest Act were amended by the Orissa Assembly in 2001 on the basis of the 2000 Orissa Forest (Amendment) Bill. The objectives of the bill were to arrest further damage to forest resources and prevent the killing of forest animals. This is to be achieved through imposing stringent punitive measures on offenders.

The state government has also declared a new policy to reduce the theft of wood from the forest. According to the new provision, seized wood is to be auctioned where it is confiscated. There are incentives for helping to apprehend the rule-breakers: 25 per cent of the proceeds from the auction go to the person who has provided information and another 25 per cent to the officer who confiscated the items.

The state has placed special emphasis on improving forest conditions in the southern so-called KBK region (covering the old districts of Koraput, Bolangir and Kalahandi) under a long-term plan since the late 1990s. However, there has been no apparent enthusiasm for this in political or bureaucratic circles. The state's forest bureaucracy carries on with its usual

Table 8.1 *Timeline of evolution of community forest management (CFM) and joint forest management (JFM) in Orissa*

Year	Point	Comment
Pre-20th century	Village communities use forests – many also manage through customary rules	
1922–1928	Land survey	Colonial attempt to survey areas of what would become Orissa from revenue records.
1936	Orissa created	
1936	Lapanga village forest protection committee (VFPC) formed in Sambalpur district	First recorded CFM in Orissa.
1959	*Forest Enquiry Report*	
1960s	Youth clubs take up forest protection as an activity	In context of social mobilization and launching of community development programmes.
1970s	Forest protection by self-initiated groups evolved in Puri (Nayagarh) and Balangir, Sambalpur district	Emergence of CFM.
1972	Orissa Forest Act	Empowered government to declare any land reserved, protected or village forest. Restricts local access to forest and treats *podu* (shifting cultivation) as inadmissible.
1980s	CFM spreading in Nayagarh, Balangir, Dhenkanal and Mayurbhanj districts	Growth of CFM.
1983–1985	Start of Social Forestry Project with financial assistance from the Swedish International Development Agency (Sida)	Promotion of woodlots on village land rather than community involvement in forest management.
1985	Orissa Village Forest Rules	Legal recognition of VFPCs and declaration of village forests.
1987–1988	National Environmental Awareness Campaign	Sensitizes the people to the need for environmental and forest protection.
1988 (August)	Resolution by Government of Orissa (GoO) to involve villages in protection of degraded reserved forest	Formation of self-initiated forest protection groups (SIFPGs). Focus on conservation and subsistence needs of local people.
1990 (May)	Government resolution to involve community to protect protected forests	Formal acceptance of JFM approach.
1993 (July)	Comprehensive resolution on JFM and formation of *Van Samrakshyan Samities* (VSS)	Strengthening participatory forest management.
1994 (December)	JFM extended to social forestry areas	
1996 (September)	Further JFM resolution	More rights to communities by declaring village forests under joint management.
1997 (November)	Process initiated at GoO level to draft a new resolution on JFM	
1998 (October)	Massive campaign to form VSS by the forest department	To fulfil the target for desired numbers of VSS.

Table 8.1 *continued*

Year	Point	Comment
2000	Orissa Forest (Amendment) Bill	More power given to forest officials to fight mafias in order to check degradation of forests.
2000 (March)	New state non-timber forest product (NTFP) policy	Ownership rights of 67 NTFPs to *gram panchayats*.
2001 (July)	Resolution of state price-fixing committee giving power to *panchayat raj* institutions to decide NTFP prices	While fixing prices, local conditions to be taken into account.
2002 (May)	Forest development agency (FDA) scheme created	All central consolidated forestry programmes to be implemented under integrated village afforestation and eco-developments (SGVYs) through FDAs and implemented through VSS.
2006	Orissa Forest Sector Development Society created with financial support from the Japan Bank for International Cooperation (JBIC) UK Department for International Development (DFID) Forest-Sector Reform Project approved	To oversee implementation of programmes on restoration of degraded forest and biodiversity conservation and to improve livelihood sources for forest-dependent populations.

Source: Kailas Sarap (see also Patnaik, 2002)

target-oriented activities because the institutional mechanism for promoting personnel is not based on innovative or creative work, but on the length of service and good conduct.

Spatial and temporal variations: Direction of forest policy

In 2003, the state embarked on an industrialization policy focusing on the mining, power (coal based) and iron and steel sectors through the exploitation of natural resources. These natural resources are found in forested regions where tribal and other poor communities live. By November 2005, more than 43 memoranda of understanding (MoUs) had been signed by companies and the state to exploit these natural resources. It has been alleged that one company, Vedanta, has violated the Ministry of Environment and Forests' (MoEF's) National Guidelines for Forest Conservation Act, and judicial proceedings have been initiated.

This industrialization strategy is promoting unregulated and *ad hoc* conversion of forest and the displacement of numerous forest-fringe villagers. With no land tenure rights and inadequate compensation, the recent direction of industrial policies is having adverse consequences for forest and forest-fringe villagers. There have been demonstrations in many areas of the state by tribal peoples and others affected by such displacement attempting to assert their rights to land and compensation. This has led to occasional protests, even to the point of clashes with the police and the death of 12 tribal protesters at Kalinga Nagar in Jajpur district in January 2006.

Land tenure and podu

The state has long been characterized by the uncertain tenurial rights of a majority of tribal and other poor people living on the forest fringe and in forest villages. Land records in Orissa are very poor, and the land rights of the majority of the state's tribal people living in hilly areas and in all categories of forest land are yet to be settled (see Box 8.1 for details). Surveys for the revenue settlement of land declared government property have excluded land with slopes of over ten degrees, despite several tribes living in these hilly areas. A large number of tribal people continue to cultivate and live without formal land titles on land declared to belong to the government. In the absence of recognized land rights, many poor cultivators, including tribal peoples, are unable to access credit from formal credit institutions to improve their farming.

Another major issue in Orissa is the presence of about 5000 'forest villages', a legacy established by different forest departments, mostly in northern and southern Orissa. The forest departments had recruited settled tribal labourers from forest areas and provided them with agricultural land in order to ensure the availability of labour for forest operations. Despite a GoI policy decision, these forest villages have not yet been converted to normal revenue villages, and, as a result, forest-dwelling communities residing in forest villages are deprived of many of the benefits of government-sponsored programmes for the tribal and other poor.

Shifting (*podu*) cultivation is a common practice in the tribal belt of Orissa, on which nearly 150,000 households belonging to different tribal groups depend. The recent policy of the state government has been to discourage it without compensating those who practised it. The forest department has persuaded the tribal peoples to stop practising *podu* cultivation and has encouraged the formation of VSS in these areas. In our Kandhamal district study site, we found that in some villages youths have persuaded *podu* practitioners to practise settled cultivation and other livelihood activities instead. Shifting cultivation practices have transformed over time, and with the increasing sedentarization of tribal groups, many so-called 'shifting cultivators' are simply landless people who depend on cultivating low-productivity forest areas because they do not own more productive agricultural lands.

Tribal populations have numerous claims to forest lands that have not been resolved, and neither have their rights been settled. The state has been supposed to settle their rights on these lands since before the recently introduced Forest Land Bill in parliament. Several civil society organizations in the state have supported tribal communities' demands for land rights on land which they have been cultivating for decades, mobilizing communities in different parts of the state in support of their demands. Even some of the organizations in the Maoist insurgent-affected areas (the so called 'naxalites' named after the naxalbari area of West Bengal, where they first emerged during the late 1960s), including our study sites, have been demanding land rights and a stop to the exploitation of tribal groups in forest-related activities.

Box 8.1 Contested landscapes: The construction of legal forests in Orissa

Kundan Kumar

According to official data, 58,135km² (38 per cent) of Orissa is categorized as 'forest land'. However, the forest covers only 48,838km² (FSI, 2003), implying that 10,000km² of forest land is not forested, thus indicating that there are serious problems with the way in which 'forest land' has been defined.

Forest tenure and access issues in Orissa can be categorized under two headings. The first is the larger issue of construction of 'legal forests', which privileges forest land use over other land uses. This issue is significant in Orissa because categorization of land as forests has been very problematic. Second is the issue of rights and access to forest products and services from land legally categorized as forests. This box focuses on the construction of 'legal forests' in view of the Scheduled Tribes and Other Traditional Forest Dwellers (Recognition of Forest Rights) Bill, which was passed on 18 December 2006.

Orissa, comprising 24 princely states and parts of three ex-colonial provinces, has a complex history of forest governance. The princely states and provinces had different forest laws and policies. In 1972, the Orissa

Forest Act (based on the 1927 Indian Forest Act) was passed, superseding prior forest laws in an attempt to bring uniformity to forest governance. Thus, the 'legal forests' of Orissa have emerged as the consequence of a complex mosaic of laws and policies over the last century. However, the non-recognition of the critical dependence of forest communities on forests and forest land, and consequent denial of their rights has been constant in this complex process.

Prima facie forest laws provide for the settlement of rights, including land rights, of local people and communities before declaring a legal forest. In the case of Orissa, this presumption often does not hold true due to a number of the following factors.

Non-recognition of rights on shifting cultivation land

Estimates of the area under shifting cultivation in Orissa range from 5298 to 37,000km^2 (Thangam, 1984; Patnaik, 1993) – about 44 per cent of the total forested area of the state, the highest percentage in India (Mishra, 1995). The *Forest Enquiry Committee Report* of 1959 records 12,000 square miles (almost 30,720km^2) of land in Orissa under shifting cultivation (GoO, 1959), equal to almost 20 per cent of the area of the state. Another report from the Orissa forest department states that 31,237.9km^2 of forest are affected to varying degrees by shifting cultivation.

After independence, the Government of Orissa (GoO) did not recognize shifting cultivation as a legitimate land use and settled all shifting cultivation areas, including forest land, as government land. The denial of rights in vast areas of tribal communities' customary land led to the *criminalization of one of the most important sources of tribal livelihoods*. Shifting cultivation on forest land continues to be practised by a large number of tribal communities, who are treated as encroachers by the Orissa forest department.

Declaration of deemed reserved forests and deemed protected forests

Most forests in Orissa are either in the merged princely states or from the area under the jurisdiction of the 1882 Madras Forest Act. The princely states declared certain forested areas reserved or other classifications, generally without any proper settlement of rights. In many states, 'wastelands' (all lands excluding reserved forests or land settled as private land) were declared forest land. Similarly, where the 1882 Madras Forest Act was applicable, large areas were declared reserved land and protected land without settlement of rights on land or forests. Many of these forest lands contained settled populations.

After independence, the GoO amended the 1927 Indian Forest Act (IFA) by adding Section 20-A(1), which provided that all reserved forest areas in the princely states would automatically be 'deemed' to be reserved forests. Another amendment of the 1927 IFA laid down that 'forests recognized in the merged territories as Khesara Forest, village forest or protected forests or forests by any other name designated or locally known, shall be deemed to be protected forests within the meaning of the act'. These blanket amendments completely ignored the fact that none of these forests have ever been properly surveyed and the rights have never been settled.

In Madras Presidency areas, reserved lands and protected lands created under Chapter III of the 1882 Madras Forest Act were 'deemed to be protected forests' under the 1972 Orissa Forest Act (Section 33(4)). These 'deemed forest' areas were neither surveyed, nor were the rights of the cultivators and inhabitants settled at any time. In many forest divisions, deemed reserve forests constitute more than 90 per cent of reserved forests.

Large numbers of people, especially tribal groups, who continue to live in these forests as they have for generations, are now considered encroachers. An analysis of the 2001 Census of India in 9 districts shows that there are 443 settlements with a population of 70,000 people located on forest land. For instance, 25 villages are located within Beheda Reserved Land in the Umerkote range of Nabarangpur district (Orissa Forest Department, 1999, p275). Another 24 villages are located within reserved and proposed reserved forests in Nabarangpur, Jharigam and the Umerkote range (Orissa Forest Department, undated, p275).

Faulty land and forest settlements

The formal recognition of rights over land used for cultivation is recorded through revenue surveys and settlements. In most tribal areas of Orissa, the first such formal recording of rights took place only after independence. Such surveys left out areas declared forest land and other remote areas. For example, almost 13,000km^2 of forested area and hillslopes were omitted from the first Koraput District Survey and Settlement (1938–1964), even though there were settlements and cultivation in these areas. In other tribal districts, large forest blocks and the cultivated areas within them were left out. From independence until the end of the 1960s, cultivators were encouraged to clear forest areas for cultivation. Much of such land cultivated in forests was never regularized. Thus, by the 1970s there were large cultivated areas of forest land.

Recognizing this problem, the GoO brought in a resolution in 1972 to identify and settle forest land already being cultivated in favour of the tribal and other poor groups working on them. However, little was done and the 1980 Forest Conservation Act put a stop to all such settlement of forest land by cultivators. Since 1980, only 27ha of forest land already under cultivation has been legally settled with the cultivators with clearance from central government.

In 1995, the Orissa Forest Department estimated that 74,380ha of forest areas have been 'encroached upon' for cultivation (GoO, 1995), although this is most likely a major underestimate as the forest department provides data only for forests under its jurisdiction, whereas most of the cultivation of forest land is on land categorized as revenue forests. This estimate also excludes the massive 3,132,700ha of forest land estimated by the same report as affected by shifting cultivation to varying degrees.

This summary briefly indicates the scope of tenurial and access issues related to forest lands in Orissa. Vast areas of land categorized as forests are under occupation by cultivators, mostly tribal groups. There are no proper estimates as official data is unreliable and underestimates the true extent of the problem. Hundreds of thousands of families all over the state are cultivating forest land without any rights. This situation has become very serious since the 1980 Forest Conservation Act, which has made the provision of legal rights on forest land extremely difficult.

The situation has been aggravated by the inaction of GoO, which is unwilling to recognize the magnitude of the problem, even now. For instance, according to the state government, only 4729ha of forest land are eligible for regularization according to guidelines issued by the Government of India (GoI) in 1990 (only forest land cultivated since before the 1980s is eligible for regularization). However, the GoO's own data show that in just one forest division of Nowrangpur there are 23,039ha of pre-1980 encroachments on forest land (Orissa Forest Department, undated).

The issue came to a head in 2002 when the Ministry of Environment and Forests (MoEF) issued a letter asking that all those cultivating government land be evicted within six months. This would have meant making millions of forest dwellers across the country homeless and would have affected hundreds of thousands of families in Orissa. It provoked massive outrage and grassroots mobilization across the whole of India, as a result of which the GoI drafted a proposed law to grant land rights to tribal populations cultivating forest land and submitted it to parliament in 2005. This proposal was sent to a Joint Parliamentary Committee (JPC), which has suggested even more additions, such as providing community rights in forests and the inclusion of non-tribal populations cultivating forest land. It has also suggested that shifting cultivators should be given land-use rights to customary shifting cultivation land. If these recommendations are incorporated within the proposed law, it will change the face of forest governance in India and could help to address the problems with the construction of 'legal forests', as discussed above.

Sample villages and their forests

Having reviewed the context in which different modes of PFM are proceeding in Orissa, we now focus on the field study which was undertaken in three districts of the state – Bolangir, Kandhamal and Sambalpur – selected purposively to focus on hinterland areas where the major proportion of forests are located and where JFM and CFM practices have been in operation for the last few years. Two blocks were selected randomly from each district, and two functional VSS and two self-initiated forest protection groups (SIFPGs) were randomly selected from each block. One village with neither JFM nor CFM – that is, no type of protection – in each district was also selected (see Table 8.2).

Primary data on household-level variables relating to JFM and SIFPG villages were collected through a sample survey of 22 per cent of all households comprising 155 JFM and 105 CFM households, and information was also collected from each of the 3 'no-protection' villages. The households were classified as very poor, poor, medium rich and rich on the basis of subjective wealth ranking. More than 80 per cent of the households in the study villages were classified in the lowest two wealth-ranking categories as either poor or very poor.

Most of the VSS villages contain three or more castes, including backward castes and scheduled tribes. The Kandhamal villages, as well as most of the SIFPG villages, are dominated by scheduled tribe categories. Farming is the dominant source of livelihood for more than 75 per cent of households, followed by wage employment in all villages. The majority of households are marginal farmers (landholdings varying between 1 to 2.5 acres, or 0.4ha to 1ha) and small farmers (landholdings of 2.5 to 5 acres, or 1ha to 2ha). Dependence on the forest by these groups for their livelihood is very high. Infrastructure facilities in the study villages are relatively poor, with poor water supply, road access, healthcare, and sanitation and school facilities. Firewood is the main source of energy for 99 per cent of households.

Households grow a variety of crops on their land and overall cropping intensity is low. Winter paddy is the dominant crop in all three regions, followed by pulses, coarse cereals and vegetables. In Kandhamal district, farmers grow turmeric on so-called *donger* land (land situated just below the forest fringe), using leaves from the forest to cover the crop's first growth. The crop output and prices realized are very low due to the lack of marketing facilities in all three districts.

Forest composition on the study sites

All three districts have mixed forest with different varieties of timber, firewood, bamboo and NTFPs including edible fruits, seeds, sal leaf for leaf cups and plates, kendu leaves, and myrobolans (the fruits of harida, bahada and amla). Forests are also used for grazing cattle.

Outcomes and impacts of PFM

Local forest management institutions

This section examines the village VSS and SIFPG institutions, particularly their initiation, processes and constitutional arrangements (see Table 8.2). The institutions were formed during the late 1980s and mid 1990s. Although some VSS were initiated in the mid 1990s, these were generally inactive until 1997; but all of the VSS we studied have been functioning since 1998 due to the gradually increasing involvement of the forest department. However, the SIFPGs had been active since their inception.

Formation

There are some differences in the formation of VSS and SIFPGs. SIFPGs were formed by local people or *Yubak Sangha* (youth groups) in response to increasing concern about the decline in the availability of fuelwood and other forest products. JFM is a government-sponsored programme with a code of conduct covering the formation of forest protection committee (FPCs) by forest officials. Some SIFPGs have converted to VSS status. Of the six VSS, four were initiated by the forest department and the other two (Khandam and Tudubali) were converted from self-initiated groups. It should be noted that two other

Table 8.2 *Sample villages, forest type and management provisions*

District		Name of village	Year of formation	Initiation type	Number of households	Ethnic group**	Forest (ha)	Forest types	Protection provision
Sambalpur	*Vana Samarakshyan Samiti* (VSS)	1 Khandam	1994–1995	Converted SIFPG/ village	90	OBC/SC	43.79	Sal, miscellaneous	Social fencing
		2 Kutasinga	1994–1995	Forest department/ village	86	ST/OBC	130.82	Miscellaneous	Social fencing
	SIFPG	3 Jharmunda	1994–1995	Self/ village	87	ST/SC/OBC	57.31	Sal, miscellaneous	Paid watcher
		4 Palokhaman*	1990–1991	Self/ hamlet	49	ST	95.39	Sal, miscellaneous	Paid watcher
	No protection	5 Dhunchali			39	ST		Sal, miscellaneous	–
Bolangir	VSS	6 Chandhan Juri	1997–1998	Forest department/ village	104	OBC/SC	93.43	Miscellaneous, bamboo	*Thengapali* (paid forest guard)
		7 Mursing	1997–1998	Forest department/ village	228	Miscell-aneous	427.81	Miscellaneous, bamboo	Paid watcher
	SIFPG	8 Junani Bahal	1987–1988	Self/village	66	ST/SC	20.54	Miscellaneous, bamboo	*Thengapali*
		9 Mandapala	1994–1995	Self/hamlet	94	OBC/ ST/SC	35.43	Sal, miscellaneous	*Thengapali*
	No protection	10 Talchakel			118	OBC		Miscellaneous	–
Kandamal	VSS	11 Tudubali	1994–1995	Converted SIFPG/village	80	ST/SC	81.47	Sal, miscellaneous	*Thengapali*
		12 Pokari	1997–1998	Forest department/ village	122	ST/SC/OBC	53.91	Sal, miscellaneous, bamboo	Social fencing
	SIFPG	13 Mandaguda	1987–1988	Self/hamlet	93	ST/SC	12.82	Sal	Social fencing
		14 Banegaon*	1988–1989	Self/village	77	ST	29.80	Sal, miscellaneous, bamboo	Social fencing
	No protection	15 Bandabaju			141	ST/SC			–

Notes: * SIFPG villages Palokhaman in Sambalpur and Banegaon in Kandhamal district have been converted into VSS during 2004 to 2005.
** BLGR = Bolangir;
KM = Kandhamal;
OBC = other backward castes (i.e. amongst the poorest caste not 'scheduled');
SBP = Sambalpur;
SC = scheduled castes;
ST = scheduled tribes.

Source: Kailas Sarap (from field survey, 2004–2005)

SIFPGs (Palokhaman and Banegaon) were converted to VSS during 2004 and 2005. VSS or SIFPG may be formed at whole-village level if the village is small, or at hamlet level if the village is large. Out of six VSS and six SIFPGs, only one SIFPG in each district was formed at hamlet level, and the others were all formed at village level (see Table 8.2).

The constitutional arrangements for VSS and SIFPGs are as follows: prior to the formation of executive committees, a general body must be created. The general body consists of one male and one female from each member household.

Those SIFPG formed by the *Yubak Sangh* had only the youths of their villages in the general body. Later on, the other villagers were assimilated within general body and forest protection affairs. Such cases were found in two VSS and four SIFPGs. Of the VSS/SIFPGs formed at hamlet level (three out of six SIFPGs), only the households of these hamlets are members of the general body. The VSS/SIFPGs formed at village level include all households, although there are important exceptions.

One case of social exclusion emerged in Chandanjuri VSS in Balangir district, where all 30 scheduled caste households have been excluded from the general body. In Mursing VSS, about 30 households do not participate in VSS affairs due to village factionalism; yet they are still collecting forest products despite not being considered members of the VSS.

The participation of members from poor backgrounds is low in comparison with richer members. It has been found that the latter not only attend meetings frequently, but also influence decisions taken in meetings, whether of the executive committee or the general body. As such, the majority of participants are unable to influence FPC decisions. The attendance of female members is lower than that of males, and although some women members do attend meetings, they generally do so sporadically rather than regularly. This is true of both VSS and SIFPGs. The participation of women in meetings is rare except in one VSS, where a women's self-help group has been formed, raising the participation of women members.

Executive committees and their composition

Out of the six SIFPGs we studied, four had been initiated through *Yubak Sangh* and the remaining two through village committees. The difference is in age – SIFPGs formed by village committees consist of elderly people, while in those initiated by the *Yubak Sangh*, youths fill the important posts, although they are supported by the village elders. There is no universal or formal procedure in the selection of SIFPGs' executive committee members. Each SIFPG follows its own procedures, although in many ways the procedures of all are similar. There is no imposition or outside interference in the selection procedure of executive committee members.

In the VSS, the executive committee members are supposed to be elected at a general body meeting in the presence of forest officials. The reality was often found to be very different in the study areas. In four of the six study VSS, the forester played a vital role in selecting executive committee members, choosing his preferred candidates. But in two VSS, the people proposed a list of members as their executive committee representatives, which was formed at the initiative of forest department officials.

The VSS elect between 9 and 14 executive committee members, including at least 3 women, while SIFPGs have between 5 and 12 members. Whether VSS or SIFPG, the selection of members is such that the rich dominate the key office positions. There are rich class representatives in five of the six VSS and in all six SIFPGs.

Different caste representation has been found in the executive committees of VSS/SIFPG; but the numerically dominant caste in each village is generally represented more in its executive committee. The key executive members enjoy multiple posts in various institutions at once. This is true of all VSS and SIFPGs. These members wield a great deal of power in influencing decisions relating to resource management and other important factors.

There is a provision for women representatives on the executive committee for VSS villages, and consequently there are women representatives in all VSS and four of the six SIFPGs. Twenty-three per cent of all VSS executive committee members are women and just over 16 per cent are in the SIFPGs. This could be one area where self-initiated processes score lower than government-initiated ones. There is some variation in different study districts; there were relatively more women on the executive committee in Kandhamal district than in Balangir and Sambalpur.

Meetings (executive committee/general body)

Frequency. Local people may have their own ideas about how frequently they need to meet. However, according to VSS rules, the executive committee should call a meeting at least once a month, whereas a general body meeting should be organized once every three months or so. All six VSS studied are meeting at least once a year, and half organize more frequent meetings, often following directions from forest department staff (see Table 8.3). For instance, in Chandhan Juri, 15 executive committee and five general body meetings were held during 2002 to 2003, and Khandam held 6 executive committee and 13 general body meetings in the same year. VSS general body meetings are supposed to be held in the presence of a forester; but several took place without one.

In SIFPGs, the frequency of meetings is comparatively low because they are organized only when required. Executive committee meetings were less and general body meetings more frequent, reflecting the fact that FPCs exemplify a more participatory decision-making system, whereas VSS operate by representative-based decision-making. For instance, Junai Bahal held five executive committee and ten general body meetings during 2002 to 2003. In Mandaguda, four executive committee and seven general body meetings were organized. In SIFPGs, there is hardly any difference between executive committee and general body meetings since non-executive committee members also participate in these meetings.

Attendance. The quorum requirement of a meeting is at least 50 per cent in VSS villages, although there is no such rule in SIFPGs. In reality, both VSS and SIFPGs see lower attendance at executive committee/general body meetings. The average attendance at VSS executive committee meetings was 6.5 and at general body meetings 29.5, and 5.0 and 24.5, respectively, at SIFPG meetings. It is clear that although both VSS and SIFPG general bodies have a large number of members, their attendance at meetings is low. The presence of women members is low or negligible.

Record-keeping. In VSS, the forester is in charge of keeping records, although some VSS maintain and keep their own. Three VSS are doing this, even if irregularly and unsystematically. All SIFPG records are maintained and kept by the office bearers. There are no outside guidelines, so in all SIFPGs the record-keeping is unsystematic and irregular. But records are freely available to the villagers.

Methods of forest protection

Two main methods of forest protection are employed by VSS and SIFPGs: *thengapali* (paid watchers) and 'social fencing'. Villages with a high chance of outside interference tend to use patrols, and some pay one or two individual *thengapalis* or forest guards (see Table 8.3). As protection becomes established and the risk of outside interference declines, villages tend to move on to 'social fencing' – that is, consensual observation of norms with each person acting as unofficial watcher. Women do not directly participate in forest protection activities in either the VSS or SIFPGs study villages; but they do participate indirectly. When they are in the forest and learn of illegal felling by outsiders they immediately inform male members. One such case occurred in Khandam village (in Sambalpur district) in 2003 when some outside mafias were cutting trees. Women members immediately informed the villagers and the thieves were caught and handed over to forest officials. However, it was found that in

Table 8.3 *Some features of executive committee and general body meetings in* Vana Samarakshyan Samiti *(VSS) and self-initiated forest protection group (SIFPG) villages*

Meetings	Protection method (%)			Frequency of meeting per year		Average number of individuals attending meeting			VSS/SIFPG record-keeping (%)	
	Thengepali (paid forest guard)	Paid watcher	Social fencing	Executive committee	General body	Executive committee	General body	Good	Fair	Bad
VSS	33.3	16.7	50	7.7	9.0	6.5	31.1	16.7	83.3	
SIFPG	50	16.7	33.3	3.0	10.3	4.0	24.5	33.3	16.7	50

Source: Kailas Sarap (from field survey, 2004–2005)

other areas of the study districts there are certain SIFPG FPCs in which women members participate in protection directly through *thengapali*.

Representation, decision-making, and relations between the village and forest department and other institutions

In all study villages, the elites dominate decision-making. Some of them also collude with lower-level forest officials who hold key positions in the villages. These people have a good relationship with the forest department compared to ordinary members, who are treated harshly by officials for committing even minor mistakes.

The forest department plays the dominant role in the total management process with the help of FPC members. Since the start of JFM, the forest department has been in constant touch with VSS villagers and was cooperating well with some VSS members. The majority of VSS villagers hardly participate in any decision-making processes. However, in the case of SIFPGs, the situation is different. Before JFM, some forest staff had cooperative relations with certain villagers, helping them in times of need. After the JFM rules were initiated, forest staff started converting self-initiated groups to VSS. This was slow to begin with but intensified later in order to meet VSS targets, as a result of which the officials no longer cooperated with self-initiated groups.

Participation of women

The participation of women in VSS and SIFPG executive committee and general body meetings and in decision-making processes is low due to their low level of literacy and awareness. Patriarchal traditions dictate that young women do not appear or speak in front of elder male members. Those who do attend meetings thus tend to be passive, silent listeners. Even in some of the VSS villages (such as Chandhanjuri), male executive committee members get the signatures of female members in the register maintained for meeting purposes despite their limited active involvement. The case of Khandam village is exceptional: women frequently participate in VSS meetings. This is due to local NGO and self-help group efforts to empower women. Women's participation is also high in Banegaon in Kandhamal district, a SIFPG until 2003/2004, and now converted to a VSS.

Women are not aware of the financial expenditure of their VSS or SIFPG committees. Although a certain amount of money has been spent on each VSS, women remain completely ignorant of the details.

Post-formation VSS support

In order to make PFM effective, the forest department has taken a number of steps to provide VSS villages with post-formation support in the form of funding from different sources for various activities. The villagers have also benefited from some waged work in this process. Post-formation support includes boundary demarcation, cleaning, coppicing, silvipastural practices, plantation, dam construction and soil conservation (see Table 8.4).

Boundary demarcation in all VSS and SIFPG villages has been initiated, and has been completed in four VSS and five SIFPG villages. The forest department demarcates forest boundaries with the help of revenue officials. Villagers help by contributing physical labour, stretching measuring chains between two forests and making the lines prominent. Occasionally, members engaged in such work are paid by the forest department. The forest department has taken more interest in this work in VSS villages than in the SIFPG villages since it has to report progress in the VSS to higher officials. However, in SIFPG villages, members have contributed voluntary labour with some guidance from forest officials, which they had been requesting for some time. In any case, the forest department's involvement is necessary in boundary demarcation, whether for VSS or SIFPG villages, because the ownership of forests rests with the state.

The forest department has taken measures to raise the capacity of VSS members through building awareness, organizing health camps, and account and fund management. Training facilities on a variety of forest issues have been provided by the divisional forest office and other related departments and by NGOs. Even in SIFPG villages, the office bearers have received such external exposure. Evidence from the field reveals that, generally, executive committee office bearers in VSS villages have received training, although mainly in the case of male members, rather than female ones. The chances of training for ordinary members are rare in both types of villages. The training has raised awareness of forest protection and management. However, it should be noted that it has cemented the relationship between forest officials and elites more than it has benefited the management of forests.

Clearly, VSS villages have received various aspects of post-formation support from the forest department; but SIFPG villages have been discriminated against. However, the effect of the forest department in providing post-formation support is marginal, and its coverage is confined to a few VSS. Similarly, capacity-building is confined to a few executive members.

Interface between VSS/SIFPG and the forest department

Under JFM rule, the forest department and the VSS are partners in managing the forest. Although a cooperative and complementary role is needed, the forest department assumes the role of senior partner in forest management, especially in decision-making and mobilization of resources, and in executing the programmes.

Relations with local government bodies

The *panchayat*, or local government body, plays a key role in the JFM system. Under the 1996 Orissa joint forest management, the *sarpanch* (village council head) is the president of the VSS committee and other village ward members represent the *panchayat*. The *gram panchayat* informs the forest office concerned about management work implemented from time to time. Under the 2002 NTFP Amendment Bill, the power to issue licences to collectors within their boundaries for extraction of 68 NTFP items has been given to the *panchayat*. Although there are representatives from local *panchayats* in all six study VSS, there is no formal coordination process between *panchyats* and VSS, as yet. In the SIFPGs, there is also no formal representation of *panchayat* office bearers, even though they form part of the groups in such villages.

Self-initiated group federations

Self-initiated groups have formed federations at block, district and state level at the initiative of NGOs and communities. There are many federations (such as the Orissa Jungle Manch) working in different parts of the state, including in our study districts. These federations coordinate themselves at the district and state levels in order to project their common interests to the state. They play the role of pressure groups for the SIFPGs by providing alternative policy measures for forest protection, marketing structures, proper prices for NTFP products, etc, and try to mobilize people into conserving forest resources by imparting knowledge, helping to resolve conflict and emphasizing forest-based livelihoods.

Participatory micro-planning for forest management

The implementation of JFM in Orissa requires the preparation of micro-plans in order that villagers' involvement in managing the forest area in question accords with norms approved by the forest department divisional forest officer (DFO). A micro-plan is formally prepared to cover the next five years. Its preparation is supposed to be the result of collaboration between the forest officials and the villagers according to Orissa JFM guidelines. But from the study villages it is clear that villagers' involvement in this is negligible. In the preparation of micro-plans, the people's participation is seen as similar to a wage labour role; only a few people participate in the process and they perform tasks such as assisting forest department staff in demarcating forest areas and providing relevant information. Forest staff seek to maintain their traditional 'scientific forestry' management procedures regardless of the villagers' needs. Local people are not made aware of the planning process or of their potential role within it.

Of the six sample VSS villages, the forest department had completed micro-plans in only three. In the other villages, the plan was 'in process' – a draft had been sent to the DFO concerned for suggestions and approval, and remained with him. This reflects the lack of completion of micro-plan preparation across the state, in which nearly 60 per cent of VSS micro-plans had not been completed by 2003.

In the SIFPGs, no micro-planning processes are followed. They follow traditional methods of management based on indigenous knowledge, which includes encouraging the growth of plants through preventing grazing at the initial growth stage, pruning and cutting branches, restricted cutting of young plants or even leaves, and regeneration of plants in open spaces. FPCs in SIFPG villages plan the use of forest resources through the decisions of collective members.

Impact on households

This section analyses the direct and indirect benefits derived by households in the study area.

Changes in household assets, entitlements to collective assets and opportunities, and distribution of landownership

The pattern of landholding (both ownership and non-privately owned cultivated land) among the VSS, SIFPG and no-protection villages revealed that the majority of very poor and poor households own and cultivate very little land (see Table 8.5). The average size of non-owned cultivated land is slightly higher than owned land because most of these households have customarily cultivated small areas of land in the forest to which they do not have ownership deeds. About 40 per cent of total land among the VSS households in Kandhmal is cultivated without ownership deeds (see Table 8.6), approximately 6 per cent in Balangir and 3 per cent in Sambalpur. In SIFPG villages the overall percentages of households with such land were 19 per cent in Kandhamal district, 12 per cent in Balangir and 7 per cent in Sambalpur; and in no-protection villages, 41 per cent in Kandhamal district, about 2 per cent

Table 8.4 *Post-formation management work in VSS and SIFPG villages in the study districts*

Districts	Forest protection status	Village	Boundary Demarcation Y/N	'Cleaning' Y/N	No. patches	Coppicing area (ha)	Check dam construction Y/N	Fund allocated (Rs.)	Silvi- Y/N/ Proposed	Gap plantation pasture Y/N	*areas	Soil conservation Y/N	Areas or fund allocated (Rs.)	Funding
Sambalpur	VSS	Khandam	Complete	Y	All	43.8	N	—	P	P	—	P	—	FD
		Kutasingha	Complete	N	—	—	Y	25,000	N	N	—	P	—	FD
	SICFMG	Jharmuda	Complete	N	—	—	N	—	N	N	—	N	—	
		Palokhaman	Complete	Y	_ of PF	40.0	Y	—	N	N	—	Y	—	Watershed Dept
	No protection	Dhunchali	–											
Bolangir	VSS	Chandanjuri	Complete	Y	2	30.0	Y	18,000	N	Y	2	Y	30,000	FD; Soil Conservation Dept.
		Mursing	Not complete	Y	2	30.0	N	—	P	P	2	N	—	FD; afforestation under Social Forestry Programme
	SICFMG	Junani bahal	Complete	Y	All	20.5	Y	—	N	N	—	N	—	–
		M.Kutenpali	Complete	N	–	—	N	—	N	N	—	N	—	–
	No protection	Talchakel	–											
Kandhamal	VSS	Tudubali	Not complete	Y	_ of PF	40.0	Y	80,000	N	Y	1	N	—	NF NIPDIT (NGO)
		Pokari	Complete	Y	All	53.9	N	—	N	N	—	Y	4 areas	FD; Watershed Department
	SICFMG	Mandaguda	Complete	N	–	—	N	—	N	Y	—	Y	—	Social Forestry Programme
		Banegaon	Complete	N	–	—	N	—	N	N	—	Y	10,000	FD

Note: Y = Yes, N = No; P = Proposed to do; PF = Protected Forest; NIPDIT = National Institute for People's Development, Investigation and Training (Phulbani (Orissa) based NGO)

Source: Kailas Sarap (from field survey, 2004–2005)

in Balangir and 3 per cent in Sambalpur. It is to be noted that in the case of very poor and poor households, the percentage of 'encroached' land was much higher than for other groups of households in all districts. Furthermore, a part of the land is fallow wasteland. As a result, the crops harvested from such tiny plots provide only a meagre subsistence. Because of this the villagers depend on the forest for a significant part of their livelihood.

It is to be noted that some VSS members and about one-third of SIFPG members have had to vacate land under *podu* cultivation, which they had been practising for a long time. This happened in Tudubali (VSS), Mandaguda (SIFPG) and Banegaon (SIFPG) villages in Kandhamal district after the formation of FPCs in these villages. The FPC members persuaded them to leave the areas in the greater interests of forest protection. About 31 SIFPG and 7 VSS members, each with about one-fifth of an acre, were affected by this. Initially, they put up strong resistance; but in the end they had to vacate the land to be included in the protected areas. No compensation was paid. However, some of the affected members shifted to areas further from the reserved forest in order to continue *podu* cultivation.

No difference was found in the average size of households' landholdings before and after PFM in the study villages since there had been few land transactions. There has been no further encroachment of land by households in recent years because PFM has brought a collective resistance on the part of the group to individual encroachments in PFM areas.

The average number of bullocks owned by different categories of households in each of the three forest management situations does not differ significantly (see Table 8.7), although the medium-rich and rich households have more goats and, hence, can benefit more from forest grazing rights. The average value of livestock rises with an increase in household status of all three forest management types. There has been little change in the number of bullocks

Table 8.5 *Landholding patterns in VSS, SIFPG and no-protection villages in the study districts*

Wealth status	Total number of households	Owned land (acres)	Land leased in (acres)	Land Encroached		Fallow wasteland (acres)	Operated land (acres)
				leased out (acres)	land (acres)		
VSS							
Very poor	60	0.80	0.07	0.08	0.41	0.05	1.31
Poor	52	2.41	0.04	0.02	0.24	0.32	2.39
Medium rich	32	5.02	0.23	0.09	0.41	0.31	5.44
Rich	11	8.62	0.09	0.07	0.54	1.02	8.30
Total	*155*	*2.75*	*0.10*	*0.07*	*0.58*	*0.25*	*3.25*
SIFPG							
Very poor	53	1.48	0.07	0.03	0.26	0.19	1.59
Poor	29	1.84	0.37	0	0.27	0.18	2.3
Medium rich	16	3.75	0.1	0.21	0.8	0.1	1.34
Rich	07	12.02	0.28	1.0	0.08	2.36	9.02
Total	*105*	*2.62*	*0.26*	*0.13*	*0.34*	*0.38*	*2.71*
No-protection							
Very poor	28	0.71	0.02	0.06	0.25	0.04	0.88
Poor	39	2.51	0.04	0.10	0.43	0.06	2.82
Medium rich	09	5.99	0	0.39	0.17	0.39	5.38
Rich	–	–	–	–	–	–	–
Total	*76*	*2.26*	*0.02*	*0.12*	*0.33*	*0.09*	*2.40*

Source: Kailas Sarap (from field survey, 2004–2005)

Table 8.6 *Percentage of land without ownership deeds (encroached) to total operated area by different groups of households in different study districts*

	Wealth status	Kandhamal	Bolangir	Sambalpur
VSS	Very poor	64.08 (1.03)	20.45 (1.76)	0
	Poor	37.95 (1.95)	4.56 (2.46)	0
	Medium rich	32.14 (3.64)	1.8 (6.10)	0 (1.59)
	Rich	25.25 (5.94)	0 (7.08)	0
	Total	*39.71 (2.01)*	*5.88 (3.57)*	*2.71 (3.70)*
SIFPG	Very poor	28.57 (1.4)	10.0 (2.0)	16.67 (1.2)
	Poor	25.0 (1.6)	6.89 (2.9)	3.45 (2.90)
	Medium rich	15.15 (3.3)	25.45 (5.5)	12.77 (4.7)
	Rich	4.08 (4.08)	0 (11.0)	0
	Total	*19.05 (39.71)*	*12.12 (3.30)*	*7.14 (2.8)*
No-protection	Very poor	69.62 (0.79)	6.33 (0.79)	18.52 (1.08)
	Poor	38.79 (2.13)	2.17 (2.76)	0.58 (5.15)
	Medium rich	21.74 (0)	0	0 (6.25)
	Rich	0	0	0
	Total	*40.88 (1.81)*	*1.97 (2.54)*	*3.39 (3.25)*

Note: Figures in brackets represent land operated per household (acres).

Source: Kailas Sarap (from field survey, 2004–2005)

and other animals in VSS villages since PFM was introduced because there has been little surplus income available to the majority of households with which to buy additional livestock.

Direct assets and entitlement changes
During the initial years of protection in VSS villages, some restrictions were applied to the collection of certain types of NTFPs and the grazing of livestock to allow forest regeneration. Poor households were affected by this, although the extent has varied in different study VSS. This practice had already been applied in the SIFPG villages as well. It was found that restrictions on grazing were strictly observed in all SIFPG villages from the initial stage of protection since the prevention of forest degradation was seen by most members as a key purpose for initiating forest protection.

The livelihood of the poorer group, whose main occupation is making baskets and other items from bamboo, was also initially affected. For instance, in Chandajuri VSS, the FPC decided not to allow the group to take or even buy bamboo because mature bamboo fetches a higher price and the FPC was seeking to raise its revenue in this way. However, after a brief initial period of growth, the restriction was not enforced. Furthermore, it was found that in most VSS villages, the collection of different NTFPs after PFM intervention has increased by 20 to 25 per cent for all groups in comparison with the pre-PFM period. Households obtain a number of products, including firewood, fodder and food items from the protected forest, both for their own use (see Table 8.8, column 8) and for sale. The increase in the use of NTFPs is due to the growth of forest, which in turn has led to increased availability of forest produce.

Income from forests
The main direct benefit to local households comes from the sale and consumption of NTFPs and wood products, such as timber and firewood. The monetary value of these items has been estimated in order to calculate household income. The intangible benefits are largely

Table 8.7 *Average number and composition of livestock per household in the study districts*

	Wealth status	Total number of households	Bullock/ buffalo	Cow	Goat/ sheep	Poultry
VSS	Very poor	60	1.6	0.2	1.4	1.3
	Poor	52	2	0.8	1.1	1.8
	Medium rich	32	2.4	1.1	2	1.6
	Rich	11	2.8	2.3	2.5	1.3
	Total	*155*	*2*	*0.8*	*1.5*	*1.5*
SIFPG	Very poor	53	1.5	0.7	1.4	2.5
	Poor	29	1.9	0.6	1.7	1.4
	Medium rich	16	2.5	1	2.5	1.5
	Rich	07	2.3	1.1	1.7	1.6
	Total	*105*	*1.8*	*0.8*	*1.7*	*1.1*
No protection	Very poor	28	1.3	0.7	2.2	4.9
	Poor	39	2	0.9	2.5	3.4
	Medium rich	09	3.2	1.3	4.7	5.1
	Rich	–	0	0	0 0	
	Total	*76*	*1.9*	*0.9*	*2.6*	*4.3*

Source: Kailas Sarap (from field survey, 2004–2005)

environmental and difficult to measure directly. However, we have obtained the households' responses concerning their perceived indirect benefits due to the growth of the forest.

Income derived from forest, agriculture and other sources for different wealth ranks in all three management situations is shown in Table 8.8. The relatively high income in VSS compared to SIFPG villages is due to the greater forest area per household and the better prices realized for forest products.

The average income earned from forest-related activities by VSS households overall was 4442 rupees. Income from the crop sector was 7365 rupees and the overall average total is 17,950 rupees. The situation for SIFPGs was lower overall: 3154 rupees from forest; 6117 rupees from cropping; and, in total, 16,236 rupees. In the villages with no forest protection, mean income from the forests was 3349 rupees, 6878 rupees from cropping, and, in total, 14,107 rupees (see Table 8.8 and Figure 8.1).

In the VSS villages, the direct annual contribution of forest to total income of overall households was 24.7 per cent, and 19.4 and 23.9 per cent in the SIFPG and no-protection villages, respectively. This contribution goes up to 30.8 to 42.7 per cent for poor and very poor groups in VSS villages. It varied form 18 to 27 per cent in SIFPGs, and 27.5 to 34.6 per cent in no-protection villages. Clearly, the contribution of forest income for poor and very poor groups in all three types of forest management is very high, and, as such, their dependence on forest is considerable. Dependence on the forest is relatively higher in the Kandhamal villages, where more than 80 per cent of households are tribal, in comparison with Sambalpur and Balangir.

Contribution of different forest products to total forest income

Mahul and toll, kendu leaf, firewood, leaf plates and edible produce contributed about 89.2 per cent of total forest income in the VSS, 92.4 per cent in the SIFPGs and 92.6 per

Table 8.8 *Composition of income from different sources in VSS, SIFPGs and no-protection villages in study regions, Orissa*

Type of group	Wealth ranking	Number of house-holds	Income from forest collection in rupees (% of total income bracketed)	Crop income in rupees (% of total income bracketed)	All other income in rupees* (% of total income bracketed)	Total income In rupees (% of total income bracketed)	% of total forest income utilized for household consumption	% of total forest income contributed by summer season
VSS	Very poor	60	4249 (42.7%)	2536 (25.5%)	3165 (31.8%)	9950	32.4	76.3
	Poor	52	4210 (30.8%)	6529 (47.8%)	2920 (21.4%)	13,659	32.6	78.7
	Medium rich	32	502 (20.0%)	13,153 (52.5%)	11,396 (27.5%)	25,051	32.8	79.7
	Rich	11	4970 (8.0%)	21,547 (34.9%)	35,266 (57.1%)	61,783	31.1	78.6
	Total	**155**	**4442 (24.7%)**	**7365 (41.0%)**	**6143 (34.3%)**	**17,950**	**32.5**	**78.1**
SIFPG	Very poor	53	2869 (27.0%)	3759 (35.4%)	3977 (37.6%)	10,605	47.1	71.7
	Poor	29	3384 (17.8%)	6196 (32.6%)	9412 (49.6%)	18,992	39.7	65.6
	Medium rich	16	3497 (20.6%)	9117 (53.8%)	4321 (25.6%)	16,935	45.8	73.5
	Rich	07	3584 (7.2%)	21,697 (43.6%)	24,531 (49.2%)	49,812	45.7	78.6
	Total	**105**	**3154 (19.4%)**	**6117 (37.7%)**	**6965 (42.9%)**	**16,236**	**44.6**	**70.7**
No-protection villages	Very poor	28	2807 (34.6%)	1997 (24.6%)	3311 (40.8%)	8115	41.1	57.6
	Poor	39	3707 (27.5%)	6254 (46.5%)	3497 (26.0%)	13,458	39.9	55.5
	Medium rich	9	3482 (10.0%)	24,771 (71.2%)	6551 (18.8%)	34,804	38.5	62.3
	Total	**76**	**3349 (23.9%)**	**6878 (49.0%)**	**3790 (27.1%)**	**14,017**	**40.1**	**57.0**

Note: * Other income includes wages and income from business and sale of livestock.

Source: Kailas Sarap (from field survey, 2004–2005)

cent in the no-protection villages (see Figure 8.2). It is interesting to note that over half of forest-related income in no-protection study villages comes from firewood, whereas it was around one-third in the self-initiated study villages and under one-quarter in VSS villages. The protection of forest in VSS and SIFPG villages has led to the controlled sale of fuel-wood by FPC members. By contrast, there is widespread use of forest for this purpose in the no-protection villages, which is fast leading to forest degradation. Our study team observed this situation. Clearly, any type of protection is better than none for forest conservation. The prices realized by these items are generally depressed by very

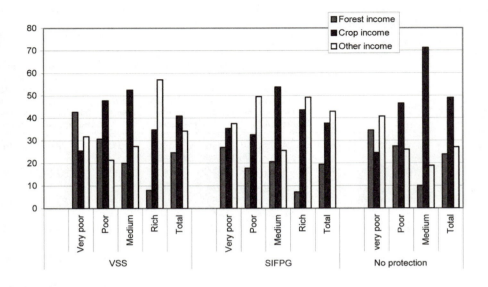

Figure 8.1 Composition of income from different sources (percentage) in different forest management situations according to wealth group

Source: Kailas Sarap (from field survey, 2004–2005)

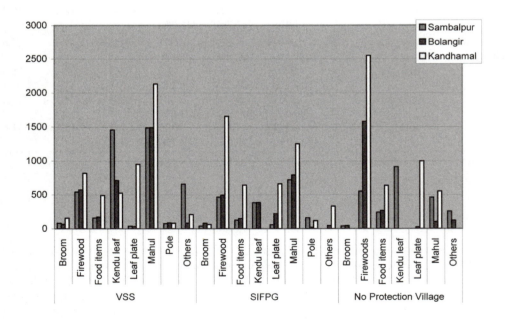

Figure 8.2 Contribution of different forest products to total forest income in three different districts of Orissa

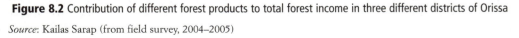

Source: Kailas Sarap (from field survey, 2004–2005)

unfavourable market relations; if fairer market prices could be achieved, income from forests would be far higher. It should be noted that no VSS in the study areas has harvested timber, and, as such, they have not generated any income from this source for their members. The contribution of PFM to wage income from forest-related work in VSS villages was negligible. The forest department generates about ten days of employment per household per year pruning, cutting and planting trees. In SIFPG villages there was no waged work; but members voluntarily contribute labour for the development of forest in a type of collective action.

Contribution of different seasons to total income from forest

VSS villages achieve the greatest and no-protection villages the smallest proportion of their total forest income in the summer season; this is reversed in the rainy season (see Table 8.8).

Of the total income generated from forest, about 32.5 per cent was utilized for direct consumption in VSS villages and 44.6 and 40 per cent in SIFPG and no-protection villages, respectively. By contrast, a major proportion of forest produce is utilized by households in all management situations in order to generate cash income, although all systems use the forest more for consumption purposes than to generate cash income in the rainy season, when it acts as an insurance mechanism against hunger for poor forest-dependent communities in the absence of alternative sources of income. On the whole, forests provide the poorest groups with fundamental insurance against starvation and hunger. They are also critically essential in the case of illness; for instance, in areas where malaria is prevalent, as in our study villages, sales of NTFPs are often the main source of income to pay for treatment.

The overall income of poor and very poor households in all three management situations is below the official poverty line. The annual forest-derived income per household and per worker is also low, partly due to the small per capita area and low productivity of forest, and the depressed prices for forest products realized by the households. The income derived per household is relatively higher in VSS villages in comparison with that of SIFPG and no-protection villages. Clearly, the positive impact of PFM on the livelihoods of forest users in the study districts is marginal.

Well-being and vulnerability

The access of VSS households to NTFPs from the forest has increased with JFM. Male and female members can freely go into protected forests to collect NTFPs. As local people's awareness of forest protection has increased, 'social fencing' practices have spread, and the time required to patrol the forest has decreased in VSS as well as SIFPG villages.

A majority of VSS and SIFPG members believe that PFM has affected their economic condition positively even though the impact is small, particularly in poor and very poor households who spend more than 40 per cent of their working time collecting and processing forest products. The mean income in VSS villages was 32 rupees per day per household, and 23 rupees and 21 rupees in SIFPG and no-protection villages, respectively (see Table 8.9). It was relatively lower in the case of poor and very poor compared with richer households, especially in the VSS villages. As a result, their livelihood conditions remain very poor. In generating even this low income, some of them faced a variety of problems, including the unhelpful attitudes of lower-ranking forest officials, marketing agents and others.

Table 8.9 *Income from forest-related activities per day per household*

	Very poor	Poor	Medium rich	Rich	Mean
VSS	29	29	34	41	32
SIFPG	23	25	24	25	23
No-protection villages	20	22	18	–	21

Source: Kailas Sarap (from field survey, 2004–2005)

Interlinked loans

One reason why the sale of forest products realizes such low prices is the prevalence of inter-linked loans. There has been no apparent improvement in the indebtedness of poor and very poor VSS or SIFPG households, or in their health and educational levels, and the interrelationship of these factors plays an important role in perpetuating the poverty of these poorer households. Prevalent killer diseases such as malaria require funds for immediate treatment, making cash-poor households vulnerable to money lenders who are, typically, also buyers of forest produce. Poor households therefore often have to borrow against future sales at rates depressed by the money lender, as well as exorbitant interest rates of 60 per cent or more per year.

Of the total VSS and SIFPG member households studied, 47 per cent of VSS and 40 per cent of SIFPG members have outstanding debts. Of these, 58 per cent of VSS, 64 per cent of SIFPG and 61 per cent of no-protection borrowers have loans interlinked with the sale of forest products, meaning that a substantial part of income from the sale of NTFPs goes to moneylenders. Although product availability has increased, 95 per cent of VSS and 86 per cent of SIFPG households said that there has been no change in their level of debt.

The FPCs are unable to generate funds to help members in need and there are no grain banks in any of the study villages. Furthermore, the sum of common assets accumulated is negligible due to lack of contributions from the forest department in VSS villages (except one). Even though, in the SIFPG villages of Junani Bahal and Mursing, the FPCs have sold bamboo to generate funds, these are not used for the benefit of members but for cultural activities, including the construction of temples in the villages.

Out-migration by local households is common practice across the study area. Individual members, as well as families, migrate to nearby or distant places. This distress migration has social costs in terms of leaving families behind or leaving the village for sudden long periods, with serious implications for education and the well-being of migrants' children. Migration has not decreased due to PFM activities in VSS and SIFPGs, although PFM has raised the expectations of members that it might provide alternative income sources in the lean season.

NTFP marketing

The marketing of forest produce, including medicinal plants, is an important problem in all three management situations. The Orissa Forest Development Corporation (OFDC) and the Tribal Development Co-operative Corporation (TDCC) purchase minor forest products and agricultural surplus from tribal gatherers. The ostensible reason that they engage in buying is to reduce the risk of exploitation of tribal groups by traders.

The OFDC and TDCC procure forest produce through procurement and collection centres, including multipurpose co-operative societies spread across tribal and other areas of the state. But they are unable to fulfil their objectives due to a lack of institutional and financial capacity, a lack of official enthusiasm for procuring the targeted amount and a shortage

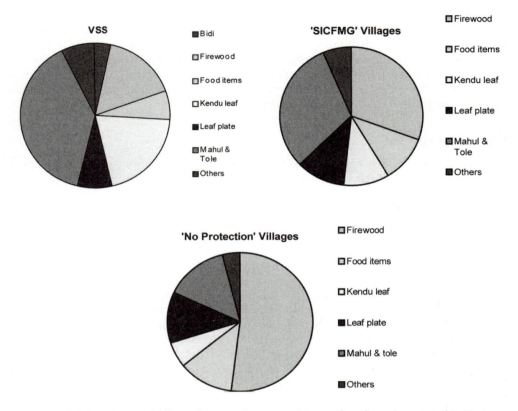

Figure 8.3 Contribution of different forest products to total income from forest (percentage) in *Vana Samarakshyan Samiti* (VSS), self-initiated forest protection groups (SIFPGs) and no-protection villages in Orissa

Source: Kailas Sarap (from field survey, 2004–2005)

of funds – therefore, they have been incurring financial losses. It has been found in the study areas that although they are entitled to buy many non-timber forest products, they collect only a few main types of NTFPs (see also Mallik, 2000). As a result, the primary collectors have to sell much of their harvest to private traders. Clearly, there are many deficiencies in these organizations' operations, including failure to collect quantities of even available NTFP items, a lack of sufficient support to primary collectors with regard to scientific procurement and irregular payment. The reasons for these failures are clandestine collection, poor marketing networks, corrupt practices by officials and exploitation by private traders.

The involvement of NGOs in the marketing of products is also negligible. Only in one VSS village, Chandanjuri, has a local NGO purchased medicinal plants (harida, amla and bahada) in limited quantities. The *gram panchayats* have now been entrusted with the procurement of forest produce. However, they have neither the financial and organizational capacity, nor the willingness to handle the marketing of the produce. The primary gatherers have to depend on traders in the village, as well as in nearby towns, as a final resort to sell their collected output. It has been found that households in all the study villages sell a variety of products both in and outside their villages. Many sell products at prices much below those fixed by the district committees for the purpose and through interlinked transactions.

Furthermore, they sell most of their products in raw form without any value addition. In such a situation the net value realized after deduction of transaction costs and implicit interest charges is far lower than the gross value. Experience in other areas shows that NTFP prices are more stable and remunerative if they are processed where they are collected. But efforts towards this end by the state, private sector or NGOs are largely lacking in the study areas to date. Investment in small-scale enterprises to add value could increase employment, as well as the income of collectors of forest products.

Impact on forests

The unambiguous evidence is that PFM has generally improved forest condition. A comprehensive forest survey of the 12 VSS and SIFPG villages was conducted in 2003 using global positioning systems (GPS), a summary of which is presented in Table 8.10. The area of forest as found in the present study is based on the indications of FPC members to the GPS team of the forest boundaries. Closed forest is more extensive in VSS and SIFPG villages. Although degraded areas were allotted to VSS groups, most now have closed forest, indicating that the forest has grown since the formation of VSS. This was corroborated by VSS members. The SIFPG forests have more closed canopy than open forest and there are growing stocks of stems. Thus, the forest is seen to be growing in both the protection systems.

In view of the absence of baseline data, it is difficult to compare pre- and post-PFM conditions. However, during our field survey FPC members in each of the study villages made some observations about the condition of their forests. A majority of respondents believed that this has improved, more so in SIFPG than in VSS villages since SIFPGs started protecting their forests earlier. A majority of VSS and SIFPG respondents believed that there was a greater presence of birds and small animals in the protected forests. The groundwater level and hydrological conditions in VSS and SIFPG villages seem to have improved, according to local people's views.

Policies, practices and impacts: Governance and the policy process

Having reviewed how PFM has affected livelihoods in Orissa, we now try to explain the outcomes.

Policy and governance processes

There is no Orissa state forest policy, although the 1972 Orissa Forest Act and 2000 amendments guide the state on forest-related matters. National guidelines and notifications relating to PFM also partially guide forest governance in the state. Orissa has not implemented all of the central policy guidelines, particularly the 2000 Central Guidelines, in JFM (such as by providing tenurial security to VSS by registering them under the 1960 Society Registration Act and according self-initiated groups legal status).

At district level, the divisional forest office (conservators, divisional forest officials and field staff) largely implement directives from the state capital. The lower officials at range and beat level exercise little independence from their DFOs in their interpretation of the rules. As a result, they are unable to take quick action at ground level when necessary. They fear that senior officers would not like them to take the initiative or innovative action. At the village level, it has been found that villagers and FPC members are ignorant of most of the PFM guidelines and look to their range officer or forester for guidance.

Table 8.10 *Forest categories in sample VSS and SIFPG villages*

District	Village and type of forest management	Forest area per house-hold (ha)	Total forest area (ha)	Forest condition	Area (ha)	Main species present	Size range of trees (dbh in cm)	Stem per ha
Sambalpur	Khandam VSS	0.49	43.79	Closed	42.00	Sal, bija, dhura, mahul, asan, sissu, amala and arjuna	05–49	93,757
				Open	1.79	Sal, dhoura and riohini fruit	05–39	678
	Kutasingha VSS	1.52	130.82	Closed	10.59	Sal, arjuna, dhoura, piasal, mahul and sissu	05–50 and above	88,940
				Open	20.23	Sal	05–50 and above	5289
	Palokhaman SIFPG	1.95	95.39	Closed	95.39	Sal, dhoura, piasal, mahul and dahtuk	05–50 and above	84,518
	Jharmunda SIFPG	0.66	57.31	Closed	51.54	Sal, char, dhoura, piasal and mahul	05–50 and above	57,650
				Open	5.73	Sal and char	05–29	8728
Bolangir	Chandanjuri VSS	0.90	93.43	Closed	80.45	Sal, sissu, dhoura, bamboo, harida, mahul and sahaj	05–50 and above	56,816
				Open	12.96	Kendu, rajmai and bamboo	05–50 and above	2937
	Mursingh VSS	1.88	427.81	Closed	397.82	Sissu, dhoura, bamboo, kendu, sal, piasal and mahul	05–29	722,343
				Open	22.52	Char, riohini fruit and bamboo	05–10	9078
	Junanibahal SIFPG	0.31	20.54	Closed	20.54	Sal, piasal, dhoura, mahul, sissu and bamboo	05–50 and above	19,823
	M. Kutenpali SIFPG	0.38	35.43	Closed	32.39	Sal, char, piasal and kendu	05–19	75,725
				Open	2.55	Sal, char and jamun	05–19	2518
Kand-mal	Tudubali VSS	1.02	81.74	Closed	74.26	Sal, mahul, dhoura and piasal	5–49	128,132
				Open	7.21	Sal, mahul, char and bhalia	10–19	2,320
	Pokari VSS	0.44	53.91	Closed	52.14	Sal, piasal, char, harada, mahul and bamboo	5–50 and above	89,834
				Open	1.77	Mahul	10–29	109
	Mandagoda SIFPG	0.14	12.82	Closed	12.72	Sal, mango, mahul, sissu and bandhan	10–50 and above	13,949
	Banegaon SIFPG	0.39	29.80	Closed	28.30	Sal, char, piasal, bahada, mahul, sissu, bamboo and mango	5–29	40,727
				Open	1.49	Sal, mahul and mango	20–50 and above	51

Note: Forest areas are areas as found by the GPS survey team.

Source: Kailas Sarap (from field survey, 2004–2005)

Major policy actors, their changing characteristics and influences at different levels

This section analyses the utilization of forest and forest products by certain important stake-holders involved in forest use, and their behaviour in relation to each other. These stakeholders are the forest department; industrial timber and forest land users, including sawmills and mining companies; the timber mafia; and NGOs. Of course, communities, in general, and forest-fringe communities, in particular, are the major users of forest products. We have already discussed the livelihood patterns of forest-dependent communities. Now we discuss the influences of certain policy actors on forest and forest-fringe people.

The forest department

The development of forest management is largely dependent on the thinking and practice of the forest department, whose duty it is to provide a responsive and transparent forest management system. Its roles need to be regularly redefined according to changing circum-stances and social priorities. The Orissa Forest Department has sought to adapt, particularly after the 1988 National Forest Policy (NFP) and the related JFM notification. In the case of extensive forests, the forest department's role is to provide protection and scientific manage-ment, and to promote people's participation in forest management and ecological development outside the forest area. In the case of forest areas within village boundaries, its crucial role is to facilitate the people's protection of the forest and to promote increased and sustainable production from the forest. The forest department has to consider the equitable distribution of forest products and to strengthen village-level institutions in order to protect and manage the forest on a sustainable basis with the involvement of the villagers. The preparation of micro-plans at village level in consultation with villagers is essential. The forest department also has to play a developmental role in improving the livelihoods of forest dwellers. The achievement of the forest department in Orissa in realizing these goals is not satisfactory. It is mainly concerned with protecting the forest and has not taken any signifi-cant measures to improve the livelihoods of forest dwellers.

There is a great deal of mutual distrust between forest department staff and forest-dependent communities, especially in areas where SIFPGs have been protecting the forest. This is due to a variety of reasons. The forest department was apprehensive of communities' use of the forests – the encroachments, grazing, illegal felling and so on – and viewed the forest-dependent communities as a threat to 'its' forests. By contrast, the communities feared the loss of access to traditionally used forests due to the greed of corrupt forest department officials. In such an antagonistic situation the effectiveness of forest protection and manage-ment is low.

Industry and forest use

A number of forest products, including bamboo, sal seeds and other NTFPs, are utilized by forest-based industries, mainly paper mills, plywood and MDF units, seed oil units, 55 small-scale industrial units and 665 sawmills (licensed and not). These industries employ an estimated 25,000 people (Singh, 1997) and require large quantities of timber and non-timber raw materials for the sustained supply of which they depend on forests. If it is assumed that 50 per cent of trees planted are used by industry, Orissa would need to plant an estimated 90 million trees a year, working at three times the present rate of planting. However, the production of services and goods from forests has decreased considerably since the 1950s for two reasons: the depletion of the forest owing to increased biotic pressures and the official diversion of forestland to non-forest use. Large-scale diversion of forest areas to non-forest use (about 27,055ha of forest area between 1982 to June 2002; author's estimate based on available sources) has taken place in the state. As a result, there is a supply shortage. This pattern of development has generated demand for certain forest products beyond supply capacity. Forest-based industrialists use their influence with the powerful decision-making state elites. For instance, the government recently decided to permit the opening of additional sawmills, allowing them at a distance of less than 10km from the forest. Such a situation would reduce access to forests by forest-dependent communities, given the current growth of forest.

Mining and deforestation

Orissa is endowed with rich mineral resources, including deposits of bauxite, chromites, coal, dolomite, fireclay, graphite, granite, iron, manganese, limestone, nickel and quartzite. These deposits are located across the districts of Kalahandi, Koraput, Keonjhar, Balangir, Dhenkanal, Sambalpur, Jajpur, Sundargarh and Phulbani. The chromites, manganese and iron extracted in Orissa in 2004 to 2005 constituted 97.37, 28.56 and 33.91 per cent, respectively, of those of the whole country (GoO, 2005–2006). There are a total of 364 leases, including 148 for iron, manganese and chromites, operating over 37,664ha (including 19,263ha of forest cover) (FSI, 1999).

In December 2003, the GoO approved a policy on the granting of mining leases and transfer of land for commercial projects in scheduled areas of the state, liberalizing permission for mining leases in forest areas without consideration of the views of local people, in lieu of which the locals would receive some cash compensation. Some of the money collected from leaseholders would be utilized for the development of the areas. The local people do not like these schemes because they are apprehensive that their livelihoods will be adversely affected due to their low human capital. Areas of the state with mineral deposits have become 'hunting grounds' for many national and international mining companies, with long-term implications for the growth of forests and the welfare of forest-dependent communities. Since the boundaries of forest lands and tribal rights have not been clearly settled, the mining issue has led to extreme tension and conflict between local people, on the one hand, and mining leaseholders and security personnel, on the other. Conflict is occurring in many parts of the state, including Langigarh, in Kalahandi district, and Kashipur, in Koraput district, where the construction of an aluminium factory has been taking place. The tribal community has been agitating for some time to assert its rights on forest land where it has been settled for a long time, although the state has not settled land rights in its favour. In the absence of recognized rights, the tribal community is being displaced without (or with paltry) compensation. As a result, these displaced people may have to look for new patches of forest to settle.

Forest mafias

There are two aspects to the illicit extraction and sale of forest products. The first is the prevalence of individuals collecting fuelwood for sale. The early morning transport of head or bicycle loads to local towns is a common sight. However, organized gangs engaging in the illicit felling of timber is an entirely different matter, and the operation of forest mafia in the state is a subject of public concern. According to the latest Forest Survey of India (FSI, 2003), the loss of forest cover in the state due to illegal timber felling is very high. Unauthorized removal of timber is rampant from vulnerable areas, such as the forest in the Mahanadi River Basin, which spreads across Sambalpur, Angul, Cuttack, Nayagarh and Boudh districts; Similipal forests in Mayurbanj district; the forests of Rayagada, Koraput and Malkangiri districts on the Orissa–Andhra Pradesh border; and the forest corridors of the Orissa–Jharkhand and Orissa–Chhatisghgarh borders, as is widely reported in local newspapers on a regular basis.

Due to the high demand for wood for furniture and house construction, particularly in urban areas and neighbouring states, a well-organized felling racket is operating involving prime timber trees in forested area. For instance, during 2003 to 2004, about 113,808 forest offences were booked, involving 107,093 offenders. The value of forest products seized in these offences was around 150 million rupees. On a smaller scale, local woodcutters reportedly gain access to nearby forests by bribing junior forest officials. This has become so normalized that the cutters say that they have to pay a toll fixed by forest officials for tree felling.

The forest department claims that measures have been taken by forest officials at the state level. For instance, during the year 1998 to 1999, the department seized 3162 vehicles with stolen forest products, although of these only 345 involved motorized vehicles and the other 2817 involved bicycles, bullock carts and other such. During seizures of illicitly felled forest products, organized mafia groups have attacked and injured many forest department staff; for instance, from 1994 to 2002, 348 personnel were assaulted by members of the mafia (Orissa Forest Department, 2004). However, the lower officials are not cooperating in implementing anti-mafia measures, partly because of a nexus between forest officers and the mafia and partly due to mafia intimidation of forest officials. Our discussions with some forest officials revealed that both these aspects are true. Some forest officials pleaded that they have to protect their own lives against the power of the mafia.

The punishment procedure for illicit timber felling and smuggling and forest encroachment is cumbersome, and at the village level the ranger or forester has limited power. These officers have to report large amounts of products seized to a higher officer or the police. The matter goes to court, which takes a long time to process it. Thus, punishing the mafia is a protracted and time-consuming process. Furthermore, there is evidence to suggest that occasionally there is connivance between forest officers and the mafia, which inevitably negatively affects the reputation of forest department staff with local people.

Involvement of NGOs in the forestry sector
Orissa has a rich tradition of institutional pluralism with a long history of NGOs, community-based organizations (CBOs) and civil society activities. Some NGOs take a keen interest in the forestry sector and engage with it in three ways. First, they have been active in policy formulation and reform in the state. Second, many NGOs have played important roles in organizing self-initiated forest protection groups. Lastly, some of those who took the initiative in the formation of forest protection groups have promoted their networking in order to strengthen these groups with a view to improving their bargaining power with other organizations and the state. Networking is seen as crucial to addressing various issues such as information dissemination, conflict resolution, creating platforms for the exchange of ideas, and building pressure for reform. In order to strengthen the state's CFM network, CFM groups have started to federate themselves at cluster, block and district levels. Networking and the evolution of federations vary at the local level, with different representation mechanisms catering for different issues and undertaking diverse activities. One of the networks' main demands is that the state should adopt a CFM policy instead of the current JFM policy, in which the forest department is the controlling agency (Vasundhara, 2002).

Decentralized local government and participatory forest management

This section considers the *panchayati raj*, institutions with particular reference to scheduled areas and the related *Panchayat* Extension to Scheduled Areas (PESA) legislation.

The state of Orissa has accepted the extended provision of Part IX of the Indian Constitution relating to *panchayati raj*, albeit with certain modifications. The 1997 Orissa *Gram Panchayat* Act provides power to the *gram sabha* (village assembly, also known as *gram sason* in Oriya, the language of Orissa) for the operation of participatory democracy and command over natural resources. Specifically, Section 4(III) of PESA in the state has conferred on *gram sason* the power to plan and approve programmes on various matters of socio-economic development.

The PESA act also provides for the control and trade of NTFPs in the state. According to clause 3(c), any individuals wishing to purchase NTFPs from primary gatherers or to trade in NTFPs must register themselves in the specific *gram panchayat* for the NTFP and pay a registration fee to the range officer concerned, along with a minimum 10 per cent royalty. Clause 3(e) allows government agencies such as the OFDC and TDCC to register to procure one or more NTFP items.

However, the Orissa State Act does not give any such power to *gram sabha* on matters of land acquisition, minor minerals, or the planning and management of minor water bodies. Instead, it has entrusted all of these powers to *zilla parisad* (the district-level tier of local government), which is not required to consult the *gram sabha* while exercising them. Despite this apparently progressive legislation, there are a number of problems with the operation of PESA in Orissa.

Our enquiry in the study villages revealed that a majority of tribal households have no idea of PESA, and the few who have heard of it do not understand what it does. Similarly, the *panchayats'* control of NTFP trade is extremely weak since they lack the capacity to exercise the power vested in them and, in any case, the majority of *panchayat* personnel are not aware of their power. Clearly, the *panchayats* in the state, and especially in tribal areas, so far lack the necessary institutional and financial capacity to assume the control and marketing of NTFPs. This is not to say that they should not, but rather that building *their* capacity is urgently required if improvements in NTFP marketing to benefit the bargaining power of local collectors are to take place.

Operation of forest development agencies

Forest development agencies (FDAs) are centrally funded by the MoEF and have been working in 28 of Orissa's forest divisions since 2003, providing a high level of financial assistance to selected VSS groups for forest development and infrastructure activities in the VSS villages. The FDA is a five-year programme, beginning in 2002, which aims to protect over 880,000ha of state forests at an estimated cost of nearly 652 million rupees. But by the end of its fourth year, only 1354 VSS were covered by FDAs, about 14 per cent of all the VSS (9677) in the state. These FDA-financed VSS protect about 50,000ha of forests at a cost of 382 million rupees. The FDAs seem to have been unsatisfactory and not participatory, and the objective for which they were formed has not been fully achieved due to lack of participation in VSS by local people; elite and forestry official domination of community forest user groups (CFUGs); lack of focus on improving the livelihoods of the poorest; and lack of transparency in the use of funds. Furthermore, there is rift between VSS villages with FDA assistance and those without. Our visits to the study sites revealed that forest officials are concentrating more of their resources and time on FDA-assisted villages than on non-FDA villages, despite VSS and SIFPG villagers' attempts to get assistance from the forest department by visiting the department time and again. The result is forest department neglect of non-FDA VSS villages.

The significant forest department funding involved is acting as an incentive to many people in self-initiated CF villages to convert to VSS. However, as we have seen from the field study, conversion from self-initiated groups to VSS comes at a cost. The self-initiated groups have often regenerated degraded forest lands independently; but on conversion to VSS they are only entitled to a share of the net proceeds of the harvest of forest products that have matured since the formation of the VSS. Therefore, they are expected to forgo any share of timber revenue from the original self-initiated group. Furthermore, they are obliged to have the local forester as secretary of their committee and to depend on his attendance to hold a meeting, a condition which has often led to the stagnation of VSS groups as local foresters lack the time and often the inclination to attend these meetings. Their views and influence are unlikely to be welcomed by forest-dependent communities anyway as they are trained to emphasize restrictions on livelihood forest use.

Inter-sectoral linkage

There are a number of state-level departments and ministries associated with the improvement of tribal livelihoods, including the Ministry of Rural Development, the Ministry of Tribal Development, co-operative societies, *panchayati raj* institutions (PRIs) and NGOs. The forest and revenue departments are two major departments comprehensively involved in forest-related activities, which, however, work independently of each other because each wishes to maintain its autonomy. There is, therefore, a lack of coordination and monitoring in the implementation of different programmes and duplication of some activities, as a result of which the impact of each of these institutions is reduced. In such a situation the utilization of resources is not optimal, and there are problems with the proper implementation of livelihood programmes associated with the state's tribal and forest-fringe people. Conflicts arise over the use of forest land and the management of forest in the absence of coordination between the departments. For instance, about 45 per cent of forest land is under the control of the revenue department and the rest is under the forest department. Even though ownership rights to certain forest land rest with the revenue department, the forest department controls its management. Recently, the revenue department gave mining leases to many companies on its own forest land; but the forest department does not recognize this because it considers it illegal to mine forest land without its permission. Such conflicts are rampant nowadays since the revenue department has given leases, especially in degraded forest areas, for mining activities and the construction of sponge iron factories.

Similarly, the coordination between the VSS/SIFPGs and the *panchayat* is weak and negligible. As a result, intra- and inter-village conflicts are not properly resolved. The *panchayat* has a wider jurisdiction, is better trusted by villagers and can offer collective protection and dispute resolution. But there is no effective coordination between the forest protection committees and *panchayats* in the study area, including in scheduled areas. Furthermore, there is no coordination between the many NGOs and other institutions working in forest-related activities; most NGOs working in the state try to maintain their own identities in terms of their *modus operandi* and project themselves as autonomous institutions, even though some work on behalf of government agencies, including the forest department. Because of the lack of coordination between institutions, progress in the livelihoods of forest-dependent communities is slow and sometimes repetitive, leading to suboptimal use of resources for forests and people dependent on forests.

Politics of information and knowledge about forest management

In all of the VSS villages, it was found that the forest officials' views on managing forests and the choice of species have been implemented at the expense of village priorities and preferences. Even in open patches of different forests, forest officials have planted their choice of species. The villagers' preferences have been ignored based on the technocratic rationale that given the soil types, particular species of trees are suitable for the patch. Furthermore, the forest officials believe that certain species, such as teak and sal, can survive better even if affected by grazing. In order to fulfil their target of progress in conservation and forest growth, they provide commercial timber species for plantation in open spaces. But given that the local people have accumulated much knowledge over long periods of time through their conservation, regeneration and development of forests, there is a need to strengthen CF management and co-learning approaches in the state.

Main opportunities and constraints to improved PFM

In Orissa the state's behaviour towards PFM has been paradoxical. It has encouraged the development of JFM, but neglected the already existing SIFPGs as a policy choice, despite the fact that SIFPGs were highly successful in protecting forests before the formation of

VSS. JFM policy is not robust since it has no legal standing or tenurial security. This offers a very poor incentive to forest protection groups. Moreover, issues relating to improving the livelihoods of forest-fringe communities have not received adequate attention from the various policy actors, especially the legislature and forest departments, as well as the general bureaucracy involved in policy formulation and implementation. The impact of PFM policies on the livelihoods of forest-fringe people and on forest development in Orissa has been marginal.

Our interaction with stakeholders at all the VSS and SIFPG study sites revealed that local institutional development is weak. The FPCs do not cooperate with other related institutions such as the *panchayati raj* and rural and tribal development departments. In such a situation the bargaining power of forest protection members in relation to other stakeholders, including forest officials, traders and others, is very weak, and the impact of individual departments on the living standards of forest-fringe people is marginal. Even though women's participation in collecting forest products is significant, their role in decision-making is negligible. There are, though, some positive developments in some villages in Sambalpur and Kandhamal districts. The participation of women has been high and members have realized reasonable prices in selling forest products due to the active presence of NGOs and self-help groups. These institutions have generated awareness concerning the conservation and development of forest resources and increased members' bargaining power with other stakeholders, even if it still remains at a relatively low level.

Given the above scenario, it is necessary to generate awareness among primary stakeholders, forest and other bureaucracy, at block and district levels, which implement forest and other policies affecting the livelihoods of forest-fringe communities. Similarly, the attitude of legislators and policy planners needs to be sensitized to the context of livelihood issues, including the land rights of tribal and other groups dependent on the forest. We found that legislators of different political parties in the state are worried about the deforestation and degradation of forests, and have been demanding strict measures against the mafia stealing forest products; but they have not shown any such concern about the improvement of the livelihoods of forest-fringe people. During its second study phase, the study team found high awareness of the necessity to improve forest dwellers' income from sales of forest products through better exchange with different buyers and value addition. But forest dwellers felt helpless due to a lack of credit to enable them to afford simple tools and training. They are unhappy with the government's recent decision to privatize the trade in sal seeds in the state since they feel that private traders will offer very low prices. The local villagers are also concerned about the improvement of the forest on which they depend. Not only civil society and NGOs, but also villagers at the study sites are concerned about deforestation and the impact of mining-based industrial activities, such as sponge iron factories being installed in tribal and other forest areas. For instance, one of the very successful forests protected by a self-initiated group in the district of Sambalpur has been deforested recently by an iron factory opening nearby. Similarly, the study team found awareness in local people of the demand for forest land rights for families who have been in possession there for many decades.

There is an urgent need for policy-makers dealing with forest and land to offer forest dwellers incentives to protect and manage their local forests for livelihood objectives through the provision of tenurial security. Policy-makers also need to improve the desperate living standards of the rural poor through NTFP exchange relations, the value addition of the NTFP processing, coordination of the activities of FPCs with *panchayati raj* institutions, and integration of the activities of different departments dealing with forest, tribal and rural development.

Given the several types of conflict in FPC villages, including the disharmony between VSS groups with and without the FDA's financial support recurrent in the absence of legal status for VSS or SIFPG groups, conflict resolution has not been satisfactory. For instance, a majority of the villagers in Khandam are in conflict with the Binapani self-help group, which leased its village pond from the *panchayat* in 2005 for catching and selling fish against the wishes of the majority of villagers. Since the president of the self-help group is the sister-in-law of the VSS president, the villagers started pressuring him to ask her to surrender the leased pond to the villagers. When he did not succeed in this they wanted him to step down from his post. Now the villagers are divided into two groups: some supporting the president and some the SHG. Since then all forest-related works have stopped, even though during the first year of the plan under the FDA the VSS had carried out some entry-point activities.

Policies and future strategies

Certain measures must be taken to allow PFM to function properly given the political culture of the state. These include improvement of PFM institutions; provision of tenurial security to forest-fringe people; resolution of conflicts; and the prevention of illegal and excessive felling of trees, illegal mining and mafia activity. There is a need to enhance economic benefits through eliminating the exploitation of sellers of forest products and increasing value addition and the productivity of the state's forests.

Improvement of institutions
PFM institutions, including the executive committee and the general body, have the potential to be dynamic, representative and democratic by proper inclusion of all groups, including the poor and women. The involvement of NGOs and SHG groups in VSS and SIFPGs will enhance women's participation. FPC members should be given training on forest-related issues, environmental awareness, record-keeping, etc. This will result in FPCs functioning democratically and transparently.

Security of tenure and conflict management
The different types of conflict present in VSS and SIFPG villages can be resolved by raising the power of FPCs by giving them legal status through the provision of tenurial security, registering them under the Society Registration Act and involving the *panchayats*. The involvement of *panchayati raj* institutions and other departments dealing with tribal and rural development will improve the livelihood of forest-fringe people through the efficient use of allocated resources, and raise the bargaining power of FPC members in different institutions in their areas. Proper coordination between FPCs and other village institutions will improve conflict management, micro-planning and the overall development of forest-fringe villages and people dependent on forests.

Prevention of illegal felling of trees, mining and mafia activity
A critical measure would be to reduce the illegal felling activities of mafias, large and small. The apprehension of mafia groups is not easy because of the patronage they enjoy from various quarters, the inadequate power of lower forest officers who deal directly with the problem and the prolonged court proceedings. Suppression of the mafia would require the elimination of channels such as illegally established sawmills and furniture shops through which the timber is sold.

The government amended some of the provisions of the 1972 Forest Act in 2000 in order to address the mafia problem; but the enforcement of the act is poor, given that 40 per cent of field-level lower forest posts are vacant and funds allocated for them are also minimal. Our enquiries of many beat officers in the course of our fieldwork revealed that they feel no zeal for their work because the size of the territory that they have to guard is so large. They also feel the extra VSS-related work as a burden. This makes fighting the mafia problematic.

Similarly, the state is unable to control illegal mining activities in the forest due to the connivance of lower bureaucracy with the leaseholders. Strong, transparent measures are required here.

Increased economic benefits

As an incentive to the people to protect the forests, increased economic returns are necessary. It is essential to plant a species mix with the needs of the villagers in mind and to include open spaces in order to grow a variety of NTFPs. Investment in the forest sector needs to be increased in yearly and medium-term planning by the state for forest regeneration.

The operation of FDAs can be improved by giving proper representation to VSS in decisions about the allocation of funds and in monitoring the development activities carried out as a result. In order to raise value addition, processing units for marketable NTFPs can be set up for each VSS/SIFPG or for a group of VSS/SIFPGs in easily accessible places. The marketing of these products should be taken care of by private and public agencies, including NGOs, who can help in ensuring reasonable prices. The operation of SHGs, grain banks and NGOs dealing in credit provision and buying forest products in the tribal areas should be strengthened.

Integration with other schemes (rural development)

Agriculture is an important source of livelihood, but is primitive near forest areas due to a lack of modern technology, irrigation and the conservation of rainwater available near the forest. Lack of households' own surplus and government apathy in the provision of infrastructure, credit and tenurial security contribute to this situation. Increased public investment to raise the productivity of existing assets, such as land and irrigation, is necessary to revitalize agriculture. Furthermore, there is a need for enhanced focus on human resource development, including skills and capacity-building among JFM or CFM households, raising their bargaining power with other agents, and awareness of information and technology. There is a need to coordinate different departments working in tribal/forest areas in order to achieve these objectives.

Above all, the current path of development that prioritizes extraction of resources by mines and the subordination of agriculture, forest and human resource development in tribal/forest areas is detrimental to forest and forest-dependent people, given their uncertain property rights on forest land and their low human capital. This must be changed in order to minimize the degradation of forests, displacing large numbers of forest-dependent communities and uprooting their livelihoods.

Conclusions

On the whole, PFM should be viewed as a comprehensive process focusing on the development of institutional dimensions such as the provision of legal status to FPCs, decentralization and devolution of power, increased participation of women, equity and economic sustainability. In order to achieve these goals, the programme must be made dynamic and flexible enough to respond to socio-cultural and institutional diversity (such as encouraging the already viable SIFPGs), along with VSS, on the one hand, and to the changes emerging on the political and economic fronts, on the other, such as the rising primacy of the market, the reduced role of the state and the increasing role of civil society and *panchayati raj* institutions. However, given the political and economic structure of the state and its lack of effective political and bureaucratic decentralization, its emphasis on the top-down approach to development, and its orientation towards mining-based industrialization instead of land-based and tribal development, it will take many years to improve the livelihoods of forest-fringe and forest-dwelling people.

References

Babu, A. (2003) 'Forest based livelihood and rural poor in Orissa', in Sarap, K. et al (eds) *Agrarian Transformation in Orissa, Vol I*, Sambalpur, Sambalpur University, P.G. Department of Economics

Bhaskar, V. (1999) 'Implementing JFM in the field: Towards an understanding of the community bureaucracy interface', in Jeffery, R. and Nandini, S. (eds) *A New Moral Economy for India's Forest? Discourses of Community and Participation*, Delhi, Sage

Das, V. (1996) 'Minor forest produce and the rights of tribals', in *Economic and Political Weekly*, 14 December, pp3227–3229

D'Silva, E. and Nagnath, B. (2002) 'Behroonguda: A rare success story in joint forest management', in *Economic and Political Weekly*, vol 37, no 6, 9 February, pp551–557

FSI (Forest Survey of India) (1997) *State of Forest Report*, Dehra Dun, Ministry of Environment and Forests

FSI (2001) *State of Forest Report*, Dehra Dun, Ministry of Environment and Forests

FSI (2003) *State of Forest Report*, Dehra Dun, Ministry of Environment and Forests

GoI (Government of India) (2001) *Census of India*, Delhi, GoI

GoI Planning Commission (2003) 'Forest resource and forest management policy in Orissa', *Orissa Development Report*, Bhubaneswar, NKCDC, Chapter 5

GoO (Government of Orissa) (1959) *Forest Enquiry Committee Report*, Bhubaneshwar, Government of Orissa

GoO (1993) *Joint Forest Management Resolution: Forest and Environment Department Resolution*, No 16700-10-F(prom)-20/93, 3 July, Bhubaneshwar, Forest and Environment Department

GoO (1995) *A Decade of Forestry*, Bhubaneshwar, Orissa Forest Department

GoO (2002) *Orissa Forestry Sector Development Project*, Bhubaneshwar, Government of Orissa

GoO (2005–2006) *Economic Survey of Orissa*, Bhubaneshwar, Government of Orissa

Kumar, K. and Choudhary, P. R. (2005) *A Socio-Economic and Legal Study of Scheduled Tribes' Land in Orissa*, Bhubaneshwar, Vasundhara, supported by the World Bank

Mallik, R. M. (2000) 'Sustainable management of non-timber forest products in Orissa: some issues and options', *Indian Journal of Agriculture Economics*, vol 55, no 3, pp383–397

Mishra, P. K. (1995) *Critical Issues in Shifting Cultivation: A Critical Appraisal of the Orissa Forestry Development Programme*, vol II, Annexure L

Orissa Forest Department (1999) *Revised Working Plan for Nowrangpur Division, Cuttack*, Bhubaneswar, Orissa Forest Department

Orissa Forest Department (2000) *PCCF Office Status Report*, Bhubaneshwar, Orissa Forest Department

Orissa Forest Department (2004) *Orissa Forest Status Report (2003–04)*, Principal Chief Conservator of Forests, Bhubaneshwar, Orissa Forest Department

Patnaik, B. K. and Brahmachari, A. (1996) 'Community-based forest management practices: Field observations from Orissa', *Economic and Political Weekly*, 13 April, pp968–975

Patnaik, M. (2002) 'Community forest management in Orissa', *Community Forestry*, RCDC, Bhubaneswer, January, vol 1, no 1&2, pp4–9

Patnaik, N. (1993) *Swidden Cultivation amongst Two Tribes of Orissa*, Bhubaneshwar, Cendernet, and New Delhi, ISO/Swedforest

PCCF (Principal Chief Conservator of Forests, Orissa) (1991) *Orissa Forest*, Bhubaneswar, Statistical Branch Office of the Principal Chief Conservator of Forests

PCCF (1999) *Orissa Forest*, Bhubaneswar, Statistical Branch Office of the Principal Chief Conservator of Forests

PCCF (2001) *Orissa Forest*, Bhubaneswar, Statistical Branch Office of the Principal Chief Conservator of Forests

PCCF (2003) *Orissa Forest*, Bhubaneswar, Statistical Branch Office of the Principal Chief Conservator of Forests

Poffenberger, M. and McGean, B. (eds) (1996) *Village Voices, Forest Choices: Joint Forest Management in India*, Delhi, Oxford University Press

Prasad, R. (1999) 'Joint forest management in India and the impact of state control over non-wood forest products', *Unasylva*, issue 198, Rome, FAO

Regional Centre for Development Cooperation (2002) *Community Forestry*, various issues, including vol 1, no 5, September

Sarap, K. (2004) *Participatory Forest Management and Livelihood among Forest Dependent People in India and Nepal: Review of Policies*, Working paper no 2, Norwich, University of East Anglia and Sambalpur University

Sarap, K. and Mahamallik, M. (2001) 'Food insecurity, coping strategy and livelihood patterns among households in tribal areas of Orissa', in Asthana, M. D. and Medrano, P. (eds) *Towards Hunger Free India: Agenda and Imperatives*, New Delhi, Manohar

Sarin, M. (1999) *Policy Goals and JFM Practice: An Analysis of Institutional Arrangements and Outcomes*, Collaborative research supported by WWF-India and IIED, New Delhi, WWF-India

Saxena, N. C. and Ballabh, V. (1995) 'Forest policy and the rural poor in Orissa', *Wastelands News,* vol 11, no 2, November–January 1996, New Delhi, pp9–13

Singh, B. P. (1997) 'Forest development in Orissa: A status paper', in *Orissa-Forest*, issue 4, December, Bhubaneshwar, Department of Forest and Environment, Government of Orissa, pp1–13

Sunder, N., Mishra, A. and Neeraj, P. (1996) 'Defending the Dalki forest: Joint forest management in Lapanga', *Economic and Political Weekly*, vol 31, no 45–46, pp3021-3025

Thangam, E. S. (1984) 'Agro-forestry in shifting cultivation control programmes in India', in Jackson, J. K. (ed) *Social, Economic and Institutional Aspects of Agroforestry*, Tokyo, United Nations University

Vasundhara (1996) *Community Forest Management in Transition: Role of the Forest Department and Need for Organizational Change*, Bhubaneswar, Mimeograph

Vasundhara (1998) *Non-timber Forest Products and Rural Livelihoods, with Special Focus on Existing Policies and Market Constraints*, London, DFID

Vasundhara (2002): 'Need to look beyond JFM: Learning from community forest management in Orissa', *Banabarata*, issue I

Winrock International India (2005) *Orissa Forest Vision Document*, New Delhi, Winrock International India

9

Participatory Forest Management in Andhra Pradesh: Implementation, Outcomes and Livelihood Impacts

V. Ratna Reddy, M. Gopinath Reddy, Madhusudan Bandi,
V. M. Ravi Kumar, M. Srinivasa Reddy and Oliver Springate-Baginski

Synopsis

Andhra Pradesh was formed in the 20th century from component regions with diverse backgrounds, similarly to Orissa, and has struggled to become a unified state. It comprises an extremely diverse range of agro-ecological areas, from (mainly tribal-populated) forested hill in the north-east to arid red sanders forests in the south. Earlier participatory forest management (PFM) initiatives in areas under the Madras Presidency areas were shunned by the state government in the 1950s, and it was only with the introduction of joint forest management (JFM) in 1992 in Andhra Pradesh that local people's forest use was again formally legitimated. JFM implementation was heavily supported by the World Bank from 1995 to 2000 and 2002 to 2007.

The Andhra Pradesh Forest Department has sought to distinguish itself by transforming its PFM programme from JFM to community forest management (CFM), although while the rhetorical claims for policy evolution seem laudable, this chapter examines whether the outcomes and impact of PFM, whether JFM or CFM, on the livelihood systems of people in Andhra Pradesh have been beneficial.

The overall conclusions are that PFM implementation has been highly problematic due to the lack of any real devolution of power to local people, and the persistence of patronage power relations between forest department field staff and local elites. In tribal areas, many *adivasis*, in the context of comprehensive and chronic social oppression and marginalization, have suffered under PFM from being excluded from their customary fallows cultivation lands, and seeing them planted with exotic tree species. Although households have received cash disbursements under the World Bank project for a few months' wage labour in lieu of the cultivated land they have been forced to give up, this is merely a temporary project-based palliative, and without improving the livelihood productivity of the forest land resource it is probable that after the project cycle ends local land-poor households will revert to fallows cultivation in the forest. Only if local people have authority to plan longer-term sustainable forest land management, based on secure tenure, is PFM likely to succeed.

The context of participatory forest management (PFM) policy in Andhra Pradesh

Socio-economic profile of the state

Andhra Pradesh consists of three distinct regions: Coastal Andhra, Rayalaseema and Telangana (see Map 6). Telangana was a part of a Muslim kingdom ruled by the Qutub Shahi and Nizam dynasties. Andhra and Rayalaseema were part of the Madras Presidency of the British Empire, separated from the Madras State in 1953 to form the separate Andhra State, the first to be formed based on linguistic nationality (the language being Telugu). In 1956, the Telangana region was amalgamated with Andhra State based on linguistic affinity, resulting in the creation of the present state of Andhra Pradesh, with Hyderabad as its capital.

Andhra Pradesh is a state of great contrast and diversity. Coastal Andhra is a prosperous area with productive agriculture. Due to assured irrigation facilities, the Green Revolution occurred in some of its districts, such as Godavari, Krishna and Guntur. Telangana is a semi-arid region with limited irrigation and, hence, is comparatively less developed, and Rayalaseema is an arid zone with low rainfall with harsh environmental conditions. During the last 50 years, the National Congress Party has mainly ruled the state.

Overview of forests in Andhra Pradesh

There are two main forest areas in the state (see Map 6). The northern tribal belt runs across northern Telangana and Coastal Andhra, and contains high-value timber trees such as teak and sal, and the central north–south belt to the eastern side of Rayalaseema consists of dry forests with valuable red sanders (*Pterocarpus santalinus* spp). About 23 per cent of the geographical area of the state is forested, of which 79 per cent is reserved as state forests (see Table 9.1).

Table 9.1 *Legal status of forests in Andhra Pradesh*

Particulars	Area (km²)	%
Total geographical area of state	2,75,068.00	–
Forest area under forest department	63,814.00	23.20
Reserved forests	*5478.63*	*79.10*
Protected forests	*12,365.34*	*19.38*
Unclassified forests	*969.76*	*1.52*

Source: Andhra Pradesh Forest Department (2004)

The area under forest has been subjected to degradation and has declined due to various factors (see Table 9.2). It is estimated that only 36 per cent of Andhra Pradesh's forests have a crown density of more than 40 per cent (Samaj, 2006). Clearance for agriculture and the diversion of forests to rehabilitate project-affected people were the main factors in this depletion, made worse by industries, mining and encroachments. The National Remote Sensing Agency (NRSA) estimated that the extent of forest area degraded in the year 1988 to 1989 was about 38 per cent, compared with 24 per cent in the whole of India.

Table 9.2 *Loss of forest in Andhra Pradesh (hectares)*

Purpose	1950 to 1983–1984 (ha)	% total area lost	1984 to 1991–1992 (ha)	% total area lost
Rehabilitation	66,759	32.18	66,767	28.30
Agriculture	87,289	42.07	104,902	44.47
Non-agriculture	18,816	09.07	19,154	08.12
Mining*	5461	02.63	15,907	06.72
Encroachments	29,160	14.05	29,160	12.36
Total	*207,485*	*100.00*	*235,889*	*100.00*

Note: * Singarani Colieries are vast coal mines located in the Khammam district of Andhra Pradesh. As the mining industry grows it requires more forest land.

Source: Andhra Pradesh Forest Department (1999)

The declining quality of forests is reflected in the reduction of state forest revenue. The contribution of forests to the net domestic product has diminished from 9.4 million rupees in 1980 to 1981, to 8.5 million rupees in 1996 to 1997. As a result, forestry has become a low priority sector in the state. This is reflected in the recruitment of staff; the total number of forest staff is 2752, of which there are 136 forest service officers and 2616 ministerial staff. Many posts are not filled since there has been no recruitment since 1993. Thus, limited manpower and budgetary constraints have compelled the forest department to seek external funding. Under the World Bank's structural adjustment policy, the Andhra Pradesh government has secured funding that has led to PFM practices such as JFM.

The role of forests in agrarian livelihoods

According to recent estimates, about 14 per cent (10 million) of the state's population are dependent on forests for various livelihood activities. Many of these are tribal people, who account for 6.59 per cent of the total population and who depend on forests as the main source of their livelihood.

Agriculture, both shifting and settled, is a primary livelihood activity of forest communities. The rural poor depend on forests for fodder, fuelwood, soap nuts, gum, tamarind, amla, honey, etc. Forest dependence varies across seasons and regions. Dependence is also conditioned by the quality of the forests and existing policies and institutional arrangements. Dependence is high in seasons with less agricultural activity, particularly in the summer months, and in regions with good availability of forest produce and markets. The three regions of Andhra Pradesh represent these variations in forest and livelihood linkages. The ecological composition and value of the forests also differ between the regions. The tribal areas in Adilabad and Visakapatnam have teak forests and non-timber forest products (NTFPs) with higher forest-livelihood linkages, while Rayalaseema has mixed forests. Forest products provide poor households not only with income, but also with domestic requirements such as wood, fuel, fodder and food.

Review of forest policies and major events with a timeline

The formal origin of the state's forest policies occurred during the mid 19th century. Before the formation of the Andhra Pradesh state, two sets of forest policies were followed. While Rayalaseema and Coastal Andhra followed British policies, Telangana was covered by the Nizam forest policies (see Table 9.3).

Table 9.3 *Timeline of forest policies in Andhra Pradesh*

Year	Policy	Details
1856	Establishment of the forest department	The British created a separate department for the management of forests: Andhra and Rayalaseema were brought under the same department.
1867	Establishment of Nizam State Forest Department (Telangana)	The Nizam government established a forest department in 1867 to manage valuable teak forests.
1882	Madras Forest Act	Under its legal framework, most of the forests in Andhra Pradesh were declared state property.
1925	Forest *panchayats*	First attempt to devolve forest management to local communities in Coastal Andhra and Rayalaseema.
1945	Hyderabad Forest Act	Forests reserved as state property.
1967	Andhra Pradesh Forest Act	Modelled on the 1882 Madras Forest Act. Provides a legal and administrative framework for forest management.
1980s	Social Forestry Programme	Introduced during the 1980s to improve forest cover by encouraging extensive tree plantations, mainly to satisfy industrial demand.
1992	Joint forest management (JFM) with the support of the World Bank	Under the framework of the 1988 Forest Policy, the Andhra Pradesh government issued the JFM scheme seeking the involvement of village communities in forest management.
2002	Community forest management (CFM)	In 2002, JFM was modified to CFM, and communities were given more autonomy to manage forests.

Before independence, both the British and the Nizam forest policies asserted state monopoly control over forests for their timber and revenue requirements. This process deprived the tribal communities who depended on forests for survival of their livelihoods. This mismatch between the state and the people's interests resulted in a constant struggle, and JFM/CFM schemes are another attempt by the state to bridge the gap between policy frameworks and the everyday use of forests by the people.

Recent policy changes

JFM was promoted in Andhra Pradesh with the aim of protecting the degraded forestlands through *Vana Samarakshyan Samiti* (VSS), which, in return for their forest protection, were entitled to 50 per cent usufruct rights to forest products such as NTFPs, grasses and dry fuelwood, and a 50 per cent share of the net proceeds from the final harvest (after forest department harvesting and marketing costs). This was modified in 1996 to allow the communities 100 per cent usufruct rights on forest products. There are currently 8343 VSS covering 2.3 million hectares of forest land and involving 1.3 million VSS members, including over 600,000 women and nearly 700,000 scheduled tribes and scheduled caste members as co-partners in forest management (Government of Andhra Pradesh, 2002). The regional variations in forest cover and JFM coverage are presented in Table 9.4.

Direction of policy

An assessment of JFM/CFM in Andhra Pradesh indicates a fundamental weakness in the legal framework due to the lack of a legal basis to the scheme. The entire programme has been built on orders issued by different executive branches of government without clear linkages to enabling legislation. It is therefore susceptible to modifications in both positive and negative directions. With this in mind, Jodha (2000) has called for the provision of legal identity and status to JFM committees (VSS), including their registration under the 1860 Societies Registration Act, to equip and strengthen them to adapt to the increasingly formalized and commercialized environment (including the formal credit market), and to protect them against the whims of forest officials prone to misusing their discretionary powers.

Table 9.4 *Regional variations in Andhra Pradesh*

Particulars	Telangana	Coastal Andhra	Rayalaseema	Total
Geographic area (km²)	114,863	92,906	67,229	**274,998**
Forest area (km²)	29,242	19,563	15,008	**63,813**
Forest area as a percentage of geographic area	25.46%	21.06%	22.30%	**23.20%**
Forest area as a percentage of state forest area	45.82%	30.68%	23.52%	**100%**
JFM area (km²)	9681	4084	3067	**16,832**
JFM area as a percentage of regional forest area	33.11%	20.88%	20.44%	**26.38%**
JFM area as a percentage of state JFM area	57. 52%	24.26%	18.22%	**100%**
Number of *Vana Samarakshyan Samiti* (VSS)	3040	2653	1296	**6989**
Mean forest area per VSS (ha)	318.45	153.94	236.65	**240.84**

Source: mid-term evaluation by Rao (2006)

Methodology

The study used quantitative and qualitative research methods for data collection in order to assess the impact of PFM on livelihoods in Andhra Pradesh. To evaluate JFM, data were collected via questionnaires at the household level, and for CFM, qualitative methods, such as focus group discussions, semi-structured interviews and transect walks, were used.

Three districts were specifically selected to reflect the three agro-climatic regions of the state. Forest density and concentration of VSS were used as selection criteria in the region. Accordingly, Visakhapatnam in Coastal Andhra, Kadapa in Rayalaseema and Adilabad in Telangana were selected.

The target population for village-level study was the range of forest-dependent communities, particularly those directly and indirectly affected by the implementation of JFM. Eighteen villages – three VSS and three non-VSS villages from each district – were selected. VSS were selected using stratified random sampling on the basis of their ranking by the forest department in terms of performance. Non-VSS villages adjacent to the VSS villages and having similar physical and social characteristics were selected as approximate 'control' sites (see Table 9.5).

A sample of 25 households from each VSS village and 15 from each non-VSS village were selected through stratified random sampling for an intensive survey. A total of 225 households from the VSS villages and 135 households from the non-VSS villages were selected. Households in the village were stratified in terms of landholding patterns (landless

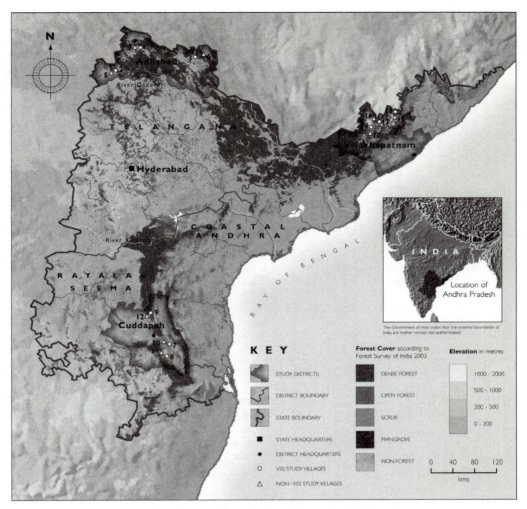

Map 6 Andhra Pradesh, including districts selected for study

Source: Jonathan Cate (original material for this book)

= no land; marginal farmers = less than 1ha; small farmers = between 1ha and 2ha; medium farmers = between 2ha and 4ha; and large farmers = more than 4ha). Although this fixed measure allows comparison between areas, it does not take into account different agro-ecological conditions or livelihoods: for instance, in dry lands, larger plots would be needed than on irrigated or hill land. This measure refers to *de jure* landholding with *pattas* (tenurial rights), even though there are many households in the study who lack formal landholdings but are *de facto* landholders in the sense that they are cultivating the land in customary ways, such as *podu*.

Table 9.5 *Details of sample* Vana Samarakshyan Samiti *(VSS) villages*

District	Village	Number of FPC households	Area of forest (ha) per FPC household	Forest type	Total forest area (ha)	Closed forest (ha)	Open forest (ha)	Plantation (ha)	Bare land (ha)
Adilabad	1 Pandhirlodhi VSS	37	6.13	Tropical evergreen forest	227.04	*191.58*	*31.32*	*2.07*	*2.07*
	2 Sainagar Thanda VSS	35	6.00	Tropical evergreen forest	208.35	*170.53*	*35.44*	–	*2.38*
	3 Heerapur VSS	67	9.54	Tropical evergreen forest	639.28	–	*616.22*	*23.06*	–
Kadapa	7 Ramachandrapuram VSS	25	15.49	Thorny bushes	387.43	*206.68*	*180.75*	–	–
	8 S.R.Palem VSS	25	17.51	Thorny bushes	437.90	*328.19*	*102.71*	*7.00*	–
	9 Mudireddypalle VSS	69	4.05	Thorny bushes	279.92	–	*279.92*	–	–
Visakhapatnam	13 Sobhakota VSS	73	3.87	Southern moist evergreen	283.21	–	*227.53*	*53.68*	–
	14 Gudlamveedi VSS	50	4.24	Southern moist evergreen	212.09	–	*48.66*	*85.96*	*68.47*
	15 Nandivalasa VSS	51	3.69	Southern moist evergreen	188.69	–	*69.44*	*30.60*	*88.65*

Source: Reddy et al (field survey, 2004–2005)

Outcomes and impacts of PFM

This section focuses on the practice of PFM implementation and demonstrates how different practices and varied local conditions lead to variations in outcomes. The implementation of PFM involves selecting sites; forming groups; transferring rights and responsibilities; drafting micro-plans; disbursing financial and other support; access; sale of forest produce; and so on. Analysis was carried out using a range of methods, including the livelihoods framework discussed in Chapter 4.

JFM processes in Andhra Pradesh: Local forest management institutions

The objective of JFM policy is 'motivating and organizing village communities for protection, afforestation and development of degraded forest land, especially in the vicinity of habitation' (GoI, 1990, p157). The programme was intended to be assisted by non-governmental organizations (NGOs), village communities (as beneficiaries) and the state forest departments. Once a VSS is formed, it becomes responsible for protecting its forest.

Constitution of a VSS

A VSS is a forest protection committee constituted under JFM in Andhra Pradesh. The section officer or range officer is expected to convene meetings at the village level in which all adult members of the village participate. The quorum of the meeting is set at 50 per cent of the households in the village. After explaining the concept of JFM, the forest section officer constitutes the VSS. In our sample villages, forest officials approached the village leaders after conducting general body meetings and convinced them of the benefits of VSS. No NGO was associated with the formation of VSS in Adilabad and Visakhapatnam, while in Kadapa, the VSS in all of the study villages were formed with the involvement of NGOs and forest department officials.

Constitution of management committees

The management committee is a representative body, elected to execute the VSS works, with 8 to 10 elected members, at least 30 per cent of whom are women. The 'election' to the management committee is said to be 'unanimous', which casts doubt on the whole point of the procedure. In general, management committees are dominated by the strongest groups in tribal villages and influential elites in non-tribal villages. This is because forest department officials, when forming VSS, contacted the village elders and influential persons, and nominated office bearers according to their advice. This process was generally accepted without protest, mainly due to fear of losing waged employment.

Social inclusion and exclusion

According to the guidelines, 50 per cent of the households in the village must be involved in VSS, membership being provided for two members of each family (in most cases the household 'head' and spouse). In practice, male members dominate (79 per cent in Visakhapatnam, 81 per cent in Kadapa and 83 per cent in Adilabad). In terms of the numbers of each category of farm household, it is the marginal and small farmers in Visakhapatnam and Adilabad and the landless in Kadapa who comprised most of our sample respondents. The scarcity of women and old people among the members means that they are excluded from their traditional access to forests.

Almost all of the households in the VSS villages are members, with only minor exceptions, such as in Heerapur and Sobhakota, where 19 and 14 households, respectively, have not become members. In Sobhakota, people have refused membership due to fear of losing *podu* land, and in Heerapur, because they live too far (1.5km) from the main village and due to the availability of six months' employment on railway works in a nearby town. The VSS at Nandivalasa began as a cluster of villages; but soon only the main Nandivalsa village members remained and the others became inactive. VSS male and female membership is not exactly equal in most of the villages, although the differences are marginal.

Governance

In all our sample villages, the general body and management committee meetings appear to have been conducted on a regular basis, with general bodies every six months and management committees every month. Interaction between the general body, management committee and forest officials is required when making operational decisions and planning activities, such as forest plantation and waged work. Regular attendance of members at meetings would ensure proper awareness and accountability. In villages such as Srirangaraju Palem, disadvantaged communities such as the Mudiraj were deliberately not invited to meetings, which resulted in conflict in the community. This, in turn, affected people's wider participation in JFM.

The attendance of VSS members at meetings was reported to be low in all three districts. To the question 'Do members attending the meetings have a say in VSS decisions?', the majority of respondents in Kadapa (88 per cent) and Adilabad (80 per cent) districts indicated that they always had a say at meetings, while only 45 per cent in Visakhapatnam had a similar experience. The main reason for this is that women and minority clan members were not allowed to express their views in management committee meetings. The VSS appears to function democratically in terms of participation in general bodies in all the villages; but some in management committee decision-making processes are dominated by their president. This was observed in Nandivalasa, Ramachnadrapuram and Sai Nagartanda VSS in discussions concerning the nature of VSS work, members to be called for VSS waged work, seasons of work, etc. Some of the villagers in Nandivalas complained to the range officer about the management committee's misappropriation of VSS funds; but no action was taken. In Mudireddypalle, the general body had removed the president on an allegation of misusing VSS funds for personal gain. This attests that JFM funds are prone to be misused in the absence of strict monitoring mechanisms.

In decisions, the VSS represents the majority view of the community. A good majority of members in all three districts agreed with this (89 per cent in Visakhapatnam, 85 per cent in Kadapa and 96 per cent in Adilabad). It was observed that decisions by the general body are honoured in the management committee, although some presidents continue to be domineering about the nature of works to be undertaken.

Forest management planning (micro-plan)

The micro-plan is a blueprint that registers the scheme of activities to be followed by VSS. JFM policy states that:

> ... *a working scheme for the area will be prepared by the range officer concerned, in consultation with the MC [management committee] after carrying out a micro-planning exercise; the exercise would focus on demand for traditional forest products from that area and identify the measures necessary to increase the productivity through natural or arterial regeneration of the forests.* (Government of Andhra Pradesh, 1992)

Participatory rural appraisal (PRA) methods were used to prepare the micro-plan. JFM policy stipulates that the micro-plan should be prepared by all stakeholders, all groups of men and women, local NGOs, the forest department and *panchayats*.

The drafting of the micro-plan has received a lot of criticism in JFM. Although it is supposed to be prepared with the participation of all general body members, in practice, forest department staff play a dominant role, as was the case in our sample villages. However, the presence of NGOs at the preparation of Kadapa's micro-plan made hardly any difference, as NGOs are tied by their donors and go by the prescriptions of the forest department'. NGO members in Kadapa stated that the forest department allotted only three days for the completion of the micro-plan, ensuring that its own forest department preferences took priority. The main objective of forest department officials in preparing micro-plans is to make them an integral part of their working plans.

Although the majority of general body members in the sample villages accepted that they were in agreement with the micro-plan model (52 per cent in Visakhapatnam, 57 per cent in Kadapa and 82 per cent in Adilabad), a sizeable proportion of respondents replied that they had not been consulted about it, showing that a substantial proportion of the population does not have a voice in VSS work. For instance, in Sainagartanda VSS in Adilabad, villagers felt that due to extensive teak plantation, the availability of the wild fruits which provide an income in the agricultural lean season has decreased. These community concerns were not incorporated within the micro-plan. In Visakhapatnam district, reclamation of *podu* land for VSS forest has taken place on a massive scale. According to the study by Winrock International, about 25,000ha of *podu* have been brought under JFM (Borgoyary, 2002). Details of *podu* land brought under JFM were not mentioned in the micro-plan in our sample villages, suggesting that micro-plans were prepared mainly according to the choices of the forest department without due consideration for people's concerns.

VSS and the forest department

The forest department's traditional control over forests was expected to be reduced by PFM. Respondents in our sample villages (44 per cent in Visakhapatnam, 41 per cent in Kadapa and 55 per cent in Adilabad) mentioned that this has not happened. In Andhra Pradesh, the involvement of the community in forest management could be seen in tribal areas of Visakhapatnam, where access to the forest and *podu* were regulated by the council of elders. Tribal festivals and lifestyle include many conservation and community ethics. In Adilabad, trees such as *muhua* were preserved as community property. Some of the NGOs (Samatha, 2005) accuse the VSS of being responsible for the demise of traditional forest management. However, our data reveal that respondents in all three districts overwhelmingly disagreed with this suggestion (92 per cent in Visakhapatnam, 90 per cent in Kadapa and 87 per cent in Adilabad).

JFM has initiated scope for regular interaction between local people and the forest department in a relationship that has changed from one resembling police and thieves to co-protectors of forests. This change of mindset, at least at policy level, has come about due to frequent interaction between forest department staff and local people. However, respondents were unanimous that the forest department should not completely give up its control over forests because they fear that the absence of the forest department would lead to village conflict and, possibly, to domination by outsiders.

Effective forest protection

One of the main roles envisaged by the JFM policy for VSS is to protect forests from local and non-local extraction, including timber and fuelwood, NTFPs and grazing, fodder and grasses. This has resulted in conflict between VSS members and outsiders who have customary rights, such as grazing their animals in the forest, a problem observed in Kadapa's Mudiraddey Pelly scheduled caste colony, where villagers could not prevent upper caste people from using VSS areas for grazing due to fear of losing their employment as agricultural labourers.

In our sample villages, forest protection appeared to be weak, particularly in Heerapur in Adilabad and Mudireddypalle in Kadapa. Two factors were behind this: first, due to the large number of households available for waged employment in VSS, work is minimal, and, second, caste conflicts (in Heerapur between backward castes and scheduled tribes) and sub-caste conflicts (between Madiga and Malas SCs) weaken possibilities for collective action. As a result, forest vegetation has not improved. The forests adjoining VSS are also becoming degraded due to incursions by outsiders who are not part of the VSS, such as upper caste members in Heerapur and a few Mudireddypalle families. Violations of VSS rules are normally related to unauthorized cutting of poles, theft of forest produce and grazing without permission. Interestingly, the sample villages have not officially recorded any such offences by members.

Markets for forest produce

The marketing of forest produce is an important income-generating activity of forest-dependent people. In Adilabad, people prefer to sell produce to middlemen, rather than to centralized government collection centres, due to the problem of transport costs. However, in Adilabad and Kadapa districts, the availability of NTFPs is less than in Visakhapatnam. Hence, the marketing systems are less developed and collectors sell off produce to the private traders who contact them at their homes. In Visakhapatnam, the Girijan Cooperative Corporation (GCC) is the sole authority for the collection of forest produce. Although there is criticism of the GCC monopoly over forest products, the majority prefer to use it for the fair price that it offers. Since JFM, the marketing of wood and some NTFPs has been carried out by the forest department in our sample villages.

Podu (shifting cultivation), encroachment and rehabilitation

Shifting cultivation, or *podu*, was seen in Visakhapatnam and is a contentious issue. Tribal people critically depended on *podu* for survival since it contributes about 20 per cent of their income, besides providing food security. There has been a continuous struggle over the practice of *podu* between the forest department and tribal people. The forest department has branded it as ecologically detrimental, although its scientific credentials are in vogue – some studies argue that *podu* is a suitable form of cultivation on the hill slopes in India. Under JFM, about 37,000ha of *podu* land was under JFM (Samaj, 2006). It is estimated that about 20,000ha of *podu* in Visakapatnam was brought under JFM treatment (Borgoyary, 2002). Although the natural cycle of *podu* had gradually become settled agriculture, tribal groups with highly diversified cropping patterns (on average, they cultivate ten crops on a *podu* patch) protect soil fertility and nutrients.

  Podu has been practised in all the three sample villages in Visakhapatnam, and most of the tribal households had *podu* land, although they had only temporary 'D-form' *patta,* granted by the revenue authority to people with the consent of the forest department for a temporary period of time. The forest department considers *podu* cultivators as *de jure* and *de facto* encroachers, and does not recognize their rights since they are not recorded in legal documents. After the inception of JFM, the forest department confiscated most of the tribal *podu* land for not having 'D' *pattas*. We were told that occasionally, the forest department had bribed VSS members to identify the land of households without 'D' *pattas*. Tribal activist groups have strongly criticised these forest department methods as divisive and exemplifying a 'divide-and-rule' mentality against them.

Podu/encroachment and JFM in the study villages

In Sobhakota village, the forest department had confiscated a few hectares of *podu* land well before the inception of PFM. But strangely, there was no serious resistance except for mild protests. This silence was mainly due to the forest department's promises of compensation and rehabilitation, although these have not been fulfilled. The forest department's promises of compensation were kept in the case of Nandivalasa, however, where affected families were given 'roof tiles' (about 2000 to 4000 rupees per household). Twenty of the 30 families were also promised the award of *pattas* to the lower part of the hill on which they practised *podu*.

Livelihood impacts

The impact of PFM was studied using a revised sustainable rural livelihoods framework, narrowed to a cluster of indicators, disaggregated across the different wealth ranks. An attempt was made to understand sustainable changes compared to the situation had the policy not been implemented. This is very challenging because the changes are complex and other factors are also variable. The following indicators are used to illustrate the impact of PFM on the livelihoods of people in Andhra Pradesh:

- changes to household assets, entitlements to collective assets and opportunities (e.g. loss or gain of forest land from cultivation, grazing and fuelwood);
- change in income;
- vulnerability and well-being;
- sustainability of impacts and causal links between JFM and impacts.

Impact on livelihoods is measured in terms of changes in various indicators due to JFM in our sample villages. Conditions pre- and post-VSS, wherever possible and in both VSS and non-VSS villages, were included. The measurement of impact is based on information collected from 25 sample VSS households and 15 non-VSS households in each district. Interestingly, VSS villages seem to have more degraded forests and non-VSS have much better forest cover. The impact of PFM is measured across different economic groups based on their landholdings in order to examine its distributional aspects.

Changes in household assets and entitlements to collective assets

This section describes how households' primary productive private assets and entitlements to collective assets have changed with JFM. The principal assets relevant to livelihoods in the study areas are private agricultural lands, entitlement to public forest and grazing lands.

Changed forest condition and changed entitlements (natural capital)

The changed regulatory framework around forest use (i.e. restrictions on extraction) and the management inputs (planting, etc) of JFM has led to changes in forest condition. About 70 per cent of the forest lands allotted to VSS are under JFM. JFM restricts the traditional access of people to neighbouring forests, legitimizing this restriction by providing alternative wage employment in VSS works. These JFM restrictions were particularly severe for those households most dependent on the forest (usually the poorest) and during the initial period when forest access is most restricted to allow regeneration. Data on the availability of various forest products to households indicate that availability has, in most cases declined (in some cases, significantly) (see Table 9.6). Significant improvement was observed only in Adilabad in the case of fodder (all farmers) and NTFPs (landless), although there has been a significant decline in the availability of fuelwood, especially for the landless and small and marginal farmers. Restrictions on grazing had a negative impact on poor households, especially at the outset of JFM, forcing them to sell off their ruminants, while small and marginal farmers were less affected because they have their own grazing land. In the case of fuelwood, small and marginal farmers used gas supplied under a separate government project. Here, again, the landless, dependent on fuelwood, suffered. The poor are the worst victims of restrictions in VSS areas.

Time spent on collecting forest products (per head load) is affected by product availability, condition and location of forests, and is also determined by whether a forest is closed or restricted by the VSS. If the forest quality is poor, households have to spend more time in collecting and transporting produce. In our sample villages, there has been a decline in the time spent collecting each unit of forest produce for all households, particularly the landless. In Visakapatnam and Kadapa, time spent on fuelwood and fodder collection has declined in the case of landless (see Table 9. 8). But in some cases, such as the collection of NTFPs in all sections of Visakhapatnam and fodder and fuel collection in Adilabad, time to collect has

Table 9.6 *Mean changes in resource availability (kilograms per household) across wealth groups in VSS villages before and after joint forest management (JFM)*

VSS area	Resource	Landless (kg per household)	Marginal and small wealth groups (kg per household)	Medium and large wealth groups (kg per household)	All (kg per household)
VISAKHATNAM	Fuel: before	289.60	408.43	4.82	234.29
	Fuel: after	284.33	294.18	5.11	194.54
	Fuel: % change	−1.82% ns	−27.97%**	5.93% ns	−16.96%**
	Fodder: before	13.33	5.47	0.79	6.53
	Fodder: after	0.10	7.15	1.07	2.77
	Fodder: % change	−99.25% ns	30.79% ns	36.36% ns	−57.51% ns
	NTFP: before	37.96	30.86	11.25	26.69
	NTFP: after	41.20	31.27	10.75	27.74
	NTFP: % change	8.54% ns	1.33% ns	−4.44% ns	3.94% ns
	Timber: before	0.00	2.87	4.82	2.56
	Timber: after	0.00	1.16	7.68	2.95
	Timber: % change	0.00%	−59.56% ns	59.26% ns	14.93% ns
KADA	Fuel: before	1784.68	3494.32	0	2639.50
	Fuel: after	874.90	868.64	0	871.77
	Fuel: % change	−50.98% ns	−75.14%**	0.00%	−66.97%**
	Fodder: before	1.11	0.68	0	0.90
	Fodder: after	0.28	0.82	0	0.55
	Fodder: % change	−75.00% ns	20.00%	ns 0.00%	−38.87% ns
	NTFP: before	29.73	93.59	0	61.66
	NTFP: after	36.62	44.18	0	40.40
	NTFP: % change	23.18% ns	−52.79% ns	0.00%	−34.48% ns
	Timber: before	16.71	45.45	0	31.08
	Timber: after	0.00	0.00	0	0.00
	Timber: % change	−100.00% ns	−100.00% ns	0.00%	−100.00% ns
ADILABAD	Fuel: before	810.00	799.27	682.38	763.88
	Fuel: after	501.42	565.10	644.29	570.27
	Fuel: % change	−38.10%**	−29.30% ns	−5.58% ns	−25.35%**
	Fodder: before	51.18	26.02	62.38	46.53
	Fodder: after	75.64	58.25	33.71	55.87
	Fodder: % change	47.80% ns	123.88% ns	−45.95% ns	20.08%*
	NTFP: before	38.63	63.97	53.10	51.90
	NTFP: after	48.72	53.97	72.90	58.53
	NTFP: % change	26.12%**	−15.63% ns	37.31% ns	12.78% ns
	Timber: before	288.24	142.96	395.24	275.48
	Timber: after	14.22	26.21	14.29	18.24
	Timber: % change	−95.07%*	−81.66%*	−96.39%*	−93.38%**

Notes: ns = not significant; * = significant at 0.01 confidence levels; ** significant at 0.1 confidence levels.

Source: Reddy et al (field survey, 2004–2005)

Table 9.7 *Mean changes in collection time in hours per year per household across wealth groups in the VSS villages before and after JFM*

VSS area	Resource	Landless (kg per household)	Marginal and small wealth groups (hours per year per household)	Medium and large wealth groups (hours per year per household)	All (hours per year per household)
VISAKHATNAM	Fuel: before	14.00	10.69	6.88	10.52
	Fuel: after	8.33	8.21	6.00	7.52
	Fuel: % change	−40.48%*	−23.16%**	−12.73% ns	−25.45%**
	Fodder: before	4.00	4.46	2.88	3.78
	Fodder: after	0.83	3.90	2.50	2.41
	Fodder: % change	−79.17% ns	−12.50% ns	−13.04% ns	−34.90% ns
	Timber: before	0.00	0.20	0.25	0.15
	Timber: after	0.00	0.13	0.88	0.34
	Timber: % change	0.00%	−33.33% ns	250.00% ns	125.23% ns
	NTFP: before	40.24	39.44	30.80	36.83
	NTFP: after	32.27	36.01	22.61	30.29
	NTFP: % change	−19.82%*	−8.70%*	−26.61% ns	−18.38%**
KADA	Fuel: before	13.50	17.11	0.00	10.20
	Fuel: after	8.80	7.05	0.00	5.28
	Fuel: % change	−34.85%**	−58.77%**	0.00%	−48.22%*
	Fodder: before	0.00	0.00	0.00	0.00
	Fodder: after	0.00	0.00	0.00	0.00
	Fodder: % change	0.00%	0.00%	0.00%	0.00%
	Timber: before	0.09	1.05	0.00	0.38
	Timber: after	0.00	0.00	0.00	0.00
	Timber: % change	−100.00% ns	−100.00% ns	0.00%	−100.00%*
	NTFP: before	34.41	48.00	0.00	27.47
	NTFP: after	30.67	27.07	0.00	19.25
	NTFP: % change	−10.85% ns	−43.61%**	0.00%	−29.93%**
ADILABAD	Fuel: before	13.77	15.56	14.60	14.64
	Fuel: after	8.73	10.13	9.40	9.42
	Fuel: % change	−36.56%**	−34.94%**	−35.62%*	−35.71%**
	Fodder: before	11.97	1.91	1.20	5.02
	Fodder: after	1.90	2.31	0.80	1.67
	Fodder: % change	−84.12% ns	21.31% ns	−33.33% ns	−32.05% ns
	Timber: before	2.93	1.81	0.00	1.58
	Timber: after	1.37	0.88	0.00	0.75
	Timber: % change	−53.41% ns	−51.72% ns	0.00%	−52.77%*
	NTFP: before	59.74	67.44	46.42	57.87
	NTFP: after	55.96	61.02	32.08	49.69
	NTFP: % change	−6.32%**	−9.53%**	−30.88%**	−15.58%**

Notes: ns = not significant; * = significant at 0.01 confidence levels; ** significant at 0.1 confidence levels.

Source: Reddy et al (field survey, 2004–2005)

Table 9.8 *Change in time spent on collecting forest products per kilogram before and after joint forest management*

District	Forest product	Wealth groups	Quantity (kg)		Hours to collect		Total collected (kg per hour)	
			Before	After	Before	After	Before	After
VISAKHATNAM	Fuel	Landless	289.60	284.33	14.00	8.33	20.68	34.13
		Marginal and small	408.43	294.18	10.69	8.21	38.20	35.83
		Medium and large	4.82	5.11	6.88	6.00	0.70	0.85
		All	*234.29*	*194.54*	*10.52*	*7.52*	*22.27*	*25.86*
	Fodder	Landless	13.33	0.10	4.00	0.83	3.33	0.12
		Marginal and small	5.47	7.15	4.46	3.90	1.22	1.83
		Medium and large	0.79	1.07	2.88	2.50	0.27	0.42
		All	*6.53*	*2.77*	*3.78*	*2.41*	*1.72*	*1.14*
	NTFP Timber	Landless	37.96	41.20	40.24	32.77	0.94	1.27
		Marginal and small	30.86	31.27	39.44	36.01	0.78	0.86
		Medium and large	11.25	10.75	30.80	22.61	0.36	0.47
		All	*26.69*	*27.74*	*36.83*	*30.29*	*0.72*	*00.91*
		Landless	0.00	0.00	0.00	0.00	0.00	0.00
		Marginal and small	2.87	1.16	0.20	0.13	14.35	8.92
		Medium and large	4.82	7.68	0.25	0.88	19.28	8.72
		All	*2.56*	*2.95*	*0.15*	*0.34*	*17.06*	*8.67*
KADA	Fuel	Landless	1784.68	874.90	13.50	8.80	132.19	99.42
		Marginal and small	3494.32	868.64	17.11	7.05	204.22	123.21
		Medium and large	0	0	0.00	0.00	0.00	0.00
		All	*2639.50*	*871.77*	*10.20*	*5.28*	*258.77*	*165.10*
	Fodder	Landless	1.11	0.28	0.00	0.00	0.00	0.00
		Marginal and small	0.68	0.82	0.00	0.00	0.00	0.00
		Medium and large	0.00	0.00	0.00	0.00	0.00	0.00
		All	*0.90*	*0.55*	*0.00*	*0.00*	*0.00*	*0.00*
	NTFP	Landless	29.73	36.62	34.41	30.67	0.86	1.19
		Marginal and small	93.59	44.18	48.00	27.07	1.94	1.63
		Medium and large	0	0	0.00	0.00	0.00	0.00
		All	*61.66*	*40.40*	*27.47*	*19.25*	*2.24*	*2.09*
	Timber	Landless	16.71	0.00	0.09	0.00	185.66	0.00
		Marginal and small	45.45	0.00	1.05	0.00	43.28	0.00
		Medium and large	0	0	0.00	0.00	0.00	0.00
		All	*31.08*	*0.00*	*0.38*	*0.00*	*81.78*	*0.00*
ADILABAD	Fuel	Landless	810.00	501.42	13.77	8.73	58.82	57.43
		Marginal and small	799.27	565.10	15.56	10.13	51.36	55.78
		Medium and large	682.38	644.29	14.60	9.40	46.73	68.54
		All	*763.88*	*570.27*	*14.64*	*9.42*	*52.17*	*60.53*
	Fodder	Landless	51.18	75.64	11.97	1.90	4.27	39.81
		Marginal and small	26.02	58.25	1.91	2.31	13.62	25.21
		Medium and large	62.38	33.71	1.20	0.80	51.98	42.13
		All	*46.53*	*55.87*	*5.02*	*1.67*	*9.26*	*33.45*
	NTFP	Landless	38.63	48.72	59.74	55.96	0.64	0.87
		Marginal and small	63.97	53.97	67.44	61.02	0.94	0.88
		Medium and large	53.10	72.90	46.42	32.08	1.14	2.27
		All	*51.90*	*58.53*	*57.87*	*49.69*	*0.89*	*1.17*
	Timber	Landless	288.24	14.22	2.93	1.37	98.37	10.37
		Marginal and small	142.96	26.21	1.81	0.88	78.98	28.78
		Medium and large	395.24	14.29	0.00	0.00	0.00	0.00
		All	*275.48*	*18.24*	*1.58*	*0.75*	*174.35*	*24.32*

Source: Reddy et al (field survey, 2004–2005)

increased due to access restrictions in JFM areas. Since JFM, it is taking more time to gather fodder and fuelwood. The main reason for this is the initial restrictions on access to VSS forests, necessitating travelling further for fuel and fodder.

Agricultural land (natural capital)

Land is the most important natural capital for households even in forest fringe areas, where access to land remains the fundamental determinant of the economic status of most households. For this reason we have categorized the households according to their landownership in order to capture variations across economic groups (see Figure 9.1). However, ownership of land is more problematic as an indicator in tribal areas such as Visakhapatnam and Adilabad districts, where households have customarily practised agriculture on forest lands for generations without formal tenure rights. While the forest department considers this land use 'encroachment', tribal people argue that it is their own customary land and demand land *pattas*. The forest department has been trying to discourage *podu* by bringing it under JFM areas. Thus, in tribal areas natural capital has declined, resulting in loss of income since JFM due to restrictions on *podu*/encroached land.

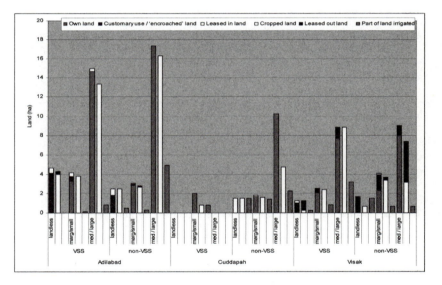

Figure 9.1 Mean landholding and land use of respondent households by wealth rank
(VSS and non VSS members)

Source: Reddy et al

Access to livestock and other assets (physical capital)

Physical capital complements natural capital in creating viable and productive livelihoods. Important physical capital assets in the sample VSS include irrigation assets, livestock and other assets. Changes in physical capital resulting from the implementation of JFM are neither the main emphasis of the policy, nor were they found to be particularly significant.

Physical infrastructure has been created via JFM in our study sites. Through entry-point activities, the forest department has created community halls, temples and water tanks to create a favourable atmosphere between itself and the local people, as well as supplying agricultural implements and other items to help reduce people's dependency on forests.

Table 9.9 *Changes in livestock holdings (numbers per household) across wealth groups in the VSS villages*

District	Animal	Numbers kept before/after JFM and % change	Landless	Marginal and small	Medium and large	All
VISAKHAPATNAM	Milk Animals	Before	2.00	1.36	0.75	1.35
		After	0.00	1.44	1.25	1.30
		% Change	*−100.00*	*5.88*	*66.67*	*−3.7*
	Draught Animals	Before	2.00	2.11	0.80	1.96
		After	0.67	1.62	1.80	1.56
		% Change	*−66.67*	*−23.08*	*125.00*	*−20*
	Small Ruminants	Before	3.00	2.88	4.33	3.04
		After	2.50	2.72	5.50	3.00
		% Change	*−16.67*	*−5.65*	*26.92*	*−1.32*
KADAPA	Milk Animals	Before	3.00	2.20	0	2.78
		After	0.00	2.00	0	0.56
		% Change	*−100.00*	*−9.09*	*0*	*−79*
	Draught Animals	Before	1.50	1.50	0	1.50
		After	1.00	1.000	0	1.0
		% Change	*−33.33*	*−33.33*	*0*	*−33.33*
	Small Ruminants	Before	13.13	12.29	0	12.89
		After	4.13	7.86	0	5.17
		% Change	*−68.57*	*−36.05*	*0*	*−60*
ADILABAD	Milk Animals	Before	2.06	2.08	1.25	2.00
		After	0.94	2.62	8.50	2.39
		% Change	*−54.05*	*25.93*	*580.00*	*19.5*
	Draught Animals	Before	1.70	1.00	2.00	1.37
		After	1.55	1.83	3.00	1.81
		% Change	*−8.82*	*83.33*	*50.00*	*31.96*
	Small Ruminants	Before	2.42	4.55	10.00	4.11
		After	6.75	4.64	30.00	7.54
		% Change	*179.31*	*2.00*	*200.00*	*83.48*

Notes: These statistics are presented as descriptive based on limited sample size in each subcategory so should not be necessarily taken to be statistically significant.

Source: Reddy et al (data from field survey, 2004–2005)

In Visakhapatnam, roof tiles were distributed to all families, resulting in good housing for the poor tribal people who had depended on grass and leaves; this also reduced the pressure on forests for leaves, grass and timber. The distribution of iron agricultural implements was aimed at reducing dependency on forests for wood. These measures helped to build a friendly atmosphere between VSS members and the forest department.

Livestock

Livestock constitutes an important part of households' productive resources. Grazing access is particularly important for the poorest households, who lack on-farm supplies of fodder. After JFM, the average livestock holdings of landless households have declined considerably in all districts. This appears to be an important negative impact on the poorest. On the other hand, livestock holdings of medium and large farmers have increased in all three districts. In Adilabad, even small and marginal farmers have improved their holdings. In Kadapa, goat-rearing, an important livelihood activity of the poor, had declined due the forest department's strict policy aimed at excluding goats altogether due to their degrading effect on forest. Overall, JFM has not had a significant impact on livestock (see Box 9.1).

Box 9.1 Joint forest management (JFM) grazing policy is anti-poor: Myth or reality?

Gopinath Reddy

Grazing is one of many contentious issues that has received much attention under the JFM programme. The former Andhra Pradesh chief minister's statement at the assembly on 1 April 2001 that 'goats are the enemy of environment and forests' has been perceived as anti-poor, anti-low caste, pro-landowning and anti-livestock.

In Visakhapatnam, people in our sample villages had high numbers of livestock, including goats, but had not faced any problems due to access to wastelands where their livestock could graze. Extra care is taken by the villagers to ensure that no goat enters newly planted VSS land, although after a certain period of time they relaxed their embargo and allowed some goats into the VSS forest area. However, villagers in Kadapa complained that due to restrictions on browsing, they were forced to sell their goats. In Sainagar Tanda in Adilabad, many people depend on livestock and naturally minor incidences of trespassing in VSS forests are reported. But, overall in our sample villages, grazing-related problems were not great. This is mainly due to the steep decrease in goat numbers as rural people opt for more reliable alternative livelihoods.

Changes in household income

JFM has had an impact on the income level of people in our study VSS. Annual gross household income ranges between 8000 rupees in Visakhapatnam and 14,000 rupees in Adilabad. Kadapa's higher gross income is mainly due to non-farm employment and remittances from abroad. In all the VSS villages, agriculture is the major source of income, with income from forest produce (NTFPs) second in Adilabad and Visakhapatnam districts, and income from non-farm labour second in Kadapa.

In most cases, the income contribution of NTFPs is greatest for poor farmers (landless, marginal and small). Average income from forests is about 2000 rupees per year for non-poor households and 1500 to 2500 rupees for poor households. In Kadapa, NTFPs make a small contribution to household income in two of the sample VSSs (see Figures 9.6 and 9.7). The contribution of VSS-related wage labour is greatest in poor households.

The comparison of before and after JFM scenarios indicates that there is an increase in households' gross income across all VSS for poor and non-poor households (see Figures 9.2 to 9.7). Part of this is due to inflation. The rise in gross income is greatest in Adilabad, followed by Kadapa and Visakhapatnam (see Table 9.10). The contribution of VSS wage income constitutes 10 per cent of total income in non-poor and 13 per cent in poor households in Visakhapatnam and Kadapa, and 8 and 10 per cent, respectively, in Adilabad. Thus, non-poor households have increased their income more than poor households. Respondents in Kadapa and Adilabad informed us that during the lean agricultural season, VSS work offered a good means of survival. Wage labour for managing the forest is only offered for a few months a year, and they know that it will eventually cease. Hence, communities do not consider it a sustainable livelihood activity. The contribution of NTFPs to household income has declined at all study sites. Of course, both poor and non-poor households have experienced the income decline from NTFPs.

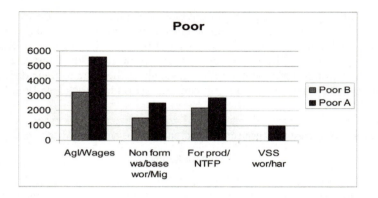

Figure 9.2 Income for poor households (before and after) in Adilabad district (rupees)

Source: Reddy et al

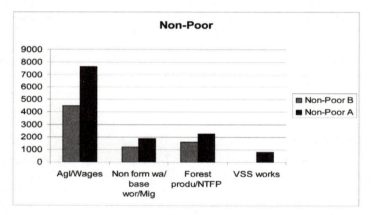

Figure 9.3 Income for non-poor households (before and after) in Adilabad district (rupees)

Source: Reddy et al

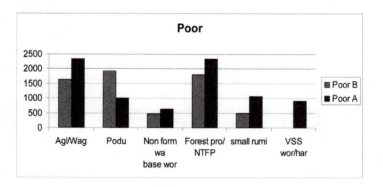

Figure 9.4 Income for poor households (before and after) in Viskhapatnam district (rupees)

Source: Reddy et al

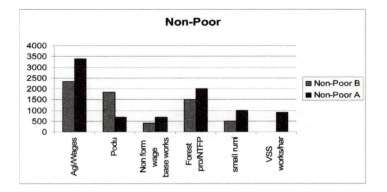

Figure 9.5 Income for non-poor households (before and after) in Viskhapatnam district (rupees)

Source: Reddy et al

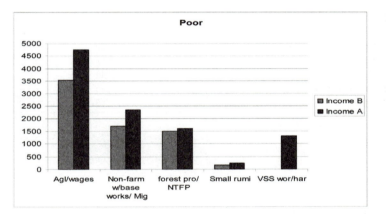

Figure 9.6 Income for poor households (before and after) in Kadapa district (rupees)

Source: Reddy et al

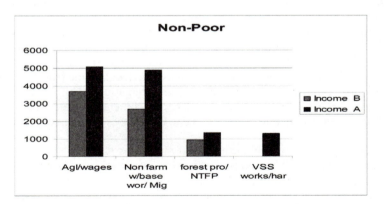

Figure 9.7 Income for non-poor households (before and after) in Kadapa district (rupees)

Source: Reddy et al

Table 9.10 *Sources of gross household income of poor and non-poor households in VSS villages before and after joint forest management (rupees)*

District	Sources	Poor		Non-poor	
		Before (rupees)	**After** (rupees)	**Before** (rupees)	**After** (rupees)
Adilabad	Agriculture/wages	3250	5617	4467	7633
	Non farm-based waged work	1500	2500	1200	1867
	Forest produce/NTFPs	2200	2875	1600	2283
	VSS works/harvesting	–	1000	–	783
	Total	*6950*	*11,992*	*7267*	*12,567*
Viskhapatnam	Agriculture/wages	1625	2350	2333	3400
	Podu	1933	1000	1833	667
	Non-farm-based waged work	467	633	400	683
	Forest produce/NTFPs	1817	2333	1517	2000
	Small ruminants	500	1067	500	1000
	VSS works/harvesting	–	900	–	900
	Total	*6342*	*8283*	*6583*	*8650*
Kadapa	Agriculture/wages	3533	4750	3700	5083
	Non-farm-based waged work/migration	1700	2342	2700	4900
	Forest produce/NTFPs	1500	1617	950	1350
	Small ruminants	167	250	–	–
	VSS works/harvesting	–	1300	–	1300
	Total	*6900*	*10,258*	*7350*	*12,633*

Source: Reddy et al (data from field survey, 2004–2005)

In Visakhapatnam, income from *podu* constitutes 30 per cent for poor and 28 per cent for non-poor households, total income having declined due to the reclamation of *podu* lands under JFM. This is one of the reasons for the relatively low growth in gross household income in this district. The decline of income from *podu* is 18 per cent for poor and 20 per cent for non-poor households. This is a decline not only in terms of income, but also of food security, although it is compensated to some extent by waged VSS work, which constitutes 11 per cent for poor and 10 per cent for non-poor. Since (of all groups) it is tribal groups who practice *podu* most, it is they who have lost their regular income and food security due to JFM.

The data reveal that forest-dependent income is less than income from agriculture, even in high-value forest regions such as Adilabad and Visakhapatnam. This explains the demand for land, which is reflected in the incidence of encroachment in these regions. The major livelihood strategy for forest and forest-adjacent dwellers is to achieve or increase their private land assets

The creation of wage employment in forestry activities is JFM's direct contribution to the household income of VSS members. Despite improved employment opportunities, unemployment continues to be on the high side in all of the sample villages (see Table 9.11). On average, two to three persons are unemployed in each household, especially in the summer season (February to April), and this is higher in VSS compared to non-VSS villages. Interestingly, unemployment is more in medium and large farming households, which could be due to greater migration by landless than from landed households.

Table 9.11 *Days of employment generated by VSS-related works across wealth groups per household*

District	Days of employment	Landless	Marginal and small wealth groups	Medium and large wealth groups	Mean
VISAKHAT-NAM	Number of days worked in last 12 months	21.67	26.30	21.00	22.99
	Number of days worked since start of JFM	161.67	162.72	119.29	147.89
KADAPA	Number of days worked in last 12 months	33.76	34.05	–	33.91
	Number of days worked since start of JFM	215.33	232.37	–	223.85
ADILABAD	Number of days worked in last 12 months	31.06	50.26	56.67	46.00

Source: Reddy et al (data from field survey, 2004–2005)

There is clear-cut seasonality in the main livelihood activities. Agricultural activity takes place from July to March. NTFPs are collected in May and June when households are free from working on their land. Landless households tend to spend more time collecting NTFPs; for instance, landless households in Nandivalasa, Visakhapatnam district's NTFP collection stretches from January to June. Dependency on forests is generally greater during the summer; however, the pressure is reduced due to the availability of VSS work in the summer (May to June).

Changes to livelihood sustainability, security and vulnerability

JFM/CFM is aimed at enhancing rural livelihood security in order to reduce vulnerability and risk. As already discussed, JFM has not significantly enhanced household livelihoods. In Kadapa and Adilabad districts we were informed that VSS work provides an income in critical seasons when the only other option is to migrate. Some said that VSS has allowed them to stay in their village and look after their families. However, JFM has had a negative impact on the tribal people in Visakhapatnam district.

Land tenure: Vacation of land due to pressure from PFM and the forest department

In Adilabad and Visakhapatnam, land used for *podu,* considered encroachment by the forest department, has been brought under JFM. Although some temporary wage labour has been offered, it may not be sustainable in the long run since it depends on external funding. As a result, the pressure on forest land may not ease. Pressure from the forest department to vacate land used for *podu* has caused farmers in the Visakhapatnam sample villages to forgo income of about 1000 rupees per annum per household, (about 20 per cent of their total income). The landless are especially badly affected; they have lost not only the income, but also the food security that *podu* provided before JFM.

Migration

Migration is a typical response to vulnerability and economic distress. Two types of migration are reported in our sample villages: permanent and seasonal, with seasonal migration more prevalent. There has been a decline in migration across all wealth ranks and among male and female members in all of the VSS villages studied. The decline in out-migration is greatest in landless households, for whom migration is a question of survival. Two villages in Kadapa have a history of at least one family member going to Kuwait or Qatar to work as unskilled labourers. This trend still continues in spite of JFM because of the comparatively better economic returns in the Middle East; but migration in Kadapa for menial work has certainly come down. In Visakhapatnam, except in Nandivalasa village, much less migration is recorded. The decline in migration is statistically significant in both male and females cases, and is more prominent among landless households.

Table 9.12 *Changes in out-migration across wealth groups in VSS study villages (days per year)*

District		Landless (days per year)	Marginal and small wealth groups (days per year)	Medium and large wealth groups (days per year)	All (days per year)
VISAKHATNAM	Male: before	12.20	17.28	17.50	15.66
	Male: after	0.00	9.72	10.00	6.57
	% change	*−100.00% ns*	*−43.78%**	*−42.86% ns*	*−58.03%**
	Female: before	0.00	9.51	9.00	6.17
	Female: after	0.00	6.89	1.88	2.92
	% change	*0.00%*	*−27.58% ns*	*−79.17%*	*−52.66% ns*
KADAPA	Male: before	34.02	5.00	0	19.51
	Male: after	5.76	2.69	0	4.23
	% change	*−83.07%**	*−46.15% ns*	*0%*	*−78.34%**
	Female: before	37.28	3.85	0	20.56
	Female: after	9.89	1.92	0	5.91
	% change	*−73.47%**	*−50.00% ns*	*0%*	*−71.27%**
ADILABAD	Male: before	19.68	14.39	6.00	13.36
	Male: after	7.74	5.45	0.00	4.40
	% change	*−60.66%*	*−62.11% ns*	*−100.00% ns*	*−67.07%**
	Female: before	16.61	13.03	6.00	11.88
	Female: after	6.45	0.00	0.00	2.15
	% change	*−61.17% ns*	*−100.00% ns*	*−100.00% ns*	*−81.90%*

Notes: ns = not significant; * = significant at 0.01 confidence levels; ** significant at 0.1 confidence levels.

Source: Reddy et al (data from field survey, 2004–2005)

Impact of JFM on forests

A comprehensive forest survey was conducted in 2004 using global positioning systems (GPS). However, we have not conducted a further GPS survey to examine the exact changes in vegetation cover. At the same time, we do not have pre-JFM data apart from micro-plans and villagers' observations. Therefore, no conclusions can be drawn yet about the impact of JFM, although some comments can be made about present characteristics and variations in vegetation across the state.

In Visakhapatnam, VSS villages have better coverage of plantations in all the ranges, while land around Paderu and Araku VSS villages is more barren. Comparison between VSS and non-VSS villages gives a rather mixed picture, even in the case of open forest and plantations, and it is not possible to attribute any differences to the policy variable. Despite the selection of degraded areas for VSS, some of the villages have less open forest than the non-VSS villages. Similarly, plantations are mostly seen around the VSS villages. JFM proposes a package for treating degraded forest land in the VSS. Activities such as extensive plantation, gap plantation, natural regeneration, singling operations for better tree growth, and soil moisture conservation works to improve water were being undertaken in the VSS, all activities that have a bearing on forest quality.

Policy, practice and impacts on governance and the policy process

PFM in India has been subject to frequent changes according to donors' preferences. After 2000, the World Bank's lending under structural adjustment schemes was subjected to severe criticism and, as a result, lending to support developmental schemes gained popularity. This thinking was reflected by the Operations Evaluation Department, which stated that 'the [World] Bank has only very recently considered the incorporation of forest policy as a part of its overall rural development and poverty reduction strategy' (WRM, 2000). In 2002, the World Bank formulated development policy lending, with emphasis on development within an efficient institutional framework (WRM, 2005), which has since come under sharp criticism.

Change to the World Bank's policy at the macro-level has reverberations in India. Recently, the World Bank announced that JFM has failed to deliver the expected results and, hence, needs modification (WRM, 2005; World Bank, 2006). In this context, JFM in Andhra Pradesh was changed to CFM, which places greater emphasis on community-driven development (see Table 9.13). The World Bank describes the objective of CFM as '"community driven" intervention that aims to reduce poverty and 'empower' communities to take autonomous decisions regarding forest management on the lands assigned to forest protection committees' (WRM, 2005). However, the mid-term assessment of the implementation of CFM has highlighted the fact that the project could not deliver its stated objectives in a genuine reform process. Issues such as people's participation in decision-making processes have remained simply rhetoric and the forest department continues to dominate proceedings (Samatha, 2005).

Table 9.13 *Major policy changes from joint forest management (JFM) to community forest management (CFM)*

	JFM	CFM
Management committee	Ten members in management committee Women's membership 30% No specific requirement for numbers of women attending Tenure two years *Sarpanch* and forest officials as members of mangement committee	Fifteen members in management committee Women's membership at least 50% 50% women should attend Tenure three years Only elected members in management committee
Financial transactions	One joint account	Dual account system with joint account and VSS account
Usufruct rights	NTFPs divided into reserved and unreserved, and communities have rights only to selected items	All NTFPs and 100% incremental volume of timber

Source: Government of Andhra Pradesh (1992, 2002, 2004)

Policies and governance processes

Although several innovative policy initiatives were promulgated with CFM, these were not translated into the field. Several tribal communities, in general, and in our sample villages, in particular, were of the opinion that they have not been empowered in terms of decisions on forest management, which are still taken by the Andhra Pradesh Forest Department. Critical issues such as land tenure and the legal status of VSS are not addressed by the 1982 policy, and it is reported that the forest department is pressuring communities to enter into contracts with private companies (WRM, 2005).

Tenure issues

Historically, successive state policy interventions for managing forests in Andhra Pradesh have aimed at exploiting forests for commercial and revenue purposes, which in the process deprive tribal populations of their livelihoods. During the mid 19th century, the colonial state prepared policies for the exploitation of forests. The 1882 Madras Forest Act imposed state control over forests in Coastal Andhra and Rayalaseema. In Telangana, the 1945 Hyderabad Forest Act imposed state ownership of forests. These acts declared forests state property and treated the customary people's access as illegal. People's access to forests was defined as a concession or privilege to be exercised with state permission. The colonial state prioritized state interests over the subsistence and longstanding interests of its people. While creating reserve forests in Kadapa in 1886, the forest settlement officer commented about forest access:

> I do not consider that either the customary practice of grazing, leaf manure cutting, grass cutting or cutting thorns for fencing, though apparently they are of longstanding, can be regarded as rights. Government constantly disturbs such customs and disposes of wastelands as it thinks fit. The claims are therefore rejected. (Government of Madras, 1886)

The imposition of state ownership resulted in perpetual contention between local people and the forest department. Several tribal revolts took place against the forest department in Andhra and Telangana (Murali, 1982). After the 1920s, the British were forced to decentralize forest management in Madras to meet the crisis that came about in the wake of the national movement. Forest *panchayats* comprising local people were created for the management of forests (Ravi, 2005). However, these were confined to the plains districts in the Madras Presidency. Little evidence of the functioning of these *panchayats* was available. After the 1950s, the forest *panchayats* were dismantled by the independent Indian government, which argued that the forests were being degraded under them. History shows that state policy measures on forests have been aimed at meeting immediate crises, rather than being long-term constructive measures. JFM and CFM are part of this historical process.

The Social Forestry Programme promoted monoculture plantations on encroached forestlands to serve the needs of industry. This negatively affected tribal livelihoods and food security in Andhra Pradesh. Participatory forest management in the form of JFM or CFM appears to be yet another attempt by the state to entrench its position and to control forest encroachment and people's access to forests. In Andhra Pradesh, JFM was funded by the World Bank. Local communities were granted usufruct rights to enjoy 25 per cent of forest products in return for protecting the forests, increased to 100 per cent on all NTFPs. However, this policy did not give communities tenurial security. As a result, PFM in Andhra Pradesh remains a wage employment generation scheme for forest-dependent communities. The entire programme is linked to external funding. Villagers and the forest department take interest and conduct meetings only when funds arrive.

Major policy actors

As with all JFM policies in different states, forest management and CFM policies envisage attitudinal change in forest department staff from 'command and control' to 'recognizing communities as equal partners'. Under JFM, Andhra Pradesh's forest department took the leading role in decision-making processes related to forest management. Under CFM, the forest department is expected to act more as a 'facilitator, regulator and provider of technical support, while the community takes the lead in forest planning and decision making' (Government of Andhra Pradesh, 2002). But at the ground level, the forest department exercises enormous powers in decision-making.

Table 9.14 *Nature of the decision-making process in VSS management*

Functional aspects	Decision-makers		
	VSS	Forest department	Other; NGO
1 Organization of general body/management committee meetings	✓	✓	✓
2 Selection of VSS forest areas		✓	
3 Micro-plan preparation	✓	✓	✓
4 Identification of works		✓	
5 Species selection	✓	✓	
6 Carrying out of works		✓	
7 Supervision of works		✓	
8 Estimation of costs		✓	
9 Fund allocation		✓	
10 Distribution of harvesting income		✓	
11 Distribution of forest products	✓		
12 Entry-point activities	✓	✓	✓
13 Keeping minutes	✓	✓	✓
14 NTFP value addition		✓	
15 Distribution of livelihood enhancement activities		✓	
16 Marketing of forest products		✓	
17 Selection of training programme		✓	

Source: Reddy et al (data from field survey, 2004–2005)

In CFM, forest department officials continue to play a dominant role in VSS management (see Table 9.14). The VSS chairman of Mudireddey Pally in Kadapa informed us that VSS works are mainly decided by VSS chairmen and forest department staff; neither management committee, nor general body members have any role in the decision-making process. We were told that due to the low educational level of management committee members in the tribal areas of Adilabad and Visakapatnam, forest department staff manipulate accounts and misappropriate VSS funds, and that sometimes VSS management committees were threatened if they did not follow forest department commands. Thus, the empowerment of communities in terms of acquiring autonomy in decision-making process cannot be achieved in CFM:

> *The CFM project is like a sugar-coated pill which is bitter inside. The forest department explains CFM as being different from the previous JFM project in Andhra Pradesh, when communities were just treated as labour to do the forest*

department works and forest protection. But what we see now after two and a half years is that CFM is just old wine in a new bottle. There are small changes, but basically this project is JFM with another name and the people do not have more power to decide how to use the forests … the forest department still dictates how the forest and land is to be used. (WRM, 2005)

NGOs played a key role in the implementation of JFM in Andhra Pradesh, especially in micro-plan preparation, organizing communities, running training programmes and building awareness and capacity. Besides implementing PFM policy, some NGOs have also influenced policy direction to some extent. Andhra Pradesh Vana Samakhya (an apex NGO network) constantly pursued the government and World Bank to sanction a resettlement action plan under which each affected family is granted 20,000 rupees, not in cash but spent on village infrastructure and the development of agriculture. The Government of Andhra Pradesh reduced the role of NGOs in the CFM programme. All financial assistance was stopped and community extension workers (CEWs), directly accountable to the forest department, were appointed to supervise and guide VSS committees. The main objective of this step was to reduce the influence of those NGOs who voiced strong objections to forest department behaviour.

'Communities' are important actors in the PFM policy process. But the concept of community is complicated. Those groups which have better negotiating and mediating access to state machinery enjoy a greater share of the benefits in almost all government policies. Our sample tribal villages have sharp socio-economic inequalities. Those who have land, cattle and awareness constitute the local elite, who play an important role in PFM policy implementation. It is their choices and interests that are prioritized in micro-plans and other VSS works. At all our study sites, elite groups held most of the important management committee posts.

Importantly, World Bank funding for JFM was a crucial factor in Andhra Pradesh. Chnadrababu Naidu, the then chief minister, who had considerable influence in the central government, succeeded in securing funding for JFM and CFM. When JFM funding ran out in 2000, for two years many deliberations took place between state government, forest department and World Bank officials. At the same time, the World Bank changed its forest policy in 2002, with new emphasis on 'development policy lending', which sought efficient institutional arrangement by involving people in forest management (WRM, 2005). Under the instructions of the World Bank, the Government of Andhra Pradesh formulated CFM without any concrete legal and tenure reforms (Samatha, 2005). Thus, World Bank policy choices have influenced the nature of CFM policy.

Politics of information and knowledge about forest management

Forest policies and their implementation are plagued with contradictions. The forest department, for instance, historically practises silvicultural management systems, standardized for long-term timber production across its estates. Foresters inherited timber production and revenue generation as their main task from colonial policies, and the livelihood concerns of forest-dependent communities are given less priority.

The problems start with the role of forests in the context of ecology. The type of species and management system that is best depends on one's overall objectives. Local people prefer short-term NTFP products and coppiced poles to meet day-to-day needs; but the forest department's priorities, in contrast, focus on green cover, long-term timber production and eviction of non-forest land uses, such as agriculture. The forest department species and management system commonly prefers fast growing and unbrowsable exotic species, or long-term sal or teak management. In PFM, the concept of the micro-plan is incorporated to focus on local people's priorities in forest management. However, as already discussed, micro-plan choices are decided by the forest department, rather than local people. The NGO network in Kadapa informed the authors that the forest department not only gave a very short time for preparing the micro-plan, but also pressured them to follows its commands. Samatha, an NGO working in Visakapatnam, criticized the forest department for forcing small NGOs to follow its commands without question.

In PFM, the forest department encouraged exotic plants, such as silver oak, to increase green cover; but communities were interested in NTFPs such as tamarind, amla, soap nut and other locally available products. Instead, the forest department imposed long gestation plantations, such as teak, in VSS areas. Communities could not afford to wait for long-maturing financial benefits since their priority is to get income in the short term. This is the prime example of the gaps between local understanding of forest management and the forest department's managerial knowledge.

As for the environmental impact of PFM, villagers have some interesting views. In both Kadapa and Adilabad districts, people said that due to soil and water conservation structures, such as check dams (which were an integral part of the forest schemes), the water flow to local tanks is obstructed and leads to a decline in water levels in the village tank. They argued that soil water conservation structures are only useful for forest growth, not for agriculture. On the other hand, these activities are promoted as the best options for soil and water conservation not only by the forest department, but also by the department of agriculture and rural development. In the absence of proper information and knowledge, these issues lead to conflicts and delays in implementation.

The case of *podu* is an example of the politicization of technical information and existing studies and local knowledge. In policy circles, *podu* is conceived of as detrimental to the environment. A range of forest department narratives exists against *podu*; but so far there is no substantive evidence to prove this argument. Some studies suggest that *podu*, in its original form of 7 to 15 years of crop rotation, is both an environmentally friendly practice and the most productive livelihood use of the ecological niche (Chandran, 2000; IFAD, 2001). While *podu* is looked on from the environmental angle, its role in livelihood protection has not been given due importance at the policy level. As a result, policies tend to obstruct livelihoods in the name of environmental protection, while livelihood protection and enhancement is being fixed as the main objective of JFM.

Major changes in actions

Several changes have taken place in the cast of different actors involved in PFM. Leadership problems emerged during elections for management committees in CFM. The Mudireddypalle scheduled caste colony in Kadapa, Pandhiriloddi village in Adilabad, and Sobhakota village in Visakapaatnam witnessed several conflicts over leadership issues. In Mudriddeypelle, sub-castes among scheduled castes fought for posts in the management committee. In Sobhakota VSS in Visakhapatnam, elections were conducted for management committee posts four times. The main reason for the conflict is that VSS management committee posts become favourable targets for dominant village groups (due to the money involved in VSS work), who wish to patronize their own group members in wage employment.

Overall, forest department dominance in forest management has not reduced over time, and there is no transparency in VSS functioning. In all our sample villages, VSS account books are not given to the management committee. In Sobhakota VSS in Visakhapatnam, in spite of several requests, account books were not handed over to the management committee. Whenever DFOs visit the village, they bring the books with them, but take them back again afterwards.

Delay in the sanctioning of funds is a big problem in CFM. People complain that although they submit proper budget estimates for works, the forest department never sanctions the full budget. In Kadapa and Adilabad, allocation of insufficient funds for VSS works resulted in work stagnating. Due to this, the management committee and villagers lost interest in forest protection, and VSS functioning was adversely affected.

Within the forest department there is a range of opinions on the future path of PFM. The DFO of Paderu division of Visakapatnam said that there is ambiguity and contestation within top-level officials of the forest department regarding the direction of CFM policy: while one section favours extensive plantation in the VSS area, the more progressive faction prefers livelihood-oriented management. It seems that the later group is prevailing and the forest department has decided to give greater importance to livelihood promotion under the patronage and instruction of the World Bank. This has resulted in the initiation of livelihood-enhancing activities, such as the distribution of honeybees, the 'adda leaves' stitching machine, (leaf plate-making) and extraction of oil from bodha grass in Kadapa (CENPAP, 2006). But the viability of these activities is open to question. The Adilabad district range officer felt that honeybees and vermin-compost may not yield desirable results since they require expertise and training, which is not forthcoming from the forest department.

Despite many flaws, certain positive outcomes of PFM have come about. Under PFM, foresters are forced to shed their professional aloofness and mingle with villagers in the implementation of VSS works. The DFO of Paderu division of Visakpatnam district informed our researchers that 'Earlier we went with sticks to control *podu*; now we are going with helping hands.'

A woman in Kadapa district expressed that 'When I was a child, whenever a forester came to the village, the entire village used to shiver; now none in the village are afraid of forest department officials.'

Inter-sectoral coordination

The primary object of the all rural development programmes has been the reduction of poverty levels with sustainable utilization of resources. CFM policy sought effective inter-sectoral coordination among various departments associated with rural development. For this, advisory bodies at state, district, divisional and village levels were created. At the village level, there is an advisory committee for each of the four VSS, comprising lower forest officials, the *panchayat sarpanch*, representatives of a tribal development agency, village administration officer and local NGOs. But the convergence of various developmental programmes and coordination within the departments is not evident in any of our study villages, where villagers informed us that *panchayat sarpanchs* did not take an interest in VSS activities. The main reason is at policy level; provisions for the *Panchayat* Extension to Scheduled Area Act (PESA) give *panchayats* and *grama sabhas* an important role in managing VSS activities; but this has not been put into force in Andhra Pradesh. Thus, VSS are not accountable to elected *panchayats*. This has resulted in the absence of an effective controlling mechanism in VSS functioning and coordination with other development activities.

After the introduction of PFM, there was a change in illegal timber trading and smuggling. In Kadapa and Adilabad districts, illegal cuttings of valuable trees and smuggling of red sanders appeared to have been reduced. In Kadapa, people caught illicit transporters of red sanders red-handed and informed the forest department, who took effective action, resulting in the reduction of smuggling. In Adilabad, the clandestine felling of teak has been reduced due to the fact that in the agricultural slack season people used to cut teak saplings to sell. With the wage employment available to them from JFM, their dependency on these activities appeared to have come down. This is an example of how alternative sources of income have (at least temporarily) reduced dependency on forests.

Main opportunities and constraints of PFM

A period of ten years may seem too little for assessing the livelihood impacts of PFM – forest management is claimed to take a long gestation period to give palpable results. However, it is a general trend in our study sites that the JFM/CFM project has not achieved the envisaged impact on the livelihoods of local people. It has, however, initiated an institutionalized organization, where people can be involved in decision-making processes concerning forest resource management.

On the negative side, CFM implementation has generated conflicts and group divisions among tribal societies (Samatha, 2005). Management committee posts are occupied by dominant village groups and result in the polarization of group politics. In Kadapa VSS sites, although the management committee did not allow goat grazing, it continued illegally. Management committees were unable to control the clandestine utilization of forest resources due to fear of serious conflicts. Villagers in Visakhapatnam found it difficult to control the entry of neighbouring villagers into their VSS areas since this led to conflicts, and the forest department did not extend its protection. This shows that VSS require proper tenurial security in order to take more concrete interest in forest management.

No significant improvements are reported in the availability of biomass products such as fuel, fodder and timber for domestic requirements after PFM. This is attributed to the lack of proper planning and training of local people. In CFM, the forest department is trying to generate income from various activities, such as a *boda* grass oil extraction plant in Kadapa, which is found to have an international market. A bio-diesel plantation in Kadapa and Adilabad district has also been initiated. The forest department is currently lobbying for permission from the government to initiate a coffee plantation in VSS areas in Visakhapatnam. Bamboo cultivation with added value through artisanal production of bamboo products is also officially encouraged by the forest department. All of these activities seem to be potential income generators for local communities.

The livelihoods approach to forest management in CFM has not been translated into reality. The main reason for this is that micro-plan preparation and execution was carried out by the forest department, which did not give importance to income-generating activities, especially in the short term. With regard to species selection and treatment practices, the forest department dictated the models, and villagers were not given a choice as to species selection. Consequently, local specificities of forest management, including NTFPs, are not being given proper attention. Since there is no active NGO involvement in CFM, things appear to be worse.

Conflicts for leadership have come up in CFM for two reasons. First, the ruling party has changed from the Telugu Dasem Party to the Congress Party, which resulted in supporters of both parties trying to become members of the management committee. Outside political interference was observed at the time of the VSS election in Kadapa. These party politics created rifts between communities and resulted in the delay of works and the misutilization of funds. Second, the posts of VSS chairperson and vice-chairperson become prestigious symbols to mobilize people under wage employment. Most of the dominant groups, especially in Kadapa district, competed for these positions; as a result, the rest of the villagers became indifferent to VSS work. In JFM, money was sanctioned immediately after estimation of works done in the VSS. But in CFM, VSS committees were asked to commence works although the sanctioning of funds was delayed. Due to delays in the disbursement of money, villagers lost interest in participating in VSS works.

Future strategies

A sizeable proportion of the poor in Andhra Pradesh are forest dependent for sustenance. The remoteness of some of these villages from the district headquarters and the absence of all-weather approach roads (physical capital at the community level) deprive people of several government-initiated poverty alleviation programmes and other facilities. This has resulted in greater dependence on forests. For instance, in many poor tribal villages in Visakhapatnam, state schemes are hardly seen at all and they remain mainly dependent on *podu* cultivation for survival. Therefore, there is a strong case to take a holistic look into the overall development of the region and the living conditions of these forest dwellers, and to initiate appropriate measures that would benefit the forest and the people living in and around it. In this regard, provision of proper infrastructure and access to markets would go a long way to supporting local people's livelihoods and reducing their dependence on *podu* and the forest.

Most of the state forest departments have not surveyed the lands declared as 'forests', which have actually been under cultivation for many generations, and treat them as state forests. Since the extremist groups, such as the Maoists, are supporting the poor cultivators, the forest administration does not want to open up these issues. At present, VSS do not have a secure tenure system and people and communities do not pay serious attention to management. The 1967 Andhra Pradesh Forest Act should be amended to provide legal status to VSS with better tenurial security for communities.

Where illegal cultivation is found, the state government can take up such land plantations initially for wage labour and, subsequently, for the usufruct rights. But this chapter indicates that this is not a sustainable model after cash for employment provided by VSS runs out. This could be altered, to some extent, through selection of appropriate species that would support the livelihoods of the poor on a sustainable basis. There are large numbers of medicinal plants with a short gestation period (six months' to two years' duration), which can be grown for immediate income streams for poor forest-dependent people. Similarly, there are clonal plantations of fast growing species that can produce a yield in five to six years. Bamboo plantation can give long-term sustainable yield from the sixth year onwards. Thus, a combination of different species can provide substantial incomes. Such afforestation schemes must be dovetailed with particular families so that they take an interest in the entire programme. There should be no attempt at eviction of families living on forest land unless there are compelling reasons. This calls for policies towards secure tenurial rights and a less expansive forest department 'land grab' policy.

Institutional structures should be strengthened by formulating various accountable mechanisms, such as effective linkages between VSS and *panchayats* and inter-departmental coordination. They should look into the present functioning of VSS and identify the channels of corruption in order to eliminate, or at least reduce, them. NGO involvement has the potential to minimize leadership conflicts at the village level. In some instances, women's involvement was undermined by their husbands. Sometimes the husband attended meetings and collected bank cheques on behalf of his wife. It should be made mandatory that women attend meetings and collect their cheques personally.

There should be proper linkages between research on forest products and VSS committees. In particular, income-generating activities such as value addition to NTFPs and medicinal plants should be encouraged. Viable market linkages should be established between industry and VSS committees.

In summary, the introduction of PFM has not brought any substantial changes to the livelihoods of forest-dependent communities. The stated objectives in the policy documents have not been translated into reality due to a number of significant and structural constraints, as discussed in this chapter. However, the most encouraging outcome of PFM is that it has opened new space for hope to bring changes to the livelihoods of people within the sustainable framework of sustainable forest resource management.

References

Andhra Pradesh Forest Department (1999) *Facts and Figures 1999*, Forest Department and Government of Andhra Pradesh

Andhra Pradesh Forest Department (2002) Andhra Pradesh Community Forest Management Project, Project Implementation, Plan Vol. I and II, Project Monitoring Unit, Andhra Pradesh Forest Department, Hyderabad

Andhra Pradesh Forest Department (2004) *Facts and Figures*, Hyderabad, Government of Andhra Pradesh

Bhatia, B. (2005): 'Competing concerns', *Economic and Political Weekly*, vol XL, no 47, pp4890–4893

Borgoyary, M. (2002) *Impact of JFM on Encroachment of Forestland: Case Study of Five Selected VSS/FPC in Visakhapatnam, Andhra Pradesh,* New Delhi, Resource Unit for Participatory Forestry (RUPFOR), Winrock International India

CENPAP (2006) *Concurrent Monitoring and Evaluation: Mid Term Review – 2006*, APCFM, AP Forest Report, Hyderabad

Chandran, M. D. and Subash, (2000) 'Shifting cultivation, sacred groves and conflicts in colonial forest policy in the Western Ghats', in Richard H. Grove, Vinita Damodaran and Satpal Sangwan (eds) *Nature and the Orient*, (ed) Oxford University Press, New Delhi

GoI (Government of India) (1989) *Ministry of Environment and Forests No. 6-21/89*, New Delhi, GoI

GoI (1990) *Ministry of Environment and Forests GO. I.LR*, 1 June 1990, New Delhi, GoI

Gopal, K. S. and Sanjay, U. (2001) *A Report on Livelihoods and Forest Management in Andhra Pradesh*, Prepared for the Natural Resources Management Programme, Andhra Pradesh, September

Government of Andhra Pradesh (1992) *Government Order M.S. No. 218*, Dt: 28-08-1992, Hyderabad, Government of Andhra Pradesh

Government of Andhra Pradesh (2002) *Government Order M.S. No. 4*, Dt: 12-02-2002, Hyderabad, Government of Andhra Pradesh

Government of Andhra Pradesh (2004) *Government Order M.S. No.13*, Dt: 12-02-2004, Hyderabad, Government of Andhra Pradesh

Government of Andhra Pradesh (2005–2006) *Economic Survey*, Hyderabad, Government of Andhra Pradesh Planning Department

Government of Madras (1886) Letter from the Forest Resettlement Officer, Cuddapah district, 8 May, no 3, F. no 319 in Board of Revenue Proceedings (BORP), Government of Madras, Madras

IFAD (2001) *Shifting Cultivation: Towards Sustainability and Resource Conservation in Asia*, International Institute of Rural Reconstructions, Philippines

Jeffery, R. and Sunder, N. (eds) (1999) *A New Moral Economy for India's Forests – Discourses of Community and Participation*, New Delhi, Sage Publications

Jodha, N. S. (2000) 'Joint forest management of forests: Small gains', *Economic and Political Weekly*, 9 December, pp4396–4399

Murali, A. (1982) 'Allrui Sitarama Raju and the Manyam Rebellion of 1922–1924', *Social Scientist,* vol 12, no 4, pp3–33

National Remote Sensing Agency (1995) *Report on Area Statistics of Land Use/Land Cover Generated using Remote Sensing Technique*, Hyderabad, Government of India

Pathak, A. (1994) *Contested Domains: The State Peasants and Forests in Contemporary India*, New Delhi, Sage

Rao, K. K. (2006) 'Community forest management and joint forest management in the Eastern Ghats, Andhra Pradesh', in N. H. Ravindranath, K. S. Murali and K. C. Malhotra (ed) *Joint Forest Management and Community Forestry in India: An Ecological and Institutional Assessment*, New Delhi, Oxford and IBH Publishing

Ratna Reddy, V., Gopinath Reddy, M., Malla Reddy, Y. V. and Soussan, J. (2004) *Sustaining Rural Livelihoods in Fragile Environments: Resource Endowments or Policy Interventions? (A Study in the Context of Participatory Watershed Development in Andhra Pradesh)*, Working Paper Series, No 58, Hyderabad, Centre for Economic and Social Studies, Nizamiah Observatory Campus, Begumpet

Ravi, K. V. M. (2005) 'Colonizing forest landscape: History of forest policies in Madras Presidency, 1880–1930', National Conference on Dalit Concerns in Forestry, Centre for People's Forestry, Hyderabad

Ravindranath, N. H., Murali, K. S. and Malhotra, K. C. (eds) (2000a) *Joint Forest Management and Community Forestry in India: An Ecological and Institutional Assessment*, New Delhi, Oxford IBH Publishers

Ravindranath, N. H., Murali, K. S. and Mulhotra, K. C. (2000b) *Joint Forest Management and Community Forestry in India: An Ecological and Institutional Assessment*, New Delhi, Oxford IBH Publishers

Reddy, V., Ratna, M., Gopinath R., Velayutham, S., Madhusudan, B. and Springate-Baginski, O. (2004) *Participatory Forest Management in Andhra Pradesh: A Review*, Working Paper, Centre for Economic and Social Studies, Series No 62, Nizamiah Observatory Campus, Begumpet, Hyderabad, Andhra Pradesh

Saberwal, V. (1999) *Pastoral Politics: Shepherds, Bureaucrats and Conservation in Western Himalaya*, Delhi, Oxford University Press

Samaj, V. M. (2006) *A Study on the Socio-Economic Status of Podu Land Cultivators in VSS Areas in Andhra Pradesh*, Hyderabad, Government of Andhra Pradesh

Samatha (2005) *Evaluation of Community Forest Management in Andhra Pradesh*, Hyderabad; Samatha

Sarin, M. (2003) *Trees Hide the Woods: Hindu Survey of the Environment 2003*, pp111–115

Shiv, S. (2001) *Officialising Clandestine Practices of Environmental Degradation in the Tropics*, Working Paper, vol II (7), New York, Harvard Centre for Population and Development Studies

Venkatraman, A. and Falconer, J. (1998) *Rejuvenating India's Decimated Forests through Joint Actions: Lessons form Andhra Pradesh*, Joint Forest Management Andhra Pradesh, www.jfmindia.org

World Bank (2006) *India: Unlocking Opportunities for Forest Dependent People*, New Delhi, Oxford University Press

WRM (World Rainforest Movement) (2000) *World Rainforest Movement Bulletin*, issue 31

WRM (2005) *World Rainforest Movement Bulletin,* issue 93

Part III

Answering the Research Questions

We have set out to analyse the process of policy-making in forest management with particular reference to participatory forest management (PFM) reform and its livelihood impacts in India and Nepal. Our analysis in Part I links historical experience, dynamic discourses and practice, and in Part II, we present detailed evidence of the policy process and the outcomes and impacts of policy in specific regions. Chapter 10 in Part III reviews and evaluates these processes and their impacts in terms of the changes in the livelihoods of rural people. The focus of the discussion is directed downwards to policy in practice, its impacts 'on the ground' and to the identification of the immediate causes of those impacts. Part I focused on the larger-scale and more general aspects of policy-making, and Part II provided evidence of policy at the state level and below. Chapter 10 is primarily focused on analysing the local and practical aspects of policy practice.

Chapter 11 reviews what might be done to reform forest policy and practice at a strategic level. The promise and critique of 'participation' in forest policy is also discussed. So far, this book has argued that local participation in the management of the forests of India and Nepal is principally justified on the grounds that it can enhance the livelihoods of rural people, especially the poor. However, although participation is claimed to offer a great deal more than material benefits, the practice of 'participation' on the terms set by forest administrations (organizations generally reluctant to concede power) has attracted criticism that it has become a 'catch word', a ritual, or, as has been suggested in the Introduction, even in some cases a 'wolf in sheep's clothing' (i.e. having significant negative consequences for local people). Some important critiques and rethinking of the notion of participation are discussed in Chapter 11.

Forest administrations' existing practices adhere to a durable and robust technocratic political culture, and have been resistant to fundamental reform. Individual reforms of particular aspects of practice necessarily require a range of complementary and mutually reinforcing reforms, and are unlikely to succeed on their own. For example, an obvious target for reform in India is the extent of local discretion in planning local forest management (i.e. the scope of so-called 'micro-plans' and the way in which they are framed, developed and approved). As has been seen in Part II, in India these micro-plans are formally subordinate to working plans, which are designed centrally according to forest department objectives, and therefore currently allow virtually no discretion to local forest users to determine their preferred forest management system. However, for micro-plans and working plans to be written in a more participatory manner to reflect the management objectives and preferred rules of access of the local forest users (an obvious recommendation), a whole range of linked reforms in forest administration would be necessary. There would need to be some handover of management authority to local people; relaxation of the silvicultural imperatives that have hitherto predominated; post-formation facilitation and provision of appropriate technical support as needed; and so on. As Chapters 2 and 3 have demonstrated, these aspects are linked in a mutually reflexive manner in which each would need to change to fulfil the reform objective of PFM. The Nepal hills here provide an instructive contrast to India, where such changes have been successfully implemented, as discussed in Chapter 5.

This view might imply a fundamental pessimism in which 'tinkering' with one aspect of PFM may seem pointless because it requires sympathetic changes in the whole discourse practice of the forest administration, which, according to both the general analysis in Part I and the empirical findings in Part II, has been largely unforthcoming in India. However, in spite of the acknowledged difficulties that arise from struggling to reform these powerful institutions (as well as the other difficulties involved with inequitable agrarian societies), there are, nevertheless, significant opportunities for reform. There are many actors involved, as Chapter 2 has described, and the working of the forest administration and the orientation of its personnel is far from monolithic. In many different cases, there are clear opportunities to move forest policy towards more just outcomes.

PFM is just one among a number of policy changes that has, as pressure for reform has grown from many different actors (international funding institutions (IFIs), forest intellectuals, social movements and political parties, as well as other government departments), required at least tactical incorporation by the forest administrations in both Nepal and India. However, Part II showed that the forest administration in Nepal has been less entrenched, and has had less field capacity to control and manage its forest estate, with the result that historically rooted local initiatives and community-based natural resource management practices have survived better than in many parts of India, and therefore PFM has found a more sympathetic environment. On the other hand, despite India having significantly more self-initiated processes, more active civil society engagement and contestation, and in many areas stronger resistance to forest department predations on the ground, there has been relatively less progressive policy development. Senior forest administrators in Nepal have been more receptive and proactive in implementing participatory reforms, and IFIs have been more dominant in financing the forest administration and have been able to push their participatory agendas more insistently than in India. At the same time, the political turmoil that has overtaken Nepal in recent years has destabilized the development of community forestry in the hills, although it is apparent that local institutions are continuing to function and to manage their livelihood needs despite the conflict, albeit with very limited official support.

10

Comparative Policy and Impact Issues

Oliver Springate-Baginski and Piers Blaikie with Ajit Banerjee,
Om Prakash Dev, Binod Bhatta, V. Ratna Reddy, M. Gopinath Reddy,
Sushil Saigal, Kailas Sarap and Madhu Sarin

These days the relationship between the forest department and people is bad. That is accepted. But ten years ago it was awful, and if we keep on going in another ten years it may even be OK. (Divisional forest officer, Vizakhapatnam, Andhra Pradesh, 2004)

Community forestry has been good for people as it provided us authority to protect, manage and use forest resources. It has improved people's relationship with district forest office staff. However ... we used to buy bamboo in winter and make many handicrafts and sell in Dharan to maintain livelihoods; but this traditional income source has not been considered in community forestry. (Chairperson Tatopani CFUG, Dhankuta district)

Regional variations in the participatory forest management (PFM) 'deal'

Wherever collective agreement is sought over participation in forest management, a 'deal' is struck among the parties at the outset, whether in self-initiated groups or state-initiated processes. The nature of PFM deals varies in important respects both between and within Nepal and India, reflecting the varying interests and bargaining power of the actors in the different regional political ecologies. Table 10.1 summarizes the different PFM deals in the Indian states and Nepalese regions studied in Part II. As we have seen, the fundamental difference has been that in Nepal, forest management has been handed over to local community forest user groups (CFUGS), whereas in India, forest departments have only given local people inducements (such as rights to non-timber forest product (NTFP) collection, wage labour and a share in the 'final harvest') to protect and work on state forest management.

In the *Nepalese hills*, after a PFM policy process that began in the 1970s, the terms of the 1993 Forest Act and associated 1995 Forest Guidelines provide for legally independent CFUGs to assume management responsibility for community forests (CFs) adjacent to settlements, to be managed according to ten-year operational plans (OPs) drafted by the local people according to their priorities. Local people can largely decide their own management practices and harvesting regulations and levels, as long as the OP is approved by the district forest officer (DFO), whose main criterion is that timber off-take is less than 50 per cent of the estimated annual timber increment (calculated based on a lengthy timber inventory conducted by district forest office field staff). CFUGs receive virtually no financial support (e.g. for wage labour, as in the World Bank project in Andhra Pradesh); but, in principle, are able to market timber and are exempted from tax on any forest-related revenue, although most CFUGs are too inaccessible to market all but the highest value timber, such as khair (*Acacia katechu* spp) cost effectively.

In the *Nepalese tarai*, despite the forests having far higher value and being more accessible, similar deals as those in the hills have been negotiated, although they have been more selectively offered by DFOs, who in the early 1990s, initially sought to create a wall of CFUGs along the southern edge of the main *tarai* forest belt in order to keep distant forest users from extracting forest products. Subsequent reluctance to hand over the remaining forests to CFUGs, or piecemeal transfer of forest patches, has led to a complex and confusing situation where immigrants from the hills, often responsible for clearing forest land, have management control of many remaining forests, yet the traditional residents (including many tribal groups) are now further from the forest frontier and have become distant users, excluded from user group membership and therefore access rights. Timber harvesting is a highly contentious issue which, under circumstances where there is often elite control, can lead to pro-rich outcomes through hidden subsidies to the richer members of the communities by timber distribution to households at subsidized prices that are, nevertheless, beyond poorer households' ability to pay (Iversen et al, 2006). The Government of Nepal had introduced a 40 per cent tax on timber marketing; but after lobbying from the Federation of

Table 10.1 *The participatory forest management (PFM) 'deal' in different study areas*

Country	Nepal		India			
Region	Hills	*Tarai*	West Bengal	Orissa		Andhra Pradesh
PFM model	**CF**	**CF**	**JFM**	**SIFPG**	**JFM**	**JFM/CFM**
Extent of total forest land under PFM	23%	12%	44.61% (81% in South-West Bengal)	15–20%	11.2%	27%
Number of local groups formed	~ 13,000	1245	3892	4928	6912	8343
Estimated % of groups active	71%	72%	50% (South-West Bengal)	65%	50%	72%
Legal status of local institution	Legally registered under 1993 Forest Act as independent body	Legally registered under 1993 Forest Act as independent body	*Ad hoc* local body without independent legal status	No independent legal identity	*Ad hoc* local body without independent legal status	*Ad hoc* local body without independent legal status
Responsibility to plan and implement forest management	With local PFM institution subject to district forest officer (DFO) approval	With local PFM institution subject to DFO approval	Remains with the forest department	Remains with the forest department	Remains with the forest department	Remains with the forest department

Table 10.1 *(continued)*

Country	Nepal		India			
Region	Hills	*Tarai*	West Bengal	Orissa		Andhra Pradesh
PFM model	**CF**	**CF**	**JFM**	**SIFPG**	**JFM**	**JFM/CFM**
Responsibility to protect forest from illicit extraction	Local PFM institution in conjunction with the forest department	Local PFM institution in conjunction with the forest department	Local PFM institution in conjunction with the forest department	Formally with the forest department; local people assume role	Local PFM institution in conjunction with the forest department	Local PFM institution in conjunction with the forest department
Timber product marketing, distribution and share	Community forest user groups (CFUGs) manage everything – tax exempt	CFUGs manage according to DFO felling and movement permit – 15% government tax on sal and khair timber marketed outside CFUGs	The forest department manages everything –25% share of final harvest net of management costs to village (predominantly 12-year sal rotation) and 25% intermediate yield	The forest department manages everything in theory – no share to the village	The forest department manages everything –50% share of final harvest (net of forest department management costs) to village, plus 100% of intermediate harvest	The forest department manages everything –100% share of harvest (net of forest department management costs) to village
Funding basis for forest department PFM implementation	Various donors	Various donors	State government, World Bank, forest development agency (FDA)	Not funded	State government, FDA	World Bank, many different central schemes
Funding for PFM local group activities	Self-funded by group	Self-funded	Limited funding	Self-funded	Limited funding	Substantial funding under five-year schemes via above sources

Source: data from Part II of this volume

Community Forest Users, Nepal (FECOFUN), it was reduced and *tarai* CFUGs became subject to only a 15 per cent tax on sal (*Shorea robusta* spp) and khair (*Acacia katechu* spp) timber sold outside the CFUG. In 2000, a co-management approach called collaborative forest management (CollFM) was piloted by the Department of Forests (DoF) with Dutch donor support in three districts (Rauthat, Bara and Parsa) in order to widen stakeholder involvement and incorporate consideration of distant users' livelihood needs along with those of local users. However, this model has been challenged through energetic lobbying by FECOFUN, who has said that it does not want to see the CF model diluted through an increased DoF role.

In *West Bengal*, after a decade of informal experiments by forest department field staff with local people, the 1989 Joint Forest Management (JFM) orders provided for the forest department to organize the forest-adjacent people into *village forest protection committees* (VFPCs) in order to participate jointly with forest department field staff in regenerating, protecting and managing blocks of forest (areas under either sal coppice or exotic plantation). This involves mutually drafting micro-plans (a formal agreement of what local people will take from the forest, agreement that they will protect the forest, and what they receive from the forest department in terms of a share of revenue from harvesting) and (at least when there was donor support) wage labour opportunities for implementing the plan (e.g. planting). The standard forest department working plans for JFM forest are not modified to reflect local needs; indeed, forest department objectives are more effectively implemented through community involvement. The main benefit to local people has been their entitlement to collect subsistence fuel, fodder and other NTFPs from the forest. With donor support, there were some wage labour opportunities. However, the right of local people to receive 25 per cent of the net income from forest department marketing of timber products has, in practice, led to inconsequential payments, which about one-third of villages have received, while in many villages this has not even been honoured. This is partly due to the forest department carrying out timber felling periodically rather than on an annual basis by block in each forest protection committee (FPC), as discussed in Chapter 7.

In *Orissa*, after the consolidation of the state in 1948 to 1949 and subsequent nationalization of the forests, the Orissa Forest Department lacked the capacity to manage and protect the forest in remote rural areas, and in these *de facto* open-access conditions many forests began degrading. Gradually, self-initiated forest protection groups (SIFPGs) emerged to protect the forests on which local people depended for livelihood product flows and services by regulating local extraction and excluding outsiders. The groups spread widely and by the 1980s there were, according to estimates, about 8000 village groups protecting some 2 million hectares of forest in the state (see Regional Centre For Development Cooperation, 2002), with many affiliated into district federations. The forest department gradually changed its orientation, and from neglecting the SIFPGs began, from 1990 onwards, to formalize these groups into JFM *Vana Samrakshyan Samities* (VSS), under the Orissa JFM notifications, on similar terms to those that had worked for the West Bengal Forest Department. Local forest department field staff were expected to organize forest-adjacent settlements into VSS to take on the responsibility of protecting the forest and supporting its management in terms defined in a micro-plan, although, again, the forest department's overarching working plans for the division remain wholly unchanged. The local forest guard becomes *ex officio* member of the VSS committee; but from field research it is apparent that many 'converted' groups are stagnant, often due to the non-attendance of local forest guards at meetings. When local self-initiated groups are converted to VSS, local people's livelihood use of forests is legitimated, they are given authority to exclude outside forest users and are supposed to receive the support of the forest department in this (although, in practice, this was found to be rare, and a number of cases have come to the attention of Orissa Forest Department of field staff complicit in illicit extraction). However, they are only entitled to a 50 per cent share of the final timber revenue (net of forest department costs) on the estimated increment from the time of formation. This is resented by self-initiated groups since many have regenerated degraded forests through their own efforts and independently of any forest department support, and now feel that the forest department is taking the fruits of their labours. There are virtually no wage labour opportunities in Orissa JFM; until 2006, there had been no donor support and forest development agency (FDA) support has occurred only in a small number of sites, although this may be expected to change. In

2006, both the UK Department for International Development (DFID) and the Japan Bank for International Cooperation (JBIC) initiated forest-sector projects in Orissa that involve an element of PFM support.

In Andhra Pradesh, the JFM notification of 1991 provided for the formation of local *Vana Samrakshyan Samities* institutions (in which the local forest guard becomes an *ex-officio* member), to whom the responsibility for protecting their local forests and regulating product extraction is delegated. Through donor support (mainly by the World Bank), wage labour incentives have been provided for local people to participate in protecting the forest and supporting the forest department's forest management, according to micro-plans that again are subsidiary to the forest department divisional working plans. There are provisions for 'entry-point' activities, such as construction of bus shelters and community halls, and wage labour opportunities over the five-year village project support cycle in order to compensate for lost incomes when forest land access becomes restricted. Additionally, there is provision for the disbursement of 100 per cent net revenues from forest product marketing to be distributed to the VSS, although with many forests under 80-year teak rotations, there will be a very long wait after the initial distribution of poles from forest thinning. However, the main emphasis on time-bound wage labour incentives has often led to the disinclination of local people to continue formal forest management activities when the funding phase has ended. This may be contrasted with the situation of self-initiated groups and user groups in Nepal, where (without interference) most of the groups tend to continue in perpetuity.

We may compare the PFM deal in the five different regions, according to Table 10.1. The percentage of land under PFM in the different regions varies from 15 per cent in the Nepal *tarai* to over 81 per cent in South-West Bengal, reflecting the different motivations of governments to share management and the different agro-ecological and settlement patterns. Whereas in South-West Bengal there is very high population density, in northern areas of the *tarai* there is very little settlement, five large protected areas (covering 7.2 per cent of the *tarai*; DoF, 2005), and a lack of interest from the forest administration to further spread handover.

We have given estimated levels for how many of the groups formed are currently active. 'Active' here is taken to mean at least a minimum level of collective activity, such as meeting periodically. It is difficult to get a clear idea of the level of activity without extensive randomized survey, although at our best guess, based on extrapolation from data collected during our survey sampling process, the levels of activity across the study areas vary between 50 and 90 per cent. It is evident in the study areas that there is a high degree of variability. In Andhra Pradesh, for instance, the level of activity is closely correlated to funding disbursement, with many groups becoming inactive after funding stops.

The key point here is that just because a group is formed does not mean it will continue, and so to claim that over 100,000 groups have been formed in India is not the same as saying that this many are active, a figure which may be closer to 50 to 70 per cent of this based on these estimates.

The legal status of the local institution differs significantly between Nepal and India. Whereas in Nepal, CFUGs are legally registered under the 1993 Forest Act, in India, JFM committees in most states have no independent legal basis beyond a memorandum of understanding (MoU) with the DFO.

Planning forest management in Nepal is the responsibility of the CFUGs and the DFO simply approves the operation plan that they develop, whereas in India, joint forest management committees (JFMCs) are not able to revise forest department working plans for the forest in question.

Institutional and forest outcomes and livelihood impacts of PFM implementation

The impacts of past forest policies have, in general, favoured increased state appropriation and negatively affected those elements of rural livelihoods based on forest land (as discussed in Chapter 1 and elaborated on in Part II). In seeking to take control of forests for timber and revenue generation, forest administrations have restricted local people's rights to use them. This delegitimization of normal livelihood-oriented forest use has contributed significantly to increasing hardship for sizeable groups, and to resentment and conflict stemming from the perceived injustice. Subsequent intensive state exploitation of forest resources, combined with the forest administration's lack of ability to police the extensive forest areas, has led both to *de facto* open access and to rapid resource degradation in many, but not all, areas, especially India and the Nepalese *tarai*. There have been community responses to this degradation, as illustrated by the Orissa self-initiated groups, as well as many other local informal initiatives primarily focused on ensuring livelihood security. The state has also responded, mainly out of its concern to improve forest condition.

Institutional outcomes

In the Nepal hills, implementation of CF has addressed the underlying cause of forest degradation – the alienation of local people from the control and management of the forest resources on which they depend. CF has also recognized and built on traditions of community resource management. However, in India, the reversal of state control has yet to occur. Local people have been mobilized by the forest administration to protect state forests and plantations, but with highly circumscribed people's 'participation' in management and decision-making.

No change in tenurial and legal rights in India
By the early 20th century, the colonial administration in India had taken over large areas of forest. Similarly, in Nepal, feudal elites had taken control of most forest areas. After independence in India, and after the later democratization in Nepal during the 1950s, nationalization of forests was extended widely without settlement of local rights in both countries. This issue is at the crux of the relationship between people, forests and the state, and whereas in Nepal the 1993 Forest Act has reversed nationalization by providing for local rights in the forests, in India, since the 1950s, there has been no fundamental reform of this iniquitous forest land tenurial status quo, and the implementation of PFM has generally not addressed it. It is evident that in India, in particular, only limited livelihood impacts are possible under the current socio-political climate without local people (particularly tribal people) having legally secure access rights to adjacent forest land. The 1996 *Panchayat* Extension to Scheduled Areas (PESA) legislation does provide for tribal rights in forest land around settlements according to livelihood use, as well as control of NTFP marketing:

> *State legislation on the Panchayats that may be made shall be in consonance with the customary law, social and religious practices and traditional management practices of community resources… State legislature shall ensure that the Panchayats at the appropriate level and the Gram Sabha are endowed specifically with … (ii) the ownership of minor forest produce.*

However, commentators have observed that state-level implementation of the act has been very limited, as yet:

> *The PESA act is one of the most potent legislative measures of the recent times, which recognizes the tribal people's mode of living, aspirations, their culture and traditions. But the fact that in most of the states the enabling rules are not in place more than eight years after the adoption of the act suggests that the state governments are reluctant to operationalize the PESA mandate... The state legislations have omitted some of the fundamental principles without which the spirit of PESA can never be realized. For instance, the premise in PESA that state legislations on Panchayats shall be in consonance with customary laws and among other things traditional management practices of community resources is ignored by most of the state laws.* (Shradha and Upadhyaya, 2004)

The Scheduled Tribes and Other Traditional Forest Dwellers (Recognition of Forest Rights) Act, which was finally passed on 18 December 2006, seeks to resolve this tenure issue through ensuring that tribal people's customary land use in and around forests is protected through the granting of secure land tenure and forest use rights (Prasad, 2005).

Thus, PFM has affected forest access in both India and Nepal, but in India through discretionary administrative provisions in an *ad hoc* manner, rather than through a strengthened rights regime.

Local institutions lack legal standing and countervailing power

In India, joint forest management (JFM), the major central government-endorsed PFM programme (which it is claimed covers over 100,000 villages and several million people), is not based on any legal provision, but rather on administrative orders both in central (1990 JFM Notification and subsequent guidelines) and state administrative notifications. Although states are gradually introducing state-level forest policies, some of which refer to PFM approaches (e.g. Government of Andhra Pradesh, 2002; Government of Himachal Pradesh, 2005), local JFMC institutions still lack legal status, and the MoU – the basis of the agreement between local people and the forest administration – have no legal credibility and cannot be enforced through legal mechanisms. In principle, there is nothing stopping the divisional forest officer (DFO) from defaulting on the 'deal' when convenient. Local people have no legal basis for challenging conduct if it does not fulfil the terms of the MoU, as, for instance, is the case in two-thirds of the study villages in West Bengal where, despite being due according to the working plans, sal coppicing had not yet been conducted. As a result, JFMC members did not receive promised revenues from timber harvesting. Whereas in India the process has been managed by the executive (i.e. the forest administration), in Nepal the legislature has deliberated over it and given it constitutional backing and legal basis through the 1993 Forest Act. It remains unclear why the legislature in India has not taken up the matter; but one may speculate that the executive (i.e. the forest administration) has sought successfully to avoid legislative oversight, and the legislature has had other priorities.

The situation is far from static, however, and more permanent solutions under the constitutional local government structure do appear to be slowly emerging in India. The contradiction between the vertical top-down authority structures of separate line agencies, such as the forest departments, and the horizontal integrated bottom-up authority structures of the local government is likely, in time, to be resolved in favour of the latter. However, line agencies are proving reluctant to devolve authority to local government bodies. The *panchayati raj* institution (PRI) legislation in most states provides for them to form standing natural resource subcommittees at village level for this purpose, and this has begun to emerge in nascent form in West Bengal, with forest guards reporting to PRIs in our study areas. Nevertheless, this process is in the early stages as the forest land around villages so far remains under the control of the West Bengal Forest Department.

Customary or self-initiated organizations in Orissa (discussed in Chapter 8) lack any legal basis, although since they are not formalized they have also not been liable to the imposition of standardized formats or unwelcome provisions by the forest department.

Changes in relations between local forest staff and forest users

PFM groups in India have generally been allocated the responsibility of protecting the forest in return for the legitimization of some aspects of their livelihood use of it. The legitimization of local people's right to collect NTFPs has been a major improvement across both Nepal and India. Curtailment of harassment by forest department staff of villagers in the forest (often by male rangers of local women) has reduced the constant tension felt by villagers and frequent demands for bribery. The relaxation of tensions was stated to be one of the major improvements perceived by local people.

In most circumstances, this is an improvement; but it must also be seen as an ambivalent impact because new frictions emerge over restrictions. In several areas, one user group has been set against another – for instance, settled populations against pastoralists or shifting cultivators – and in this way the forest department is getting them to do the 'dirty work' to police the new PFM dispensation. This has particularly been the case in areas of Andhra Pradesh, where some villages have been given responsibility for 'protecting' forest from their customary cultivators or graziers, resident in neighbouring villages, and in Nepal's *tarai*, where groups on the forest fringe have been excluding 'distant' forest users.

More responsibilities but only limited rights for forest user groups

PFM implementation has changed the distribution of costs and benefits of forest protection, management and use. As discussed above, the lack of legal basis for local PFM groups in India has meant that their position in terms of rights and responsibilities has been tenuous, compared to the situation in Nepal where local groups' legal basis means that they cannot be closed down, and challenges from either local people or the forest department require the observation of due legal process, and, if necessary, ultimately settlement in court.

Insofar as livelihoods are constrained to some extent by initial closure of the forest for regeneration, they may be compensated through employment opportunities (i.e. to plant, protect and thin the new forest). This was one of the bases of the original 'Arabari' model in West Bengal, although it was not subsequently adopted by the West Bengal government. The World Bank projects in Andhra Pradesh have, however, adopted this model, although here, this period of financial support is five years, shorter than the rotation of the new species. In addition, the wages earned are only for a few months per year – only just enough to last poor families through the dry/lean season. The way in which employment opportunities are distributed among different groups with different wealth and power becomes crucial. The Andhra Pradesh Forest Department has sought to organize poorer groups into forest protection committees in order to ensure that wage labour opportunities go to them.

The forest departments have generally formally undertaken to support the protection activities of villagers against incursions by outsiders and forest mafia. Local people may be able to repel neighbouring villagers; but repelling organized smuggling gangs can be hazardous and usually requires 'strong-arm' methods. When illegal gangs arrive, sometimes armed and with motorized transport, support from the forest administration only sometimes materializes. Numerous cases emerged in this study, particularly in Andhra Pradesh and Orissa, where local people were unable to repel well-organized timber mafias, and the forest departments did not support them. In Mudireddypalle VSS village in Kadapa district, local people even reported to us that on one occasion when they telephoned the ranger to request assistance in obstructing illicit felling, he commanded them to let the felling continue, and told them that if they interrupted it they would be implicated.

Elite capture of user groups

Elite capture of local PFM groups has been a prevalent pattern, and this is hardly surprising given the generally inegalitarian nature of agrarian society in South Asia, the hierarchical nature of the state machinery, and the *ad hoc* nature of most PFM groups in India. The manner in which PFM groups were originally set up has also tended to promote this. The general pattern of initial contacts between forest staff and local elites has widely led to elites' *de facto* control of local forest user groups.

This has led to elites being in a position to direct groups' decisions in their own interests, as well as acting as a relatively unproblematic conduit for the implementation of forest department plans. In some cases, including in the Nepal *tarai*, this appears to have led to a nexus with forest staff and contractors for the generation and sharing of timber revenues. Poorer groups and women commonly have different priorities in forest management from elite men – for instance, to reduce time taken to collect fuel, fodder and NTFPS, rather than construction timber, cash or conservation. They may also prioritize mobilization of group funds for pro-poor issues such as micro-credit or emergency payment schemes (e.g. in cases of sickness, death or natural disasters such as landslides). However, elite domination has generally marginalized these priorities. In many groups, elites have shown an aversion to pro-poor activities and have preferred a more conservative approach. Conflicts of interest have also arisen where local elites may be employers or lenders, and so poorer households, sometimes dependent on them, avoid confrontation.

Elite domination is by no means inevitable, and a number of approaches have been successfully used to avoid or redress it. One frequent approach (used in the Nepal hills and in Andhra Pradesh) has been to ensure that groups are small enough to be fairly homogeneous. In Andhra Pradesh, many VSS of poorer households have been formed at hamlet level. However, this can lead to anti-poor outcomes if the forest department then imposes unpopular regulations on the poor group, such as in tribal areas of Visakhapatnam district, as discussed in Chapter 9. Splitting groups where elite domination is a problem is another approach. In Nepal, where poorer hamlets have complained to DFOs that they are suffering from the domination of elite hamlets, DFOs have agreed to split the CFUGs (e.g. in Sakhuwasabha district). Outside facilitators and social mobilizers have also sometimes been employed to try to ensure that all community members have a say in deliberations.

Economic sustainability of local PFM groups

The *economic sustainability* of the current state-led PFM implementation in India remains tenuous, whereas in Nepal its sustainability is virtually beyond question. As we have seen, in India states are generally implementing JFM as a time-bound government scheme involving fund disbursement to *ad hoc* local groups and depending on local people to go along with this. International funding institution (IFI) projects in India have typically funded this mode. Once the funding stops, there are inevitable disruptions, discontinuities and disinterest. In South-West Bengal, sustainability beyond the Word Bank funding phase has been partly achieved due to the specific ecological/silvicultural conditions present. The short rotation management of sal has promised the incentive of revenues from harvesting; but it has been insufficient and not available to many FPCs as harvesting has been done only in one-third of our sample sites and this too periodically. It is apparent that local interest is waning in the absence of significant benefits or a more democratic approach. In Andhra Pradesh, continuity of community involvement in local PFM arrangements appears highly unlikely after the project village funding cycle and derived wage labour cease – there is little for the institution to deliberate over if it cannot control forest management. Participation (meaning, as in the majority of cases in India, agreement of local people to protect the forest for the forest administration) here is merely being purchased on a temporary basis. However, there is a high level of identification with the local forest as 'our' forest, and motivation to continue protecting it from outside extraction.

On the other hand, the model of Nepal's CFUGs and the Orissa SIFPGs indicate that sustainable local groups are possible if they are based on the premise of control over and management of the forest on a recurrent basis by a locally legitimate group, motivated by protecting the security of important forest product and service flows.

Forest outcomes: Changes in forest management and forest condition

The institutional arrangements created in PFM implementation give rise to particular forms of forest management that are considered here.

Persistence of technocratic silviculture

PFM, in practice, has involved little change, as yet, in forest management techniques, particularly in India where forest departments generally continue to implement working plan provisions with virtually no concession to local needs and preferences. Forest departments in India have historically separated forests into reserved forests for highest value timber where no local rights are permitted, demarcated protected forests, in high-value forests adjacent to habitation, where some local use rights is allowed, and undemarcated protected forests in lower-value forests where non-timber use rights were not restricted. Working plans have been focused on managing the higher value reserve and demarcated protected forests, and it has generally been in the undemarcated protected forest areas (at least up until the 2000 JFM revised guidelines), of lower value, often neglected and degraded, that forest departments have sought to extend their effective management through PFM in order to achieve plantation.

In Nepal, because local groups have the authority to design their own operational plans, they can incorporate innovative practices, selective felling and varying the age and species mix of the forest. However, there has been little attempt by the forest department in Nepal to develop new PFM management systems combining the best of both local and scientific techniques (e.g. of the sort of 'new silviculture' proposed by Campbell in Hobley, 1996). Forest administrations in India and Nepal have largely failed, as yet, to facilitate co-learning approaches to forestry research that might lead to an incorporation of local knowledge, local management objectives and local management practices within plans.

Forest planting, management and harvesting

The implementation of PFM often leads to the implementation of more intensive silvicultural management in village-adjacent forests, leading, in turn, to ecological changes (in terms of species and age class composition) from whatever was there before towards a generally simplified silvicultural species mix. This can have significant implications for the livelihoods of forest users. Often (but not always) a more biodiverse species mix was present, although frequently in a deteriorating condition. In India, local people are rarely consulted about their priorities for planting and these are virtually never incorporated within working plans, which lay down the planting or selective felling of specific species in different areas and different periods of rotation (e.g. 'improvement' or 'enrichment' planting of species preferred by the forest departments – often even working towards a monoculture, such as sal in South-West Bengal and teak in northern Andhra Pradesh, by removing the variety of different species, many of which may be valued by local people, but which are lumped by foresters together as valueless and anonymous 'miscellaneous' species.

In turn, this means that diverse uses of the forest are reduced, such as petty commodity production for sale (e.g. fruit, wild fungi, rope and medicinal plants), as well as materials for artisanal production (e.g. charcoal and bamboo), although in circumstances where PFM has been introduced in already degraded forest, this change does not apply in the same way. Nevertheless, as Figure 4.1 in Chapter 4 shows, the diversity of livelihood activities from pre-PFM forests is often radically reduced where local species diversity is dimished. Eucalyptus plantations in West Bengal and Nepal's *tarai* have, for instance, drastically reduced livelihood forest product availability from fuelwood, sal leaf plates and seeds to simply low-value eucalyptus leaf collection.

Forest condition change

It is a general finding that, in most cases, degraded forests have improved in condition through PFM. In the Nepal hills at almost all study sites, forest condition had improved, particularly in the regeneration of young saplings. In the Nepal *tarai*, regeneration has also been prevalent, and even though there have been relatively high levels of timber extraction, the condition of comparable national forests has been deteriorating in the absence of effective regulation. In West Bengal, sal regeneration has been dramatic across all village forest protection committees (VFPCs). In Orissa, forest regeneration or maintenance of condition has also been a general finding. In Andhra Pradesh, research findings are more ambiguous; although regeneration has been observed in most of the study areas, in some cases this has been of exotic plantations on previously cultivated land. Thus, overall forest condition has improved, although this has not always translated into positive livelihood impacts (particularly for the poorest) because forest management has emphasized traditional forest department timber production objectives.

This finding must be qualified, however, because under conditions of high-value forests and weak governance (such as are present in the *tarai* areas of Nepal), there have been a number of cases where the over-extraction of timber trees has not been stopped as a result of nominal handover to local communities. However, with the changes in institutional arrangements, a larger part of the benefits are likely to reach local people rather than 'timber mafias'. This is the only regional case where introduction of PFM hasn't led to improvement in sustainable forest management. A second qualification is that whether the actual condition of forests has improved depends on which criteria are applied. In India, forests will have changed in composition through the implementation of forest department working plans; but in many cases the species are not relevant to local forest users. In some areas (West Bengal and Andhra Pradesh), monocultures of exotic plantations (e.g. eucalyptus and silver oak) have been introduced that are of virtually no benefit to the local forest users, and often replace useful species, forest fallows cultivation or grazing opportunities. Elsewhere, forest department working plan provisions lead to the promotion of indigenous single species (e.g. sal in West Bengal and teak in northern areas of Andhra Pradesh) that systematically reduce biodiversity and, therefore, the diversity of products available to local people. Third, through silvicultural management practices, such as cleaning and thinning the understorey, NTFPs that are valuable for livelihood use can be removed (e.g.vines for rope-making, as observed in Orissa and West Bengal).

Livelihood impacts

Here we review the *livelihood impacts* for each of the five study areas. A major finding of this research has been that, particularly in India, contemporary PFM has, as yet, had remarkably limited positive livelihood impacts. There has generally been improved and more secure access to collection of forest products, and certainly in most cases less antagonistic and exploitative relations between forest users and local forestry staff. Furthermore, there have been significant improvements in forest quality and, in many cases, increased availability of

NTFPs. However, in India this has not yet translated into the (quite reasonably) anticipated pro-poor livelihood impacts. Whatever benefits there have been have come about as a result of orthodox timber production practices, rather than due to any adaptation to a pro-poor and livelihood-oriented forest management system. The Indian forest departments have persisted in controlling the forest resources, their management and the benefit distribution, and have generally been able to perpetuate their historic timber production orientation despite the fact that social priorities have since changed. Even when the centre has put severe restrictions on timber harvesting (i.e. through the Supreme Court Centrally Empowered Committee), forest department working plan preparation still routinely includes timber harvesting provisions despite the fact that these are unlikely to be permitted. The National Working Plan Code (MoEF, undated) the guidelines for preparing working plans, has not been revised to reflect the changed social priorities.

In the *Nepal hills*, CF has led to generally improved forest condition. However, this has been achieved through strict regulation of forest use according to conservative forest management systems oriented towards timber production (as per the forest department's traditional training and elites priorities), rather than towards multiple uses (as per medium) and poor households' needs). For rich forest users who depend on the forest only to a limited degree, this has led to a slight general increase in the availability of forest products, although there has been a reduction in pasture for grazing. However, for the poorest house-holds who have larger and more diverse forest product needs, including grazing and fodder, forest product flows have actually reduced. Significant time savings in fuelwood collection have been achieved in many areas; however, again collecting time has actually increased for the poorest households because charges for collection introduced through CF are prohibi-tively expensive for them and so they can only, for instance, collect dry wood from the forest floor rather than participating in organized cutting and distribution. Income from forest activities has declined as traditional artisanal occupations have become restricted by new regulations, although there has, at the same time, been a decline of the *bista* system, (a caste-based system of in-kind payments to artisans for services) due to manufactured imports particularly affecting blacksmiths, leather workers, tailors and potters (Blaikie, Cameron and Seddon, 1976). Some new opportunities have emerged through CF fund support, such as vegetable growing, school building development and buffalo purchase loans. Before CF, collection of wild fruits, herbs, vegetables and root crops provided the poor with some income and subsistence nutrition during the lean season and in crises; but since CF, regula-tions and restrictions on the collection of these products have obstructed this safety net function of forests. On the other hand, CFUG funds have been generated through member-ship, product royalties and fines and, in some cases, donors matching funds. These funds have been mobilized for community development – often for infrastructure such as path and road building, drinking water, small-scale irrigation facilities, electrification, school building development, community halls, emergency relief payments, and so on. Pro-poor micro-credit has also begun in a few places.

In the *Nepalese tarai*, the implementation of CF has certainly been a positive turning point in the livelihoods of many CFUG members, although it has also been a negative turning point for many livelihood forest users who have not become CFUG members. The cost–benefit distribution from the introduction of CF has generally benefited all CFUG members, but has clearly favoured rich member households the most and led to significant restrictions on non-CFUG member households. Household physical assets (house, cattle sheds and other) have improved for all wealth categories, although a much larger proportion of wealthier house-holds (84 per cent) have improved their physical assets compared to poor households (34 per cent) (see Chapter 6, 'Change in household physical assets'). Prior to CF, 11.5 per cent of member households did not own permanent houses and lived in simple huts, whereas now less than 1 per cent of households do not own a permanent house. This is because timber, one of the most expensive inputs in house construction, has become easily available. In Dang, CFUG timber was made available free for house construction in the study site district, reflecting the abundance of timber in the district; thus, housing conditions of the poor improved after the introduction of CF. However, in Saptari, timber was offered at subsidized prices, and whereas the rich were in a position to take advantage of this, the poorest households could not afford the price, and in Gangajali and Churiamai CFUGs, poorer groups such as the Mushahar ethnic group still live in basic huts because they have not been given subsidies on timber (see Chapter 6).

Through the mobilization of community funds generated from timber sales, *tarai* CFUGs have been able to pay for infrastructure development, including improving their drinking water supply, building schools, irrigation, electrification and road-building. Prior to CF, only 15.5 per cent of the households had access to water taps; but this has increased to 44 per cent after CF, although here, too, the rich have received more benefits.

Under new CF regulations, many households have been obliged to shift their extraction of forest products, such as fuelwood, from the CF to neighbouring non-community forests, indicating that although their own CFs may be more sustainably managed, this can be at the cost of neighbouring open-access forests, bringing the overall sustainability of PFM into question in the more affected areas.

In the study areas of southern *West Bengal*, the introduction of JFM has generally had a positive impact on livelihoods. The average annual income per household (cash and kind) of sampled FPCs across all wealth ranks totalled 45,523 rupees, of which 5090 rupees are forest derived. Although the mean contribution of forest income sources is only 11.18 per cent overall from forest, it is skewed by the fact that it is only 0.88 per cent of the total income of the rich families. It is much higher, at 14.9 per cent, for the medium rich, higher still at 26.8 per cent for the poor, and highest, at 32.1 per cent (almost one-third) for 'poor and landless' households. Sal leaves, sabai (*Eulaliopsis binata* spp) and fuelwood are the main contributors to the livelihoods of FPC households, while wage employment is a smaller component and the share of the net income from final forest harvesting is negligible. Here, as we have seen in Chapter 7, the introduction of JFM has given the local people the feeling of 'waking up from a nightmare', particularly for the poorest families, because pre-JFM families had to collect firewood for home and income and in doing so face harassment from forest department officials who treated them as thieves. Post-JFM, all FPC members are entitled to collect NTFPs from the forests. Time taken to collect forest products has reduced substantially. For fuelwood, it has reduced from a mean of four hours per day to around half to one hour per household. Time does have to be spent patrolling the forest as well, however, averaging very roughly between 0.5 and 1.5 hours per day. Collecting and processing forest products, particularly sal plates and sabai ropes, are important livelihood activities for households, especially the poorer, and the forest condition improvement that has occurred through JFM has led to increased availability of these products, improving the livelihood security of poorer households. Wholesale prices, however, remain depressed, and so collection and processing activities offer a very low level of remuneration.

In *Orissa*, the impact of PFM on improving forest users' livelihoods in the study districts has been small. However, it does appear to have averted forest resource degradation through over-extraction, which is evident in the non-PFM villages studied. A majority of VSS and SIFPG members believed that PFM has affected their livelihoods positively, even though the impact is small, especially among the poor and very poor households.

One of the most important impacts has been the restriction imposed on shifting cultivation on forest land, a common practice among tribal groups, as they have not been granted land rights on their customary land. This has reduced the extent of cultivation of around one third of households in the SIFPGs.

The direct annual contribution of forest to total income of households was between one-fifth and one-quarter of overall income (e.g. 24.7 per cent in the study VSS villages, 19.4 per cent in the SIFPGs, and 23.9 per cent in the 'no-protection' villages, although these variations are not statistically significant). But for poorer households, the dependence on the forest is very high, the highest in all the study areas in India and Nepal: the contribution from forest use in their overall income is as high as 30.8 to 42.7 per cent, respectively, in VSS villages, 18 to 27 per cent in SICFPGs, and 27.5 to 34.6 per cent in no-protection villages. The dependence on forests is relatively higher in Kandhamal district, where more than fourth-fifths of households are tribal.

Five main forest products – fuelwood, *mahul* (flower of mahua tree used for preparing pickle and alcoholic drink), *toll* (fruit from mahua tree), *kendu* leaf (for rolling *bidi* cigarettes), leaf plates and other food items – contributed virtually all of the forest income in all types of village. In no-protection study villages, over a half of forest-related income came solely from firewood (compared to one-third in the SIFPG villages and under one-quarter in VSS villages), reflecting regulations on extraction of fuelwood for sale in forest protection groups. In the no-protection villages, fuelwood extraction for sale is at an unsustainable level and is leading to rapid degradation of surrounding forests. No VSS in the study areas has harvested timber, and as such PFM has not generated any income from this source for the FPC members. Similarly, the contribution of PFM to wage income in forest-related work in VSS villages has been negligible: the forest department has generated only about ten days of employment pruning, cutting and planting trees per year per household. In SIFPG villages, there was no paid wage work at all; but members have voluntarily contributed labour for the development of forests through collective action.

In summary, although forests provide a significant income stream to households, particularly the poorest (who are well below official poverty lines), the actual amount of annual income derived per household and per worker from forests is low, due partly to the small size of forest areas per capita, the low productivity of forests for livelihood products and the unremunerative prices realized by collectors. Forest product prices are generally depressed by very unfavourable market relations. The poor and very poor VSS and SIFPG households spend more than two-fifths of their labour time during the year in collecting and processing forest produce; but the mean income generated per household is very low (32 rupees, or about US$0.50) per day per household in VSS villages; in SIFPG and no-protection villages, it is even less, at 23 rupees and 21 rupees per day per household, respectively.

One reason why the sale of forest products realizes such low incomes is the prevalence of interlinked loans. Poor households often have to borrow against the future sale of their forest product collection, the rates for which are then depressed by the money lender who is commonly also the local NTFP buyer. Additionally, loans from private money lenders carry exorbitant rates of interest of about 60 per cent per year. Forty-seven per cent of VSS and 40 per cent of SIFPG households studied have outstanding debts, and the majority have loans interlinked with the sale of forest products since such a substantial part of the income generated from the sale of NTFPs goes to money lenders. Although product availability has increased with self-initiated groups and VSS of the study villages, 95 per cent of VSS and 86 per cent of SIFPG households said that there has been no change in their indebtedness.

In Andhra Pradesh, finally, the PFM livelihood impact situation is particularly complex due to the wide variety of different agro-ecological and cultural circumstances. There has been a significant decline in the availability of forest products (especially fuel and fodder availability) to households after the implementation of JFM/community forest management (CFM), mainly due to the restrictions on access to forests for regeneration. Although there was a small decline for small and medium farmers, the poor have been particularly affected due to their high dependency on forest products. Especially during the initial years of JFM, restrictions on grazing have had a severe impact on poor households. Grazing restrictions did not have a severe effect on most farmers because they had private land for grazing; but many of the poorest have been forced to sell off cattle. In the case of fuelwood, many farmers were able to switch to subsidized liquid petroleum gas (LPG) supplied under government projects, although here, again, the poorest could not afford even the subsidized rates and instead experienced increased restrictions on fuelwood access.

The temporarily offered VSS wage works have, however, become a new and important source of household income, comprising a mean of 8 to 10 per cent of total income during the project implementation cycle, particularly important in the hot season when there are few labour opportunities. Of course, this wage labour is only offered for a few months per year over the five-year 'treatment' of the village, so is more of a welfare payment than a sustainable livelihoods component. In Kadapa and Adilabad districts, VSS wage labour has provided an alternative for the poorest to out-migration in search of work.

In Visakhapatnam, income from *podu* (shifting cultivation), which constitutes 30 per cent for poor and 28 per cent for non-poor total income, has declined due to the reclamation of *podu* lands by forest departments through JFM. This is one of the reasons for the relatively low growth in gross household income in this district. Decline of income from lost *podu* opportunities represents 18 per cent for poor households and, even more, 20 per cent for non-poor households, affecting not only income but also food security. The decline of *podu* income has been only partly compensated (temporarily) by VSS wage work, which has contributed 11 per cent of total annual income for poor and 10 per cent for non-poor households.

Restrictions and exclusions from the forest

The impact of restrictions on forest use and other constraints (such as species change) falls unequally on households of different wealth rank. Those households who used the forest most intensively and lack sufficient privately owned land have usually suffered most from restrictions. The poorest, and particularly those of oppressed ethnic and caste origins, have, in many cases, found themselves either left out by omission or actively discriminated against at the time of formation. Two examples illustrate this. In Bolangir district of Orissa in Jarmunda, VSS, basket-makers, who had hitherto been collecting bamboo from the forest under the previous customary system, were excluded from membership of the VSS; therefore, one of the main sources of their livelihoods was interrupted. In the middle hills districts in Nepal, the formation of CFUGs has often provided the opportunity to exclude groups from using the forest that had hitherto depended heavily on it for their livelihoods. For instance, in Jalkini CFUG in Dhankuta district of Nepal, *Kamis* (the blacksmith occupational caste), who had previously produced ploughshares for the local community and had consumed significant amounts of charcoal, and who had gone on strike over levels of remuneration in the *bista* system (*jajmani,* a reciprocal in-kind and labour exchange based on caste), were excluded. These examples show how institutional or legal change in the control or distribution of resources may allow particular groups or leading factions to take advantage of the disturbance provided in accepted norms and practice.

Another important issue here is the implication of exclusion and restriction from the forest of farming systems and pastoralism. Forests can represent a nutrient bank in farming systems. PFM can, in some circumstances, secure the maintenance of these flows (e.g. fodder leaves to feed cattle to create the farmyard manure that fertilizes the fields). However, increased regulations can also disrupt these flows. The exclusion of unsustainable use practices, particularly in degraded forests, may be necessary given that the management objectives of a particular forest are reached by negotiation with forest users in an open and transparent manner. But there is undoubtedly a heavier burden on poorer groups and their needs are often marginalized in deliberations. More wealthy groups with more cash income can substitute other nutrient sources, including chemical fertilizers, or shift fulfilment of timber needs and nutrient-producing plants to their own private fields. PFM has also, in some cases, reduced access to forests. Across the middle hills of Nepal, many marginal farmers used forests for seed beds for millet cultivation until the introduction of CF prohibited this. Shifting cultivation and turmeric cultivation in the forests in Andhra Pradesh and Orissa have also been obstructed by the introduction of PFM there, as discussed in Chapters 8 and 9.

The period of time over which income streams are suspended

The impact of closing forests depends on the management system and, specifically, the period during which the forest is closed for regeneration, as well as the rotation period of the chosen species. The rotation period and the generation of subsequent income at the time of felling can vary between 12 years (sal coppice in West Bengal) and up to 80 to 120 years for teak. These rotations are derived from silvicultural textbooks of some antiquity (e.g. Jerram, 1936) and are usually adhered to in a mechanical and inflexible manner. It would be quite feasible, if large-size timber were not the sole priority, to adopt shorter rotations which maximize the opportunities for sustainable incomes, as well as sustainable forests, or even mixed-age stands for selective felling (as many private foresters do). Again, the poor's time discount rate is highest since they cannot afford to wait long for cash payouts on harvesting after a number of years, whereas the forest administration's time discount rate appears to be very low. In the Nepalese hills, five-year operational plans have been introduced through CF, regardless of the age class of the forest, and selective felling is practised within the allowable timber cut calculated by the DFO.

The *final payout* by the forest department to the village *on harvesting* may not be equitable and is seldom based on need, either in India or in Nepal. In any case, the sum is, in most cases, an insignificant amount (as discussed in Chapter 7, in South-West Bengal, in the less than one-third of sampled villages which actually received payment, the share per family per year varied between 40 rupees to 207 rupees (mean of 78 rupees), which is only 2.16 to 4.13 per cent of the total income from forests of different wealth ranks). Nonetheless, the promise of cash payments from timber harvesting is, in almost all cases, a persuasive inducement for local compliance with the forest administration's plan for participation – particularly in cases where PFM threatens to reduce the income of specific groups – for instance, *podu* cultivators and others mentioned above. The percentage of share that is supposed to be returned to the group may be as high as 100 per cent. However, this is net of forest department harvesting costs (a major source of possible malpractice and denial of income).

Implementation of PFM, through more significant devolution of secure rights, more livelihood-oriented forest management and a more pro-poor focus, clearly holds the potential for more significant (positive) livelihood and poverty impacts. However, the path for achieving this is fragile and is contingent on a number of factors. It requires institutional and cultural change in the forest administrations in order that they are willing to devolve more significant authority to independent local bodies, and to support them in their capacity development as equal partners and to promote a pro-poor orientation. National five-year plans give ambitious pro-poor provisions; yet these hardly seem to have permeated down to the district and field practice of foresters, who still persist in perceiving their job as forest management for timber production. Adapting to a livelihood and pro-poor forest management regime would involve moving to multi-product silviculture, emphasizing rapid production of NTFPs and adapting to the local needs and priorities of the poor through local deliberation, consultation and decision-making.

However, as we have seen, a historically authoritarian culture within forest administrations, along with the significant wider problems with governance in South Asia, has led so far to a general dilution of the potential for a more pro-livelihood and pro-poor process, as well as the impacts of PFM, particularly (although not exclusively) in India. This dilution has been achieved through two main processes. First, forest administrations have sought to accommodate PFM processes within their normal practice (e.g. through the perpetuation of working plans in JFM areas), rather than introduce more fundamental adaptations. Second, forestry field staff, acculturated within a semi-paramilitary hierarchical and patriarchal organization staffed almost exclusively by men, have replicated and, in many cases, extended the forest department bureaucratic authority structure to local areas through formation of PFM institutions and in their interactions with local people. These interactions have overwhelmingly been with elite men, excluding ethnic minorities, poorer households and women (this finding accords with those of Hobley, 1996, p147; Jeffrey and Sundar, 1999, p198; Jackson and Chattopadhyay, 2000; and Agarwal, 2001).

Incomes (cash and kind) from forests

If increasing income from the forest for local people is the main objective, then one of the biggest potentials for this is marketing timber. Internationally, there are a number of countries, with Mexico perhaps the foremost, where local timber management has become successfully institutionalized (Barton-Bray and Merino-Perez, 2002), although in South Asia the process is at an earlier stage. In only one of our five study areas – the Nepalese *tarai*, where there is high-value standing forest, access to markets and a permissive regulatory environment – is timber a major source of revenue for local groups. In the Nepal hills the inaccessibility and less valuable timber present obstructions and timber trading has not emerged significantly despite the permissive policy environment. This, however, is in stark contrast to India, where the forest administration still maintains a monopoly on timber trading from state land. Revenue-sharing from forest department timber marketing does come to PFM groups in India, although, in practice, this has turned out to be a very small amount (the three out of the ten study groups who received a share in West Bengal were given around 78 rupees per family per year; see Chapter 7).

Contribution of non-timber forest products to livelihoods

In recent decades, regulatory arrangements over NTFP marketing have increased, often with the tendency to favour the interests and bargaining power of merchants, contractors and forest departments. PFM has not challenged this state of affairs and often merely continues this trend of ineffective and, often corrupt, state regulation. In Nepal and in most of India's local institutions, CFUGs and *panchayats* have the right to license NTFP export out of the area. However, very few local institutions are either aware of this right or able to manage the trade; as a result, for higher-value medicinal plants, open-access over-extraction is commonplace and prices continue to be depressed. In Andhra Pradesh, the Girijan Co-operative Corporation is a para-statal NTFP buyer that has monopoly control of the trade in the state. However, it has not improved the prices that it pays collectors, and due to its limited coverage of the state middlemen, it continues to play an important and often exploitative role in NTFP trade.

Improved NTFP collection and marketing holds the greatest potential for strengthening livelihood benefits through PFM. Across the study areas, activists and NGOs (e.g. the Regional Centre for Development Cooperation, or RCDC, in Orissa and the Asia Network for Sustainable Agriculture and Bioresources, or ANSAB, in Nepal) have been working intensively to improve collection, processing and marketing conditions, although, as yet, the forest administrations have tended to continue treating NTFPs as the 'minor forest produce' they labelled them during the 19th century.

From this review of the livelihood impacts of PFM, the figures clearly indicate that forests play a major part in the livelihoods of poor people, particularly in the lean season when subsistence resources are insufficient from private land or from wage-earning opportunities. However, the contribution declines for medium-rich and rich households. It must be borne in mind that the current proportion of subsistence needs met from the forest is a result of long-term historical expropriation of the forest by the state. If the figures seem low (in some people's eyes), this cannot logically justify further expropriation to an even lower figure.

Proximate causes of policy impacts and recommendations

The summarized impacts of PFM have been listed above, and in this section their proximate causes are examined: that is the local and day-to-day practice, patterns of behaviour and institutions that have shaped these impacts. Recommendations for reforms are provided.

Scheme basis for PFM implementation

In India, JFM has been implemented on a 'scheme' basis for disbursement of funds over the medium term, rather than as a permanent reform. *Ad hoc* local institutions that lack a permanent basis are liable to become defunct. While a temporary basis may suit the forest administration, particularly if their aim is simply to enlist local people to protect plantations until they are established, it is not likely to resolve the long-term issue of lack of local rights in forest management. Local people will become less inclined to invest their time and effort where the security of the return is in doubt:

- PFM implementation needs to be treated as a permanent arrangement, not a time-bound fund disbursement scheme, if it is to lead to long-term institutional development, resource improvement and livelihood development impacts.

Selection of locations for PFM promotion: Identification of 'forests' and forest boundaries

A major strategic policy decision is identifying which areas are to come under PFM. There is a potential range of forest categories: all forest land, forest land adjacent to settlements, or only degraded forests adjacent to settlement. This is a crucial first decision for forest administrations in both Nepal and India. In India, only degraded forest was deemed suitable for shared management up until the 2000 JFM revised guidelines circular.

There may be a range of different land uses and land users in India, such as grazing land and forest fallows cultivation on lands that have become classified as 'forest' and become subject to JFM regulations. However, 'degraded' is not defined and therefore remains open to the discretion of the forest department. Often, non-legal categories such as 'scrub' and 'open forest' (as used by the Forest Survey of India for its biennial *State of Forest* reports) are used by the forest departments. In Chapter 9 we have seen how areas used for livelihood-oriented forest fallows cultivation were targeted for PFM in order to obstruct this use. On the other hand, in the Nepalese middle hills all forest has been deemed suitable for handover and even many areas of the extensive and undegraded high hills forests have been handed over.

In practice, boundary conflicts often arise after formation (especially in both the Nepal hills and *tarai*), both because boundaries are not clarified at the outset with all local users consensually, and because outdated maps or unusable chain survey maps have often been used. In Nepal, CFUGs have then had the onerous responsibility of contesting any encroachments that may occur on the old boundaries of what remains state property, leading to legal costs becoming incumbent on local forest users rather than the forest department:

- Applying PFM in tribal 'forest' land where poorest households are cultivating is anti-poor and forest departments should desist from doing this.
- Greater livelihood benefits are likely from PFM in standing forests.
- The forest in question should be resurveyed and mapped at the time of formation in order to ensure that the boundaries are up to date and commonly recognized by all.

Identification of 'forest users'

All of the *actual* users of the forest need to be identified and consulted, including local residents, distant users and seasonal users. In practice, not all users are included, partly because of the lack of clarification of boundaries discussed above, and also because of the micro-politics of the village and the associated status of some users. Finally, due to the difficulties in resolving the problem of how a user is defined (daily?; monthly?; seasonally?; every five years?), the term 'users' becomes synonymous with regular and local users. A further issue here is in- and out-migration, which require revision of the membership list, and which bring about problems for incomers who may be excluded or who may have to pay prohibitive new membership fees or bribes. Those excluded particularly include distant users, the poor and the more forest dependent. In the Nepal *tarai*, distant users have been widely excluded, often deliberately. For instance, the transhumance system in Karnali has declined because CF in the lower hills did not permit grazing for their animals, as granted by the government in the past. In Andhra Pradesh, the forest department has deliberately excluded *podu*-cultivating forest users and claimed success in persuading local tribal cultivators to 'surrender' 37,000ha from *podu* cultivation:

> ... *by providing viable alternatives during the implementation of joint forest management, nearly 37,000ha of forestland under the possession and cultivation of local people have been reclaimed through afforestation and put under productive treecrops through VSS.* (World Bank, 2004, p2)

Even with the best intentions, the inclusion of all users is very time consuming. Participatory rural appraisal (PRA), which has been officially adopted in the World Bank project in Andhra Pradesh as a means of data collection for this complex process, is a highly time-efficient way of collecting this kind of data, although PRA requires extensive training and practice with local people. In this case, only three to five days were allotted to set up the requisite group. However, in other parts of India, PRA processes have not been followed at all. While it is not absolutely necessary that PRA techniques are used, it is certainly a great help, both for data collection and for local forest staff or associated service providers, to discuss the formation of the group with the village. PRA has been extensively critiqued (Waddington and Mohan, 2004; Francis et al, 2005), and it is not difficult to recognize some of the characteristics of PRA being performed in a perfunctory manner. These include non-representation at the PRA 'performances', play-acting by participants, elite capture, pseudo-shamanism, the phoney 'feel good' factor on the part of outsiders, and so on (Hickey and Mohan, 2004). However, as we discuss in Chapter 11, many of these shortcomings can, at least, be palliated by the way in which PRA is facilitated, the linking of the initial PRA with immediate action and planning with local people, and leaving the data on which planning is based with the village in some public place (e.g. on the wall of the *panchayat* buildings). Again, as always, it is difficult to regulate for better performance of PFM and it takes expertise, time and commitment. In Andhra Pradesh, the World Bank funded the JFM/CFM programme with a plan to carry out thorough PRA, but followed unrealistic imposed schedules:

- Legitimate resource users need to be included in groups to avoid exclusion, marginalization and conflict. Where livelihood resource users are not local, co-management mechanisms may be needed to involve them.

Maintenance of forest department hegemony, prioritized over autonomy of local institutions

A recurring point in this book is that the forest administrations of both India and Nepal have sought to maintain their traditional powers as far as possible. Although in Nepal, especially in the hills, powers have been devolved to local groups based on recognition of traditional practices, in India, rather than legitimating, working with or building on existing local institutional arrangements, the forest administrations have instead undermined or confused existing practice by creating new institutions. A wide range of customary and informal institutions have thereby been undermined (such as self-initiated groups in Orissa and local governance institutions such as *gram sabhas* and *van panchayats* in Uttaranchal; see Sarin et al, 2003).

There is a fundamental contradiction between the bureaucratic and regulatory nature of the forest administration tasked with establishing PFM and the diverse location-specific landscapes of both the forest and the political ecologies of the local people who use it. In the context of asymmetric power relations, the forest administration has generally sought to retain hegemony over local people and their organizations. There are a number of related reasons for why this is so widespread. First, there are well-established bureaucratic repertoires (that is to say, a range of learned and repeated practices) on the part of forestry officials at all levels. When they confront a potentially autonomous institution, there are bound to be frustrations and differences of view, which officials then strategize to resolve. In some cases, individual officers see the inherent contradictions between repertoire and field experience and can be adroit and sensitive in negotiating a consensual outcome or even making tactical concessions. In other cases, forest staff maintain an authoritarian style, and while the letter of participation may be formally observed (e.g. targets achieved), the spirit of participation is entirely absent. Some of the actions of forest staff in negating the spirit of participation may be strategic and consciously thought out. Volunteering to devolve executive power, as PFM invites, where fulfilment of managerial objectives becomes more uncertain and out of direct control, simply does not come easily to executive officers. Finally, there may well be private interests, such as promotion prospects, avoiding censure and protecting opportunities for illegal gain – but not necessarily so. The key point here is that PFM as currently practised subordinates or, in some cases, even replaces viable local institutions, with new institutions extending a hierarchical structure down to the village level. Democratic decentralization of resource management has not happened to any great degree, leading to the contradictory situation in which there is a claimed increase of PFM; but by attempting to facilitate it the very institutions that are supposed to deliver it are undermined or made 'perfunctory', in Lele's term (Lele, 2000).

An example from Andhra Pradesh illustrates a conscious strategy by forest department staff to achieve their 'green cover' objectives, while avoiding direct confrontation by setting up JFM user groups using a 'divide-and-rule' policy – a malicious strategy contrary to the successful development of local forest user groups. For example, in Visakhapatnam district, the forest department has set one village against another by giving management responsibility for forests used by one village for shifting cultivation to neighbouring villages – which are then given the task of protecting exotic forest department plantations. In Kurnool district, many VSS have been formed on degraded forest lands used by landless grazers, and in some cases the responsibility to exclude grazing has been given to neighbouring villages – not to the grazers themselves. The predictable ensuing inter-village conflicts are damaging to the long-term prospects of the poorer households and the forest user groups.

In West Bengal, the forest department has taken over as the major partner, subordinating the FPCs to minor partner, particularly in decision-making. The micro-plan, although supposed to be written jointly, is, in practice, a product of the forest department. The forest administration treats the local institution not as a democratic organization, but instead seeks to induct to the management committee only those local people who will defer to it. During research for this book the authors experienced first-hand the hostility of rangers to 'outsiders' visiting 'their' JFMC villages without their prior consent. A previously arranged meeting was obstructed and cancelled, and local people felt that they had to obey the ranger and that they had no say as to whether they could interact with the outsiders. There are other examples where private interests were protected in such a way that the forest department purposely failed to assist local institution members in finding and punishing powerful offenders:

- Ultimately, the autonomy of local institutions must be legally assured through constitutional local self-governance structures.

Inadequate preparations and support for local PFM

A major complaint made by forest users themselves is that state-initiated local PFM groups were formed very hastily and not given support during the formation process, from the initial identification of location through to the post-formation stage. There have been a number of cases where 'target chasing' was initiated by forest departments seeking to form unrealistic numbers of groups in a very short time (e.g. the World Bank project in Andhra Pradesh, where this hurried scramble was due to unrealistic project schedules). A serious application of participation must include field traverses and negotiation with forest staff, sometimes combined with PRA, and demands time. DFOs and rangers often lack the resources, skills and incentives to fulfil this challenging task in anything more than a token manner.

In the Nepalese hills, over the last decade, an average of over 1000 groups per year have been formed. The pressure on field staff to fulfil high targets in remote areas has inevitably led to shortcuts in the formation processes, leading to weak institutionalization.

In the Nepalese tarai, the handover process was not well developed and occurred on the provisional criterion of whether potential groups were 'willing and able'. Some groups of over 2000 households were formed. In another case in Rupandehi district, a 1600ha forest was handed over to a CFUG, composed of 12 villages across a number of different VDCs and municipalities. In both cases, local people showed considerable initiative in trying to make these unwieldy institutions work, despite difficulty in publicizing meetings and inducing people to turn up in sufficient numbers to achieve a quorum. In one case, 7 buses were hired to try and bring the group together and 12 local sub-committees were formed, each of which sent a delegate (according to official guidelines). The local DFO eventually split one of such large groups into three smaller ones after a lengthy negotiating process, and this has slowly led to improvements (see Chapter 6).

In many states in India there has been a difference of expectations over how much time and effort is really required to form local participatory groups between foresters and local people. For foresters, their silvicultural working plan for the forest in question is not going to be revised through local negotiation anyway. Therefore, negotiating the micro-plan is often treated only as a perfunctory process, and one that can be dispensed with altogether or delegated to NGOs simply to secure village compliance:

- The policy implication here is that for PFM to evolve, the future formation of local groups requires much more careful consideration and more thorough process, and for those groups already formed, a consolidation or reorganization effort is required.

Raising awareness of rights and responsibilities in PFM

As a result of rapid PFM group formation, many local people are unaware of their rights, responsibilities and roles under the new system. This creates a non-transparent atmosphere where the forest administration and local elites can easily manage the local process in their favour. However, in Nepal, social mobilization by donor-funded projects (such as the DFID-funded Livelihoods and Forests Programme) has increased general awareness levels in villages, thereby moving the balance of power from elites and foresters to the general body of members:

- Raising awareness of the rights and responsibilities of users is crucial and helps to build social capital and the necessary knowledge and skills for collective action. Ideally, this implies repeated collective discussions with local people at open meetings, down to hamlet level, as well as local leadership of the process. In practice, these are often rushed or avoided altogether, with field staff simply paying a brief visit to village elites. On the other hand, in a very small number of cases, funding resources, commitment of staff and local motivation all come together and lead to successful awareness-raising.
- It is not an easy change of role for forest department field staff to go from policing to acting as development facilitators, and this cannot be expected overnight. Independent NGOs, community-based organizations (CBOs) and civil society groups can therefore be crucial as service providers if they are given a proper long-term role and not just enlisted to deliver compliant groups to the forest departments.

Development of local forest management plan and/or micro-plan

One of the most fundamental differences between PFM in India and Nepal is the level of power that has been devolved to local people in order to plan their own forest management according to their needs and priorities. In India, there are no local-level forest management plans. The micro-plan document negotiated at formation does not address the detail of forest management, which has already been written in divisional working plans. Therefore, it is really little more than a public relations exercise – the audience being both local people (offering inducements for their involvement) and donors (indicating groups formed and money disbursement agreed). In this sense, participation in micro-plan preparation often represents little more than a confidence trick.

In Andhra Pradesh, NGOs have been used to facilitate the formation of JFM groups by the forest department, which paid them on a per group basis, although the forest users themselves were generally not informed of this arrangement. Once a group is formed by an NGO, it is delivered to the forest department and the NGO's role ends. At this time, users experience a distinct change of tone and intention. One group whom the authors visited in Vizakhapatnam district, Andhra Pradesh, had contributed to the facilitating NGO detailed and considered requests for diverse forest species that would provide tree crops for livelihood needs; but when the forest department field staff arrived to implement the plan, they brought only silver oak species, reflecting their preference. This example shows the extent to which the Andhra Pradesh Forest Department has treated local needs seriously. In Adilabad, the authors also visited Beheronguda VSS, where the local people were not informed at the time of the micro-plan preparation that they would need to wait 80 years for revenue from timber harvesting in the teak forest that they were protecting. These examples show how forest departments have not yet emphasized pro-livelihood forest management techniques and technical support, but remain stuck in a pro-timber and pro-industry mode of functioning.

In Nepal, the local CFUG has the legal right to define its own forest management plan after a forest inventory has been conducted by the range office. The inventory process required under the revised 2004 guidelines remains technically inappropriate for hill CFs, offering a lengthy enumeration of valuable timber trees and only superficially addressing NTFPs. The 1993 Forest Act provides for CFUGs to write their own operational plans with the assistance of forest department field staff, although, in practice, DFOs generally reject operational plans not written by foresters. There have been many innovations in locality-specific, needs-based and multiproduct forest management planning and practice by CFUGs; but few of these have enjoyed wider uptake or been adopted by DFOs in the writing or approval of operational plans. DFOs have to assess and approve the standard operational plan format for each CFUG, the contents of which are prescribed in the CF guidelines. DFOs remain in a hierarchical organization that emphasizes their regulatory role (i.e. in avoiding excessive timber off-take), rather than a developmental role (e.g. how best to sustainably mobilize available resources to reduce poverty).

Development of local capacity to protect and/or manage forest after formation

A very different situation prevails in India from that in Nepal. In India, JFMCs are generally responsible only for protecting, and not managing, the forest. In some cases they appear to do this fairly successfully, although where organized criminal mafia groups are involved, local groups need the support of the forest department, which is often not forthcoming. A further issue here is the involvement of forest department staff in illicit timber extraction, where JFMCs can find themselves threatened. In Nepal, where in law local groups have responsibility for managing the forests, this requires substantial local capacity development that has not yet been fully achieved. Thus, in practice, the DFOs often play a predominant role in operational plan development. However, in some cases, as a result of the local DFO exercising skill and flexibility, there has been a quite remarkable level of activity and capacity development indicating local desire, capability and interest in both protecting and managing their own forests.

In conclusion, it is obvious that each of these stages in PFM formation requires time and concerted effort, and shortcuts, target-chasing and tokenism can lead to all sorts of diverse problems for local people and their institutions.

In Nepal, both in the middle hills and the *tarai* (as discussed in Chapters 5 and 6), the DoF has issued frequent circulars, directives and policy amendments since 1993. The concern is not only that there have been changes, but that these were introduced in a non-deliberative manner. They neither went through a parliamentary democratic process, nor was there any major opportunity for local people to contribute. Whereas at the time of forest handover local people began to identify it as 'their' forest, after the many revisions and dilutions of policy some local people have begun to feel that it may not be theirs after all. Furthermore, the legal responsibility for the performance of local forest management institutions has been laid on the shoulders of the DFO, who therefore tends to be loath to depart from blueprint guidelines and use his or her discretion to accommodate the variety of local needs. There have been a few high-profile legal cases where the conduct of DFOs has been challenged in court. For example, the DFO in Khavrepalanchok district was persuaded to approve the operation of the only CFUG sawmill in Nepal by the Nepal Australia Community Forest Project (Singh, 2005); but when the level of felling was believed to exceed the CFUG's approved off-take level, the DFO was suspended (although the sawmill continues to be something of a cause célèbre. The four CFUGs involved have now contracted the sawmill's management to private timber contractors. This example reflects the lack of trust between central government and the DFO, particularly in relation to local innovations and flexible use of personal discretion. As DFOs are under pressure to ensure

---

Box 10.1 Raotara village in West Bengal: A worst-case scenario of the perversion of the local user groups' capability to protect the forest (narrative of a forest protection committee member)

Raotara village forest protection committee's micro-plan was 'negotiated' in one visit by the DFO with the village *panchayat* leader and the main money lender. Tribal people who were indebted to the money lender were forced to cut the JFM forest illegally and sell timber at local markets to pay back these loans. The market-place was close to the DFO's office; but he apparently looked the other way. The land illegally cleared in this way was to be sold by the money lender to new incomers for agriculture.

---

that CFUGs do not exceed their legitimate felling levels, they are therefore hesitant to exercise discretion and flexibility in signing off operational plans.

In *West Bengal*, the top-down, non-representative nature of the JFM process is best exemplified in the standard sal coppice management system that is prevalent across the south-west study regions. There is no incentive for staff to vary this, despite the fact that village communities are interested in a range of forest products that would be better produced with other cycles. For instance, sal leaves for plate-making would be more accessible to collectors if the sal were coppiced at three years; poles for house and farm use could be generated from a six-year coppice; and a range of different tree and shrub species would give a wider range of livelihood products. Nevertheless, at present the standard working plan provisions are not varied.

Pressures for reform versus institutional inertia in forest administrations

After the lengthy list of problems with the current PFM implementation discussed above, we may reasonably ask whether the shortcomings are indicative of the forest administrations' struggles to change or their struggles to resist change. It is well-recognized in management texts that institutions are continuously required to adapt to their dynamic, changing environments, and that they commonly seek to reduce the required change by trying to adapt the environment itself in order to maintain continuity, predictability and often, also, their members' vested interests (Scott, 1999). The modernist forest administration model emerged from the 19th century, when social and imperial priorities were very different. Since the post-war period, modernization and industrialization policies have intensified pressures on this modernist model to deliver revenues and timber for the state. However, from the 1970s onwards, a change of social priorities has led to the emergence of a number of systematic factors promoting inertia and decline in the modernist forestry model.

Green felling bans

In the wake of high-profile public agitation against the industrial bias in modernist forest management in India (e.g. the Chipko movement), green felling bans were introduced by the central and state governments from 1980 onwards (as discussed in Chapter 1). Similarly, during 1990 in Nepal, a green felling ban in national forests was introduced that still persists today in national (but not community) forests. In both India and Nepal, green felling bans took management decisions out of the hands of the forest administration, fundamentally disrupting the modernist model. Forest departments have been in a sort of decline since these green felling bans were introduced, particularly because their self-generation of revenue flows was curtailed. Although these restrictions forced the forest administrations to reflect on their priorities, this has not necessarily led to fundamental reform, and most forest departments in India continue to manage forests in the hope that felling will, again, be permitted.

Working plans have traditionally been prepared for the high-value forests classified and settled as reserved forests and demarcated protected forests in India according to the National Working Plan Code, and annual plans of operations (APOs) for forest working (i.e. cutting and planting) were systematically derived from these working plans. But with bans on felling, the working plans became largely irrelevant. The predictably generated revenues were also no longer available to fund annual working and plantation, so many forest departments' normal working plans were put aside and APOs were developed by DFOs on an *ad hoc* basis. The post of working plan officer (responsible for drafting working plans) even became seen in some states as a 'punishment posting'. With felling bans in place, foresters in many areas have been out of the habit of using traditional forest management skills.

In Nepal, there have never been proper working plans, at least ones that included regeneration provisions. In the hill areas, these were never developed because the focus of forest management has been the higher-value *tarai* sal forests. Here, there has been systematic felling since the 1920s simply based on felling plans drafted by the Timber Corporation of Nepal with district forest offices. *Tarai* forests allocated for clearance for settlement (e.g. for conversion to agriculture, resettlement schemes, industrial areas, towns and highways) have also provided extensive areas for clear felling without need for replanting. A number of donor projects during the 1990s, most recently by the Finns, have sought to introduce sustainable technical forest management plans in the *tarai* to include regeneration provisions (e.g. the Bara Forest Management Plan). At the same time, the Ministry of Forests and Soil Conservation drafted similar operational forest management plans for the other *tarai* districts. However, the non-transparent and non-participatory development of these plans led to civil society groups (primarily FECOFUN) objecting to, and successfully mobilizing, public opposition to stop them (even in Finland).

Fiscal crisis

Government fiscal crises emerged during the 1980s, largely as a result of changes in the global economy, including oil price rises. As Indian states and the Nepalese government entered these crises at the same time as timber revenues dropped, funds to maintain forest departments became severely limited. The same pattern occurring internationally led to many forest departments in many countries being closed or merged with other departments, and this in itself has been a factor in promoting devolution of forest management to local people. However, in South Asia, forest departments have been able to continue with government support; but finance departments have become eager to augment their budget through donor grants or soft-loan financial assistance, and in some cases donors have obliged and projects have palliated the funding crisis. External financial support to forest departments in India is typically offset from their normal state budget, disrupting normal forest department systems and putting their funding allocations under the influence of donor projects. This can lead to an *ad hoc*, short-term orientation. While donors may have the best intentions, the disruption to the system may have serious consequences for the institutional continuity of the forest administration. Donors can influence states using financial support as leverage, particularly in the case of smaller states such as Nepal, where donor support makes up a significant part of the national development budget (although much less so in India, where it forms a much smaller proportion of the overall budget).

Human resource management, transparency and corrupt practice

A third aspect contributing to inertia in the traditional technical forest management system is the generally observed decline in human resources in South Asian bureaucracies. It has been recognized in India that the civil services have been in decline for decades. As the ex-principal secretary of Bihar, P. S. Appu, states:

> As a result of the perverse personnel policies followed by successive central and state governments, public administration has become dysfunctional. The governments no longer govern. Law and order has broken down in large parts of the country. Citizens no longer enjoy protection of life and property. The prime purpose for which the state was set up is not being served. The social contract has collapsed... An important development during the last three decades has been the growth of a malignant syndrome embodying pervasive corruption, criminalization and electoral malpractice. (Appu, 2005)

This decline may be attributed to a number of factors. Allocation of postings is often based on patronage and political influence, rather than on merit in India. Senior foresters during this study have complained that in the confidential annual reviews of performance, except in cases of extreme misconduct, full marks are routinely awarded in many forest services in an act of solidarity across the Indian Forest Service (IFS), which is in competition with the Indian Adminsitrative Service (IAS) for postings in the central administration. Complainants also said that the service consequently contains many 'duffers' whom, in the absence of a meritocratic promotion system, end up in senior positions. There are also remarkably regular staff reshuffles due to political influence disrupting continuity of work. Furthermore, with lifetime job security and a non-meritocratic promotion system, there is only limited incentive to perform well, and non-attendance, particularly of field staff, can be a frequent problem. There have been few studies of this aspect of forest administration; but the weakness of human resource development in many forest departments is a significant problem.

The prevalence of corruption has been a growing problem in India, with some notorious high-level cases such as the *Bastar malik makbuja* case, which illustrates how rampant it is. This case, one of the best-known corruption scandals, involved senior IAS and IFS officials (including a divisional commissioner) being prosecuted in 1996 for personally profiting from the exploitation of tribal people and their tree wealth (Acharya, undated).

The Supreme Court and Tribal Forest Rights Bill processes

A factor further undermining the forest administrations' control has been the judiciary, which has become increasingly central to the forest policy process in India. Since what has become known as the Forest Case began in 1995 (T. N. Godavarman Thirumulkad *v* Union of India), the Supreme Court and the Centrally Empowered Committee that it has constituted have issued a wide range of directives affecting forest land tenure, felling, etc. In 2003, orders were issued to evict anyone resident on state forest land without tenure within six months. Due to the fact that most tribal people have never had their land tenure formalized, this would have expropriated millions of India's poorest, along with the opportunistic land 'encroachers' whom it was targeting. Partly as an outcome of the chaos and public protest created by the threatened evictions, there has been a concerted effort over the last few years to resolve the anomaly of millions of tribal people cultivating in their ancestral domains without secure land tenure, and vulnerable to pauperization through arbitrary eviction. The

Scheduled Tribes and Other Traditional Forest Dwellers (Recognition of Forest Rights) Act was finally passed in 2006 as discussed above, although its passage has been beset by argument and conflict.

Implementation of the Panchayati Raj Act

The gradual rolling out of decentralized government in both India and Nepal promises a fundamental reorganization of authority structures at local level. In 1992, the 73rd Amendment to the Constitution granted constitutional status to *panchayati raj* institutions in India (accompanied by the later *Panchayat* Extension to Scheduled Areas legislation in 1996). These measures made constitutional provision for state governments to devolve 'funds functions and functionaries' to three levels of local *panchayat* bodies. Similarly, in Nepal, the 1997 Local Self-Governance Act provided for a wide range of responsibilities to be devolved to district and village development committees. Although the scope for reform is vast, in practice, devolution has been resisted at every step, and the process has been stalled by the Maobadi conflict until very recently. However, ultimately, these legislations challenge the legitimacy of local PFM bodies constituted by line agencies such as the forest department, and are likely to subsume them under constitutionally based local government bodies.

Mobilization of PFM networks

Forest departments are facing increasing civil society mobilization across the study areas, particularly as PFM groups begin to network to protect and further their interests. In Nepal, FECOFUN has been remarkably successful in negotiating a number of policy reforms through mass mobilization and lobbying. Although the organization was initially established with donor support and had to endure claims that it had partisan political affiliations, it has developed into a powerful mass organization. Through intercession with the finance ministry in 2003, proposed tax increases on CFUG timber trade were diluted. Furthermore, in 2006, through FECOFUN pressure, the future of the collaborative forest management model piloted in the *tarai* was brought into question. In India, such mobilization has been less concerted and effective. Nevertheless, evidence of emerging PFM networking can be seen in Orissa, where NTFP marketing has been moderately improved through pressure from a number of networks.

Industrialization processes in forest areas

The Government of India has set ambitious economic growth targets (specifically the continuation of the current level of 8 per cent), and in order to achieve these the cutting of corners in terms of environmental protection, and even justice, seems to be inevitable. Orissa contains major mineral reserves in forested areas, and although conversion of forest land to other uses is formally proscribed under the 1980 Forest Conservation Act, without Ministry of Environment and Forests (MoEF) approval, the Government of Orissa has been issuing MoUs to both national and international private companies to engage in mining and other industrial development plans in forest areas. Most of these areas are densely populated with *Adivasis*, who are threatened with forced eviction. Many of the numerous demonstrations have ended in killings of tribals by police, the death of 12 such tribals in Kalinganagar in January 2006 being only the most recent example.

Central funding to JFMCs through forest development agencies

The forest development agencies' scheme was begun in 2001 under the National Afforestation Board in an attempt to fund JFMCs independently of donor support. Substantial funds are made available for five years to favoured *panchayats*, although only a small number are selected. The effect in some areas, including Orissa, has been a clamour by local people to become registered as JFMCs in order to have the chance of this support.

Conclusions

This chapter has reviewed the outcomes and impacts of PFM implementation across the study areas of India and Nepal, and has found that although forest condition has generally improved, the livelihood impacts of PFM, as currently practised, have been very much constrained, with poorer households often suffering negative impacts overall.

Examining why this has occurred, we can see that although forest administrations have come under pressure to reform, they have so far been able to accommodate these demands, while often perpetuating a 'business as normal' orientation.

In the Nepal hills, the devolution of control has gone furthest. However, this may not represent a model for other areas, as here, too, poorer groups have suffered from being marginalized.

References

Acharya, K. (undated) *Teak and Trickery: An Investigation of the Malik Makbuja Case in Bastar*, Chattisgarhnet, www.cgnet.in/3636garh/FT/T2

Agarwal, B. (2001) 'Participatory exclusions, community forestry and gender: An analysis and conceptual framework' *World Development*, vol 29, no 10, pp1623–1648

Appu, P. S. (2005) 'Decline, debasement and devastation in the All India Services', *Economic and Political Weekly,* 25 February, Mumbai, Sameeksha Trust, pp826–832

Barton-Bray, D. and Merino-Perez, L. (2002) *Community Forests of Mexico: Achievements and Challenges*, Mexico City, Sierra Madre Publishers

Blaikie, P. M., Cameron, J. and Seddon, J. D. (1976) *Nepal in Crisis: Growth and Stagnation at the Periphery*, Delhi, Adroit

Chambers, R., Saxena, N. C. and Shah, T. (eds) (1989) *To the Hands of the Poor: Water and Trees*, London, Intermediate Technology

DoF (Department of Forests) (2005) *Forest Cover Change Analysis of the Terai Districts*, Kathmandu, Ministry of Forests and Soil Conservation, Government of Nepal

Government of Andhra Pradesh (2002) *Andhra Pradesh State Forest Policy*, Hyderabad, Government of Andhra Pradesh

Government of Himachal Pradesh (2005) *Himachal Pradesh Forest Sector Policy and Strategy*, Shimla, Himachal Pradesh Forest Department

Hickey, S. and Mohan, G. (eds) (2004) *Participation: From Tyranny to Transformation? Exploring New Approaches to Participation*, London, Zed

Hobley, M. (1996) *Participatory Forestry in India and Nepal: The Process of Change*, London, ODI

Iversen, V., Chettry, B., Francis, P., Gurung, M., Kafle, G., Pain, A. and Seeley, J. (2006) 'High value forests, hidden economies and elite capture: Evidence from forest user groups in Nepal's *terai*', *Ecological Economics,* vol 58, no 1, pp93–107

Jackson, C. and Chattopadhyay, M. (2000) 'Identities and livelihoods: Gender, ethnicity and nature in a south Bihar village', in Agrawal, A. and Sivaramakrishnan, K. (eds) *Agrarian Environments: Negotiating Conflicts over Resources and Identities in India*, Durham, Duke University Press

Jeffrey, R. and Sundar, N. (eds) (1999) *A New Moral Economy for India's Forests*, Delhi, Sage

Jerram, M. R. K. (1936) *A Text-book on Forest Management*, Cleveland, Ohio, Sherwood Press

Prasad, A. (2005) 'Survival at stake', *Frontline*, vol 23, no 26, Dec 30, Chennai, pp4–10

Lele, S. (2000) *Godsend, Sleight of Hand or Just Muddling Through: Joint Water and Forest Management in India*, Natural Resource Perspectives No 53, London, ODI

MoEF (Ministry of Environment and Forests) (1990) *Joint Forest Management Notification*, Delhi, MoEF, Government of India

MoEF (2000) *Memorandum No 2-8/2000-JFM (FPD) Subject: Strengthening of Joint Forest Management Programme*, Delhi, JFM Cell, MoEF, Government of India

MoEF (2006) *Report of the National Forests Commission*, Delhi, MoEF, Government of India

MoEF (undated) *National Working Plan Code*, Delhi, MoEF, Government of India

Regional Centre for Development Cooperation (2002) *Community Forestry*, vol 1, issue 5, September, Bhubaneshwar, RCDC

Sarin, M. with Singh, N. M., Sundar, N. and Bhogal, R. K. (2003) *Devolution as a Threat to Democratic Decision-making in Forestry: Findings from Three States in India*, London, ODI

Scott, J. (1999) *Seeing Like a State: Why Some Schemes to Improve the Human Condition Have Failed*, Durham, Duke University Press

Shradha, K. and Upadhyaya, S. (2004) 'Grassroots democracy: Local governance watch', *SaharaTime*, 28 December, Social Watch India, www.socialwatchindia.com/news/GRASSROOTS%20 DEMOCRACYLOCAL%20GOVERNANCE%20WATCH.htm

Singh, H. B. (2005) 'Chaubad-Bhumlu community sawmill: Empowering local people', in Durst, P. B., Brown, C., Tacio, H. D. and Ishikawa, M. (eds) *In Search of Excellence: Exemplary Forest Management in Asia and the Pacific*, Bangkok, FAO, pp135–144

Waddington, M. and Mohan, G. (2004) 'Failing forward: Going beyond imposed forms of participation', in Hickey, S. and Mohan, G. (eds) (2004) *Participation: From Tyranny to Transformation? Exploring New Approaches to Participation*, London, Zed

World Bank (2004) *Andhra Pradesh Community Forest Management Project Resettlement Action Plan (RP87)*, Washington, DC, World Bank

11

Participation or Democratic Decentralization: Strategic Issues in Local Forest Management

Piers Blaikie and Oliver Springate-Baginski

The key argument

Part I made the case that a discursive alliance exists around the popular reform narrative of forest management in both India and Nepal, although there are some differences between the two countries. The popular narrative is not static and there has been a dynamic process of adaptation and strategizing between the main actors in both countries. Nonetheless, it has had considerable long-term stability and centres on the related issues of participation in management by local forest users and redistributive and pro-poor justice and human rights. The opposing narrative of classic, state-led, top-down forest management has arguably had a longer and more stable history and is deployed to counter the popular narrative, to adapt, and even to cede ground to it when politically expedient. However, while this book advocates the former, the popular narrative contains important contradictions in the context of Indian and Nepalese forest policy reform. It is important to discuss these so that participatory forest management (PFM) is not, once more, uncritically recommended when there are unexamined problems at the heart of participation and the wider narrative of which it is a part.

The popular and classic narratives are shaped and promoted by the set of actors discussed in Chapter 2. The extent to which the various actors in the policy process can get others to do what they want (a central aspect of power) is due not only to the persuasiveness of their narratives, but also to their structural position. Here, too, there are opportunities and constraints to their positions. The popular reform narrative and the policy actors who advocate it face structural opposition, both covert and open (e.g. from the mining and commercial timber interests, as well as from the majority of the forest departments at federal and state levels). However, the actors supporting the popular reform of forestry in various ways (e.g. political activism, in the popular and academic presses, financial provision for participatory projects, project design and implementation, and international counter-parting) form a loose alliance with a variety of agendas and often with divergent goals. The case is examined, in Chapter 2, of international funding institutions (IFIs) with reform agendas and forest departments that can invoke the principles of national sovereignty in order to limit or divert reform, especially in India where forest departments are not so beholden financially to IFIs as they are in Nepal. Thus, it is important to examine the limitations of both the narrative and the alliances of those who promote it.

While the popular reform narrative is a persuasive one on rational and ideological grounds, it suffers from a degree of vagueness and inconsistency. This is not surprising considering the wide variety of actors who promote it, and the fact that local people have been disempowered from involvement in forest management for so long. It suffers from inconsistency in the *terms* used; the *assumptions* made both tacitly and explicitly about a wide variety of processes (such as trends in forest cover and quality, the causes of deforestation, the impacts of shifting cultivation and the role of the state in forest management); the *'theory'* or *models* that link human behaviour (including different styles of policy-making) to actual forest use and ecological change on the ground under different political ecological conditions; the *goals* of forest management (e.g. for national economic development or as a public good); and in *outcomes* of forest management under different political ecological conditions (e.g. issues of fact, choice of indicators, methods of measurement and use of statistics). Underlying these issues is the idealization of the 'popular' itself, de-emphasizing the stratification of most local societies and the internal politics that keep some households poorer than others. Devolving power to local people without reciprocal accountability structures risks elite domination, as we saw in all the study areas, and most of all where the forest is of highest value: in Nepal's *tarai*.

Table 11.1 provides a summary of the main contrasts between two stereotypical styles of forest policy: a classic state forest model and a popular narrative.

The controversial aspects of policy reform (terms used, assumptions, theory or models and goals) will be discussed in the four major arena of forest policy reform. These are:

1   participation in forest management;
2   the production and sharing of forest knowledge;
3   land reform and associated legal aspects of forest ownership and use; and
4   forest policy as a development process.

Participation in forest management

Participation has been a central theme of this book, but the term has often been used critically with inverted commas (indicating that the term means whatever the user has in mind), and the claims made in its name should be critically examined. Hickey and Mohan (2004b) provide a brief history of participation in development, where they lay out a number of broad and varied approaches within the context of development. These include 'political', 'emancipatory', 'liberation' and 'populist' participation, as well as 'participatory governance', among others. It is not surprising that interpretations of the meaning, goals and methods of bringing participation about are complex and often contradictory. The theorizations of each of these (and other) conceptualizations of participation are diverse and frequently overlap. It is one of the fuzziest discourses in development today, and the purpose of this book is not to enter into debate over wider issues, but to focus, instead, on the issues of participation relevant to forest policy in India and Nepal. Participatory approaches have been subject to a range of critiques (Rahnema, 1992; Cooke and Kothari, 2001; and in a less dismissive style, Hickey and Mohan, 2004a). Some of these critiques are examined later in this chapter. However, as Part II in this book has shown, the shortfalls of reaching even the most modest goals of 'participation' in forest management have been shown to be disappointing in most, but not all, instances. The practical prerequisites for any participation of local people in the planning and management of forests are not usually in place, especially in the Indian states studied in this book. Our concerns here in current debates about participation in forest management are threefold.

First, one of the key issues is the form and extent of participation of local forest users in

Table 11.1 *What is and what could be: Two stereotypes*

Aspects of administration forest	'What is': State forestry hegemony	'What could be': Popular reform narrative for democratic PFM
Forest land and the 'forest estate'	*India:* The expansion of a territorial forest estate and historical creation of forest land as an administrative category under the control of the forest department: • Composed of standing forest land and also, often, non-forest grazing land, (so-called 'wasteland', now re-categorized as 'degraded forests'). • Composition of forest-based estate has historically involved competitive 'turf-wars' with other departments (especially revenue departments) and with local people. *Nepal:* Virtually all non-private forest land nationalized in 1957.	Participatory, user-led, democratic land-use planning. Different land management according to national and local user needs: • Forest-based livelihood needs met mainly from village forests under village control. • Environmental services/biodiversity conservation from both local forests (many CFM groups are protecting forests for the environmental services that forests provide) and remote natural forests. In India, there is virtually no 'remote' forest that people do not use. Therefore, all forests should be available for livelihood needs and be under village control. • Production forestry mainly on private land, but also on public forests where it is possible to do so sustainably, and from village forests.
Forest management	*India:* Objectives set at central and state levels 'in the national interest': • Top-down centralized planning. • Production of timber and revenue supply (now changing mainly due to judicial interventions). • Long-term management emphasized. • Designed to protect, maintain and extend forest estate. • Green cover, and environmental and biodiversity conservation prioritized. *Nepal:* • Community needs-based in theory, but, in practice, there are conservative restrictions on harvesting. • *Tarai* forest management is more commercially orientated: problems of rights of distant users.	Public interest management objectives and bottom-up planning at local level as far as possible: • Short- and medium-term livelihood needs are prioritized. • Intensification options for higher incomes. • Livelihood needs, biodiversity management and environmental services are incorporated. • Forest products and services are combined with other land-use benefits. • Cyclical planning process works in conjunction with participatory monitoring.

Table 11.1 (*continued*)

Aspects of administration forest	'What is': State forestry hegemony	'What could be': Popular reform narrative for democratic PFM
Staff, organization and service provision	Main roles of forest staff: • technical and silvicultural roles for forest management planning; • field implementation – forest management, harvesting, processing and marketing; and • patrolling and enforcement.	Main roles of forest department staff (would differ by location): • In remote forests managed for environmental services: patrolling and enforcement, production for local consumption and, where possible, sustainable production. • In village forests: monitoring and support. Community-based organizations (CBOs)/non-governmental organizations (NGOs) can facilitate planning. Private service providers might provide technical service support to communities on a competitive tender basis. • Private production forests: technical support on competitive provision basis. • Retraining necessary for forest service providers staff in PFM-relevant forest management techniques.
Fiscal system and forest product marketing	• Main sources of revenue have declined due to resource degradation and green felling bans (some switching to ecotourism). • Licence system for non-timber forest product (NTFP) collection – forest corporations and state co-operatives, and involvement of contractors and traders in rent-seeking/exploitative relationship with primary collectors. • Revenue distribution: much revenue lost through illegal logging. • Perverse incentives and institutional culture – often informally encourage corruption and rent-seeking.	Fiscal system should be efficient, feasible and supportive of broader social and environmental objectives, accountable and transparent: • Liberalize marketing structures. • Provide local primary co-operatives for NTFPs. • Local institutions should control licensing and regulate off-take. • Local revenues are largely held and allocated locally in conjunction with democratic local government. • Incentives provided to motivate enforcement and combat corruption. • Remove transit obstructions, such as transit rules.
Legal powers	Forest department generally uses paramilitary powers to exclude other users, to a greater or lesser extent: • Forest department generally has poor accountability and transparency. • Non-accountable discretionary powers allow rent-seeking. • Local participatory forest management institutions are often created on a very weak legal basis – lack separate legal identity from the forest department or constitutional basis, as with local government.	Need for democratic and accountable processes: • Settlement of land tenure in equitable manner: forest estate often constituted under feudal/colonial executive fiat – historic injustice to be addressed and resolved. • Local forest users' needs cannot be excluded in practice. Giving rights to use forest leads to more sustainable use patterns – avoids 'tragedy of open-access commons'. • Forest management planning requires multi-stakeholder involvement, with real rights. Best under auspice of democratically run local government system.

Table 11.1 *(continued)*

Aspects of administration forest	'What is': State forestry hegemony	'What could be': Popular reform narrative for democratic PFM
Technical and socio-economic knowledge	• Silviculture: simplification of field reality to fit technical modelling – clear felling, mono-cropping, etc. • In practice, productivity very low, compared with NTFP production and the value of non-traded consumption goods from the forest. • Technical research on timber productivity conducted. • Forest users' needs and technical knowledge not considered.	Recognition of local knowledge, and acceptance of complexity and diversity: • Technical aspect grounded in social, as well as natural, science and outmoded technical aspects of so-called scientific forestry are upgraded, where appropriate. • Multi-cropping and management of multi-species forests prevails. • Research is needed on location-specific multi-cropping systems in which locals are usually technical experts. • Intensified, not extensive, production systems with high value for local forest users are favoured. • NTFP focus predominates. • Indigenous technical knowledge is incorporated within local planning.
Information-generation and sharing	• Long-term blueprint forest management planning according to silvicultural principles. • Monitoring of timber production according to blueprint. • Rudimentary monitoring of local institutions in some cases. • Little monitoring of livelihood impacts. • Poor dissemination of information about rights and obligations of forest users.	• Participatory monitoring of livelihood impact, resource and institutional performance. • Transparency and knowledge-sharing are highlighted. • Forest management information systems (MIS) are based on bottom-up needs. • Training is necessary.
Local institutional arrangements	*India:* • Co-opted joint forest management committees (JFMCs) with compliant elite members are inducted by district/divisional forest officers (DFOs). *India & Nepal:* • Arrangements are non-transparent and closed to a great majority of forest users. • Poor ethnic minorities and women are often ignored.	• Independent powers are highlighted. • Women are empowered with reserved positions on the executive committee. • Poor and marginal groups have reserved positions. • Pro-poor emphasis in management planning and distribution emphasized.

forest management plans that enable them to enhance their livelihoods through use of the forest. The almost universal finding in the three Indian states is that local people have very little influence on local forest management plans. The key issues of rotations, tree species, encouragement of non-timber forest products (NTFPs) and management techniques, including coppicing and cleaning the understorey, have immediate impacts on livelihoods. There are some examples where the district or divisional forest officer (DFO) or other local officer made an attempt to listen to and accommodate local priorities; but these are exceptions. They contrast with Nepal's CF, where local people have the legal right to prepare forest management plans for themselves. The overall outcome is that forest department field staff dominate the preparation of the micro-plan, which so far as forest management is concerned simply repeats in summary form what the divisional working plan proscribes for the area, to which it is subordinate. The goals of these working plans are technical silvicultural forest management for green cover and commercial objectives. They are not concerned with the needs and wishes of local people. At this point (which has been reached many times before by other professional foresters and authors), the reader is commonly urged to make the micro-plan or its equivalent more participatory to better accommodate and assist the livelihood needs of forest users. However, as Chapters 2 and 3 have explained, few of the preconditions for the planning and delivery of a participatory plan are in place. PFM requires of the forest department new skills of negotiation, new knowledge of multiple species of plants and their uses, and greatly increased resources of time, assistance, transportation and career incentives. There is no incentive to be participatory; but there *is* an incentive to launch a target number of joint forest management (JFM) groups within a very tight time budget. It is no wonder that all case studies reported the almost universal tendency of the DFO and the rangers to pay a quick visit to village leaders, organize a poorly advertised meeting or two, and then attempt to enforce the plan without its publication, leaving a map and simple outline of the plan pinned to the wall of the *panchayat* or village development committee (VDC) building. Without far more strategic and profound reform, all of these urgings to be 'more participatory' will simply not have any effect.

Second, the extent to which poorer sections and women can be successfully represented in the planning process at all levels, especially the local, is a key criterion of the success of PFM. A rather dismal record is revealed by this research and is supported by many other studies. PFM group meetings in West Bengal, Orissa, Andhra Pradesh and both regions of Nepal 'were rare', 'poorly attended' and 'not transparent', and other phrases the reader will have picked up in Part II. The representation of women of all groups, the ethnic minorities, low castes and the very poor was almost universally abysmal. Women forest users in the sample villages of Orissa 'had no idea' of JFM. Women hardly dared speak in front of men and state their own gender-specific needs, and their participation was found to be below prescribed 'quota' levels (the Nepalese hills). While there are reserved places for women which must nominally be filled, the role of women in the executive committee of the PFM scheme was usually minimal (Sarin et al, 1998, reviewed the situation in several states in India; see Locke, 1999, for an exploration of gender issues in JFM and Agarwal, 2001, p1627, for a listing of women's representation in different Indian states). Similar conclusions about the representation of poorer groups and lower castes were drawn in all case studies. It is important to note that there were variations between states in India and between villages across the case study areas.

Calls for justice are seldom a waste of breath; but calling is not enough unless there is mobilization and organization. In the agrarian political economies of South Asia, there is huge variation in the extent and form of marginalization (e.g. see Gulhati, 2000, who reports on the complete exclusion of women from the executive committee in sample villages in Jammu and Kashmir); but the call for better representation and an end to discrimination in these groups by a more participatory forest policy is 'asking the tail to wag the dog'. The issue of mobilization and organization of local forest users for collective action to manage forests will tend to reflect the inequalities inherent in the agrarian political economy. However, there are ways by which insistence on institutional reform may be able to improve the position of marginalized groups and women across all groups. Reserving places for women by insisting on one male and one female from each household for the executive committee, and by overall quotas, can help, but can be evaded in discriminatory, male-dominated villages facilitated by gender and by equity-insensitive male forest department staff. Also, who is to police such rules? The DFO does not see policing the reduction of social exclusion as a priority, except in the indirect sense that the local plan for the forest should discriminate as little as possible against disadvantaged groups. Instead, it is largely a matter of the personal style of the DFO, the ranger and the forest guard. There are many cases of dedicated and socially aware forestry officials who have achieved a democratic, open and widely representative PFM; but it usually takes a great deal of effort and flair and a long-term relationship to build up trust. In turn, this is frustrated by short periods of service in any one location and frequent transfers. The authors, on a visit to some JFM schemes in a hilly area of Andhra Pradesh where *podu* cultivation was a major issue, came across a DFO in Visakhapatnam division of Andhra Pradesh who was *too* participatory. He was building a level of openness and accountability in his dealings with local forest users which threatened illegal timber felling in which the forest department staff were apparently involved. He was transferred rapidly to another area where he could do less harm to the dysfunctional *status quo*. Thus, the professional incentive structure to be participatory and see that no major discrimination is being practiced is often negative.

Participation: The tarnishing of an idea

Participation has had a long history both in India and in Nepal. Pre-colonial local livelihood-oriented resource management involved customary community-based natural resource management arrangements in which the term 'participation' would apply *ex post*. In India, subsequent to the initial implementation of state forest management, local resistance to the demands of the colonial state led to tactical accommodation on the part of colonial forest administration to 'local participation' (e.g. *van panchayats*, Kangra forest co-operatives, Madras forest co-operatives and others; see Chapter 1). Community development was also established in the last years of the colonial administration and continued after independence. In the post-colonial period, civil demands for reform in local control and access to forest resources initially clashed with state-building and industrialization imperatives, but gradually led to a new phase of 'participation' from which social forestry and joint forest management emerged. Many examples of self-initiated groups date from this period (particularly in Orissa, as well as in Jharkhand and other states). Thus, there was plenty of 'participation' in India before the use of the term by IFIs and the Indian forest administration (e.g. in the 1988 Forest Policy). However, the element of participation in Indian forest policy and in projects financed by foreign donors, has, as we have claimed in Chapter 3, been 'oil on a squeaky wheel' to accommodate and appease further periods of unrest and loss of forest, and to reduce conflict that had led to injury and loss of forest personnel. Worse still, participation has been, in many cases, a 'wolf in sheep's clothing' in that it has enabled the forest administration to assume management control over new areas of 'forest land' and to further extend management control – including in areas where there were already self-initiated groups (see Hildyard et al, 2001, for a critique of JFM which takes a similar view to those here, and Sarin et al, 2003). This underlines the point that a major focus on participation alone misses other essential and linked considerations, especially the expansion of control of new lands by forest departments through land reclassification. It is blinkered in the extreme to maintain a narrow focus on the quality of participation in forest management in a particular sample village if large tracts of land all around it have been claimed by the forest department, who then impose overall control over what is negotiable and what remains the unexamined prerogative of the department.

However, there are deep-seated and structural reasons why the implementation of PFM has failed to fulfil the claims of participation in India, a view echoed by other commentators for decades. In Nepal, the progress of participatory forestry has been greater for a number of historical and contemporary reasons. First, there has not been the heritage of direct colonial forest management either on the state or on local institutions. There has been an indirect colonial influence through the training of forest officers in the colonial forest syllabus at Dehra Dun Forest Academy in India, and in the servicing of colonial timber demand from *tarai* forests. The state is weak and could not effectively police and manage the national forest after nationalization in 1957. Second, this left many community-based natural resource management (CBNRM) and local institutional arrangements more or less intact. In addition, in the middle hills, the opportunities for commercial felling (both legal and illegal) are limited. Third, the state's attempts to assert policing of national forests have been very circumscribed due to its limited capacity and the difficult terrain. Thus, these initial conditions were conducive for the successful handover of forest management responsibility to local people. As a result of these historical reasons in the Nepalese hills, the general level of local institutional autonomy in rural areas is generally higher than in India. The comparison between India and Nepal is an instructive tale for the Indian Forest Administration seeking to extend and deepen state control. It is that expansion and deepening of control which is likely to destroy valuable local institutional capacity for PFM (the case of the self-initiated forest user groups in Orissa in Chapter 8 is relevant here).

There has been increasing reappraisal and interrogation of 'participation' in development practice (Rahnema, 1992; Mosse, 1996; Cooke and Kothari, 2001; Hickey and Mohan, 2004a). Much of the criticism of 'participation' that has become increasingly central to donor-funded projects and policy assistance in CBNRM focuses both on the donors themselves and their simplistic or naive conceptualization of the process, as well as on structural and political conditions in the recipient country. A number of commentators have pointed out that 'participatory' projects are designed and presented to a number of different international audiences, as well as to policy elites in recipient countries – and not to the people who are supposed to 'participate' at all. For example, the World Bank launched a Global Environment Fund (GEF) 'participatory project' called the Conservation Forest Area Protection, Management and Development project in Thailand, which called for the eviction of thousands of Karen people from the protected area. The project document had 'never been translated into Thai, much less Karen: much less communicated to, much less discussed with, much less agreed to by the local Karen people' (Lohmann, 1994, cited in Hildyard et al, 2001, p59). Another example of participation as practised by the World Bank is the Andhra Pradesh Forestry Project researched in this book.

The 'performance' of participation is often designed for audiences other than the local people themselves (donor visits and evaluations, researchers, etc; see Blaikie, forthcoming). Performance can become a tokenistic practice – a small number of 'model' sites are established near roads, even airports. IFIs (e.g. the World Bank) reviews depend on limited visits by busy and highly paid consultants, who seldom have the time to visit inaccessible sites, understand complex situations or enter into time-consuming discussions with local people, especially over controversial issues – reminiscent of Chambers' (1987) memorable phrase 'rural development tourism'. The World Bank carries on 'doing participation' in the same way despite intensive criticism. For example, Cooke's (2004) first 'rule of thumb' for participatory change agents is 'Don't work for the World Bank'! Also, yawning gaps appear between claims made about participation and outcomes. There has been the assumption that the rhetoric of participation allows (or, more instrumentally, is *designed to* allow) increase control by the state and large IFIs (hence the metaphor of the 'wolf in sheep's clothing').

What can be expected from national institutional reform when international agencies such as the World Bank are not themselves democratic, transparent or responsive to voluble and sustained critique (Cooke, 2004)? As Chapter 3 has discussed, the characterization of progressive IFIs bringing the participation agenda to reluctant forest administrations is an oversimplification. It has been noted in this book, as well as in a number of cases (and this point can only be based on repeated hearsay), that donor projects have limited scope for applying leverage to promote the 'enabling environment' for participation when governments act in a manner perceived to be contradictory to a participatory approach. This has certainly been the case in India. There is considerable ongoing public debate over the impact of donors on participatory policy. Leverage by donors often fades away if their bluff is called since withdrawal from a project has considerable reputation and financial costs. On the other hand, donors have occasionally withdrawn altogether from forest-sector projects – for example, the World Bank, the Danish International Development Agency (DANIDA), the German Agency for Technical Cooperation (GTZ), the Australian Agency for International Development (Ausaid) projects from Nepal, and the UK Department for International Development (DFID) from Karnataka. Donor country office staff confide that donors have desisted from large-scale support to forest-sector projects since the late 1990s due to the intransigence of forest departments combined with grassroots protests in the case of India. This can be interpreted as evidence that leverage by donors is usually ineffective – that is, they desist from funding or leave during the project when attempted leverage fails. Additionally, major forest-sector funding donors (such as DFID) have partly shifted their support strategy to supporting state budgets directly.

The theories of participation are as much shaped by the structural constraints to practise as vice versa. Outcomes are the result of a reflexive process between what is supposed to happen under 'participation' and its actual daily practice. Reform of the forest administration's practice typically involves the repeated dilution of policy commitments in the face of institutional inertia. Implementing 'participation' at field level tends to become doing what is possible under the severe constraints to actual reform. These constraints variously include divergent objectives between different stakeholders, lack of resources to do the job properly, and lack of incentives for forest staff to risk one's job security and promotion prospects by going against the whole ethos of the forest department and its discourse/practice (as discussed in Chapter 3 and in this chapter).

Nonetheless, individual actions and innovations, as long as they are incremental and relatively small and not too radical, can sometimes be undertaken. These may include negotiating micro-plans with wide representation in specific villages by a particular DFO; providing relevant extension information as requested by local forest users; making available species relevant to local needs and providing for suitable forestry techniques; giving effective back-up for protection; facilitating the resolution of boundary disputes; and dealing with offenders in a just and equitable manner. However, even to deliver these modest measures involves much hard work and professional risk.

Is participatory forest management (PFM) a confidence trick?

A confidence trick persuades one to accept something under false pretences. PFM may be, in principle, a confidence trick if the 'deal' is misleading or allows the forest administration to default on the terms subsequently. The implication of 'participation' is transmitted through two words in the programmes of India and Nepal. The first is 'joint', as in joint forest management. If management is 'joint', then it implies that there is some participation of local people across the 'join' between civil society and representatives of the local state. In Nepal, participation is implied by the word 'community', as in CF. However, when words become palpable sets of rules that shape what local people can and cannot do, the notion of 'jointness' and the participation which ensures a fair negotiation largely evaporate in India. JFM is not forest handover, but rather a delegation of forest protection and restrictions on access along preordained lines in return for benefits, some of which are notional. Evidence suggests that whereas in Nepal the 'deal' has been largely explicit, in India there have been significant areas where it has not.

The first drawback for members of newly formed user committees in India is that there is no legally enforceable security for the deal in the long-rotation forest management provisions that are generally set by the forest departments in JFM forests. They have no option but to accept it on trust since there is no legally binding contract. In practice, forest departments' seldom have the trust of local people anyway due to a long history of painful experience. The benefits from 'final felling' may be only realizable in the remote future, and for poor people who discount future incomes very highly, this is irrelevant. In West Bengal, this has meant 12-year sal rotations, and in Andhra Pradesh, 80-year teak rotations.

Second, the marketing of the 'final' timber harvest in India is entirely under the non-accountable control of forest departments and is generally liable to be captured by a mutually self-rewarding network consisting of the forest departments, contractors and merchants. The revenue that might be distributed to the local 'beneficiaries' is *net* of the forest department's felling and marketing costs, the calculation of which is totally opaque. Furthermore, auctions of forest products are conducted by the forest departments in ways that are often liable to price-fixing by cartels of merchants – particularly evident in West Bengal, resulting in artificially deflated prices below the value on the open market, and therefore in reduced income flows to local people. At this stage of a JFM scheme, 'jointness' and 'participation' are distant goals that have long since disappeared for those who agreed, or at least acquiesced, to the scheme.

The charge that JFM is a confidence trick not only stems from the fact that forest users do not know what the future costs and benefits of a new JFM scheme are, as we have discussed above, but is also due to the way in which the 'deal' is presented under conditions of very unequal power and knowledge. A number of coercive and pressurizing arguments are employed by forest department field staff. The first gambit is that unless the village forms into the group promptly, then it is possible that other neighbouring villages will be given land that is customarily 'theirs'. The second is that no protection can be given from neighbouring villagers and forest mafias unless they become 'officially recognized' JFM groups. Third, in cases where donor projects provide lucrative wage-earning opportunities, these will be foregone unless agreement is reached quickly. This typically comes accompanied with 'conditions' that must be accepted – for example, that villagers must make a prior commitment to stop all grazing in their JFM forest, which hits the poorest dependent on raising small livestock the hardest.

This account of JFM in India is negative; but there are some mitigating outcomes and, as always, exceptions. The mitigating outcomes have been signalled at various points in this book, and in summary they are as follows:

- Regularization of livelihood-oriented forest use and presence in the forest, which no longer needs to be undertaken under cover of darkness: this is especially important for women, who use the forest on a daily basis.
- Local people now have the authority and, perhaps, the incentive (although to what extent it is fulfilled is open to doubt) to exclude outsiders from designated JFM forests.
- There has been widespread forest regeneration in most JFM forests.

In Nepal, our examination in Part II shows that the 'PFM deal' has been much more favourable, and participation is less of a tired cliché or a confidence trick. The empirical evidence from the Nepalese middle hills (see Chapter 5) is encouraging. This is not to say that the CF programme does not suffer from serious flaws (elite capture, leading to non-inclusiveness, neglect of pro-poor criteria and the exclusion of poorer groups who prove troublesome to the village elite); but the devolution of power over forest management has been much more substantial, both in tenurial terms and in the process of handover of forests to CFUGs.

Is participation salvageable?

Participation of local users in forest management in any except the most perfunctory and tokenistic way cannot be an effective part of forest policy in India and Nepal unless other prior and more structural causes are addressed. First, there has to be a widely accepted balance between state power and citizens' rights. In practical and applied terms, local people have to know what they can legitimately expect, what the limits of a DFO's and a forest ranger's power are, and the conditions under which the personal discretion of the DFO is lawful. They also need knowledge of the means by which they can object and to whom if they feel that government officers have acted unreasonably. Here, participation as a right of citizenship is a common theme running through development theory and participation (Hickey and Mohan, 2004a). Participation also depends on the agrarian political economy and the degree to which representation is possible across gender, class, caste and ethnic divides. It can be a contributor to a much broader and more radical movement, but that is all. Some movements, such as the People's Campaign for Decentralizing Planning in Kerala, launched in 1996, the National Forum of Forest People and Forest Workers in New Delhi, and other apex organizations such as the Federation of Community Forest Users, Nepal (FECOFUN), are among many examples.

Second, the responsibility of the forest departments for facilitating participation in forestry requires a great deal of skill and the acquisition of a range of political, social and economic information about how the local political ecology works. Participatory rural appraisal (PRA) has been widely practised in both India and Nepal, mostly by non-governmental organizations (NGOs) and IFI-funded programmes. However, it has often been made routine, and numerical targets for PRA are set, which encourages a perfunctory approach. Added to this is the lack of time, resources and skills to perform PRA to anything more than a perfunctory level.

The methods of participatory development, including in forestry, have been well developed but increasingly criticized, and PRA has been the target of recent criticism. Sundar (2000) illustrates some of the pitfalls of PRA as practised in JFM. One of them is that villagers mould their demands according to what they feel the project can deliver, with the result that it requires a very skilled negotiator to understand discursive strategies used by sections of the local population and other dissenting and sometimes submerged voices, and to accommodate these into a micro-plan that also pays attention to the goals of national forest policy. This is not easy, a fact that many critical accounts do not acknowledge. If the political context of participation is hostile to PRA (as inegalitarian and not politically aware of citizens' rights), and sufficient resources are not in position for most DFOs and rangers to carry out PRA with due care anyway, then critics must come up with something better and more feasible than PRA – for all its mistranslations, unacknowledged exercise of power by outside facilitators and spurious claims on the presentation of local knowledge in 'participatory development'. In more general terms, Mosse (2001b, p16) claims:

> *Indeed, a good understanding of local configurations of power – local leadership styles, factions and alliances and gender relations – is a prerequisite for the organization of effective community-based PRA and for the interpretation of its outputs.*

Therefore, PRA as practised in the real world cannot be seen to be a feasible method, and, instead, manufactures consent and constructs social and botanical categories according to the forest department, leaving alternatives unheard and undiscussed. In view of the challenge in acquiring essential information about local political ecology and, specifically, the material links between forest land and livelihoods and the distribution of access across groups, the question may be asked: 'Can PRA ever be made to provide a basis for the exchange of information and open discussion by a broadly represented local community?' There is a strategic choice here. There is either an effort to put the necessary resources in place for a tolerably effective PRA, or a retreat to *less participation* through a partial withdrawal of forestry administrators to a support and extension role.

Let us look at the latter strategy first: admitting that participation in its current form is not salvageable, and that there should be *less participation*. In its place could be the long overdue granting of security of tenure for forest users (including in currently formed JFM groups), and a proper and accountable legal basis for the interaction between local people and the state. As Chapter 1 and other chapters in this book have discussed, there has been a major land grab by the forest administration of India, and in many areas the prescribed process for the settlement (i.e. recognition or compensation for extinguishment) of customary rights has not been observed (see, for example, Sarin, 2003, 2005). In other areas in which self-initiated forest protection groups had been established, customary tenure has been encroached on or undermined by the state, with new maps showing boundaries of exclusions and residual privileges. More secure and unambiguous rights would allow local users more space to manage their own forests without the need for 'participation' and the enforcement of protection. If forestry administrators retreated from their hegemonic role and their historic mission to impose their policy goals (as listed in Table 11.1), particularly in village-adjacent areas, there would be space for more devolution of decentralized forest management. Foresters could then, instead, take a facilitative extension, support and monitoring role where local forest users could form a variety of protection groups and plan their own management strategies. This issue is central to PFM and why the forest administration is currently able to determine the terms for PFM.

Turning now to the second strategy, which seeks to improve the practice of participation, training in 'political ecology' (as defined in the Introduction) by all of the main forestry training facilities and new courses developed in all aspects of PFM are essential. It is at the stage of early training at forestry colleges that forest officers and other field staff become acculturated and learn their technical repertoire, as well as the regulatory framework to deliver it. The textbooks currently used are dated and heavily biased towards silviculture, and do not adequately handle the management of mixed natural forests, grasslands, and the wide range of NTFPs, including medicinal and aromatic plants (Roy, 1995). A revision of syllabi is required for the training of DFOs, working plan officers at circle level, forest rangers and guards (see numerous other recommendations, such as Roy, 1995; Shaikh and Kurian, 1995; World Bank, 2006). This is a key issue, but must be accompanied by precise job descriptions for all officers, setting out professional goals, objectives, indicators, reportage systems, and monitoring commitments with evaluation criteria that will be used in job performance evaluation. New syllabi would include rural sociology (issues of elite capture, marginalization of poorer groups, village politics, and issues of political strategy in agrarian situations); livelihood analysis; gender issues; local usage and local names of biota on forest land; training on how to facilitate local institutions; practical exercises; field trips for facilitation of PFM

plans; problem-solving; review and critiques; etc (see Campbell, 1995, for practical suggestions about new elements in the syllabus and new, practical learning-by-doing experiences;
and World Bank, 2006, promoting the need for facilitative rather than regulatory knowledge). These social facilitation courses need to be given the same weight as the more
traditional courses on silviculture.

Furthermore, activities other than PRA should be included as part of DFO and ranger
training and should feature field trips with forest users, participatory inventories, forest
traverses with different groups (e.g. women only), meetings with different groups, map-
making and so on. There will be many who are cynical about calls for extra training. Indeed,
one of the writing team (who will remain nameless) felt from personal experience that short
training courses were often attended primarily for their shopping potential in the host town!
There are certainly many grounds for cynicism; but they can be reduced (but never entirely
set at rest) if the training always has a test, the passing of which is mandatory for professional
qualification or promotion, and the application of the training is written into job descriptions.

It may seem paradoxical, but both strategies (of retreating from participation, but
improving its practice where it remains) can be promoted at the same time in order for the
forest administrations to provide less 'participation', but more relevant support and advice.

Knowledge creation

The second underlying issue behind the policy measure outlined in Chapter 10 and in Table
11.1 is knowledge creation in forest management, comprising technical knowledge about
biota found on 'forest land' and mostly referring to trees and shrubs. It also includes the
categorization of biota and land uses and their representation on maps. Forest department
knowledge, as used at state level and below, in the field, is characterized by Latin names
(usually excluding local names), although there are lists in general publications (see, for
example, Singh, 2004, p248f). There is a focus on silviculture and commercial potential,
blueprint and traditional repertoire, which is indifferent to knowledge produced by different people (e.g. women, ethnic minorities and historically long-term forest dwellers). Bearing
in mind the pitfalls of a stereotypical differentiation of 'scientific'/official knowledge and
indigenous technical knowledge (Agrawal, 1995), local knowledge is site specific, subsistence or small-scale commodity production-orientated, culturally interpreted (particularly in
the case of medicinal herbs), multi-species, and differentiated by gender, class, caste and
ethnic origin.

The extent to which local knowledge and management practices that support livelihoods
are used and incorporated within local plans is another criterion for successful PFM, which
will be in a better position to deliver enhanced livelihoods. This topic is discussed further in
the wider context of knowledge creation and power below. There are a number of cases
discussed in Part II (e.g. a leading farmer of Sindhupalchok district in Nepal, who shared
indigenous knowledge with the DFO and project staff in Chapter 5, and other instances in
West Bengal listed in Chapter 7). This issue is much written about, and there have been
cautions against a naive populism which glorifies indigenous technical knowledge (Agrawal,
1995; Blaikie et al, 1997; Lettmayer, 2000). The term indigenous technical knowledge (ITK)
itself is problematic and difficult to define. However, there is a wide range of indigenous
practices for both the management of multi-species forests, including what have come to be
called NTFPs, as well as plantations, which have been ignored by forest departments in
India and Nepal but which contribute significantly to livelihoods. Examples of sacred
groves, sacred corridors, agro-forests, cutting and coppicing, and rhizome planting have
been documented (see a comprehensive listing in Pandey, 2004). At best, inventories have

been made of NTFPs and medicinal plants, and in a few states in south India a network for the conservation and sustainable utilization of medicinal plants has been set up, although the areas covered are very small (200ha to 300ha in each state) (Srivastava, 2004). The main point is that the issue of the value of ethno-forestry is well known and gets good coverage in official publications (e.g. Bahuguna and Upadhyay, 2004). The systemic institutional reforms needed for the integration of indigenous knowledge and formal forestry have been published (Pandey, 2004, p203). The same kind of broad transitions advocated here (but not yet implemented) are reminiscent of the farming systems research of two decades or more ago. Technocratic agricultural innovations produced by research stations were seldom adopted by farmers because production conditions were different and much more limited on-farm than at research stations. The ways in which the innovations meshed (or failed to mesh) with the goals of different farmers were simply not known. On-farm research then became the norm, and later gender issues, discrimination and bottlenecks for inputs and credit came within the remit of farming systems research. Participation in agricultural research and the relocation of the production of new knowledge to the farm itself is now so familiar as to be unremarkable. The practical revolution (as opposed to academic and official publications that do not form part of policy) in forestry knowledge in India and Nepal has hardly begun. The rhetoric is now in place and attracts attention at all levels – central and state government, IFIs and national NGOs. However, the institutional, professional, financial, organizational and educational changes have not taken place, except in very small pockets and isolated projects.

Here, again, there are a number of linked reforms that have to be undertaken before a widespread incorporation, development and dissemination of local forest management practices can take place. At present in India, the Ministry of Environment and Forests (MoEF) has six regional forestry and forest product research stations at which new technologies and techniques are researched, although their record in providing useful innovations for the subsistence and small-scale commodity production (especially in NTFPs) of local users is insubstantial (Chawii, 2000). Their main recent research programmes are cloning of seedlings through tissue culture (particularly exotics such as eucalyptus, teak and bamboo), identification of new uses for different species, such as Jatropha for biodiesel, and medicinal plants. The problem is that there is very little economic incentive to reorient research away from directly commercial research for industry to the alternative research that is required for PFM. In Nepal, forestry research into multi-species forests and their livelihood uses has simply not been a priority area at either the Department of Forest Research and Survey (DFRS) within the Ministry of Forests and Soil Conservation (MoFSC) or at the Institute of Forestry, Pokhara, although the latter has recently begun to embrace work on NTFP issues.

Research funding for indigenous practices at *ex situ* sites must increase and be linked to regional centres of research. *In situ* 'in-forest' research still appears confined to isolated PhDs, project-initiated studies, and NGOs, although the need for this is, again, well known. It is worth repeating: the path which farming systems research took to relocate focus and location from centralized, technocratic institutions of knowledge production to a decentralized, site-specific and social situation-specific system of research is very relevant to India and Nepal today (although in India this transformation hasn't happened even in agricultural research, otherwise so many farmers would perhaps not be committing suicide today). Rastogi (2004), for example, states that the present emphasis of JFM in India is on 'development' and 'application', rather than on research into different silvicultural models for different areas, extending beyond timber to other products, as well by 'going beyond classical systems and approaches' and undertaking linked socio-economic research (Rastogi, 2004, p227f). The general case for the incorporation of local knowledge is, of course, well established.

There is another, different area of knowledge production that is closely linked to participation and the power to designate what is legitimate knowledge and what is dismissed as worthless hearsay and prejudice. The production of maps shows the geographical extent of different ecological, legal and tenurial details of land claimed by forest departments and local people. Maps are a form of claim by the forest department, and the 'map as fact' is a form of knowledge creation through claim-making. All maps serve interests and are embedded in a history that they help to construct. As Sundar (2000) points out in her discussion of PRA and mapping in JFM, it is of little surprise that official maps prepared by forest departments can often become effective claim-making devices to extend control of land by classifying it as forest and, therefore, under the control of the department (see Chapters 2 and 3). In other cases, maps of 'degraded forest' (which is officially classed as having less than 40 per cent canopy cover) are prepared and then JFM is located in some of these forests – but these degraded forests may well be natural grasslands serving as pasture or yield all sorts of tubers, roots and grasses, which are not part of the criteria for designating forest as degraded (Sundar, 2000, pp88–89).

These forest administration maps may be very different from the mental maps of customary use, rights, obligations and forest type held by village-level users, and official maps (even if lacking legal basis) as used by the forest department are treated as the only authoritative source. Maps are considered an essential and irreplaceable tool of forest planning at the working and micro-plan level. They are invariably drawn up at district headquarters and conform to standard cartographic procedure by the exclusive categorization of different areas of 'forest land' (even if there is no tree standing at present and there have been no trees there in living memory, as discussed in Chapter 1). But in India, in many states there aren't even proper maps since the areas have never been surveyed – 40 per cent of reserved forests in Orissa, for example, and most states in north-east India. There are two important considerations when adapting these practices for PFM. While PRA has been introduced in a number of districts, it has also often been converted into a routine and enfeebled exercise through target-chasing and a lack of incentives for field staff to conduct PRA properly, as this chapter has already discussed. A more participatory mapping exercise would expand the ways in which the boundaries are drawn between different forest types, and their uses and rights. The conventional way organizes space into discrete micro-regions, while local practice often involves a wide range of 'boundaries' that are rendered permeable by different categories of users and different uses in different seasons (see Messerschmidt, 1985, for Nepal). Local maps drawn in a participatory way would produce a much more complex space–time map. Current working plans, with their categorization of forest, land use and rights of access, are superimpositions on a much more complex management of land and forests through space (boundaries) and time (seasons and longer periods of tree rotations). Maps such as these undoubtedly exist as mental maps in various self-initiated forest user groups throughout India and Nepal. The micro-plan usually does violence to pre-existing mental maps, which, in a more self-determined policy framework, provide an alternative source of local knowledge about forest management. Another related challenge is the extent to which computerized databases and satellite imagery for mapping land use, species distribution, biodiversity and endangered species, forest cover and forest quality can be used in a participatory way and mesh with local people's mental maps.

Lastly, one vital aspect of knowledge creation and dissemination has hitherto been neglected. This is the issue of regular monitoring of PFM (Bahuguna and Upadhyay, 2004; Roy, 2004; World Bank, 2006). As these authors point out, monitoring provides an essential loop of new and continuously updated knowledge between state and civil society; but there are no mechanisms in place to activate the loop and allow rapid and timely local information to reach national policy-makers, the media, the general public and IFIs involved in financing PFMs on a continuous basis. Continuous monitoring of PFM has advantages of being able to see how forest management is actually going and to indicate emerging problems and corrective action, whether information and practice is relevant to users and getting their views about what they themselves are doing (Roy, 2004). Bahuguna and Upadhyay (2004) present a generalized model of a feedback loop between monitoring and policy. The authors have useful social monitoring suggestions about the working of local institutions (Bahuguna and Upadhyay, 2004, p313); but even this could become formulaic and not sufficiently 'process orientated'. The list of indicators for monitoring has also no mention of locally generated and relevant indicators arrived at through participatory processes, and is still silvi-culturally orientated. Monitoring should be done *continuously* in a participatory manner (ideally self-monitoring), and used as a basis for needs-based support provision, not as an occasional *ex post* activity (Springate-Baginski et al, 2003). Most monitoring studies are really *ex post* evaluations and only done once, and most were 'done largely from the perspective of the forestry department' (Springate-Baginski et al, 2003, p30). Other criticisms include lack of transparency, no reportage of sample size and techniques, little discussion or choice of indicators, little discussion of actual impact and, importantly, virtually no evidence of these studies having any leverage and impact within the forestry department. Again, this is another example of the disconnection between suggestions about participatory practice that are in the public domain and the actual day-to-day practice of forestry administrations.

A range of views from different (and often contesting) parties is useful for the monitoring process. Public auditing of state-controlled programmes in India has been initiated by some social movements, and have strengthened and spread. There is also an important role for the voluntary sector and independent NGOs to initiate and facilitate public auditing. The public audit of JFM should follow experience gained in Gujarat in a number of sectors other than forestry, and from the Nepalese experience where this process has been taken into the forestry sector. Additionally, financial and tactical support from donors may be necessary to promote such a process. The activities of the Forest Integrity Network, internationally, and the Transparency International India (TII) West Bengal chapter's activities in examining transparency or lack thereof in forest management are other examples. Furthermore, the initiation of independent apex federations of forest users and the strengthening of existing forest user federations and their networking into the political process at central and state levels would broaden the scope for views other than those of the forest administration to be heard and openly debated. FECOFUN in Nepal has demonstrated ably that PFM-based networks can play an important role, not only in advocating for CF, but even for promoting wider democracy, as has occurred over recent years during the conflict. However, in India, instead of nurturing and respecting similar independent federations, the Indian forest bureaucracy's response has been to create pseudo federations of JFM committees totally controlled by the forest departments themselves in the form of forest development agencies.

Prospects for reforming key underlying problem areas

Here is a quotation about forestry in India made over ten years ago by two distinguished authors. It posits an outdated administration and outdated ideas that are demonstrably unravelling. Many others have made similar points, and this book reiterates several of them. Yet, calls for reform and specific policy changes, as in this quotation, have not led to a distinct new policy direction:

> *The structure for the administration of public (including forest) lands remains essentially colonial in nature. While reform of agricultural land was pressed forward following independence, the management of public lands has remained frozen. Obviously it too needs a radical reorientation. These lands should be divided into two categories: (a) lands devoted to ecological security, and (b) community-managed lands devoted to providing livelihood security. The*

commercial plant production function should be fully shifted to private agricul-
tural lands. Given such an outlook, the foresters would play the role of joint
managers with people of lands devoted to ecological security or to livelihood secu-
rity and an extension machinery serving tree farmers. (Gadgil and Guha, 1995,
p174)

'The management of public lands has remained frozen.' This resonates with our own analysis of India, in which PFM (despite providing some economic and monetary concessions in
the form of entitlements to forest produce) remains largely a rhetorical device unable to deci-
sively thaw the long-established classic forestry narrative and its closely knit set of practices.
Movement on reform and a thaw on the frozen style of management of forests requires a
more self-reflective and open process of policy-making and implementation. This can be
called a 'learning environment'. A learning environment for PFM would require well-
thought out and widely negotiated structural changes. Wiersum (2000) reviews these
changes in an adaptive framework and, in distinction, a paradigm shift. Although the author
does not commit to one or other framework, the conceptual model of forest management as
a 'knowledge and action system' has all the characteristics of paradigm change – easy enough
to advocate, difficult to bring about. The main difficulties which successful reform faces are
not local people, nor even elite capture and forest mafias (although these remain serious
obstacles in some cases), but the bureaucratic structures and culture within which PFM must
thrive or which it can contribute to reforming. Here, a process-orientated approach to policy
is helpful, especially where there is complexity at different scales (see the multi-scale policy
map, Figure 2.1, in Chapter 2), between different agencies (forest administrations, revenue
departments, the Tribal Welfare Department and local self-government institutions, as well
as customary community institutions and the *gram sabha*), other actors and cross-sectoral
concerns (see Chapter 2), cross-cutting issues (e.g. trees, environmental protection, poverty,
gender and human rights) and political ecological diversity (see the Introduction), as well as
contradictions in the production of forest knowledge between scientific and indigenous
forest management. These considerations of complexity and sustained and unresolved
conflict and uncertainty all suggest the necessity for a move away from blueprint approaches
(as currently employed by the forest department) and blueprint recommendations for piecemeal (and therefore doomed) reform. Instead, this book approaches the question 'what is to
be done?' not by answering with a list of recommendations backed by pious hopes, but by
an attempt to understand the *process of policy-making and the underlying discourses and practices.* However, to adapt another quotation from a similar historical background to the one
above: 'It is not only our business to understand the world, but to change it.' Understanding
and reform of policy are mutually reinforcing. More understanding (e.g. of policy narratives
and the actors who tell them, or the ways in which 'participation' in the field actually
happens) hopefully leads to reflection and change. At the same time, discrete changes in how
policy is made and to its content lead to further reflection and new information. If this new
information and experience are monitored, recorded, widely disseminated to stakeholders
and discussed in an open way, 'learning by doing' becomes less of a cliché and more of an
essential path to continuous reform. A learning institution will therefore accept complexity
(and the issue of forest policy in India and Nepal could hardly be more complex), monitor
the policy process at all levels, and keep the focus of policy and its diverse impacts wide to
include those who fall outside the direct control of the forest administrations and outside its
administrative remit. Ultimately, in democratic societies it is for the democratically elected
representatives to direct the executive. The sort of ongoing and sustained reform identified
as necessary above would require the articulation of local aspirations for change, through
civil society mobilization, into the democratic expression of 'political will' in order to redi-
rect the forest administration towards the current social priorities. Such civil society

mobilization is seen in a number of social movements, both legal (e.g. the Chipko and Ekta Parishad movements, and the Campaign for Survival and Dignity, which lobbied for the forest rights act passed in December 2006 by the Indian Parliament), as well as extra legal (various Maoist groups in both Nepal and India, many of whom have expressed land tenure reform, particularly in tribal areas, as part of their manifestos). However, with the democratic systems in contemporary South Asia under severe strain from feudalistic patronage structures, abuse by criminal opportunism and, most recently, by corporate-led 'economic reform' entailing massive resource transfers to the corporate sector, the challenge to achieving reform remains as great as ever.

A final word. Policy reform was never a matter of 'truth talking to power'. This book has made a number of truth claims, some of them supported by direct measurement and applied statistical tests. Others have been supported by the evidence of documents and interviews, and appeals to reason or other persuasive strategies – we have told narratives in the same way as our actors on forest policy and management have done. However, the amount of evidence presented from so many diverse actors over such a period of history has the sheer weight of persuasiveness, even if we do not make a final claim to truth.

References

Agrawal, A. (1995) 'Dismantling the divide between indigenous and scientific knowledge', *Development and Change*, vol 26, pp413–439

Agarwal, B. (2001) 'Participatory exclusions, community forestry, and gender: An analysis for South Asia and a conceptual framework', *World Development*, vol 29, no 10, pp1623–1648

Bahuguna, V. K. and Upadhyay, A. (2004) 'Monitoring needs for JFM: The perspective for policy makers', in Bahuguna, V. K., Capistrano, D., Mitra, K. and Saigal, S. (eds) *Root to Canopy: Regenerating Forests through Community–State Partnerships*, New Delhi, Commonwealth Forestry Association (India Chapter) and Winrock International India, Chapter 28, pp309–316

Banerjee, A. K. (2004) 'Tracing social initiatives towards JFM', in Bahuguna, V. K., Capistrano, D., Mitra, K. and Saigal, S. (eds) *Root to Canopy: Regenerating Forests through Community–State Partnerships*, New Delhi, Commonwealth Forestry Association (India Chapter) and Winrock International India, Chapter 3, pp45–56

Baviskar, A. (2000) 'Claims to knowledge, claims to control: Environmental conflict in the Great Himalayan National Park, India', in Ellen, R., Parkes, P. and Bicker, A. (eds) *Indigenous Environmental Knowledge and its Transformations*, London and New York, Routledge, Chapter 4, pp101–120

Blaikie, P. M. (forthcoming) 'Is small really beautiful: Community-based natural resources management in Malawi and Botswana', *World Development*

Blaikie, P. M., Brown, K., Dixon, P., Sillitoe, P., Stocking, M. A. and Tang, L. (1997) 'Knowledge in action: Local knowledge and development paradigms', Special Issue of *Farming Systems*, vol 4, pp217–237

Campbell, J. (1995) 'Tailoring forest management systems to people's needs', in Roy, S. B. (ed) (1995) *Enabling Environment for Joint Forest Management*, New Delhi, Inter-India Publications, Chapter 4, pp59–92

Chambers, R. (1987) *Rural Development: Putting the Last First*, London, Longman

Chawii, L. (2000) 'Choking research' in *Down To Earth*, 30 November

Cleaver, F. (2005) 'The social embeddedness of agency and decision-making', in Hickey, S. and Mohan, G. (eds) (2004) *Participation: From Tyranny to Transformation*, London and New York, Zed Books, pp271–282

Cooke, B. (2004) 'Rules of thumb for participatory change agents', in Hickey, S. and Mohan, S. (eds) *Participation: From Tyranny to Transformation?*, London and New York, Zed Books, Chapter 3, pp42–55

Cooke, B. and Kothari, U. (2001) *Participation: the New Tyranny?*, London and New York, Zed Books

Edmunds, D. and Wollenberg, E. (2005) 'Strategic approach to multi-stakeholder negotiations', in Sayer, J. (ed) (2005) *The Earthscan Reader in Forestry and Development*, London and Sterling, VA, Earthscan, Chapter 20, pp395–414

Ellen, R., Parkes, P. and Bicker, A. (eds) (2000) *Indigenous Environmental Knowledge and its Transformations*, London and New York, Routledge

Gadgil, M., and Guha, R. (1995) *Ecology and Equity: Use and Abuse of Nature Contemporary India*, London, Routledge

Gulhati, M. (2000) 'Evolution of forest management practices in Shivalik Hills, Udhampur District, Jammu and Kashmir', in Ravindranath, N. H., Murali, K. S. and Malhotra, K. C. (eds) *Joint Forest Management and Community Forestry in India*, New Delhi and Calcutta, Oxford and IBH Publishing Co Pvt Ltd, Chapter 7, pp151–170

Hickey, S. and Mohan, G. (eds) (2004a) *Participation: From Tyranny to Transformation*, London and New York, Zed Books

Hickey, S. and Mohan, G. (2004b) 'Towards participation as transformation: Critical themes and challenges', in Hickey, S. and Mohan, C. S. (eds) *Participation: From Tyranny to Transformation?*, London and New York, Zed Books, Chapter 1, pp3–24

Hildyard, N., Hegde, P., Wolvekamp, P. and Reddy, S. (2001) 'Pluralism, participation and power: Joint forest management in India', in Cooke, B. and Kothari, U. (eds) *Participation: The New Tyranny?*, London and New York, Zed Books

Lettmayer, G. (2000) 'Learning to respect about cooperation with resource users', Paper presented at Workshop on Learning from Resource Users – a Paradigm Shift in Tropical Forestry, 28–29 April, Austria

Locke, C. (1999) 'Constructing a gender policy for joint forest management in India', *Development and Change,* vol 30, pp265–285

Lohmann, L. (2001) 'Pluralism, participation and power: Joint forest management in India', in Cooke, B. and Kothari, U. (eds) (2001) *Participation: The New Tyranny?*, London and New York, Zed Books

Messerschmidt, A. D. (1985) *People and Resources in Nepal: Customary Resource Management Systems of the Upper Kali Gandaki,* Proceeding of the Conference on Common Property Resource Management, Annapolis, MA, 21–26 April

Mosse, D. (1996) 'The social construction of "people's knowledge" in participatory rural development', in Bastian, S. and Bastian, N. (eds) (1996) *Assessing Participation: a Debate from South Asia*, New Delhi, Konark Publishers

Mosse, D. (2001a) 'Process documentation research and process monitoring', in Mosse, D., Farrington, J. and Rew, A. (eds) (2001) *Development as Process: Concepts and Methods for Working with Complexity*, India Research Press, Delhi, Chapter 2, pp31–53

Mosse, D. (2001b) 'People's knowledge, participation and patronage: Operations and representations in rural development', in Cooke, B. and Kothari, U. (2001) *Participation: The New Tyranny?*, London and New York, Zed Books, Chapter 2, pp16–35

Mosse, D., Farrington, J. and Rew, S. (eds) (2001) *Development as Process: Concepts and Methods for Working with Complexity.*, New Delhi, India Research Press

Pandey, D. N. (2004) 'Ethnoforestry and sustainability science for JFM', in Bahuguna, V. K., Capistrano, D., Mitra, K. and Saigal, S. (eds) *Root to Canopy: Regenerating Forests through Community-State Partnership*, New Delhi, Winrock and Commonwealth Forestry Association, Chapter 19, pp195–210

Pottier, J., Bicker, A. and Sillitoe, P. (eds) (2003) *Negotiating Local Knowledge; Power and Identity in Development*, London, Pluto Press

Rahnema, M. (1992) 'Participation', in Sachs, W. (ed) *The Development Dictionary: A Guide to Knowledge as Power*, London, Zed Books, pp116–131

Rao, J., Murali, K. S. and Murthy, I. K. (2004) 'Joint forest management studies in India: A review of the monitoring and evaluation methods', in Ravindranath, N. H. and Sudha, P. (eds) (2004) *Joint Forest Management in India: Spread, Performance, Impact*, Hyderabad, Universities Press, Chapter 2, pp26–65

Rastogi, A. (2004) 'Research, development and application imperatives under JFM', in Bahuguna, V. K., Capistrano, D., Mitra, K. and Saigal, S. (eds) *Root to Canopy: Regenerating Forests through Community–State Partnership*, New Delhi, Winrock and Commonwealth Forestry Association, Chapter 22, pp227–234

Roy, S. B. (ed) (1995) *Enabling Environment for Joint Forest Management*, New Delhi, Inter-India Publications

Roy, S. B. (2004) 'Participatory vegetation monitoring; examples from West Bengal', in Bahuguna, V. K., Capistrano, D., Mitra, K. and Saigal, S. (eds) *Root to Canopy: Regenerating Forests through*

Community–State Partnership, New Delhi, Winrock and Commonwealth Forestry Association, Chapter 25, pp267–273

Sarin, M. (2003) 'Bad in law', *Down to Earth*, 15 July

Sarin, M. (2005) *Laws, Lore and Logjams: Critical Issues in Indian Forest Conservation,* Gatekeeper Series, 116, London, IIED

Sarin, M. with Ray, L., Raju, M. S., Chatterjee, M., Banerjee, N. and Hiremath, S. (1998) *Who is Gaining? Who is Losing? Gender and Equity Concerns in Joint Forest Management*, New Delhi, SPWD

Sarin, M., Singh, N. M., Sundar, N. and Bhogal, R. K. (2003) *Devolution as a Threat to Democratic Decision-making in Forestry? Findings from Three States in India,* ODI Working Paper 197, February, London, ODI, pp36–40

Shaikh, M. H. A. and Kurian, J. (1995), 'Training needs in joint forest management', in Roy, S. B. (ed) *Enabling Environment for Joint Forest Management*, New Delhi, Inter-India Publications, Chapter 14, pp161–166

Singh, N. K. (2004) 'NTFP development: Constraints and potential', in Bahuguna V. K., Capistrano, D., Mitra, K. and Saigal, S. (eds) *Root to Canopy: Regenerating Forests through Community–State Partnership*, New Delhi, Winrock and Commonwealth Forestry Association, Chapter 21, pp221–226

Springate-Baginski, O., Yadav, N., Dev, O. P. and Soussan, J. (2003) *Institutional Development of Forest User Groups in Nepal: Processes and Indicators*, Rural Development Forestry Network Newsletter, London, ODI

Srivastava, A. K. (2004) 'Sustaining JFM through medicinal plants', in Bahuguna, V. K., Capistrano, D., Mitra, K. and Saigal, S. (eds) *Root to Canopy: Regenerating Forests through Community–State Partnership*, New Delhi, Winrock and Commonwealth Forestry Association, Chapter 20, pp211–220

Sundar, N. (2000) 'The construction and destruction of "indigenous" knowledge in India's Joint Forest Management Programme', in Ellen, R., Parkes, P. and Bicker, A. (eds) *Indigenous Environmental Knowledge and its Transformations*, London and New York, Routledge, Chapter 3, pp79–101

Wiersum, K. F. (2000) 'Incorporating indigenous knowledge in formal forest management: Adaptation or paradigm change in tropical forestry?' in Lawrence, A. (ed) *Forestry, Forest Users and Research: New Ways of Learning*, Wageningen, The Netherlands, European Tropical Forest Research Network

World Bank (2006) *India: Unlocking Opportunities for Forest Dependent People*, Delhi, World Bank

Glossary of Indian and Nepalese Terms

Adivasis: tribal or indigenous peoples (although the term is contested).

Ailani: barren or unregistered land.

Ban-butyan: forest bush.

Bari: non-irrigated land where mainly maize and millet are grown.

Bhari: backload; approximately 40kg of fuelwood.

Birta: traditional type of land grant given to high-caste Hindus. A grant of land to a noble as a reward for service to the state, this led to the emergence of *Birta* land tenure. It was usually both tax free and heritable, and without a set time limit. It was valid until it was recalled or confiscated.

Bista: a reciprocal in-kind and labour exchange based on caste.

Chautara: resting place with a tree.

Churia/Siwaliks: the first ranges arising north of the Indo-Gangetic plains up to 1000m. The term *Siwaliks* is used throughout the Himalayan region. *Churia* (or *Chure*) is a Nepalese word for the Siwalik range.

Dalit: so-called 'lower' castes, including Kami, Damai and Sarki groups.

Gram Sabha: village assembly.

Jagir: land grant paid as payment for the work; tax needed to be paid.

Janjati: ethnic group historically indigenous to Nepal, such as Gurungs, Magars and *Rais Janajatis* (those belonging to non-Hindu groups, such as Gurungs, Magars and Rais).

Jimmawal: village headman, who was given, in some cases through government decree, the responsibility to manage the forest.

Kamis: local blacksmith making charcoal to make and repair iron tools.

Kharbari: grassland.

Kipat: ancient type of communal land tenure. On *Kipat* land the headman gave individual households the right to cultivate and to collect forest products.

Kamaiya: agricultural bonded labour in far and mid-western region of Nepal.

Kulo: locally developed irrigation system in the village.

Manapathi: system of payment in kind according to local measure (*mana*).

Mohi: government tenure arrangement whereby tenants cultivate others' lands.

Mukhiya: chief, leader, headman, tax collector during Rana regime.

Mundari Khuntkatti: inalienable land rights granted to settled tribals in the 1908 Chhotanagpur Tenancy Act (CNT) that gave perpetual title of the lands to the group as a whole, and also prohibited the sale or transfer of such lands.

Pakho-Pakhera: sloping land with scattered trees and bushes.

Panchayat: Local village committees. In India *panchayats* (also called *gram panchayats*) have become the basis of decentralized local government. See the note on page xvii for a fuller explanation of the term. In Nepal, the quasi-feudal single-party '*panchayat* system' was introduced by King Mahendra in 1960 and endured until 1990. *Panchayats* have since been replaced by Village Development Committees as the basis for local government in the country.

Panchayati Raj: The principle of local self-government espoused by the Indian nationalist movement and constitutionally mandated in 1992 under the 73rd Amendment. Please refer to note on page xvii.

Pradhan Panch: the elected leader of the village council (*panchayat*).

Rana: a family regime of *de facto* rulers who ruled Nepal from 1846 to 1950.

Sukumbasi: landless people.

Talukdari: system of local (usually hereditary) function whereby the functionary of the state collects taxes. Grant of land and forest to local leaders for their control, use and management, especially in Brahmin-Chettri villages. Existed until the 1950s.

Tarai/Bhabar: Nepal's southern flat land territory along the Indian border, the *tarai* region is broadly divisible into the southern stretch of the alluvial plains (the *tarai* proper) and the northern colluvial deposits (the *Bhabar*) along the southern foothills of the Siwalik range (100–1500m).

Zamindar: landlord.

Index